# Logic Colloquium '03

LECTURE NOTES IN LOGIC

A Publication of
THE ASSOCIATION FOR SYMBOLIC LOGIC

LECTURE NOTES IN LOGIC 24

# Logic Colloquium '03

*Proceedings of the Annual European Summer Meeting of the Association for Symbolic Logic, held in Helsinki, Finland, August 14-20, 2003.*

Edited by

Viggo Stoltenberg-Hansen
Department of Mathematics
Uppsala University

Jouko Väänänen
Department of Mathematics and Statistics
University of Helsinki

CRC Press
Taylor & Francis Group
Boca Raton London New York

CRC Press is an imprint of the
Taylor & Francis Group, an informa business

Addresses of the Editors of Lecture Notes in Logic and a Statement of Editorial Policy may be found at the back of this book.

**Sales and Customer Service:**
A K Peters, Ltd.
888 Worcester Street, Suite 230
Wellesley, Massachusetts 02482, USA
http://www.akpeters.com/

**Association for Symbolic Logic:**
Sam Buss, Publisher
Department of Mathematics
University of California, San Diego
La Jolla, California 92093-0112, USA
http://www.aslonline.org/



**Library of Congress Cataloging-in-Publication Data**

Logic Colloquium (2003 : Helsinki, Finland)
Logic Colloquium '03 : proceedings of the Annual European Summer Meeting of the Association of Symbolic Logic, held in Helsinki, Finland, August 14-20, 2003 / edited by Viggo Stoltenberg-Hansen, Jouko Väänänen.
p. cm. – (Lecture notes in logic ; 24)
Includes bibliographical references.
ISBN-13: 978-1-56881-293-9 (hbk. : acid-free paper)
ISBN-13: 978-1-56881-294-6 (pbk. : acid-free paper)
ISBN-10: 1-56881-293-0 (hbk : acid-free paper)
ISBN-10: 1-56881-294-9 (pbk. : acid-free paper)
1. Logic, Symbolic and mathematical–Congresses. I. Stoltenberg-Hansen, Viggo. II. Väänänen, J. (Jouko), 1950- III. Association for Symbolic Logic. IV. Title. V. Series.

QA9.A1L64 2003
511.3–dc22 2005057435

CRC Press
6000 Broken Sound Parkway, NW
Suite 300, Boca Raton, FL 33487
270 Madison Avenue
New York, NY 10016
2 Park Square, Milton Park
Abingdon, Oxon OX14 4RN, UK

# TABLE OF CONTENTS

# COLLOQUIUM PROGRAM

**Tutorial speakers**

Michael Benedikt, *Model theory and complexity theory*
Bell Labs, Lisle, USA.

Stevo Todorcevic, *Set-theoretic methods in Ramsey theory*
C.N.R.S. – UMR 7056, Paris, France.

**Plenary Speakers**

Alexandru Baltag, *Doing things to models: Trans-structural dynamic logics*
Oxford University (Computing Laboratory), Oxford, United Kingdom.

Howard S. Becker, *Generalized model theory and Polish group actions*
University of South Carolina, Columbia, USA.

Matthew Foreman, *Has the CH been settled?*
University of California, Irvine, USA.

Jean-Yves Girard, *From operator algebras to logic*
IML, UPR 9016, Marseille, France.

Martin Grohe, *The succinctness of monadic logics on finite trees*
Humboldt-University of Berlin, Berlin, Germany.

Peter T. Johnstone, *A survey of realizability toposes*
DPMM, Cambridge, United Kingdom.

Simo Knuuttila, *On the history of modality as alternativeness*
University of Helsinki, Helsinki, Finland.

Menachem Kojman, *On the role of infinite cardinals*
Ben Gurion University, Beer Sheva, Israel.

Michael C. Laskowski, *Applications of descriptive set theory to uncountable model theory*
University of Maryland, College Park, USA.

Larisa Maksimova, *Decidable properties of logical calculi and of varieties of algebras*
Siberian Branch of Russian Academy of Sciences, Novosibirsk, Russia.

Per Martin-Löf, *Are the objects of propositional attitudes propositions in the sense of propositional and predicate logic?*
Stockholm University, Stockholm, Sweden.

Ludomir Newelski, *Lascar strong types and topology*
Wroclaw University, Wroclaw, Poland.

Dag Normann, *The continuous functionals in the perspective of domain theory*
University of Oslo, Oslo, Norway.

Graham Priest,*Inconsistent arithmetics and inconsistent computation*
University of Melbourne, Melbourne, Australia.

Pavel Pudlák, *Consistency and games*
Academy of Sciences, Prague, Czech Republic.

Michael Rathjen, *Realizability, Forcing, and independence results for constructive theories*
University of Leeds, Leeds, United Kingdom, and Ohio State University, Columbus, USA.

Saharon Shelah, *Good frames, for what are they good?*
The Hebrew University, Jerusalem, Israel.

Richard A. Shore, *The boundary between decidability and undecidability in degree structures*
Cornell University, Ithaca NY, USA.

**Special Sessions**

**Model Theory**

Tuna Altinel, *Simple groups of finite Morley rank*
Universite Claude Bernard Lyon-1, Villeurbanne, France.

Tapani Hyttinen, *Random logarithm: An example of a homogeneous class of structures*
University of Helsinki, Helsinki, Finland.

Wai Yan Pong, *The differential order*
California State University, Carson, USA.

**Set Theory**

Arthur W. Apter, *Indestructibility and strong compactness*
Baruch College of CUNY, New York, USA.

Steve Jackson, *Some applications of cubical markers*
University of North Texas, Denton, USA.

Heike Mildenberger, *New canonization theorems and dense free subsets in* $\aleph_\omega$
University of Vienna, Vienna, Austria.

Maurice C. Stanley, *Outer models and genericity*
San Jose State University, Tracy, USA.

**Recursion Theory and Arithmetic**

Klaus Ambos-Spies, *Generators of the computably enumerable degrees*
University of Heidelberg, Heidelberg, Germany.

Oleg V. Kudinov, *Computable listings and autodimensions of rings and fields*
Siberian Division of RAS, Novosibirsk, Russia.

Chris Pollett, *Weak arithmetics and unrelativised independence results*
San Jose State University, San Jose, USA.

John V. Tucker, *Specification and computation on topological algebras*
University of Wales Swansea, Swansea, United Kingdom.

**Proof Theory and Non-classical Logic**

Ralph Matthes, *Inductive constructions for classical natural deduction*
University of Munich, Munich, Germany.

Erik Palmgren, *Problems of predicativity in constructive topology*
Uppsala University, Uppsala, Sweden.

Jeff Paris, *Uncertain reasoning*
Manchester University, Manchester, United Kingdom.

Tarmo Uustalu, *Type-based termination of recursive definitions*
Institute of Cybernetics, Tallinn, Estonia.

# TUTORIAL

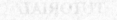

# GENERALIZING FINITE MODEL THEORY

MICHAEL BENEDIKT

**Abstract.** We overview work on languages for querying finite structures, where the finite structures "live" inside some infinite interpreted structure. The goal is to take the query languages that have been investigated in finite model theory — first-order logic, second-order logic, fixpoint logic, etc. — and find natural extensions of them that take into account the ambient structure. Most of our discussion will be on first-order query languages. From the point of view of model theory, this means we fix an infinite structure and look at first-order formulas with additional relational predicates that range over finite sets; thus our languages lie between first-order logic and weak second order logic. The bulk of the survey consists of characterization theorems stating what sorts of queries can be defined in these logics. Not surprisingly, the expressiveness of first-order queries is determined by the model theory of the infinite structure. When the structure has a decision procedure, one can ask how complex it is to evaluate these formulas, in terms of the size of the finite structure that is plugged in for the free relational parameters. That is, we can consider formulas in these logics as templates for standard formulas, and ask how decision procedures behave as the templates are instantiated differently. The model-theoretic perspective and the symbolic computation perspective turn out to be closely connected. We also discuss applications of the results to spatial databases and to abstract complexity.

**§1. Introduction.** Classical Finite Model Theory and Descriptive Complexity Theory [25, 17] give results about the expressiveness and complexity of logics over finite relational structures. This paper surveys work generalizing finite model theory to the situation where there is some infinite interpreted structure $M$ where the finite relations live. We want to talk about logics that can access both the finite structure and the infinite background model; what classes are definable within such a logic, and what can we say about them with respect to computability and complexity? Answering these questions requires a suitable model of definability in the presence of built-in structure, as well as a notion of complexity appropriate when one works in an abstract setting, possibly over noneffective domains. There are several candidate approaches to logics over finite models "with a background structure", and for each of these there are multiple complexity models one could compare against. Not surprisingly, the emphasis in this survey will be on the approaches the author has worked on, namely Embedded Finite Model Theory and "generic complexity". In this introduction, we describe this approach and give a brief acknowledgement of the alternatives.

**Logic Colloquium '03**
Edited by V. Stoltenberg-Hansen and J. Väänänen
Lecture Notes in Logic, 24

Probably the first research explicitly aimed at studying the situation where finite models sit within an infinite structure was the "Metafinite model theory" of Grädel and Gurevich [22]. They consider a many-sorted model consisting of a finite relational structure $S$, an infinite structure $M$, and an embedding of the former into the latter. In [22] several logics for querying these "mixed" structures are defined; these do not allow quantification over the infinite model $M$ — instead they deal with first-order and higher-order logics working either over elements in the finite part or over functions from the finite part to the infinite part. The resulting theory, applied with $M$ being the real field, provides a tight connection between definability and complexity within the BSS computation model (see below). This relationship is explored in [23]. Metafinite model theory has not been much explored outside of the reals. Since first-order quantification over the infinite part is prohibited one would not expect that the computational power of the resulting query languages will align with the model theory of the infinite structure in the general case.

In this paper, we take the approach of Embedded Finite Model Theory (EFMT), which originates in work within the database community [26]. Here one examines logics that mix first-order quantification over the infinite structure $M$ with first and higher-order quantification over a finite relational structure $S$ whose domain is a subset of $M$. Because the logics of Embedded Finite Model Theory allow first-order quantification over the infinite part, even the weakest of them are fairly powerful — effective evaluation of a formula will involve applying a decision procedure for $M$. We will see that in the resulting theory the expressive power of first-order query languages closely reflects the underlying model theory. The drawback is that in this setting we do not know any logic that precisely captures a complexity class.

EFMT was developed with two particular infinite models in mind, the real field and the real ordered group. The initial goal was not to give insight into either complexity theory or model theory. Rather, logical query languages were devised to specify queries on collections of points in the plane, and more generally, semi-linear or semi-algebraic regions in Euclidean space. For example, given a collection of labeled points in the plane, one can use first-order queries to ask whether the green points are all collinear, whether a polygon defined by one collection of points contains another collection, etc. The work of Kanellakis, Kuper, and Revesz [26] and its successor papers explore the analogies between evaluation of these query languages and traditional relational database query languages, and develop spatial database systems based on these languages. For details on the spatial query languages coming out of this research, a recent survey is [28].

Although the spur for EFMT was database applications, the perspective of this paper will be model-theoretic. EFMT gives another way to gauge the strength of a first-order theory, by measuring how much computation the corresponding query languages can express. Currently those connections are understood only in a crude way. For example, we know that stable structures behave differently from arbitrary structures (see Section 4). But even at this point it is clear that by classifying models based on their query languages we arrive at new dividing lines.

To talk about how expressive a query language is, we need to talk about complexity over a first-order structure. A natural notion of computability and complexity over a model was developed by Poizat [32], generalizing the abstract algebraic computability of Blum, Shub, and Smale [11]. Roughly speaking this "BSS complexity" classifies a problem by the logical complexity of uniformly-generated sequences of formulae needed to define the problem. For example, a problem is in *NP* over a structure *M* if it has a nicely-generated collection of polynomially-growing existential formulas that define it. BSS complexity gives a good way to compare the complexity of problems that live within the same structure. We will sometimes want to look at complexity measures closer to concrete computation, so that we can make direct comparisons between the expressiveness of first-order logic in two radically-different structures. That is, we will want to talk about what "classical" computable functions are expressible making use of structure, regardless of the domain of the structure. One way to do this is by means of the Boolean Part of an abstract complexity class [11]. This is obtained by restricting the inputs and outputs of functions in the class to two distinguished constants from the structure. Over rings this is a natural notion, with the two constants picked to be 0 and 1, and the characterization of Boolean Parts has played a key role in obtaining transfer results between classical and algebraic complexity [29, 20]. Over an arbitrary structure there may be no natural choice, and the complexity may vary wildly as different constants are chosen. Hence we look at a different notion (again inspired by work in the database community) called "generic complexity", which measures the pure or "homogeneous" computation that can be done in any structure. We will then discuss the complexity (in the BSS and generic sense) of the queries defined in first and second order logic over structures.

**Organization.** Section 2 reviews the classical connection between finite model theory and complexity, including some results one might aim to generalize. It then gives the basic definitions of the paper, including the notion of first-order query and the complexity measures that can be applied to it. Section 3 studies which models have all of their first-order queries of low complexity. Section 4 deals with models that have bounded complexity of first-order queries in a weaker sense: first-order queries may be expressive, but cannot do any new "pure computation". Section 5 connects these first-order results with the more general question of complexity theory over a model. Section 6 gives conclusions.

**§2. Notations.**

**Complexity Classes.** We assume familiarity basic complexity classes $P$, $NP$, etc. (see, for example, [21]), which for us define collections of decision problems on strings. Most of our interest will lie within the polynomial hierarchy $PH$, which is formed starting at $P$ and closing under complementation and polynomial nondeterminism: i.e. if $T(x,y) \in PH$ and $p$ is a polynomial with integer coefficients, then $\{x : \exists y\ |y| < p(|x|)\ T(x,y)\} \in PH$. It is well-known that $P \subseteq NP \subseteq PH \subseteq PSPACE \subseteq EXPTIME$. We will also look at "smaller" complexity classes defined by *circuits* [13]. A circuit is a dag-structured presentation of a propositional formula, where we have propositions giving the value of each bit of the input string. A decidable problem $Q$ can always be "calculated by" a family of circuits $C_n$ defining the restriction of $Q$ to strings of length $n$. If the problem is in polynomial time, then it can always be given by a family of circuits of polynomial size, and in fact by a "$P$-uniform family" — $C_n$ is polytime computable from $n$. This motivates the definition of subsets of $P$ by restricting the circuits further:

- $Q \in UAC^0$ if it can be computed by a family of circuits $C_n$ whose depth is bounded independently of $n$, and where $C_n$ can be "easily computed" from $n$.
- $Q \in UTC^0$ if it can be computed by a family as above, where the circuits can have "majority gates".

The notion of "easily computable" here means logtime uniform — for details see [3]. The *non-uniform* analogs are given by removing the requirement that $C_n$ be easy to compute.

- The non-uniform variant of $P$ is the class $\mathcal{P} = P/Poly$, problems definable by a circuit family $C_n$ whose size is polynomial in $n$.
- $AC^0, TC^0$ are defined by removing the uniformity condition on $UAC^0$, $UTC^0$.
- A problem $Q$ is in $\mathcal{NP} = NP/Poly$ if there is a polynomially growing family of circuits $C_n(x_1 \dots x_n, y_1 \dots y_{p(n)})$ such that

$$x_1 \dots x_n \in Q \leftrightarrow \exists \vec{y}\, C_n(\vec{x}_n, \vec{y}_n).$$

**Finite Model Theory.** What we call "Classical Finite Model Theory" deals with characterizing what we know about a relational problem if it is definable in some logic. A *relational problem*, which we also call a *query*, is a problem (i.e. Boolean function) whose instances are sets of finite structures for some relational language. You can think of your favorite graph problems (is the graph connected, 3-colorable, etc.) as being canonical examples. To define the computational complexity of a query we will use the standard coding of a finite structure $(\{1 \dots n\}, R(x_1 \dots x_k), \dots)$ as a string: $1^n 0\ \chi_R \dots$ where $\chi_R$ is a string of size $n^k$ that has a 1 at the $m$th place exactly when $R$ holds on the $m$th $k$-tuple.

Given a sentence $\phi$ in some logic over signature $S$, we can then talk about the complexity of $\{F : F \models \phi\}$. The first thing that is clear is that first-order problems are in $UAC^0$ — this follows from just "converting quantifiers to conjunctions and disjunctions". First-order logic over a pure relational signature $S$ gives a small subset of the class $UAC^0$, and thus is computationally extremely weak. Therefore we often look at first-order languages with built-in predicates on the coding:

- $FO(<)$ = first-order logic over $S$ plus the ordering on the code domain $\{1 \ldots n\}$
- $FO(+, *, <)$ = first-order logic over $S$ plus ordering plus arithmetic on domain $\{1 \ldots n\}$.

An issue with this sort of coding is that given an $S$ structure, there are many ways to order it. We often restrict attention to problems that are *generic*: the answer depends only on the $S$-structure, not on the ordering. We define the query classes $FO_{GEN}(<)$, $FO_{GEN}(+, *, <)$ to be the subsets of $FO(<)$, $FO(+, *, <)$ that are generic. It is known [25] that:

$$FO \subsetneq FO_{GEN}(<) \subsetneq FO_{GEN}(<, +, *) \subsetneq UAC^0.$$

In some situations, classes given by definability are exactly classical complexity classes. An example is the theorem of Fagin, stating that *NP* queries are the same as those definable in Existential Second Order Logic. In other cases, classes given by logics are "finer" than complexity classes:

- Existential Monadic Second Order Logic $\subset NP = ESO$
- First-Order Logic $\subseteq P$

The following table gives some exact relationships (i.e. for queries that are closed under isomorphism, membership in the complexity class is equivalent to definability in the logic). The proofs can be found in [25]:

| Class | Logic |
|---|---|
| PH | Second Order |
| NP | Existential Second Order |
| $UTC^0$ | First-Order plus Counting quantifiers plus order |
| $UAC^0$ | First-Order with Arithmetic plus order |

**Embedded Finite Model Theory.** We now fix a model $M = \langle \mathbb{U}, \ldots \rangle$ in a finite language $L$, and extend the previous notation.

Let $S$ be a relational signature disjoint from $L$, and consider an $S$ structure $D$ with domain a *finite* subset of $\mathbb{U}$. Any such $D$ determines an $L \cup S$ structure $(M, D)$, the expansion of $M$ in which $S$ is interpreted as in $D$. We call such a $D$, an *M-embedded finite model.* For example, if $M$ is the real field $\langle \mathbb{R}, +, *, 0, 1, < \rangle$ and $S = \{P\}$, unary, an embedded finite model is just a finite set of reals with a distinguished subset.

An *M-query* is a collection of embedded finite $M$-models. We want to look at $M$-queries definable by logics. For example, *First-Order Logic* is (syntactically) first-order logic over $L \cup S$. There are multiple semantics for an FO logic query $\phi$. By default, we look at *natural semantics*: to evaluate $\phi$ on embedded $D$ we evaluate it in the standard sense on the expansion $(M, D)$. For example, if our model is $\langle \mathbb{R}, +, *, 0, 1, < \rangle$, $S = \{G(x, y)\}$, then a first-order query is:

$$CoLin = \exists a\, \exists b\, \forall x\, \forall y (G(x, y) \leftrightarrow y = ax + b).$$

This evaluates to true on $D$ iff all points of $G^D$ are collinear. Since quantification is over the infinite structure, first-order logic in this default sense may not be weak at all. We use $FO(M)$ to be the set of queries defined using first-order logic under the natural semantics.

An alternative semantics is to restrict quantification to the finite structure. This gives us the notions of restricted and active query, which play a key role in the remainder.

DEFINITION 2.1.

- For $D$ an embedded finite model, the *active domain* of $D$, denoted $adom(D)$, is the union of all the projections of relation symbols in $D$.
- *Restricted First-Order Logic*, denoted $FO_r(M)$, is built from the first-order definable formulae of $L$ using the Boolean operators and the quantifiers $\exists x \in adom\, \phi(x, \vec{y})$ and $\forall x \in adom\, \phi(x, \vec{y})$.
- A formula $\phi \in FO_r(M)$ will be called a *restricted query*.
- An $FO_r(M)$ query $Q$ built up from only quantifier-free formulae using active domain quantifiers is called an *active query*. We write $Q \in FO_{act}(M)$.

$(M, D) \models \exists x \in adom\, \phi(x, \dots)$ iff there is an $a \in adom(D)$ such that $(M, D) \models \phi(a, \dots)$, and similarly for $\forall x \in adom$.

For example, $CoLin$ can be rewritten as an $FO_{act}$-query (roughly):

$$\forall u_1, u_2, v_1, v_2, w_1, w_2 \in adom$$
$$G(u_1, u_2) \wedge G(v_1, v_2) \wedge G(w_1, w_2) \rightarrow \frac{v_2 - v_1}{u_2 - u_1} = \frac{w_2 - w_1}{u_2 - u_1}.$$

If $M$ has QE, every restricted query can be converted to an active query. Even if $M$ itself is not decidable, we can still evaluate an active query effectively over any effectively presented $D$.

*Restricted Second-Order Logic* is built up similarly using $\exists X \subset adom$ and $\forall X \subset adom$, where $X$ is a new relation symbol. $(M, D) \models \exists X \subset adom\, \phi$ iff there is a subset $X_0$ of the $adom(D)^{arity(X)}$ such that $(M, D, X_0) \models \phi$. *Active Second Order Logic* is built up as above, but from quantifier-free formulae.

Note that none of the logics we deal with here can distinguish between two embedded finite structures $D$ and $D'$ based on the same underlying structure

$M$ which have the same interpretation for all the predicates in the relational signature $S$, even if the domains of $D$ and $D'$ are different. Hence we will generally assume that an embedded finite structure has domain equal to its active domain.

**Complexity Models.** How do we give the complexity of a logical formula over embedded finite structures? The most naive measure is *bit complexity*. If $M$ is decidable, we can restrict attention to queries without parameters from $M$ and to embedded finite models whose domain consists of definable elements of $M$, looking then at traditional complexity. For example, over the real field, we can look at formulae with rational parameters and embedded finite models whose active domain is in $Q$, and code both the formulas and the models as bit strings. Of course, bit complexity depends on the choice of a binary coding of definable elements, and different codings can yield radically different complexities. Although bit complexity reflects a concrete computation model, it doesn't match the level of abstraction of our query languages, which take operations of the model $M$ as primitive. A better fit is *BSS complexity* [11], which we review now.

Still fixing first-order structure $M$ in a finite language, we say that a *problem instance* over $M$ is a sequence of elements from $M$. The *size* of the input is just the number of elements in the sequence. A *BSS machine* $T$ over $M$ has a finite state control, an input tape that initially contains a sequence of elements from $M$, a fixed set of work tapes, each of which will contain a sequence of elements from $M$, and a transition function. During a transition $T$ can set a tape element by applying a function from $M$ to arguments from other work tapes or the input tape, can move heads on one of the tapes or can change state by applying a Boolean combination of $M$-predicates.[1] These transition functions of $T$ can make use of a fixed set of parameters from $M$ in the predicates or functions.

The time complexity of a machine $T$ is the number of steps as a function of the size of the input: i.e. each step costs one unit. As an example, if $M = (\mathbb{R}, +, *, 0, 1, <)$, one can consider the problem: Given a set of reals, determine if one of them is a product of the others. This can be done very easily in cubic time on a BSS machine. We say that a problem is $P_M$, $EXPTIME_M$ etc. if the number of steps is bounded by a polynomial (resp. exponential) function of the number of elements in the input. A problem $Q$ is in $NP_M$ iff there is a predicate $T(\vec{x}, \vec{y})$ in $P_M$ and polynomial $p(n)$ with:

$$\vec{x} \in Q \leftrightarrow \exists \vec{y} \in M^{<p(|\vec{x}|)}\ T(\vec{x}, \vec{y})$$

i.e. by default, non-determinism means we choose an element from $M$.

---

[1] In other equivalent formulations, one has a set of registers with a single element per register, and one has a current register instead of a head.

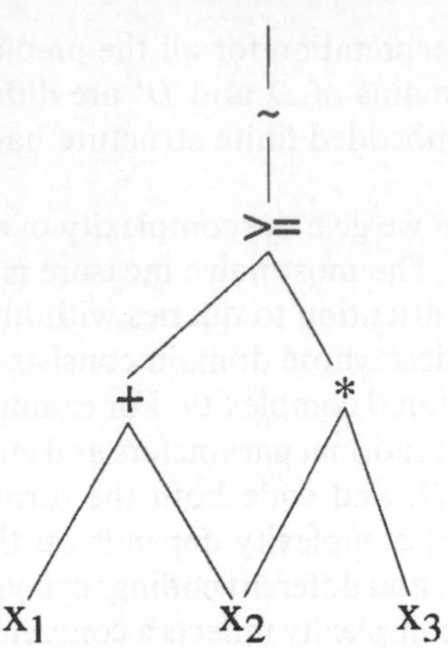


FIGURE 1. A circuit for $\neg(x_1 + x_2) \geq (x_1 * x_3)$.

If $M$ does not have QE, non-deterministic classes are badly-behaved (e.g. may have no complexity bound). Hence in talking about BSS complexity here we will normally restrict to models that have QE. However, we could extend the definitions above to allow the machines to use a fixed set of definable functions and predicates, and many of the statements in the remainder would then extend to the non-QE case.

We now look at the analog of circuit complexity. Any problem $Q$ solvable in time bounded by the arity of its arguments is given by a computable sequence of circuits. An $M$-circuit is a circuit with gates for every function and relation of $M$.

A problem is in $AC^0_M$ (or "has BSS complexity in $AC^0$") if it can be solved by a family of $M$-circuits of polynomial size and constant depth, and likewise for $UAC^0_M$.

In a model with QE, any restricted logical formula defines a quantifier-free problem. We have the following relationship of logics to complexity (we leave out the $M$ subscript on the right):

- $FO_{act} \subseteq UAC^0$
- $SO_{act} \subseteq EXPTIME$.

These inclusions are provably strict for most interesting models. One can lift the restriction on QE in this result by working in the extension of the BSS model with definable primitives, as mentioned above.

**Generic Complexity.** Finally, it is helpful to have a notion of complexity that looks at problems that 'make sense' in any structure. A class of embedded finite models for relational signature $S$ is *generic* if it is closed under $S$-isomorphism.

i.e. the query depends only on isomorphism type of the interpretations of the relational predicates $G \in S$, not the embedding in $M$. Consider the natural analogs of two "classic" queries from finite model theory:

- Parity = $\{V$ finite subset of $M$: $|V|$ is even$\}$.
- Connectivity = $\{G$ finite graph in $M^2$: $G$ is connected$\}$.

It is clear that Parity and Connectivity are both generic. In contrast, the query *CoLin* mentioned earlier is not generic. We let $FO_{GEN}(M)$ := set of generic queries first-order definable over $M$. We say a generic query is in complexity class $C$ if the corresponding collection of isomorphism types is in $C$, and say the generic complexity of $M$ is in $C$ if $FO_{GEN}(M) \subseteq C$.

**§3. Tame behavior of first-order queries.** We will now study the complexity of $FO(M)$ queries for various structures $M$. It is easy to see that if $M$ is extremely simple (e.g. the trivial infinite structure), then $FO(M)$ allows you to express only the standard FO queries of finite model theory. We abbreviate this phenomenon by saying $FO(M) = FO$ for these models. In contrast, if $M = (\mathbb{N}, +, *, <)$ one can use a simple coding argument to show that all arithmetic queries over finite structures are in $FO(M)$. This leads to the question of characterizing what $FO(M)$ is for well-known decidable structures $M$.

The main technique for getting bounds on the complexity of first-order queries is to show that every first-order query is equivalent to a restricted query. An example of this equivalence is the conversion of the *CoLin* query to a restricted query in the previous section. We say a structure has *first-order restricted-quantifier collapse* if every first-order query is equivalent to a restricted query. Likewise, we say a structure has *first-order active collapse* if every first-order query is equivalent to an active query. A simple argument using indiscernibles (see [18]) shows that:

THEOREM 3.1. *If $M$ has first-order restricted-quantifier collapse, then the generic complexity and the BSS complexity of first order queries is in $UAC^0$. In addition, we have the following finer information about generic complexity*:

- $FO_{GEN}(M) \subset FO_{GEN}(<)$.
- *If an elementary extension of $M$ contains a totally indiscernible set, then* $FO_{GEN}(M) = FO$.

So we consider how to get restricted-quantifier collapse. Recall that a model $M$ *does not have the finite cover property* ($M \in NFCP$) iff for every $\phi(x, y_1 \dots y_m)$ there is $k$ such that for every finite $A \subset M^m$,

$$\forall A_0 \in [A]^{<k}\ \exists x \bigwedge_{\vec{a} \in A_0} \phi(x, \vec{a}) \Rightarrow \exists x \bigwedge_{\vec{a} \in A} \phi(x, \vec{a}).$$

NFCP structures are stable [24], and hence do not carry definable linear orders. The significance of NFCP for complexity of FO derives from the following observation:

PROPOSITION 3.2 ([18]). *If $M \in NFCP$ then $M$ has restricted-quantifier collapse.*

From this it follows that the complex field has restricted-quantifier collapse, and hence for this model (and any algebraically closed field) first-order queries have generic complexity in FO and BSS complexity in $UAC^0$. We see from this in turn that Parity and Connectivity are not expressible as first-order queries over the complex field (since it is well-known that they are not in the finite model theory class FO [17]). Note that Proposition 3.2 gives *no bound* for the size of the corresponding restricted formula in terms of the original formula.

NFCP is much stronger than restricted-quantifier collapse. To (partially) characterize models with collapse, we need some non-classical model-theoretic properties.

A set $A \subset M$ is *pseudo-finite* if for some index set $I$, some sequence of models $M_i$ indexed by $I$ and sequence of finite sets $A_i \subset dom(M_i)$, and some nonprincipal ultrafilter $U$ over $I$, $(M, A) \equiv \Pi_U(M_i, A_i)$, with $M_i \equiv M$. Here $\Pi_U(M_i, A_i)$ is the ultraproduct. A pseudo-finite set satisfies every first-order property that holds of every finite $A$ in $M$. Theory $T$ is *pseudo-finitely homogeneous* if for every $M \models T$, pseudo-finite $A, B \subset M$, elementary bijection $h : A \to B$, and $a \in M$, we have that $h$ extends to an elementary map on $A \cup \{a\}$.

It's easy to check that NFCP implies pseudo-finite homogeneity. Pseudo-finite homogeneity characterizes collapse for stable theories.

PROPOSITION 3.3 ([18]). *For $M$ stable, Restricted Quantifier Collapse is equivalent to Pseudo-finite Homogeneity.*

Unfortunately, this characterization has not proved particularly helpful for showing collapse, even for stable theories. A particularly interesting open question is to characterize which stable structures have expansions with Restricted Quantifier Collapse. The proposition above is also of no assistance for unstable structures (although the equivalence is open in general). We now discuss sufficient conditions for restricted-quantifier collapse for unstable structures.

A model $M$ has the *Isolation Property* if for every $\gamma(x, y_1 \dots y_k)$ there is $\lambda_\gamma(x, w_1 \dots w_l)$ such that: For any finite set $A$ and $b \in M$ $\exists a_1 \dots a_l \in A$ with

$$\lambda_\gamma(b, a_1 \dots a_l) \wedge \forall x\, (\lambda_\gamma(x, a_1 \dots a_l) \to \bigwedge_{c_1 \dots c_k \in A} \gamma(x, c_1 \dots c_k) \leftrightarrow \gamma(b, c_1 \dots c_k)).$$

That is, every $\phi$-type over a finite set is isolated by some $\gamma$-type.

Isolation is a "local version" of Schmerl's "nuclearity" [36], defined by Belagradek, Stolboushkin, Taitslin, who showed:

THEOREM 3.4 ([5]). *Any model with the isolation property has first-order restricted quantifier collapse.*

All structures with isolation are unstable. It turns out that most "interesting" ordered structures have the isolation property.

THEOREM 3.5. *The following structures have the isolation property:*

- *The real ordered group and real ordered field*
- *Presburger arithmetic* $(\mathbb{N}, +, 0, <)$
- *Any o-minimal structure.*

Recall that an ordered structure $M = (\mathbb{U}, <, \ldots)$ is *o*-minimal if every definable $A \subset M$ is a finite union of intervals. Examples include real closed fields, the real field with exponentiation, the reals with all restricted analytic functions [37].

COROLLARY 3.6. *Each o-minimal structure has generic complexity in* $FO(<)$ *and BSS complexity in* $UAC^0$.

Theorem 3.5 gives a uniform method of getting restricted-quantifier collapse for ordered structures. It does not match the historical development, which saw a series of results proved by more model-specific techniques. The first step towards Corollary 3.6 was a proof for the real-ordered group in [31]; the extension to the real-ordered field and *o*-minimal structures came in [6]. The isolation technique was formalized in [5].

Isolation has not been studied in its own right, and there are many simple model-theoretic questions that arise concerning it. For example, which structures have an expansion with isolation? Can every structure with isolation be expanded with a linear order, while preserving isolation?

If we consider the concrete case of the real field, the intuition for how to prove restricted-quantifier collapse is quite basic. Inductively, we need to process queries of the form $\exists x\, Q_1(w_1) \ldots Q_n(w_n)\, \phi$, where the $Q_i$ are quantifiers. For a given embedded finite $D$, $\gamma_i(x, \vec{w}, \vec{y}) \in \phi, \vec{w} \in adom(D)$ and $\vec{y} \in \mathbb{R}$, we know that $\gamma_i(x, \vec{w}, \vec{y})$ is a union of $K$ intervals with endpoints definable from $\vec{w}, \vec{y}$. A witnessing $x$ can be taken to be midpoint of two such endpoints, and hence

$$Q = \exists \vec{z} \in adom \bigvee_j Q_1(w_1) \ldots Q_n(w_n)\, \phi'_j$$

where $\phi'_j$ is formed by replacing each $\gamma_i(x, \vec{w}, \vec{y})$ with $\gamma_i(\mathrm{Sample}_j(\vec{z}, \vec{y}), \vec{w}, \vec{y})$, where $\mathrm{Sample}_j$ samples the $j$th interval defined by two endpoints.

One can get a more efficient version using some refinements:

- Eliminate blocks of quantifiers at once.
- Find a small number of $\mathrm{Sample}_j$ covering sample points for all cells.

See [4] for a discussion of these issues in the case of signatures with only unary predicates.

What is the significance of restricted-quantifier collapse over the fields $R$ and $C$? Some formal consequences are as follows:

COROLLARY 3.7. *Every first-order query over the real field has*:

- *BSS complexity in* $UAC^0$
- *bit complexity in* $UTC^0$

The second item follows from the fact that integer multiplication can be done with $UTC^0$ circuits [34]. To get a more intuitive idea of where the interest in collapse lies, consider an FO query $Q$ over signature $S \cup L_{RCF}$. Given a finite interpretation $I$ for $S$ we can plug in to get an FO formula $Q_I$. Thus the naive evaluation of $Q$ is to plug in $I$ and apply a standard QE algorithm. For example, using [35] one would need to evaluate a quantifier-free formula of size: $|Q_I|d^{\prod_{i<w} n_i}$, where $d$ is a bound on degree of polynomials in $Q$, $Q$ has $w$ natural quantifier blocks, and block $i$ has $n_i$ variables in it. In contrast, an algorithm using Restricted Quantifier Collapse would proceed in two phases. First, get an active query $Q'$ equivalent to $Q$, and second evaluate $Q'$ in $TC^0$. Although the two steps together are no more efficient than the earlier technique, the expensive first part depends only on $Q$ and once performed can be applied cheaply to any $I$. So $Q'$ can be thought of as a quantifier-elimination schema, which can be applied "in advance" of the data.

For the real field and complex field, converting to an active query is expensive. The best current bound is on the order of $(|Q| * d)^{k^n}$, where $d$ is a bound on the degree of the polynomials, $n$ is the number of variables, and $k$ is a constant. On the other hand, it is not known if conversion to a restricted query can be done in polynomial time.

Restricted Quantifier Collapse has also been an important tool in understanding the expressiveness of Query Languages for Geographic Databases. Here one considers a first-order $\phi(G_1, \ldots, G_n)$ with $G_i$ predicates, and we let $G_i$ vary over a class of regions in Euclidean space. Example classes studied include the class of polygonal semi-linear or semi-algebraic regions. For example, using the fact that the graph connectivity query is not first-order definable over embedded finite models in the real field, one can easily show there is no $\phi(G)$ that defines *Topological Connectivity* as $G$ varies over any of the families above. See the book [28] for details.

**§4. Tame behavior of generic queries.** We have seen that for structures satisfying strong model-theoretic conditions (e.g. NFCP, o-minimality) quantification over a model can be eliminated, and first-order queries collapse to first-order logic in the sense of finite model theory. We now look at the generic complexity of FO queries. We will see that much weaker model-theoretic properties suffice to get bounds. These weaker hypotheses are "local" versions of previous properties.

It is easy to show that models without NFCP may not have queries well-behaved. Consider the standard FCP model $M$ consisting of an equivalence relation with a single class of every finite size, and the query "$U$ contains an equivalence class". This query is not equivalent to any restricted query, and it has "no complexity" in any of our complexity models. However, the class of generic queries $FO_{GEN}(M)$ is still trivial for this model. This follows from stability:

THEOREM 4.1 ([2]). *For any $M$ stable, $FO_{GEN}(M) = FO$.*

COROLLARY 4.2. *Parity and Connectivity are not definable as first-order queries over any stable structure.*

We outline a proof of Theorem 4.1 due to Casanovas and Ziegler. The key to the generic collapse result is relative restricted collapse. For structure $M$, $I \subset M$, say $M$ has *restricted-quantifier collapse on $I$* if for first-order query $\phi$ there is restricted query $\phi'$ such that

$$\forall D \subset I \quad M(D) \models \phi \leftrightarrow \phi'.$$

It is easy to see that if $M$ has restricted-quantifier collapse on some infinite totally indiscernible set $I$, then $FO_{GEN}(M) = FO$. If $M$ has restricted-quantifier collapse on some infinite indiscernible ordered set $(I, <)$, then $FO_{GEN}(M) = FO_{GEN}(<)$.

To prove relativized restricted collapse, one needs to relativize notions form stability. *$M$ does not have fcp over $I$* (we write $M \in NFCP/I$) iff:

For every $\phi(x, y_1 \dots y_m)$ there is $k$ such that for every finite $A \subset I^m$, if for every $A_0 \in [A]^{<k}$, $M \models \exists x \bigwedge_{\vec{a} \in A_0} \phi(x, \vec{a})$ then $M \models \exists x \bigwedge_{\vec{a} \in A} \phi(x, \vec{a})$.

A simple generalization of Proposition 3.2 gives the following.

PROPOSITION 4.3 ([15]). *If $M \in NFCP/I$ for infinite $I \subset M$, then $M$ has restricted-quantifier collapse over $I$.*

For $I \subset |M|$ we say $M$ is *Stable over $I$* iff there is no $\phi(\vec{x}, \vec{y})$ that defines a linear order on arbitrarily large finite subsets of $I$. Relative stability can also be defined using the relativization of ranks to $I$. Given $I \subset M$, $I_{ind}$ is the structure with domain $I$ and predicates for every relatively definable subset of $I$.

THEOREM 4.4 ([15]). *If $M$ is stable over $I$, and $I_{ind} \in NFCP$, then $M \in NFCP/I$.*

The generic collapse result, Theorem 4.1, now follows, since if $M$ is stable it is stable over indiscernible set $I$, and the structure $I_{ind}$ is trivial.

An analog of the "restricted quantification" result for stable theories also holds when allowing the additional relation to be infinite but small.

THEOREM 4.5 ([15]). *If $M \in NFCP/I$, and $I$ is small (i.e. $M$ is $|I|^+$-saturated), then every $L \cup S$ formula $\phi(\vec{x})$ is equivalent to a formula with quantifiers restricted to $I$. Furthermore, $M(I)$ is stable.*

COROLLARY 4.6. *If $M$ is stable, $I$ small and indiscernible, then every $L \cup S$ formula $\phi(\vec{x})$ is equivalent to a restricted formula, and hence $I$ is strongly minimal relative to $M(I)$.*

Note that these last results are about open formulas with an extra predicate interpreted by a fixed small infinite set, not by a varying finite set. It suggests that results about uniform definability with a free relational predicate ranging over finite subsets can be carried over to the case where the predicate ranges over small subsets, and even (with more drastic restrictions on the ambient structure) to arbitrary subsets. Further generalizations in this direction can be found in [1].

So we know that for stable theories, for every first-order query $Q$, for any indiscernible set $I$, there is a restricted query $Q'$ equivalent to $Q$ for instances that come from $I$. How complex can $Q'$ be? This is the analog to the question of complexity of active-quantifier collapse mentioned in the previous section. For general stable theories it's unlikely that anything further can be said. For strongly minimal structures more quantitative results are possible. Fournier [19] gives a technique for getting bounds on the expressivity of first-order queries over an infinite indiscernible set $I$ in an algebraically closed field. Roughly speaking, he shows that over embedded finite structures with domain in $I$ a first-order query of quantifier rank $n$ can be converted to an active query of quantifier rank at most $2^n$. As a consequence, Fournier is able to get more quantitative results on sequences of formulae needed to define generic queries. For example, we know $FO_{GEN}(M) = FO$ for $M$ stable, so therefore the Parity query is not expressible as a first-order query in a stable structure. In finite model theory, we know a stronger fact: Parity on structures of size $n$ cannot be expressed by a formula with quantifier rank less than $n$. Fournier shows an analogous bound for embedded finite models in algebraically closed fields.

THEOREM 4.7 ([19]). *If $M$ is algebraically closed, then there is no sequence of formulae $\phi_n$ of quantifier rank below $\log(n) + 1$ such that $\phi_n$ expresses Parity over embedded models of size at most $n$.*

Fournier's paper [19] has similar results for real closed fields, with both upper and lower bounds.

One can ask what classes of unstable theories have low generic complexity. Recall that a model $M$ *fails to have the Independence Property* if there is no formula $\gamma(\vec{x}, \vec{y})$ for which there are arbitrarily large finite $S$ such that

$$\forall X \in P(S)\, \exists \vec{m} \in M \bigwedge_{\vec{c} \in X} (\gamma(\vec{m}, \vec{c}) \leftrightarrow \vec{c} \in X)$$

We write $M \in NIP$. It is known that $o$-minimal structures are in NIP [37], as are stable structures [24].

THEOREM 4.8 ([2]). *If $M \in NIP$, then generic queries have complexity in $FO(<) \subset UAC^0$.*

The converse is not the case. In the next section, we will see an example of $M$ with $IP$ (that is $M \notin NIP$) with generic queries in $UAC^0$. However, it is known [9] that if $M \in IP$ then $M$ cannot have restricted-quantifier collapse. The proof of Theorem 4.8 again goes through the idea of relativization:

THEOREM 4.9. *If $M \in NIP$, $(I, <)$ is order-indiscernible in $M$ with complete dense order type, $M$ is $|I|^+$-saturated, then $M$ has restricted-quantifier collapse over $I$.*

The proof idea can be considered as extending the relativization idea of Theorem 4.1. If $M \in NIP$, $(I, <)$ is order-indiscernible in $M$ with complete dense order type, then types over $I$ are definable (a kind of relative stability). Hence $I_{ind}$ is just a linear orderering. Under these conditions, $M$ has 'restricted NFCP over $I$'. Let $\phi(\vec{x}, \vec{w}, \vec{y})$ have defining scheme $\delta_\phi(\vec{w}, \vec{u})$ over $I$. An $\vec{i} \in I$ gives a *potential* $\phi$-type: $\{\phi(\vec{x}, \vec{w}, \vec{y}) : \delta_\phi(\vec{i}, \vec{w})\}$. What is shown in [2] is that there is $k$ such that every potential $\phi$-type that is $k$-consistent is consistent, and from this relative NFCP collapse follows.

What can we say when first-order generic queries do not collapse down to active first-order logic?

$M$ has *Second-Order collapse* if every FO query is equivalent to a restricted second-order query. Similarly for Monadic Second Order. Recall that $PH$ is the polynomial hierarchy, obtained from $P$ by closing under complement and polynomial non-determinism.

If $M$ has Second Order collapse we know that first-order queries have:

- generic complexity in $SO = PH$
- BSS complexity in $PH$.

If $M$ has Monadic Second Order collapse we know the generic complexity is a proper subset of $PH$. In [7], it is shown that the random graph has monadic second-order collapse but not first-order collapse, and $FO_{GEN}(M) = MSO$. Hence generic queries expressible over the random graph are in $MSO \subsetneq PH$. Extensions of this technique have been used to get random structures $M$ with $FO_{GEN}(M) = PH$.

**Open Question:** Obtain general criteria for second-order collapse. It is conceivable that all Simple Theories have generic complexity in second-order.

What can $FO_{GEN}(M)$ be for a model $M$? So far we know that it can be:

- $FO$ (e.g. $M$ stable)
- $FO_{GEN}(<)$ (e.g. $M \in NIP$)
- $MSO$, $SO$ (Random Graph, Random $n$-ary relations)
- All arithmetic queries (full arithmetic $(\mathbb{N}, +, *, <)$).

**Open Question(s):** Are there models where every query in $FO_{GEN}$ is computable, but $FO_{GEN}$ contains more than $PH$? Computable models? Are there models where $FO_{GEN}$ is strictly between $FO_{GEN}(<)$ and $AC^0$?

By going to nonstandard models of arithmetic, it is easy to get generic queries that are "arbitrarily complex". What about decidable structures that are model-theoretically badly behaved? The best known example studied in EFMT is *Büchi Arithmetic*. That is, the structure $(\mathbb{N}, +, <, V_2(x))$, where $V_2(x)$ = highest power of 2 in binary expansion of $x$.

An equivalent structure can be obtained by identifying $\mathbb{N}$ with set of strings over two-element alphabet $\Sigma = \{a, b\}$. Consider

$$BA = (\Sigma^*, \prec, top_a(x), top_b(x), el(x, y))$$

where

- $\prec$ is the prefix relation
- $top_a(x)$ holds iff the last letter of $x$ is $a$ (and similarly for $top_b(x), b$)
- $el(x, y)$ holds iff $|x| = |y|$.

It has been shown [14] that definable relations $\phi(x_1 \ldots x_n)$ over $BA$ are given by a finite automaton with $n$ heads moving synchronously down $x_1 \ldots x_n$. As a consequence, $M$ is decidable (see also [12]). However, $BA$ is not in any standard model-theoretically well-behaved class: $BA$ has IP (that is $\notin$ *NIP*), has the strict order property, and does not have QE in a finite language. What can we say about first-order queries over $BA$?

The automaton representation of formula shows that first-order $\phi(x_1 \ldots x_n)$ can be converted to one where quantification is over strings of size less than the max size of $x_i$. The analogous result is true for queries:

PROPOSITION 4.10 ([9]). *Let $\phi \in L \cup S$ be a first-order query over $BA$. $\phi$ can be converted to a query where quantifiers are restricted to range over strings whose size is no greater than the largest-sized string in the active domain.*

This result can be used to bound the complexity of arbitrary $BA$-queries:

COROLLARY 4.11 ([9]). *Every first-order query over $BA$ has bit complexity in PH.*

We can show this is best possible; there are FO queries capturing problems complete for every level of the polynomial hierarchy [9]. One can show that the BSS complexity of first-order queries over $M$ "does not exist": that is, that there is no time bound solely in terms of the number of parameters of the input structure.

So far the results about the BSS complexity of general queries over this model are roughly similar to stable structures with the FCP. What about generic queries? These turn out to be much better behaved.

THEOREM 4.12 ([9]). $FO_{GEN}(BA) \subset UAC^0$.

The proof idea is simple, and gives an idea of how one can get bounds on the complexity of generic queries without restricting to embedded models that live inside an indiscernible sequence. Consider the model as $(\mathbb{N}, +, <, V_2(x))$ again. If $Q$ is generic, it is determined by its behavior on embedded models whose domain is an initial segment. For such $D$

$$\max\{|x| : |x| \in D\} \leq 2|D|.$$

So we can restrict quantification to be over $2|D|$. Hence evaluation of $Q$ over an ordered input $s$ coding the size $n$ of the domain and then the values of the relations on $n$ can be implemented with a first-order formula with arithmetic over $n$. It is known that such a formula can be computed in $UAC^0$.

In analogy with the results on $o$-minimal structures, one might expect that in fact $FO_{GEN}(M) = FO_{GEN}(<)$. However, this is open. In fact, the complexity of generic queries for many other decidable structures is unknown, at least to this author. For the universal tree automatic structure, results analogous to Büchi arithmetic have been shown [8]: generic complexity lies somewhere within $UAC^0$, but is not known to collapse all the way down to $FO_{GEN}(<)$. Bit complexity is high, and BSS complexity is undefined. For pseudo-finite fields, generic complexity is at least $SO$ [10], but no upper bound is known.

**§5. Embedded finite model theory and abstract complexity.** In this section we mention some connections between EFMT questions and fundamental issues in BSS complexity. First of all we review what the results of the previous section say (or don't say) with respect to the $P_M = NP_M$ question for a model $M$. Recall that the class $P_M$ is the set of problems given by a BSS machine running in polynomial time. Each such problem is given by a sequence of circuits $C_n(x_1 \dots x_n)$ where $C_n$ can be generated in polynomial time (in the traditional sense) from $n$. The class $NP_M$ is the set of problems $Q \subset \bigcup_n |M|^n$ such that there are circuits $C_n(x_1 \dots x_n, y_1 \dots y_{p(n)})$ using only finitely many parameters from $M$ such that

$$x_1 \dots x_n \in Q \leftrightarrow \exists \vec{y}\; C_n(\vec{x}, \vec{y})$$

and such that $C_n$ can be constructed in $P$ from $n$. So $P_M = NP_M$ only if there is a fast quantifier elimination algorithm for $M$.

Active Quantifier collapse is a special case where we can convert an $NP_M$ problem to a $P_M$ problem. More specifically, consider the class $EFO(M) =$ first-order $M$ queries with just existential natural quantifiers when put into prenex form. That is, queries of the form $\exists x_1 \dots x_n \forall w_1 \in adom \dots \exists w_m \in adom\ \phi(\vec{x}, \vec{w})$. If $M$ has QE, then it is always the case that $EFO(M) \subset NP_M$. But for all the models such that we can change $EFO(M)$ queries to active ones (the real field, etc. from the results in Section 3) we have a stronger inclusion: $EFO(M) \subsetneq UAC^0_M \subseteq P_M$.

In fact, $EFO$ problems are very special $NP_M$ problems. Let $NP^b_M$ ("bounded $NP$") be the subset of $NP_M$ defined by polynomial circuits

$$C_n(x_1 \dots x_n, y_1 \dots y_k).$$

That is, the number of guesses is fixed independent of input. Then we have $EFO \subset NP^b_M$ always. For the real field one has a still stronger statement then the inclusion of $EFO$ in $P_M$, namely $NP^b_M = P_M$ (this follows, for example, from results of Ben-Or and Kozen, see e.g. [35]). It would be interesting to know if active-quantifier collapse implies that $NP^b_M = P_M$. We expect that the collapse techniques can be used to show membership in $P_M$ for other problems given by "slow growing" formulas.

Of course, the fact that $EFO$ problems are very special citizens of $NP_M$ means that query collapse results are not a step towards proving $P_M = NP_M$ for any models with collapse. In fact, what we have seen is that when $FO(M)$ is in $P_M$, expressiveness of $EFO$ is very low. However, if we had $M$ with QE for which $EFO_{GEN}(M)$ is above $EXPTIME_M$, then obviously $P_M \neq NP_M$ for that model. Hence getting structures that fill in the gaps in generic complexity mentioned in the previous section might give us new examples where $P_M \neq NP_M$, ones where this holds because FO queries are intractable.

At the time of this writing, there are no structures known in a finite language for which $P = NP$ or its non-uniform version has been proved, although there are many for which provably $P \neq NP$. It is clear that for well-known structures this question is not going to be easy to solve. For the real ordered group Fournier and Koiran [20] show that deciding $P = NP$ is as hard as deciding $\mathcal{P} = \mathcal{NP}$ classically. For the complex and real fields, results of Blum, Cucker, Shub, Smale [11] and by Koiran (see [27]) show that this problem also connects to classical questions.

Since deciding $P \neq NP$ over a model may be overly ambitious, an a priori more tractable problem is to prove $P_M = NP_M$ holds iff $P_N = NP_N$ for some (every) pair of models of a given theory. [30, 16] thus deal with the question: what general conditions on $M$ or Theory($M$) allow one to conclude that $P_M = NP_M$ depends only on theory of $M$? That is, when is $P_M = NP_M$ first-order? There is another connection to EFMT here, since a similar question is open for query classes. One can ask whether the property "$M$ has restricted-quantifier collapse" depends only on the theory of $M$ and similarly for "Generic complexity of FO queries over $M$ is in $FO$" (or $SO$). For each of these properties, where it is known to hold, it depends only on the theory. Thus, unlike for $P_M = NP_M$, the issue of transfer for EFMT questions is settled for most decidable structures — what is lacking is a general model-theoretic result to this effect.

We list some techniques due to Michaux, Koiran, and Chapuis that are relevant to the question of whether the complexity of FO queries depends only on the theory. In the context of BSS complexity, Michaux observed that collapse of complexity classes easily transfers "upward":

THEOREM 5.1 ([30]). *If* $P_M = NP_M$ *and* $M \prec N$, *then* $P_N = NP_N$.

The proof is a simple application of elementary equivalence. Michaux's argument applies to any other reasonable complexity class (e.g. $\mathcal{P} = \mathcal{NP}$

transfers upward) and also to the FO query collapse properties from previous sections; $C \subset FO_{GEN}(M)$ for reasonable complexity class $C$ transfers upward.

The main question is how to transfer collapse of complexity classes "downward". In the context of Michaux's work, the question was if $P_N = NP_N$ implies that $P_M = NP_M$ holds for every elementary submodel $M$ of $N$. The obstacle is as follows: consider a problem $Q$ of $NP_M$, given by the projection of $C_n(x_1 \ldots, y_1 \ldots y_{p(n)})$.

Since $P_N = NP_N$, there is an efficiently computable sequence of $N$-circuits $D_n(x_1 \ldots x_n, z_1 \ldots z_l, \alpha)$ such that in $N$:

$$x_1 \ldots x_n \in Q \leftrightarrow D_n(x_1 \ldots x_n, \vec{\alpha})$$

But how does one eliminate $\vec{\alpha} \in N$?

Chapuis and Koiran give a 'quantitative version of stability' that is designed to resolve this question. They define a structure $M$ to be $P$-stable if given any uniform polynomially-growing sequence of circuits $D_n(x_1 \ldots x_n, \alpha)$ with $\alpha$ in an elementary extension of $M$, there is a uniform polynomially-growing $D'_n(x_1 \ldots x_n, \vec{\beta})$ with $\beta \in M$ and

$$M \models D'_n(x_1 \ldots x_n, \beta) \leftrightarrow D_n(x_1 \ldots x_n, \alpha)$$

$P$-stability is known to imply stability [16]. If every model of a theory $T$ is $P$-stable, then the $P = NP$ problem has the same answer in every model of $T$. A similar result pertains to containments of query classes in $P_M$. For example, if every model of a theory is $P$-stable, then whether $FO_{GEN}$ is contained in $P_M$ is independent of the model $M$. Unfortunately, $P$-stability is known for very few structures and these are ones for which query classes are already well-understood. For example, results of Blum, Cucker, Shub, and Smale [11] show that the theory of algebraically closed fields of characteristic zero is $P$-stable, and similar results are known for differentially closed fields [33]. One could also define the more specific elimination-of-parameters property for queries — looking specifically for models in which parameters from an elementary extension are eliminable from FO queries. Following the argument of Michaux above, such models would allow generic or restricted collapse results to transfer upward to elementary extensions.

**§6. Conclusions.** EFMT aims to relate the model-theoretic complexity of a structure to computational properties of problems defined by first-order queries. Thus far we know only that there is some relation between the model theory and the computational power of FO queries, but the range of possibilities in the first order case are not well-understood, and what happens beyond first order logic is almost completely unexplored. The analysis we have so far holds out the possibility of relating model-theoretic and complexity-theoretic nastiness for other classes of structures (e.g. simple) but this is purely speculative at this point. In the case of decidable structures, the upper bounds on query expressivity yield methods for evaluating sentences that are parameterized by relations, algorithms that behave better as the size of the parameter increases than prior methods. However, a convincing practical case for these algorithms has yet to be made.

EFMT and generalized complexity both aim to study what can be expressed with a sequence of formulas $\phi_n(x_1 \dots x_n)$. The query languages of EFMT give rise to special kinds of sequences where the size and quantifier rank of $\phi_n$ grow only moderately with $n$. BSS complexity deals with general polynomially-growing $\phi_n$, but with a focus on particular prenex classes (e.g. sequences of $\Sigma_1$ formulae). For general polynomially growing $\phi_n$, getting bounds on expressiveness is considerably more difficult, and even to resolve these questions relative to classical complexity-theoretic conjectures requires new tools. In particular, it appears to require more "quantitative" versions of notions from stability.

## REFERENCES


[1] J. BALDWIN and B. BAIZHANOV, *Local homogeneity*, Submitted for publication.

[2] J. BALDWIN and M. BENEDIKT, *Stability theory, permutations of indiscernibles, and embedded finite models*, ***Transactions of the American Mathematical Society***, vol. 352 (2000), pp. 4937–4969.

[3] D.A. BARRINGTON, N. IMMERMAN, and H. STRAUBING, *On uniformity within $NC^1$*, ***Journal of Computer and System Sciences***, vol. 41 (1990), pp. 274–306.

[4] S. BASU, *New results on quantifier elimination over real closed fields and applications to constraint databases*, ***Journal of the ACM***, vol. 46 (1999), pp. 537–555.

[5] O. BELAGRADEK, A. STOLBOUSHKIN, and M. TAITSLIN, *Extended order-generic queries*, ***Annals of Pure and Applied Logic***, vol. 97 (1999), pp. 85–125.

[6] M. BENEDIKT, G. DONG, L. LIBKIN, and L. WONG, *Relational expressive power of constraint query languages*, ***Journal of the ACM***, vol. 45 (1998), pp. 1–34.

[7] M. BENEDIKT and L. LIBKIN, *Unpublished notes*, 1999.

[8] ———, *Tree extension algebras: logics, automata, and query languages*, ***IEEE symposium on logic in computer science***, 2002, pp. 203–214.

[9] M. BENEDIKT, L. LIBKIN, T. SCHWENTICK, and L. SEGOUFIN, *String operations in query languages*, ***ACM symposium on principles of database systems***, 2001, pp. 183–194.

[10] O. BEYARSLAN and D. MARKER, *Personal communication*, 2002.

[11] L. BLUM, F. CUCKER, M. SHUB, and S. SMALE, ***Complexity and real computation***, Springer Verlag, 1997.

[12] A. BLUMENSATH and E. GRÄDEL, *Automatic structures*, ***IEEE symposium on logic in computer science***, 2000, pp. 51–62.

[13] R.B. BOPPANA and M. SIPSER, *The complexity of finite functions*, ***Handbook of theoretical computer science, volume A***, Elsevier, 1990, pp. 757–804.

[14] V. BRUYÈRE, G. HANSEL, C. MICHAUX, and R. VILLEMAIRE, *Logic and p-recognizable sets of integers*, ***Bulletin of the Belgian Mathematical Society***, vol. 1 (1994), pp. 191–238.

[15] E. CASANOVAS and M. ZIEGLER, *Stable theories with an extra predicate*, ***The Journal of Symbolic Logic***, vol. 66 (2001), pp. 1127–1140.

[16] O. CHAPUIS and P. KOIRAN, *Stability and saturation in the theory of computation over the reals*, ***Annals of Pure and Applied Logic***, vol. 99 (1999), pp. 1–49.

[17] H.-D. EBBINGHAUS and J. FLUM, ***Finite model theory***, Springer Verlag, 1995.

[18] J. FLUM and M. ZIEGLER, *Pseudo-finite homogeneity and saturation*, ***The Journal of Symbolic Logic***, vol. 64 (1999), pp. 1689–1699.

[19] H. FOURNIER, *Quantifier rank for parity of embedded finite models*, ***Theoretical Computer Science***, vol. 295 (2003), pp. 153–169.

[20] H. FOURNIER and P. KOIRAN, *Lower bounds are not easier over the reals: Inside PH*, ***ICALP***, 2000, pp. 832–843.

[21] M. GAREY and D. JOHNSON, ***Computers and intractability***, W.H. Freeman, 1979.

[22] E. GRÄDEL and Y. GUREVICH, *Metafinite model theory*, ***Information and Computation***, vol. 140 (1998), pp. 26–81.

[23] E. GRÄDEL and K. MEER, *Descriptive complexity theory over the real numbers*, ***ACM symposium on theory of computing***, 1995, pp. 315–324.

[24] W. HODGES, ***Model theory***, Cambridge, 1993.

[25] N. IMMERMAN, ***Descriptive complexity***, Springer Verlag, 1999.

[26] P. KANELLAKIS, G. KUPER, and P. REVESZ, *Constraint query languages*, ***Journal of Computer and System Sciences***, vol. 51 (1995), pp. 26–52, Extended abstract in *ACM Symposium on Principles of Database Systems*, 1990, pages 299–313.

[27] P. KOIRAN, *Transfer theorems via sign conditions*, ***Information Processing Letters***, vol. 81 (2002), pp. 65–69.

[28] G. Kuper, L. Libkin, and J. Paredaens (editors), ***Constraint databases***, Springer Verlag, 2000.

[29] K. MEER and C. MICHAUX, *A survey on real structural complexity theory*, ***Bulletin of the Belgian Mathematical Society***, (1997), pp. 113–148.

[30] C. MICHAUX, *$P \neq NP$ over the nonstandard reals implies $P \not= NP$ over $\mathbb{R}$*, ***Theoretical Computer Science***, vol. 133 (1994), pp. 95–104.

[31] J. PAREDAENS, J. VAN DEN BUSSCHE, and D. VAN GUCHT, *First-order queries on finite structures over the reals*, ***SIAM Journal on Computing***, vol. 27 (1998), pp. 1747–1763.

[32] B. POIZAT, *Les petites caillous*, 1995.

[33] N. PORTIER, *Stabilité polynômiale des corps différentiels*, ***The Journal of Symbolic Logic***, vol. 64 (1999), no. 2, pp. 803–816.

[34] J.H. REIF, *On threshold circuits and polynomial computation*, ***IEEE conference on structure in complexity theory***, (1987), pp. 118–123.

[35] J. RENEGAR, *On the computational complexity and geometry of the first-order theory of the reals*, ***Journal of Symbolic Computation***, (1992), pp. 255–352.

[36] J. SCHMERL, *On $\aleph_0$-categoricity and the theory of trees*, ***Fundamenta Mathematicae***, vol. 94 (1977), pp. 121–128.

[37] L. VAN DEN DRIES, ***Tame topology and O-minimal structures***, Cambridge, 1998.

BELL LABORATORIES
2701 LUCENT LANE
LISLE, IL 60532, USA
*E-mail*: benedikt@research.bell-labs.com

# RESEARCH ARTICLES

# INDESTRUCTIBILITY AND STRONG COMPACTNESS

ARTHUR W. APTER

**Abstract.** We construct a model in which the first two strongly compact cardinals aren't supercompact yet satisfy significant indestructibility properties for their strong compactness.

**§1. Introduction and preliminaries.** The study of indestructibility for non-supercompact strongly compact cardinals is one which has been the subject of a great deal of investigation over the last few years, most notably in the papers [1, 6, 7, 3], and [15]. We refer readers to the introductory section of [3] for a thorough discussion of the relevant history. We note, however, that in spite of all of the work done, the basic question of whether it is consistent, relative to anything, for the first two strongly compact cardinals to be non-supercompact yet to satisfy significant indestructibility properties for their strong compactness had heretofore been left unanswered. This should be contrasted to the relative ease with which Laver's forcing of [18] iterates, to produce models such as the one given in [1] in which there is a proper class of supercompact cardinals and every supercompact cardinal $\kappa$ has its supercompactness indestructible under $\kappa$-directed closed forcing.

The purpose of this paper is to provide an affirmative answer to the above question. Specifically, we will prove the following theorem.

THEOREM 1.1. *It is consistent, relative to the existence of two supercompact cardinals, for the first two strongly compact cardinals $\kappa_1$ and $\kappa_2$ to be non-supercompact yet to satisfy significant indestructibility properties for their strong compactness. Specifically, in our final model $V^{\mathbb{P}}$, $\kappa_1$'s strong compactness is indestructible under arbitrary $\kappa_1$-directed closed forcing, and $\kappa_2$'s strong*

2000 *Mathematics Subject Classification.* 03E35, 03E55.

*Key words and phrases.* Supercompact cardinal, strongly compact cardinal, indestructibility, non-reflecting stationary set of ordinals.

The contents of this paper were presented in the Special Session in Set Theory of Logic Colloquim 2003, held August 14–20, 2003 in Helsinki, Finland. The author wishes to thank the organizers for having invited him to speak at and participate in a very stimulating conference. The author also wishes to thank Hy-Line Cruises, for having provided a very relaxing setting under which the main result of this paper was proven.

*compactness is indestructible under either trivial forcing or $\kappa_2$-directed closed forcing that can be written in the form* $\text{Add}(\kappa_2, 1) * \dot{\mathbb{Q}}'$.

We note that Theorem 1.1 is a generalization of a sort of Theorem 1 of [3]. In the model constructed for this theorem, the first two strongly compact cardinals $\kappa_1$ and $\kappa_2$ aren't supercompact (and in fact, are the first two measurable cardinals), $\kappa_1$'s strong compactness is fully indestructible under $\kappa_1$-directed closed forcing, yet $\kappa_2$'s measurability, but not necessarily its strong compactness, is indestructible under arbitrary $\kappa_2$-directed closed forcing.[1] Also, for our Theorem 1.1, as for Theorem 1 of [3], it is impossible simply to iterate the forcing conditions given in [6], where a model in which the least strongly compact cardinal $\kappa$ is both the least measurable cardinal and has its strong compactness fully indestructible under $\kappa$-directed closed forcing is constructed. The reason is that these forcing conditions, being a Gitik-style iteration of Prikry-like forcings as introduced in [13], add bona fide Prikry sequences, which in turn by Theorem 11.1(1) of [11] then add non-reflecting stationary sets of ordinals of cofinality $\omega$. By Theorem 4.8 of [19] and the succeeding remarks, these non-reflecting stationary sets of ordinals can't exist above a strongly compact cardinal.

Before beginning the proof of Theorem 1.1, we briefly mention some preliminary information. Essentially, our notation and terminology are standard, and when this is not the case, this will be clearly noted. For $\alpha < \beta$ ordinals, $[\alpha, \beta]$, $[\alpha, \beta)$, $(\alpha, \beta]$, and $(\alpha, \beta)$ are as in standard interval notation.

When forcing, $q \geq p$ will mean that $q$ is stronger than $p$. If $G$ is $V$-generic over $\mathbb{P}$, we will abuse notation slightly and use both $V[G]$ and $V^{\mathbb{P}}$ to indicate the universe obtained by forcing with $\mathbb{P}$. If $x \in V[G]$, then $\dot{x}$ will be a term in $V$ for $x$. We may, from time to time, confuse terms with the sets they denote and write $x$ when we actually mean $\dot{x}$ or $\check{x}$, especially when $x$ is some variant of the generic set $G$, or $x$ is in the ground model $V$.

If $\kappa$ is a regular cardinal and $\lambda$ is an ordinal, $\text{Add}(\kappa, \lambda)$ is the standard partial ordering for adding $\lambda$ Cohen subsets of $\kappa$. If $\mathbb{P}$ is an arbitrary partial ordering, $\mathbb{P}$ is *$\kappa$-directed closed* if for every cardinal $\delta < \kappa$ and every directed set $\langle p_\alpha : \alpha < \delta \rangle$ of elements of $\mathbb{P}$ (where $\langle p_\alpha : \alpha < \delta \rangle$ is *directed* if every two elements $p_\rho$ and $p_\nu$ have a common upper bound of the form $p_\sigma$) there is an upper bound $p \in \mathbb{P}$. $\mathbb{P}$ is *$\kappa$-strategically closed* if in the two person game in which the players construct an increasing sequence $\langle p_\alpha : \alpha \leq \kappa \rangle$, where player I plays odd stages and player II plays even and limit stages (choosing the trivial condition at stage 0), then player II has a strategy which ensures the game can always be continued. Note that if $\mathbb{P}$ is $\kappa^+$-directed closed, then $\mathbb{P}$ is

[1]Theorem 1 of [3] is of course a generalization of a celebrated result of Magidor, who showed in [19] that it is consistent, relative to the existence of a strongly compact cardinal, for the least strongly compact cardinal to be the least measurable cardinal. Readers are urged to consult [19] and [4] for further details on both this result and some generalizations.

$\kappa$-strategically closed. Also, if $\mathbb{P}$ is $\kappa$-strategically closed and $f : \kappa \to V$ is a function in $V^{\mathbb{P}}$, then $f \in V$. $\mathbb{P}$ is *$\prec\kappa$-strategically closed* if in the two person game in which the players construct an increasing sequence $\langle p_\alpha : \alpha < \kappa \rangle$, where player I plays odd stages and player II plays even and limit stages (again choosing the trivial condition at stage 0), then player II has a strategy which ensures the game can always be continued.

In this paper, we will use non-reflecting stationary set forcing $\mathbb{P}_{\eta,\lambda}$. Specifically, if $\eta < \lambda$ are both regular cardinals, then conditions in $\mathbb{P}_{\eta,\lambda}$ are bounded subsets $s \subseteq \lambda$ consisting of ordinals of cofinality $\eta$ such that for every $\alpha < \lambda$, the initial segment $s \cap \alpha$ is non-stationary in $\alpha$, ordered by end-extension. It is well-known that if $G$ is $V$-generic over $\mathbb{P}_{\eta,\lambda}$ (see [9] or [2]) and GCH holds in $V$, then in $V[G]$, the set $S = S[G] = \bigcup G \subseteq \lambda$ is a non-reflecting stationary set of ordinals of cofinality $\eta$, the bounded subsets of $\lambda$ are the same as those in $V$, and cardinals, cofinalities and GCH have been preserved. It is virtually immediate that $\mathbb{P}_{\eta,\lambda}$ is $\eta$-directed closed. It follows from work of Solovay (Theorem 4.8 of [20] and the succeeding remarks) that the existence of a non-reflecting stationary subset of $\lambda$, consisting of ordinals of confinality $\eta$, implies that no cardinal $\delta \in (\eta, \lambda]$ is $\lambda$ strongly compact. Thus, iterations of this forcing provide a way to destroy all strongly compact cardinals in an interval.

We recall for the benefit of readers the definition given by Hamkins in Section 3 of [15] of the lottery sum of a collection of partial orderings. If $\mathfrak{A}$ is a collection of partial orderings, then the *lottery sum* is the partial ordering $\oplus\mathfrak{A} = \{\langle \mathbb{P}, p\rangle : \mathbb{P} \in \mathfrak{A} \text{ and } p \in \mathbb{P}\} \cup \{0\}$, ordered with 0 below everything and $\langle \mathbb{P}, p\rangle \le \langle \mathbb{P}', p'\rangle$ iff $\mathbb{P} = \mathbb{P}'$ and $p \le p'$. Intuitively, if $G$ is $V$-generic over $\oplus\mathfrak{A}$, then $G$ first selects an element of $\mathfrak{A}$ (or as Hamkins says in [15], "holds a lottery among the posets in $\mathfrak{A}$") and then forces with it.[2]

A result which will be used in the proof of Theorem 1.1 is Hamkins' Gap Forcing Theorem of [16] and [14]. We therefore state this theorem now, along with some associated terminology, quoting freely from [16] and [14]. Suppose $\mathbb{P}$ is a partial ordering which can be written as $\mathbb{Q} * \dot{\mathbb{R}}$, where $|\mathbb{Q}| < \delta$ and $\Vdash_{\mathbb{Q}}$ "$\dot{\mathbb{R}}$ is $\delta$-strategically closed". In Hamkins' terminology of [16] and [14], $\mathbb{P}$ *admits a gap at $\delta$*. Also, as in the terminology of [16] and [14] (and elsewhere), an embedding $j : \overline{V} \to \overline{M}$ is *amenable to $\overline{V}$* when $j \restriction A \in \overline{V}$ for any $A \in \overline{V}$. The Gap Forcing Theorem is then the following.

THEOREM 1.2 (Hamkins' Gap Forcing Theorem). *Suppose that $V[G]$ is a forcing extension obtained by forcing that admits a gap at some $\delta < \kappa$ and $j : V[G] \to M[j(G)]$ is an embedding with critical point $\kappa$ for which $M[j(G)] \subseteq V[G]$ and $M[j(G)]^\delta \subseteq M[j(G)]$ in $V[G]$. Then $M \subseteq V$; indeed,*

[2]The terminology "lottery sum" is due to Hamkins, although the concept of the lottery sum of partial orderings has been around for quite some time and has been referred to at different junctures via the names "disjoint sum of partial orderings", "side-by-side forcing", and "choosing which partial ordering to force with generically".

$M = V \cap M[j(G)]$. *If the full embedding $j$ is amenable to $V[G]$, then the restricted embedding $j \upharpoonright V : V \to M$ is amenable to $V$. If $j$ is definable from parameters (such as a measure or extender) in $V[G]$, then the restricted embedding $j \upharpoonright V$ is definable from the names of those parameters in $V$.*

Finally, we mention that we are assuming familiarity with the large cardinal notions of measurability, strongness, strong compactness, and supercompactness. Interested readers may consult [17] or [20] for further details. We do wish to note, however, that a measurable cardinal $\kappa$ has *trivial Mitchell rank* if there is no embedding $j : V \to M$ for which $\mathrm{cp}(j) = \kappa$ and $M \models$ "$\kappa$ is measurable". An ultrafilter $\mathcal{U}$ generating this sort of embedding will be said to have trivial Mitchell rank as well. Ultrafilters of trivial Mitchell rank always exist for every measurable cardinal. Also, if $\kappa$ is $2^\kappa$ supercompact, then $\kappa$ has non-trivial Mitchell rank. This implies that $\kappa$ is the $\kappa$th measurable cardinal, which further implies that the least measurable cardinal $\delta$ isn't $2^\delta$ supercompact.

**§2. The proof of Theorem 1.1.** We turn now to the proof of Theorem 1.1.

PROOF. Let $V^* \models$ "ZFC + $\kappa_1$ and $\kappa_2$ are supercompact". Without loss of generality, by first forcing GCH and then forcing with the partial ordering given in Theorem 1 of [6] (which can be defined so as to have cardinality $\kappa_1$), we assume that $V^*$ has been generically extended to a model $V$ such that $V \models$ "$\kappa_1$ is both the least strongly compact and least measurable cardinal + $\kappa_1$'s strong compactness is indestructible under $\kappa_1$-directed closed forcing + GCH holds for all cardinals greater than or equal to $\kappa_1$ + $\kappa_2$ is supercompact".

The partial ordering $\mathbb{P}$ to be used in the proof of Theorem 1.1 is now defined as follows. For any ordinal $\delta$, let $\delta'$ be the least $V$-strong cardinal above $\delta$. $\mathbb{P}$ begins by adding a Cohen subset of $\kappa_1$. The remainder of $\mathbb{P}$ is the reverse Easton iteration having length $\kappa_2$ which does a non-trivial forcing only at those cardinals in the open interval $(\kappa_1, \kappa_2)$ which are in $V$ measurable limits of strong cardinals. At such a stage $\delta$, the forcing done is $\mathbb{Q}^\delta * \dot{\mathbb{P}}_{\kappa_1, \delta'}$, where $\Vdash_{\mathbb{P}_\delta}$ "$\dot{\mathbb{Q}}^\delta$ is the lottery sum of all $\delta$-directed closed partial orderings having rank below $\delta'$ which can be written in the form $\mathrm{Add}(\delta, 1) * \dot{\mathbb{Q}}'$".

The intuition behind the above definition of $\mathbb{P}$ is as follows. The non-reflecting stationary set of ordinals of cofinality $\kappa_1$ added at each non-trivial stage of forcing is used to destroy all strongly compact cardinals in the open interval $(\kappa_1, \kappa_2)$. The lottery sum employed at each non-trivial stage of forcing is used to force indestructibility for $\kappa_2$. The Cohen subset added at each non-trivial stage of forcing is used to ensure that $\kappa_2$ becomes a non-supercompact strongly compact cardinal. The entire iteration $\mathbb{P}$ is defined in a way so as to be $\kappa_1$-directed closed, which means that after forcing with $\mathbb{P}$, $\kappa_1$ remains as both the least strongly compact and least measurable cardinal and retains the indestructibility of its strong compactness under $\kappa_1$-directed closed forcing.

The following lemmas show that $\mathbb{P}$ is as desired. Throughout, we assume that $G$ is $V$-generic over $\mathbb{P}$.

LEMMA 2.1. *In $V^{\mathbb{P}}$, there are no strongly compact cardinals in the open interval $(\kappa_1, \kappa_2)$.*

PROOF. By Lemma 2.1 of [5], the supercompactness of $\kappa_2$ in $V$ implies that in $V$, there are unboundedly many in $\kappa_2$ measurable limits of strong cardinals in the open interval $(\kappa_1, \kappa_2)$. Therefore, by its definition, after forcing with $\mathbb{P}$, unboundedly many in $\kappa_2$ cardinals in the open interval $(\kappa_1, \kappa_2)$ will contain non-reflecting stationary sets of ordinals of cofinality $\kappa_1$. As mentioned in Section 1, by work of Solovay (Theorem 4.8 of [20] and the succeeding remarks), this means that in $V^{\mathbb{P}}$, no cardinal $\delta \in (\kappa_1, \kappa_2)$ is strongly compact. This completes the proof of Lemma 2.1. ⊣

LEMMA 2.2. *In $V^{\mathbb{P}}$, $\kappa_2$ is strongly compact.*

PROOF. The proof of Lemma 2.2 uses a technique of Magidor for preserving strong compactness, which is unpublished by him but is used, e.g., in [2], [4], [5], and [8]. Let $\lambda > 2^{\kappa_2}$ be a regular cardinal, and let $k_1 : V \to M$ be an elementary embedding witnessing the $\lambda$ supercompactness of $\kappa_2$ such that $M \models$ "$\kappa_2$ isn't $\lambda$ supercompact". $\lambda$ is large enough so that we may assume by selecting a normal ultrafilter of trivial Mitchell rank over $\kappa_2$ that $k_2 : M \to N$ is an embedding witnessing the measurability of $\kappa_2$ definable in $M$ such that $N \models$ "$\kappa_2$ isn't measurable". In addition, as $\lambda > 2^{\kappa_2}$, Lemma 2.1 of [5] and the succeeding remarks imply that in both $V$ and $M$, $\kappa_2$ is a measurable cardinal which is a limit of strong cardinals. This means by the definition of $\mathbb{P}$ it is possible to opt for $\mathrm{Add}(\kappa_2, 1)$ in the stage $\kappa_2$ lottery held in $M$ in the definition of $k_1(\mathbb{P})$. We therefore assume that this has been done, meaning that in $M$, above the appropriate condition, $k_1(\mathbb{P})$ is forcing equivalent to $\mathbb{P} * \dot{\mathbb{S}} * \dot{\mathbb{T}}$, where $\dot{\mathbb{S}}$ is a term for $\mathrm{Add}(\kappa_2, 1)$, and $\dot{\mathbb{T}}$ is a term for the rest of $k_1(\mathbb{P})$. We consequently assume for the rest of the proof of Lemma 2.2 that we are forcing above such a condition. Since as in Lemma 2.4 of [5], $M \models$ "No cardinal $\delta \in (\kappa_2, \lambda]$ is strong" (this is since otherwise, $\kappa_2$ would be supercompact up to a strong cardinal and hence be fully supercompact), the first ordinal in the realization of $\dot{\mathbb{T}}$ occurs above $\lambda$.

Suppose that $k : V \to N$ is an elementary embedding definable in $V$ with critical point $\kappa_2$ and for any $x \subseteq N$ with $|x| \leq \lambda$, there is some $y \in N$ such that $x \subseteq y$ and $N \models$ "$|y| < k(\kappa_2)$". By Theorem 22.17 of [17], $k$ witnesses the $\lambda$ strong compactness of $\kappa_2$ in the sense that the existence of this sort of embedding implies that $\kappa_2$ is $\lambda$ strongly compact. Using this fact, it is easily verifiable that $j = k_2 \circ k_1$ is an elementary embedding witnessing the $\lambda$ strong compactness of $\kappa_2$. We show that $j$ lifts to $j : V^{\mathbb{P}} \to N^{j(\mathbb{P})}$. Since this lifted embedding witnesses the $\lambda$ strong compactness of $\kappa_2$ in $V^{\mathbb{P}}$ and $\lambda > 2^{\kappa_2}$ is an arbitrary regular cardinal, this proves Lemma 2.2.

To do this, write $j(\mathbb{P})$ as $\mathbb{P} * \dot{\mathbb{Q}} * \dot{\mathbb{R}}$, where $\dot{\mathbb{Q}}$ is a term for the portion of $j(\mathbb{P})$ between $\kappa_2$ and $k_2(\kappa_2)$ and $\dot{\mathbb{R}}$ is a term for the rest of $j(\mathbb{P})$, i.e., the part above $k_2(\kappa_2)$. Note that since $N \models$ "$\kappa_2$ isn't measurable", the realization of $\dot{\mathbb{Q}}$ is trivial at stage $\kappa_2$. Thus, the ordinals at which the realization of $\dot{\mathbb{Q}}$ does a non-trivial forcing lie in the interval $(\kappa_2, k_2(\kappa_2)]$ (the endpoint $k_2(\kappa_2)$ is included since by elementarity, $k_2(\kappa_2)$ is a measurable cardinal which is a limit of strong cardinals in $N$), and the ordinals at which the realization of $\dot{\mathbb{R}}$ does a non-trivial forcing lie in the interval $(k_2(\kappa_2), k_2(k_1(\kappa_2)))$. Since we have assumed that we have opted for $\mathrm{Add}(\kappa_2, 1)$ at stage $\kappa_2$ in $M$, we may infer that the forcing done at stage $k_2(\kappa_2)$ in $N$ is $\mathrm{Add}(k_2(\kappa_2), 1)$.

We construct in $V[G]$ an $N[G]$-generic object $G_1$ over $\mathbb{Q}$ and an $N[G][G_1]$-generic object $G_2$ over $\mathbb{R}$. Since $\mathbb{P}$ is a reverse Easton iteration of length $\kappa_2$, a direct limit is taken at stage $\kappa_2$, and no forcing is done at stage $\kappa_2$, the construction of $G_1$ and $G_2$ automatically guarantees that $j''G \subseteq G * G_1 * G_2$. This means that $j : V \to N$ lifts to $j : V[G] \to N[G][G_1][G_2]$ in $V[G]$.

To build $G_1$, note that since $k_2$ is generated by an ultrafilter over $\kappa_2$ and since in both $V$ and $M$, $2^{\kappa_2} = \kappa_2^+$, $|k_2(\kappa_2^+)| = |k_2(2^{\kappa_2})| = |\{f : f : \kappa_2 \to \kappa_2^+$ is a function$\}| = |[\kappa_2^+]^{\kappa_2}| = \kappa_2^+$. Thus, as $N[G] \models$ "$|\wp(\mathbb{Q})| = k_2(2^{\kappa_2})$", we can let $\langle D_\alpha : \alpha < \kappa_2^+ \rangle$ enumerate in $V[G]$ the dense open subsets of $\mathbb{Q}$ present in $N[G]$. For the purpose of the construction of $G_1$ to be given below, we further assume without loss of generality that for every dense open subset $D \subseteq \mathbb{Q}$ found in $N[G]$, for some odd ordinal $\gamma + 1$, $D = D_{\gamma+1}$. Since the $\kappa_2$ closure of $N$ with respect to either $M$ or $V$ implies the least ordinal at which $\mathbb{Q}$ does a non-trivial forcing is above $\kappa_2^+$, the definition of $\mathbb{Q}$ implies that $N[G] \models$ "$\mathbb{Q}$ is $\prec\kappa_2^+$-strategically closed". By the fact the standard arguments show that forcing with the $\kappa_2$-c.c. partial ordering $\mathbb{P}$ preserves that $N[G]$ remains $\kappa_2$-closed with respect to either $M[G]$ or $V[G]$, $\mathbb{Q}$ is $\prec\kappa_2^+$-strategically closed in both $M[G]$ and $V[G]$ as well.

We can now construct $G_1$ in either $M[G]$ or $V[G]$ as follows. Players I and II play a game of length $\kappa_2^+$. The initial pair of moves is generated by player II choosing the trivial condition $q_0$ and player I responding by choosing $q_1 \in D_1$. Then, at an even stage $\alpha + 2$, player II picks $q_{\alpha+2} \geq q_{\alpha+1}$ by using some fixed strategy $\mathcal{S}$, where $q_{\alpha+1}$ was chosen by player I to be such that $q_{\alpha+1} \in D_{\alpha+1}$ and $q_{\alpha+1} \geq q_\alpha$. If $\alpha$ is a limit ordinal, player II uses $\mathcal{S}$ to pick $q_\alpha$ extending each $q_\beta$ for $\beta < \alpha$. By the $\prec\kappa_2^+$-strategic closure of $\mathbb{Q}$ in both $M[G]$ and $V[G]$, the sequence $\langle q_\alpha : \alpha < \kappa_2^+ \rangle$ as just described exists. By construction, $G_1 = \{p \in \mathbb{Q} : \exists \alpha < \kappa_2^+ [q_\alpha \geq p]\}$ is our $N[G]$-generic object over $\mathbb{Q}$.

It remains to construct in $V[G]$ the desired $N[G][G_1]$-generic object $G_2$ over $\mathbb{R}$. To do this, we recall that by the forcing equivalence in $M$ given in the first paragraph of the proof of this lemma of $k_1(\mathbb{P})$ with $\mathbb{P} * \dot{\mathbb{S}} * \dot{\mathbb{T}}$, the ordinals at which the realization of $\dot{\mathbb{T}}$ does a non-trivial forcing lie in the interval $(\lambda, k_1(\kappa_2))$. This implies that in $M$, $\Vdash_{\mathbb{P} * \dot{\mathbb{S}}}$ "$\dot{\mathbb{T}}$ is $\prec\lambda^+$-strategically closed".

Further, since $\lambda$ is a regular cardinal and GCH holds in $V$ above $\kappa_2$, $|[\lambda]^{<\kappa_2}| = \lambda$, and $2^\lambda = \lambda^+$. Therefore, as $k_1$ can be assumed to be generated by an ultrafilter over $P_{\kappa_2}(\lambda)$,

$$|k_1(\kappa_2^+)| = |k_1(2^{\kappa_2})| = |2^{k_1(\kappa_2)}|$$
$$= |\{f : f : P_{\kappa_2}(\lambda) \to \kappa_2^+ \text{ is a function}\}| = |[\kappa_2^+]^\lambda| = \lambda^+.$$

Work until otherwise specified in $M$. Consider the "term forcing" partial ordering $\mathbb{T}^*$ (see [12] for the first published account of term forcing or [10], Section 1.2.5, page 8; the notion is originally due to Laver) associated with $\dot{\mathbb{T}}$, i.e., $\tau \in \mathbb{T}^*$ iff $\tau$ is a term in the forcing language with respect to $\mathbb{P} * \dot{\mathbb{S}}$ and $\Vdash_{\mathbb{P}*\dot{\mathbb{S}}}$ "$\tau \in \dot{\mathbb{T}}$", ordered by $\tau \geq \sigma$ iff $\Vdash_{\mathbb{P}*\dot{\mathbb{S}}}$ "$\tau \geq \sigma$". Although $\mathbb{T}^*$ as defined is technically a proper class, it is possible to restrict the terms appearing in it to a sufficiently large set-sized collection. As we will show below, this can be done in such a way that $M \models$ "$|\mathbb{T}^*| = k_1(\kappa_2)$".

Clearly, $\mathbb{T}^* \in M$. Also, since $\Vdash_{\mathbb{P}*\dot{\mathbb{S}}}$ "$\dot{\mathbb{T}}$ is $\prec\lambda^+$-strategically closed", it can easily be verified that $\mathbb{T}^*$ itself is $\prec\lambda^+$-strategically closed in $M$ and, since $M^\lambda \subseteq M$, in $V$ as well.

To show that we may restrict the number of terms so that $M \models$ "$|\mathbb{T}^*| = k_1(\kappa_2)$", we observe that since $\Vdash_{\mathbb{P}*\dot{\mathbb{S}}}$ "$|\dot{\mathbb{T}}| = k_1(\kappa_2)$", there is a set $\{\tau_\alpha : \alpha < k_1(\kappa_2)\}$ of terms such that for any other term $\tau$, if $\Vdash_{\mathbb{P}*\dot{\mathbb{S}}}$ "$\tau \in \dot{\mathbb{T}}$", then there is a dense set of conditions in $\mathbb{P} * \dot{\mathbb{S}}$ forcing "$\tau = \tau_\alpha$" for various $\alpha$. While $\{\tau_\alpha : \alpha < k_1(\kappa_2)\}$ may not itself be adequate, we enlarge it by choosing, for each maximal antichain $A \subset \mathbb{P} * \dot{\mathbb{S}}$ and each function $s : A \to \{\tau_\alpha : \alpha < k_1(\kappa_2)\}$, a term $\tau_s$ such that $p \Vdash$ "$\tau_\alpha = \tau_{s(p)}$" for each $p \in A$. $\tau_s$ exists using arguments from elementary forcing. Let $\mathbb{T}^*$ be the collection of all such terms $\tau_s$, ranging over all maximal antichains of $\mathbb{P} * \dot{\mathbb{S}}$. Since $M \models$ "$|\mathbb{P} * \dot{\mathbb{S}}| < k_1(\kappa_2)$ and $k_1(\kappa_2)$ is measurable", the number of such terms in $M$ is $k_1(\kappa_2)$. Finally, if $\Vdash_{\mathbb{P}*\dot{\mathbb{S}}}$ "$\tau \in \dot{\mathbb{T}}$", then once again, elementary forcing arguments establish that for some $s$, $\Vdash_{\mathbb{P}*\dot{\mathbb{S}}}$ "$\tau = \tau_s$".

Since we now know that $M \models$ "$|\mathbb{T}^*| = k_1(\kappa_2)$", and since $M \models$ "$2^{k_1(\kappa_2)} = (k_1(\kappa_2))^+ = k_1(\kappa_2^+)$", we can let $\langle D_\alpha : \alpha < \lambda^+\rangle$ enumerate in $V$ the dense open subsets of $\mathbb{T}^*$ found in $M$, such that as before, for every dense open subset $D \subseteq \mathbb{T}^*$ present in $M$, for some odd ordinal $\gamma + 1$, $D = D_{\gamma+1}$. We may then argue as we did when constructing $G_1$ to build in $V$ an $M$-generic object $H_2$ over $\mathbb{T}^*$.

Note now that since $N$ is given by an ultrapower of $M$ via a normal ultrafilter over $\kappa_2$ which is an element of $M$, Fact 2 of Section 1.2.2 of [10] (see also [12]) tells us that $k_2'' H_2$ generates an $N$-generic object $G_2^*$ over $k_2(\mathbb{T}^*)$. By elementarity, $k_2(\mathbb{T}^*)$ is the term forcing in $N$ defined with respect to $k_2(k_1(\mathbb{P})_{\kappa_2+1}) = \mathbb{P} * \dot{\mathbb{Q}}$. Therefore, since $j(\mathbb{P}) = k_2(k_1(\mathbb{P})) = \mathbb{P} * \dot{\mathbb{Q}} * \dot{\mathbb{R}}$, $G_2^*$ is $N$-generic over $k_2(\mathbb{T}^*)$, and $G * G_1$ is $k_2(\mathbb{P} * \dot{\mathbb{S}})$-generic over $N$, Fact 1 of Section 1.2.5 of [10] (see also [12]) tells us that for $G_2 = \{i_{G*G_1}(\tau) : \tau \in G_2^*\}$, $G_2$ is $N[G][G_1]$-generic over $\mathbb{R}$. Thus, in $V[G]$, $j : V \to N$ lifts to $j : V[G] \to N[G][G_1][G_2]$, i.e., $V[G] \models$ "$\kappa_2$ is $\lambda$ strongly compact". This completes the proof of Lemma 2.2. $\dashv$

LEMMA 2.3. *In $V^{\mathbb{P}}$, $\kappa_2$ isn't supercompact. In fact, in $V^{\mathbb{P}}$, $\kappa_2$ has trivial Mitchell rank.*

PROOF. The proof uses an argument due to Hamkins, which is given in Lemma 2.4 of [8]. If $V[G] \models$ "$\kappa_2$ does not have trivial Mitchell rank", then let $j : V[G] \to M[j(G)]$ be an embedding generated by a normal measure over $\kappa_2$ in $V[G]$ witnessing this fact. Write $\mathbb{P} = \mathbb{P}_0 * \dot{\mathbb{Q}}$, where $|\mathbb{P}_0| = \kappa_1$ and $\Vdash_{\mathbb{P}_0}$ "$\dot{\mathbb{Q}}$ is $\kappa_1^+$-strategically closed". $\mathbb{P}$ thus admits a gap at $\kappa_1$, so by Theorem 1.2 (the Gap Forcing Theorem of [16] and [14]), $j$ must lift an embedding $j : V \to M$ that is definable in $V$. Also by the Gap Forcing Theorem applied in $M$, $\kappa_2$ is measurable in $M$, and by Lemma 2.1 of [5], in $V$, $\kappa_2$ is a limit of strong cardinals. Since $\mathrm{cp}(j) = \kappa_2$, if $V \models$ "$\delta < \kappa_2$ is a strong cardinal", $M \models$ "$j(\delta) = \delta$ is a strong cardinal". Therefore, in $M$, $\kappa_2$ is a measurable cardinal which is a limit of strong cardinals, so it follows that there is a non-trivial forcing done at stage $\kappa_2$ in $M$. This means $j(G) = G * S * H$, where $S$ is a Cohen subset of $\kappa_2$ added by forcing over $M[G]$ with $\mathrm{Add}(\kappa_2, 1)^{M[G]}$ at stage $\kappa_2$ in $M[G]$, and $H$ is $M[G][S]$-generic for the rest of the forcing $j(\mathbb{P})$. Since $V^V_{\kappa_2+1} \subseteq M \subseteq V$, it follows that $V^{V[G]}_{\kappa_2+1} = V^{M[G]}_{\kappa_2+1}$. From this it follows that $\mathrm{Add}(\kappa_2, 1)^{M[G]} = \mathrm{Add}(\kappa_2, 1)^{V[G]}$, and the dense open subsets of what we can now unambiguously write as $\mathrm{Add}(\kappa_2, 1)$ are the same in both $M[G]$ and $V[G]$. Thus, the set $S$, which is an element of $V[G]$, is $V[G]$-generic over $\mathrm{Add}(\kappa_2, 1)$, a contradiction. This completes the proof of Lemma 2.3. $\dashv$

LEMMA 2.4. *Suppose $\mathbb{Q} \in V^{\mathbb{P}}$ is a partial ordering which is $\kappa_2$-directed closed and can be written in the form $\mathrm{Add}(\kappa_2, 1) * \dot{\mathbb{Q}}'$. Then $V^{\mathbb{P}*\dot{\mathbb{Q}}} \models$ "$\kappa_2$ is strongly compact". In fact, $V^{\mathbb{P}*\dot{\mathbb{Q}}} \models$ "$\kappa_2$ is supercompact".*

PROOF. Suppose $\mathbb{Q} \in V^{\mathbb{P}}$ is such a partial ordering. Let

$$\lambda > \max(|\mathrm{TC}(\dot{\mathbb{Q}})|, 2^{\kappa_2})$$

be an arbitrary regular cardinal, and let $\gamma = |2^{[\lambda]^{<\kappa_2}}|$. Let $j : V \to M$ be an elementary embedding witnessing the $\gamma$ supercompactness of $\kappa_2$ such that $M \models$ "$\kappa_2$ isn't $\gamma$ supercompact". As in Lemma 2.2, $M \models$ "No cardinal $\delta \in (\kappa_2, \gamma]$ is strong". Therefore, by the choice of $\gamma$, it is possible to opt for $\mathbb{Q}$ in the stage $\kappa_2$ lottery held in $M$ in the definition of $j(\mathbb{P})$. In addition, the next non-trivial forcing in the definition of $j(\mathbb{P})$ takes place well beyond $\gamma$. Thus, above the appropriate condition, $j(\mathbb{P})$ is forcing equivalent in $M$ to $\mathbb{P} * \dot{\mathbb{Q}} * \dot{\mathbb{R}} * j(\dot{\mathbb{Q}})$. We therefore assume we are forcing above such a condition.

We are now in a position to be able to use the standard reverse Easton techniques to show that $V^{\mathbb{P}*\dot{\mathbb{Q}}} \models$ "$\kappa_2$ is $\lambda$ supercompact". We argue in analogy to, e.g., the proof found in the last paragraph of page 679 of [8]. Specifically, let $G$ be $V$-generic over $\mathbb{P}$ and $G_1$ be $V[G]$-generic over $\mathbb{Q}$. The usual arguments show that $M[G][G_1]$ remains $\gamma$ closed with respect to $V[G][G_1]$. Thus, by the definitions of $\mathbb{P}$ and $\mathbb{R}$, since every ordinal in the realization of $\dot{\mathbb{R}}$ is above $\gamma$, $\mathbb{R}$ is $\gamma$-strategically closed in $M[G][G_1]$ and hence by closure in $V[G][G_1]$ as well. Also, since we are dealing with the usual sort of reverse Easton iterations throughout, if $G_2$ is $V[G][G_1]$-generic over $\mathbb{R}$, $M[G][G_1][G_2]$ remains $\gamma$ closed with respect to $V[G][G_1][G_2]$, and $j$ lifts in $V[G][G_1][G_2]$ to $j : V[G] \to M[G][G_1][G_2]$. This means we can now by closure and the definition of $j(\mathbb{Q})$ find in $V[G][G_1][G_2]$ a master condition $q$ for $j''G_1$ and let $G_3$ be a $V[G][G_1][G_2]$-generic object over $j(\mathbb{Q})$ containing $q$. Again since we are dealing with reverse Easton iterations, $j$ lifts in $V[G][G_1][G_2][G_3]$ to $j : V[G][G_1] \to M[G][G_1][G_2][G_3]$, so $V[G][G_1][G_2][G_3] \vDash$ "$\kappa_2$ is $\lambda$ supercompact". As $G_2 * G_3$ is $V[G][G_1]$-generic over $\mathbb{R} * j(\mathbb{Q})$, a partial ordering which is $\gamma$-strategically closed in $V[G][G_1]$, $V[G][G_1] \vDash$ "$\kappa_2$ is $\lambda$ supercompact". Since $\lambda$ was arbitrary, $V^{\mathbb{P}*\dot{\mathbb{Q}}} \vDash$ "$\kappa_2$ is supercompact". This completes the proof of Lemma 2.4. ⊣

Theorem 1.1 now follows from Lemmas 2.1–2.4. By Lemma 2.1, there are no strongly compact cardinals in $V^{\mathbb{P}}$ in the interval $(\kappa_1, \kappa_2)$. Therefore, by Lemma 2.2 and the fact that $\mathbb{P}$ is defined so as to be $\kappa_1$-directed closed, in $V^{\mathbb{P}}$, $\kappa_1$ is both the least strongly compact and least measurable cardinal and retains the indestructibility of its strong compactness under $\kappa_1$-directed closed forcing, and $\kappa_2$ is the second strongly compact cardinal. By Lemma 2.3, $\kappa_2$ isn't supercompact. By Lemma 2.4, $\kappa_2$ remains strongly compact after forcing with a $\kappa_2$-directed closed partial ordering $\mathbb{Q}$ that can be written in the form $\mathrm{Add}(\kappa_2, 1) * \dot{\mathbb{Q}}'$, and $\kappa_2$ is clearly strongly compact after doing a trivial forcing over $V^{\mathbb{P}}$. This completes the proof of Theorem 1.1. ⊣

We remark that in the proof of Theorem 1.1 just given, it is not necessary for $\kappa_1$ to be both the least strongly compact and least measurable cardinal. By Theorem 2 of [6], we may assume, e.g., that our model $V$ has been constructed to be so that $V \vDash$ "$\kappa_1$ is both the least strongly compact cardinal and least cardinal $\delta$ which is $\delta^+$ supercompact $+ \kappa_1$'s strong compactness and $\kappa_1^+$ supercompactness are indestructible under $\kappa_1$-directed closed forcing + GCH holds for all cardinals greater than or equal to $\kappa_1$ + $\kappa_2$ is supercompact". We may then define the partial ordering $\mathbb{P}$ used in the proof of Theorem 1.1 as before. Since $\mathbb{P}$ is defined so as to be $\kappa_1$-directed closed, in $V^{\mathbb{P}}$, $\kappa_1$ remains as both the least strongly compact and least cardinal $\delta$ which is $\delta^+$ supercompact and retains the indestructibility of its strong compactness and its $\kappa_1^+$ supercompactness under $\kappa_1$-directed closed forcing. The remainder of Theorem 1.1 then follows as earlier.

**§3. Concluding remarks.** In some sense, Theorem 1.1 raises more questions than it answers. We conclude this paper by listing some of them. In particular:

1. Is it possible to prove Theorem 1.1 without the requirement that indestructibility for $\kappa_2$ for non-trivial forcing be under partial orderings which begin by adding a Cohen subset of $\kappa_2$? (With the current proof of Theorem 1.1, if we drop this requirement in the definition of $\mathbb{P}$, then $\kappa_2$ remains a supercompact cardinal after forcing with $\mathbb{P}$. This is since the proof of Lemma 2.3 becomes invalid without this requirement, and the proof of Lemma 2.4 becomes valid for trivial forcing.) Note, however, that in practice, when indestructibility is used in the context of supercompactness, beginning a construction by adding a Cohen subset to a supercompact cardinal is essentially harmless.
2. Is it possible to prove Theorem 1.1 such that after doing a non-trivial $\kappa_2$-directed closed forcing, $\kappa_2$'s supercompactness isn't resurrected, i.e., such that after doing a non-trivial $\kappa_2$-directed closed forcing, $\kappa_2$ remains a non-supercompact strongly compact cardinal? In particular, is it possible to prove a version of Theorem 1.1 in which $\kappa_2$ is both the second strongly compact and second measurable cardinal?
3. In general, for $\alpha > 2$, is it consistent, relative to anything, for the first $\alpha$ strongly compact cardinals $\kappa_\alpha$ all to be non-supercompact and to have their strong compactness indestructible under (some version of) $\kappa_\alpha$-directed closed forcing?

Note that as the proof of Theorem 1.1 currently stands, the forcing $\mathbb{P}$ can't be iterated, in the sense that if we define a similar type of partial ordering above $\kappa_2$, forcing with it resurrects the supercompactness of $\kappa_2$. Thus, the techniques used in proving Theorem 1.1 are inadequate for answering the last of the above questions.

## REFERENCES


[1] A. Apter, *Laver indestructibility and the class of compact cardinals*, ***The Journal of Symbolic Logic***, vol. 63 (1998), pp. 149–157.

[2] ———, *Strong compactness, measurability, and the class of supercompact cardinals*, ***Fundamenta Mathematicae***, vol. 167 (2001), pp. 65–78.

[3] ———, *Aspects of strong compactness, measurability, and indestructibility*, ***Archive for Mathematical Logic***, vol. 41 (2002), pp. 705–719.

[4] A. Apter and J. Cummings, *Identity crises and strong compactness*, ***The Journal of Symbolic Logic***, vol. 65 (2000), pp. 1895–1910.

[5] ———, *Identity crises and strong compactness II: Strong cardinals*, ***Archive for Mathematical Logic***, vol. 40 (2001), pp. 25–38.

[6] A. Apter and M. Gitik, *The least measurable can be strongly compact and indestructible*, ***The Journal of Symbolic Logic***, vol. 63 (1998), pp. 1404–1412.

[7] A. Apter and J. D. Hamkins, *Universal indestructibility*, ***Kobe Journal of Mathematics***, vol. 16 (1999), pp. 119–130.

[8] ———, *Exactly controlling the non-supercompact strongly compact cardinals*, ***The Journal of Symbolic Logic***, vol. 68 (2003), pp. 669–688.

[9] J. Burgess, *Forcing*, ***Handbook of Mathematical Logic*** (J. Barwise, editor), North-Holland, Amsterdam, 1977, pp. 403–452.

[10] J. Cummings, *A model in which GCH holds at successors but fails at limits*, ***Transactions of the American Mathematical Society***, vol. 329 (1992), pp. 1–39.

[11] J. Cummings, M. Foreman, and M. Magidor, *Squares, scales, and stationary reflection*, ***Journal of Mathematical Logic***, vol. 1 (2001), pp. 35–98.

[12] M. Foreman, *More saturated ideals*, ***Cabal Seminar 79–81***, Lecture Notes in Mathematics, vol. 1019, Springer-Verlag, Berlin and New York, 1983, pp. 1–27.

[13] M. Gitik, *Changing cofinalities and the nonstationary ideal*, ***Israel Journal of Mathematics***, vol. 56 (1986), pp. 280–314.

[14] J. D. Hamkins, *Gap forcing: Generalizing the Lévy-Solovay theorem*, ***The Bulletin of Symbolic Logic***, vol. 5 (1999), pp. 264–272.

[15] ———, *The lottery preparation*, ***Annals of Pure and Applied Logic***, vol. 101 (2000), pp. 103–146.

[16] ———, *Gap forcing*, ***Israel Journal of Mathematics***, vol. 125 (2001), pp. 237–252.

[17] A. Kanamori, ***The Higher Infinite***, Springer-Verlag, Berlin and New York, 1994.

[18] R. Laver, *Making the supercompactness of κ indestructible under κ-directed closed forcing*, ***Israel Journal of Mathematics***, vol. 29 (1978), pp. 385–388.

[19] M. Magidor, *How large is the first strongly compact cardinal?*, ***Annals of Mathematical Logic***, vol. 10 (1976), pp. 33–57.

[20] R. Solovay, W. Reinhardt, and A. Kanamori, *Strong axioms of infinity and elementary embeddings*, ***Annals of Mathematical Logic***, vol. 13 (1978), pp. 73–116.

DEPARTMENT OF MATHEMATICS
BARUCH COLLEGE OF CUNY
NEW YORK, NEW YORK 10010, USA
*E-mail*: awabb@cunyvm.cuny.edu
*URL*: http://faculty.baruch.cuny.edu/apter

# SOME APPLICATIONS OF REGULAR MARKERS

CHARLES M. BOYKIN AND STEVE JACKSON[†]

**Abstract.** We show that Borel sets of markers with regular geometric properties exist for the translation equivalence relation on $2^{\mathbb{Z}^n}$, and use this method to obtain results such as a new proof of the hyperfiniteness of $2^{\mathbb{Z}^n}$ and the existence of a continuous embedding from $2^{\mathbb{Z}^n}$ into $E_0$.

This paper is a contribution to the study of countable Borel equivalence relations on standard Borel spaces. By a countable Borel equivalence relation we mean a Borel equivalence relation on a standard Borel space with every equivalence class countable. For $E, F$ Borel equivalence relations on standard Borel spaces $X$, $Y$, we say $E$ is reducible to $F$, written $E \leq F$, if there is a Borel $f : X \to Y$ such that $\forall x, y \in X(xEy \Leftrightarrow f(x)\,F\,f(y))$. If in addition $f$ is one-to-one then we say $E$ is embeddable into $F$, written $E \sqsubseteq F$. Recall also $E_0$ is the equivalence relation on $2^\omega$ defined by $xE_0y$ iff $\exists n\, \forall m \geq n\, (x(m) = y(m))$. The equivalence relation $E_0$ is generated by the action of the *odometer* $\theta$ on $2^\omega$ (roughly speaking, $\theta(x)$ is obtained by adding 1 to $x$ thought of as an infinite base 2 representation). We refer the reader to [1, 3] for more of the basic theory of countable Borel equivalence relations. We thank the referee and Su Gao for helpful suggestions and conversations. In particular, we thank the referee for suggesting that we prove Theorems 20, 21, and 24.

We concentrate here mainly on the specific equivalence relations $(2^{\mathbb{Z}^n}, E_T^n)$, where $E_T^n$ denotes translation equivalence on the Polish space $2^{\mathbb{Z}^n}$: for $x, y \in 2^{\mathbb{Z}^n}$, $xE_T^ny$ iff $\exists \vec{a} \in \mathbb{Z}^n\, (\vec{a} \cdot x = y)$, where the natural action of $\mathbb{Z}^n$ on $2^{\mathbb{Z}^n}$ is given by

$$(\pi_{\vec{a}}(x))(b_0, \dots, b_{n-1}) = (\vec{a} \cdot x)(b_0, \dots, b_{n-1}) = x(a_0 + b_0, \dots, a_{n-1} + b_{n-1}).$$

We study these spaces to gain more insight into more complicated equivalence relations, specifically $2^{\mathbb{Z}^{<\omega}}$ ($\mathbb{Z}^{<\omega}$ denotes the infinite direct sum of $\mathbb{Z}$), for which many basic open questions (such as the hyperfiniteness) remain open. Recall that a countable Borel equivalence relation $E$ is said to be hyperfinite if $E = \bigcup E_n$ is the increasing union of subequivalence relations $E_n$ which

*Key words and phrases.* Borel, hyperfinite, countable Borel equivalence relations.
[†]Research supported by NSF Grant DMS-0097181.

**Logic Colloquium '03**
Edited by V. Stoltenberg-Hansen and J. Väänänen
Lecture Notes in Logic, 24

are finite, that is, each $E_n$ equivalence class is finite. It is open whether an increasing union of hyperfinite equivalence relations is necessarily hyperfinite, and the special case of $2^{\mathbb{Z}^{<\omega}}$ (with translation equivalence) is the simplest case of this. A result of Weiss (see [3]) shows that the $2^{\mathbb{Z}^n}$ relations are all hyperfinite. That argument works by showing that $2^{\mathbb{Z}^n}$ has a hyperfinite subequivalence relation $G$ of finite index (that is, each $2^{\mathbb{Z}^n}$ class contains only finitely many $G$ classes), and then appealing to the general fact that if $G$ is hyperfinite and $G$ has finite index in $E$, then $E$ is hyperfinite (see [3]). In Weiss' proof, the existence of Borel markers with a regular geometric structure is established, and this in turn produces the hyperfinite subequivalence relation $G_n$ of $2^{\mathbb{Z}^n}$ with $|2^{\mathbb{Z}^n}/G_n|$ finite. Unfortunately, the indices $|2^{\mathbb{Z}^n}/G_n|$ go to infinity with $n$, and thus this seems to provide no information for $2^{\mathbb{Z}^{<\omega}}$. We present here a new proof of the hyperfiniteness of $2^{\mathbb{Z}^n}$ which does not pass through the finite index result. Unfortunately, this proof too encounters difficulties when attempting to extend to $2^{\mathbb{Z}^{<\omega}}$, but they seem to be of a different nature. Our methods here lead to results about markers with stronger regularity properties which have other applications. For example, we use these methods to show that there is a continuous embedding from $2^{\mathbb{Z}}$ into $E_0$.

In §1, a general construction of marker sets is developed. In §2, we will give a new proof that $2^{\mathbb{Z}\times\mathbb{Z}}$ is hyperfinite. Lastly in §3, we show that there is a continuous embedding from $2^{\mathbb{Z}}$ into $E_0$.

**§1. Marker sets.** Let $F_n$ denote the free part of $2^{\mathbb{Z}^n}$. That is, $x \in F_n$ iff $\forall \vec{a} \neq (0,\dots,0)\ (\vec{a}\cdot x \neq x)$. We first show the existence of marker sets having regular geometric structure which are relatively clopen on $F_n$. The existence of Borel marker sets with these same properties played an important role in Weiss' proof of the hyperfiniteness of $2^{\mathbb{Z}^n}$. For $x, y \in F_n$ and $xE_Ty$, there is a unique $\vec{a} \in \mathbb{Z}^n$ with $\vec{a}\cdot x = y$, and we let $\operatorname{dist}(x,y) = \|\vec{a}\| = \sqrt{a_0^2 + \cdots a_{n-1}^2}$. If $x$ is not $E_T$ equivalent to $y$, we let $\operatorname{dist}(x,y) = \infty$.

Theorem 1. *Let $D > 0$ be a positive integer. There is an open set $S \subseteq 2^{\mathbb{Z}^n}$ which is relatively clopen on the free part $F_n$ satisfying the following.*

(1) $\forall x \neq y \in S\ (\operatorname{dist}(x,y) > D)$.

(2) $\forall x \in F_n\ \exists y \in S\ (\operatorname{dist}(x,y) \leq D)$.

Proof. It is enough to construct $S \subseteq F_n$ which is relatively clopen in $F_n$ and satisfies (1) and (2). For in this case we let $S'$ be an open set in $2^{\mathbb{Z}^n}$ with $S' \cap F_n = S$. $S'$ still satisfies (1) by an easy argument using the fact that $F_n$ is dense in $2^{\mathbb{Z}^n}$ [if $x, y \in S'$ and $\operatorname{dist}(x,y) \leq D$, let $\vec{a}\cdot x = y$, and let $U$ be a neighborhood of $x$ such that for all $z \in U$, $z \in S$ and $\vec{a}\cdot z \in S$. Then take $z \in U \cap F_n$ to get a contradiction.]

Consider basic neighborhoods $U = U_C$ in $F_n$ determined by $C : [-N,N]^n \to \{0,1\}$. That is $U_C = \{x \in F_n : x \restriction [-N,N]^n = C\}$. We consider the basic neighborhoods to be ordered first by $N$, and then by some fixed ordering on $2^{[-N,N]^n}$. We say the basic neighborhood $U$ is $D$-good if for all $\vec{a} \in \mathbb{Z}^n$ with $\|\vec{a}\| \leq D$ we have $U \cap \pi_{\vec{a}}(U) = \emptyset$. Thus, if $x, y \in U$ and $x \neq y$, then $\text{dist}(x, y) > D$. For every $x \in F_n$ there is an integer $N(x)$ such that $x \upharpoonright [-N(x), N(x)]^n$ is $D$-good. This follows from the fact that there are only finitely many $\vec{a}$ with $\|\vec{a}\| \leq D$, and for each such $\vec{a}$, $x \neq \pi_{\vec{a}}(x)$, and there is a large enough $m$ so that $x \upharpoonright [-m, m]^n$ witnesses this.

Let $M_1 = U_1$ be the least $D$-good basic neighborhood. Clearly for $x \neq y$ in $M_1$, $\text{dist}(x, y) > D$. Let $R_1 = \{x : \text{dist}(x, M_1) \leq D\}$. Let $U_2 > U_1$ be the least $D$-good neighborhood such that $U_2 - R_1 \neq \emptyset$. Let $M_2 = U_1 \cup (U_2 - R_1)$. Clearly if $x \neq y \in M_2$, then $\text{dist}(x, y) > D$. Let $R_2 = \{x : \text{dist}(x, M_1 \cup M_2) \leq D\}$. In general, let $U_{n+1} > U_n$ be the least $D$-good neighborhood such that $U_{n+1} - R_n \neq \emptyset$, where $R_n = \{x : \text{dist}(x, M_1 \cup \cdots \cup M_n) \leq D\}$. Let $M_{n+1} = U_{n+1} - R_n$.

Let $M = \bigcup_k M_k$. Clearly $M$ is open in $F_n$ and for all $x \neq y$ in $M$ we have $\text{dist}(x, y) > D$. Suppose $x \in F_n - M$. Let $U$ be a basic neighborhood of $x$ which is $D$-good. Let $k$ be such that $U_k < U \leq U_{k+1}$. Since $x \notin M$ and $U \leq U_{k+1}$, it follows that $x \in R_k$. That is, $\text{dist}(x, M_1 \cup \cdots \cup M_k) \leq D$, and so $\text{dist}(x, M) \leq D$. This shows (2) of the theorem and also shows that $M$ is clopen in $F_n$, for $F_n - M = \bigcup_{0<\|\vec{a}\| \leq D} \pi_{\vec{a}}(M)$, a finite union of open sets. ⊣

**§2. Hyperfiniteness of $2^{\mathbb{Z}^n}$.** In this section, we give a new proof that $(2^{\mathbb{Z}^n}, E_T^n)$ is hyperfinite. This proof will directly give a countable increasing union of finite subequivalence relations which equals $E_T^n$. Actually, we just show that the free part $F_n$ is hyperfinite as the general case follows from this [sketch: for $w = (\vec{w}_1, \ldots, \vec{w}_k)$ an independent set of vectors in $\mathbb{Z}^n$ (equivalently, independent as elements of $\mathbb{R}^n$), let $X_w \subseteq F_n$ be those $x$ such that $\{g \in \mathbb{Z}^n : g \cdot x = x\}$ is the group generated by the members of $w$. It suffices to show that each $X_w$ is hyperfinite. Let $\vec{u}_{k+1}, \ldots, \vec{u}_n \in \mathbb{Z}^n$ be such that $\vec{w}_1, \ldots, \vec{w}_k, \vec{u}_{k+1}, \ldots, \vec{u}_n$ form a basis for $\mathbb{R}^n$. Let $G$ be the group generated by $\vec{u}_{k+1}, \ldots, \vec{u}_n$, so $G \cong \mathbb{Z}^{n-k}$. $G$ acts freely on $X_w$, and the induced subequivalence relation has finite index in $E_T$. So it suffices to show that every free action of $\mathbb{Z}^{n-k}$ on a Polish space is hyperfinite. But any such action easily embeds into $(2^\omega)^{\mathbb{Z}^{n-k}}$ (as $\mathbb{Z}^{n-k}$ actions) which in turn easily embeds into $F_{n+1}$.]

For $x, y \in 2^{\mathbb{Z}^n}$ with $xE_Ty$, recall $\text{dist}(x, y) = \min\{\|\vec{a}\| : \pi_{\vec{a}}(x) = y\}$. For $x, y \in \mathbb{R}^n$, let $\rho(x, y)$ be the usual Euclidean distance between $x$ and $y$. We let $B(\alpha, \varepsilon)$ denote the open ball about $\alpha \in \mathbb{R}^n$ of radius $\varepsilon$.

Theorem 2. *$(2^{\mathbb{Z}^n}, E_T^n)$ is hyperfinite.*

Proof. To ease notation, let $X = F_n \subseteq 2^{\mathbb{Z}^n}$ denote the free part of $2^{\mathbb{Z}^n}$, and let $E_T$ denote translation equivalence on $X$. We show that $(X, E_T)$ is hyperfinite (which suffices by the above comments).

Fix a sequence $D_1 < D_2 < \cdots$ of positive integers such that $D_{n+1} > 2^{n+1}(\sum_{i \leq n} D_i)^2$. Let $M_n \subseteq X$ be a clopen marker set for $D_n$ as in Theorem 1.

Thus, the $M_n, D_n$ satisfy:

(1) $\forall x \neq y \in M_n\,(\operatorname{dist}(x,y) > D_n)$.
(2) $\forall x \in X\, \exists y \in M_n\,(\operatorname{dist}(x,y) \leq D_n)$.
(3) $\frac{(\sum_{i<n} D_i)^2}{D_{n+1}} < \frac{1}{2^{n+1}}$.

DEFINITION 3. A *marker sequence* is a sequence of $E_T$ equivalent points $(x_n)_{n\geq 0}$ in $X$ with the following properties:

(1) $\forall n \geq 1\,(x_n \in M_n)$.
(2) $\forall n \geq 1\,(\operatorname{dist}(x_{n-1}, x_n) \leq 10D_n)$.

We say marker sequences $(x_n), (y_n)$ are *equivalent* if $x_0 E_T y_0$.

We say a sequence $(x_n)_{n\geq 0} \in X^\omega$ is *slowly growing* if $\forall n \geq 1$ $\operatorname{dist}(x_{n-1}, x_n) \leq 10D_n$. In particular, this implies that all of the $x_n$ are $E_T$ equivalent.

For $x, y \in X$ with $xE_Ty$. Let $x \diamond y$ be the unique $\vec{a} \in \mathbb{Z}^n$ such that $\pi_{\vec{a}}(x) = y$. We regard $x \diamond y$ as an element of $\mathbb{R}^n$.

If $\vec{x} \in X^\omega$, define for each $k \in \omega$, $C_k^{\vec{x}} = \{\frac{x_k \diamond y}{10D_{k+1}} : y \in M_{k+1}\} \subseteq \mathbb{R}^n$.

CLAIM 4. For each sequence $\vec{x} \in X^\omega$ there is a point $\alpha \in \bar{B}_{\frac{1}{2}}$ (the closed ball of radius $\frac{1}{2}$) such that $\forall \varepsilon > 0\, \forall m\, \exists k \geq m\; C_k^{\vec{x}} \cap B(\alpha, \varepsilon) \neq \emptyset$.

PROOF. For each $k$, $C_k^{\vec{x}} \cap \bar{B}_{\frac{1}{2}} \neq \emptyset$ from property (2) above. The result follows from compactness. ⊣

For $\vec{x} \in X^\omega$, let $B_{\vec{x}}$ be the set of all $\alpha \in \bar{B}_{\frac{1}{2}}$ satisfying the above claim for $\vec{x}$. Clearly $B_{\vec{x}}$ is a closed subset of $\bar{B}_{\frac{1}{2}}$.

CLAIM 5. If $(\vec{x}), (\vec{y})$ are slowly growing and $x_0 E_T y_0$, then $B_{\vec{x}} = B_{\vec{y}}$.

PROOF. Let $\alpha \in B_{\vec{x}}$. Fix $\varepsilon > 0$ and $m \in \omega$. Let $k_1 \geq m$ be large enough so that for all $k \geq k_1$

$$\left(\operatorname{dist}(x_0, y_0) + 20 \sum_{i\leq k} D_i\right) < 10\left(\frac{\varepsilon}{4}\right) D_{k+1}.$$

Let $k \geq k_1$ be such that $C_k^{\vec{x}} \cap B(\alpha, \varepsilon/4) \neq \emptyset$. Let $z \in M_{k+1}$ be such that

$$\rho\left(\frac{x_k \diamond z}{10D_{k+1}}, \alpha\right) < \frac{\varepsilon}{4}.$$

Since $\vec{x}$ and $\vec{y}$ are slowly growing and $k \geq k_1$,

$$\|x_k \diamond y_k\| \leq \left(\operatorname{dist}(x_0, y_0) + 20 \sum_{i\leq k} D_i\right) < 10\left(\frac{\varepsilon}{4}\right) D_{k+1}.$$

Thus $\rho(\frac{x_k \diamond z}{10D_{k+1}}, \frac{y_k \diamond z}{10D_{k+1}}) < \frac{\varepsilon}{4}$ and so $\rho(\frac{y_k \diamond z}{10D_{k+1}}, \alpha) < \frac{\varepsilon}{2}$. This shows $\alpha \in B_{\vec{y}}$, proving that $B_{\vec{x}} = B_{\vec{y}}$. ⊣

For $x \in X$, let $B_x = B_{\vec{x}}$ where $\vec{x}$ is the constant sequence $x$. From Claim 5 we have that if $xE_Ty$, then $B_x = B_y$. The map $x \to B_x$ is easily a Borel map from $X$ to the standard Borel space $\mathcal{F}$ of non-empty closed subsets of $\mathbb{R}^n$ (with the Effros Borel structure, or equivalently the Beer topology, see [4]). Let $g : \mathcal{F} \to \mathbb{R}^n$ be a Borel selector, that is, $g(F) \in F$ for all non-empty closed subsets $F \subseteq \mathbb{R}^n$ (see [4]). Let $\alpha(x) = g(B_x)$, so $x \to \alpha(x)$ is Borel and $E_T$ invariant.

For each $x \in X$ define now the *canonical marker sequence* $CMS(x) = (x_n)$ associated to $x$ in the following manner. Let $x_0 = x$. For $n \geq 1$ choose $x_n \in M_n$ in a Borel manner satisfying the following:

(1) $\text{dist}(x_{n-1}, x_n) < 10D_n$.
(2) $\rho(\alpha(x), \frac{x_{n-1} \diamond x_n}{10D_n})$ is a minimum for the points satisfying (1).

CLAIM 6. For all $x, y \in X$, $xE_Ty$ iff $CMS(x)$ agrees with $CMS(y)$ on a tail, that is $\exists k\ \forall l \geq k\ (CMS(x)(l) = CMS(y)(l))$.

PROOF. Clearly if $CMS(x)$ agrees with $CMS(y)$ on a tail then $xE_Ty$, since every point of $CMS(x)$ is equivalent to $x$. Suppose then that $xE_Ty$. Thus $CMS(x) = (x_n)$ and $CMS(y) = (y_n)$ are equivalent marker sequences which are slow growing. Let $\alpha = \alpha(x) = \alpha(y) \in \mathbb{R}^n$. Note that if $s_1 \neq s_2 \in M_{N+1} \cap [x]$, then $\text{dist}(s_1, s_2) > D_{N+1}$ and so for any $z \in [x]$, $\rho(\frac{z \diamond s_1}{10D_{N+1}}, \frac{z \diamond s_2}{10D_{N+1}}) > \frac{1}{10}$. Since $\alpha \in B_{CMS(x)}$ by Claim 5, we may choose $N$ so that the following hold:

(1) $\rho(\alpha, \frac{x_N \diamond x_{N+1}}{10D_{N+1}}) < \frac{1}{40}$.
(2) $(\text{dist}(x_0, y_0) + 20 \sum_{i \leq N} D_i) < \frac{D_{N+1}}{4}$.

By (2),

$$\rho\left(\frac{x_N \diamond x_{N+1}}{10D_{N+1}}, \frac{y_N \diamond x_{N+1}}{10D_{N+1}}\right) < \frac{1}{40}.$$

Thus $\rho(\alpha, \frac{y_N \diamond x_{N+1}}{10D_{N+1}}) < \frac{1}{20}$ and there is no $z \in M_{N+1}, z \neq x_{N+1}$, which could satisfy

$$\rho\left(\alpha, \frac{y_N \diamond z}{10D_{N+1}}\right) < \frac{1}{20}.$$

So $y_{N+1} = x_{N+1}$. Now by the way the canonical marker sequences are defined ($z_{n+1}$ only depends on $z_n$) it follows that $\forall k \geq N+1\ (y_k = x_k)$. Therefore $CSM(x)$ and $CSM(y)$ agree on a tail. ⊣

We have just shown that $(X, E_T)$ Borel embeds into $(X^\omega, E_0^X)$, where $E_0^X$ denote the equivalence relation of eventual equality on $X^\omega$. In the terminology of [1] this relation is hypersmooth, and it follows that $(X, E_T)$ is hyperfinite (see [1]). However, we can see this directly as follows. For each $x \in M_k$ there are at most $(20D_k)^n$ points $y$ equivalent to $x$ within distance $10D_k$ of $x$ (i.e., $\text{dist}(y, x) < 10D_k$). Therefore there are at most $(\sum_{i=1}^k 20D_i)^n$ points that choose any particular $x$ as the $k$th term in their canonical marker sequence.

Define $xE_{T_k}y$ if $CMS(x)(k) = CMS(y)(k)$ (and hence these sequences agree past $k$). Thus, each $E_{T_k}$ equivalence class is finite, and $E_T = \bigcup_{k=1}^{\infty} E_{T_k}$. ⊣

**§3. A continuous injective embedding from $2^{\mathbb{Z}}$ into $E_0$.** In this section we prove the following theorem. Recall $E_T$ is the equivalence relation on $2^{\mathbb{Z}}$ generated by the shift map $\phi$ (i.e., $\phi(x)(n) = x(n+1)$).

THEOREM 7. *There is a continuous embedding $f : (2^{\mathbb{Z}}, E_T) \to (2^{\omega}, E_0)$.*

This result is an extension of the well-known result of Slaman-Steel that a Borel embedding exists from $2^{\mathbb{Z}}$ to $E_0$. Let $F$ be the free part of $2^{\mathbb{Z}}$.

LEMMA 8. *Let $t$, $D$ be positive integers. There are pairwise disjoint relatively clopen (in $F$) sets $(T_i)_{i=1}^{N}$ such that $M = \bigcup_{i=1}^{N} T_i \subseteq F$ satisfies:*

(1) $\forall x \neq y \in M\ (\mathrm{dist}(x, y) > D)$.
(2) $\forall x \in F\ \exists y \in M\ (\mathrm{dist}(x, y) \leq D)$.
(3) $\forall x \neq y \in T_i\ (\mathrm{dist}(x, y) > 10 \cdot t \cdot D + 1)$.

PROOF. From Theorem 1 there is a clopen marker set $M' \subseteq F$ for marker distance $10 \cdot t \cdot D + 1$, that is:

(1) $\forall x \neq y \in M'\ (\mathrm{dist}(x, y) > 10 \cdot t \cdot D + 1)$.
(2) $\forall x \in F\ \exists y \in M'\ (\mathrm{dist}(x, y) \leq 10 \cdot t \cdot D + 1)$.

For each $x \in M'$, let $s(x)$ be the largest integer $s$ such that

$$\{\phi^{\alpha}(x) : 1 \leq \alpha < (s+1)(D+1)\} \cap M' = \emptyset.$$

Note that $s(x)$ is no larger than $22 \cdot t$. Let $T_0 = M'$, and for $i \geq 1$ let

$$T_i = \{\phi^{i(D+1)}(x) : x \in M' \wedge s(x) \geq i\}.$$

Let $N$ be the largest $i$ such that $T_i \neq \emptyset$ and let $M = \bigcup_{i=1}^{N} T_i$. Each $T_i$ is clopen in $F$ since $\phi$ is a homeomorphism and $\{x \in M' : s(x) \geq i\}$ is clopen for each $i$ (being a finite Boolean combination of clopen sets). Thus, $M$ is clopen in $F$. Also, (2) of Lemma 8 is easily verified by the definition of $s(x)$. ⊣

LEMMA 9 (Disjoint Markers). *Let $t$ be a positive integer. Let $1 > \varepsilon > 0$ and $(D_i)_{i \geq 1}$ be a sequence of integers such that $D_1 \cdot \varepsilon > 2\frac{(2t+1)^2}{1-\varepsilon}$ and for all $n \geq 2$,*

$$\sum_{i=1}^{n-1} \left( \frac{D_n}{D_i(1-\varepsilon)} + 1 \right) < \frac{D_n \cdot \varepsilon}{(2t+1)^2}.$$

*There are clopen Marker Sets $(M_i, D_i)_{i \in \omega}$ in $F$ with the following properties:*

(1) $\forall x \neq y \in M_i\ (\mathrm{dist}(x, y) > D_i(1 - \varepsilon))$.
(2) $\forall x \in F\ \exists y \in M_i\ (\mathrm{dist}(x, y) \leq D_i(1 + \varepsilon))$.
(3) *The following collection of sets is pairwise disjoint:*

$$\{\phi^{iD_j}(M_j) : j \geq 1,\ |i| \leq t\}.$$

PROOF. Let $\hat{M}_i$ be given by Lemma 8, using $t$ and $D = D_i$. We will adjust the points in $\hat{M}_i$ by no more than $D_i \cdot \varepsilon$ to obtain $M_i$. Let $\hat{M}_1 = \bigcup_{i=1}^N \hat{T}_i$ be the decomposition of $\hat{M}_1$ as in Lemma 8. We successively adjust the points in $\hat{T}_i$ to obtain $T_i$, and we then let $M_1 = \bigcup_{i=1}^N T_i$. Let $T_1 = \hat{T}_1$. Note that $\{\phi^{iD_1}(T_1) : |i| \leq t\}$ are all pairwise disjoint since points in $T_1$ are at least $10 \cdot t \cdot D_1 + 1$ apart. Suppose that the sets $T_1, T_2, \dots, T_{l-1}$ have been defined and satisfy the following properties:

(1) $\{\phi^{iD_1}(T_j) : 0 < j \leq l-1, |i| \leq t\}$ are all pairwise disjoint. Let

$$B = \bigcup_{0<j\leq(l-1)} \bigcup_{|i|\leq t} \phi^{iD_1}(T_j).$$

(2) Each $x \in T_i$ is a right shift by no more than $D_1 \cdot \varepsilon$ of a point in $\hat{T}_i$.

CLAIM 10. For each point $\hat{x} \in \hat{T}_l$ there exists a $\alpha < D_1 \cdot \varepsilon$ such that

$$\{\phi^{\alpha+iD_n}(\hat{x}) : |i| \leq t\} \cap B = \emptyset.$$

PROOF. For each point $\hat{x} \in \hat{T}_l$,

$$|\{\phi^s(\hat{x}) : 0 \leq s \leq D_1\} \cap B| \leq 2\left(\frac{2t+1}{1-\varepsilon}\right).$$

This is true since all pairs of points in $T = \bigcup_{u=1}^{l-1} T_u$ are at least $D_1 \cdot (1-\varepsilon)$ apart and we have $2t+1$ translations of this set $T$ in $B$. Thus

$$\left|\{\phi^s(\hat{x}) : 0 \leq s \leq D_1 \cdot \varepsilon\} \cap \left(\bigcup_{|i|\leq t} \phi^{iD_1}(B)\right)\right| \leq 2\frac{(2t+1)^2}{1-\varepsilon}.$$

Since $(D_1 \cdot \varepsilon) > 2\frac{(2t+1)^2}{1-\varepsilon}$, there is an integer $\alpha$ with $0 \leq \alpha < D_1 \cdot \varepsilon$ such that $\phi^\alpha(\hat{x}) \notin \bigcup_{|i|\leq t} \phi^{iD_1}(B)$. Hence for $|i| \leq t, \phi^{\alpha+iD_1}(\hat{x}) \notin B$. ⊣

For each $\hat{x} \in \hat{T}_l$, let $s(\hat{x})$ be the least non-negative integer $\alpha < D_1 \cdot \varepsilon$ satisfying the claim. Let $T_l = \{\phi^{s(\hat{x})}(\hat{x}) : \hat{x} \in \hat{T}_l\}$. Clearly $T_1, \dots, T_l$ still satisfies (1) and (2) above. Let $M_1 = \bigcup_{i=1}^N T_i$. Note that $\{\phi^{iD_1}(M_1) : |i| \leq t\}$ are all pairwise disjoint.

Now we can inductively adjust the rest of the $\hat{M}_n$ in a similar manner also avoiding $\{\phi^{iD_j}(M_j) : |i| \leq t, 1 \leq j < n\}$. Suppose that $M_1, \dots, M_{n-1}$ have been chosen satisfying the lemma. Let

$$B = \bigcup_{0<j<n} \bigcup_{|i|\leq t} \phi^{iD_j}(M_j).$$

Notice for any set $C \subseteq F$ and distance $D$ with the property $x, y \in C$ implies

$$\text{dist}(x, y) > D \cdot (1-\varepsilon).$$

It follows that

$$\left|\{\phi^s(\hat{x}) : 0 \le s \le D_n \cdot \varepsilon\} \cap \left(\bigcup_{|i|\le t} \phi^{iD_n}(C)\right)\right| \le \left(\frac{D_n}{D(1-\varepsilon)} + 1\right)(2t+1).$$

Hence for each $\hat{x} \in \hat{M}_n$

$$\left|\{\phi^s(\hat{x}) : 0 \le s \le D_n \cdot \varepsilon\} \cap \left(\bigcup_{|i|\le t} \phi^{iD_n}(B)\right)\right|$$
$$\le (2t+1)^2\left(\sum_{i=1}^{n-1}\left(\frac{D_n}{D_i(1-\varepsilon)} + 1\right)\right) < D_n\varepsilon.$$

For each point $\hat{x} \in \hat{M}_n$, let $s(\hat{x}) < D_n \cdot \varepsilon$ be the least non-negative integer such that $\phi^{s(\hat{x})} \notin \bigcup_{|i|\le t} \phi^{iD_n}(B)$. Let $M_n = \{\phi^{s(\hat{x})}(\hat{x}) : \hat{x} \in \hat{M}_n\}$. Since $\hat{M}_n$ satisfies (1) and (2) of Lemma 8 for $D_n$, and each point of $\hat{M}_n$ was moved no more that $D_n \cdot \varepsilon$ to the right in constructing $M_n$, it is clear that $M_n$ satisfies (1) and (2) of Lemma 9. For each $x \in M_n, x \notin \bigcup_{|i|\le t} \phi^{iD_n}(B)$ and so for $|i| \le t$, $\phi^{iD_n}(x) \notin B = \bigcup_{0<j<n} \bigcup_{|i|\le t} \phi^{iD_j}(M_j)$. This gives property (3) of Lemma 9. ⊣

We now apply Lemma 9 for $t = 3$ and $(D_i)_{i\ge1}$ as in Lemma 9 and also satisfying the extra stipulation that $n!$ divides $D_n$. This produces relatively clopen sets $M_i' \subseteq F$. We let $M_i \subseteq 2^{\mathbb{Z}}$ be open with $M_i' = M_i \cap F$. These "extended" marker sets satisfy the following:

(1) For each $n$,
  (a) $\forall x \ne y \in M_n\ (\mathrm{dist}(x, y) > D_n(1-\varepsilon))$.
  (b) $\forall x \in F\ \exists y \in M_n\ (\mathrm{dist}(x, y) \le D_n(1+\varepsilon))$.

(2) The following sets are all pairwise disjoint:

$$M_1, \phi^{D_1}(M_1), \phi^{-D_1}(M_1), \phi^{2D_1}(M_1), \phi^{-2D_1}(M_1), \phi^{3D_1}(M_1), \phi^{-3D_1}(M_1),$$
$$M_2, \phi^{D_2}(M_2), \phi^{-D_2}(M_2), \phi^{2D_2}(M_2), \phi^{-2D_2}(M_2), \phi^{3D_2}(M_2), \phi^{-3D_2}(M_2),$$
$$\ldots,$$
$$M_k, \phi^{D_k}(M_k), \phi^{-D_k}(M_k), \phi^{2D_k}(M_k), \phi^{-2D_k}(M_k), \phi^{3D_k}(M_k), \phi^{-3D_k}(M_k),$$
$$\ldots$$

[(1a) and (2) follow since they are true for the $M_i'$, and $F$ is dense is $2^{\mathbb{Z}}$.]

(3) Each $M_i$ is open and $M_i \cap F$ is relatively clopen in $F$.

(4) $n!$ divides $D_n$.

We fix some conventions. We regard the finite binary strings $2^{<\omega}$ to be ordered first by their lengths, and then lexicographically. We implicitly refer to this order when we use the phrase "least string." If $s \in 2^{<\omega}$, we let

$s^* \in 2^\omega$ be the result of concatenating infinitely many copies of $s$ forwards and backwards. That is $s^*(i) = s(i \bmod n)$, where $s = (s(0), \dots, s(n-1))$ has length $n = \mathrm{lh}(s)$.

Fix for the moment $x \in 2^{\mathbb{Z}}$. We define in stages some triples of the form $(I_a, k_1^a, k_2^a)$, for $a \in \omega$. At stage $n$, for all $a \leq n$ if $(I_a, k_1^a, k_2^a)$ has not yet been defined, then we (perhaps) define $(I_a, k_1^a, k_2^a)$ according to the following:

(1) Let $\hat{k}_1^a$ be the least integer $k$ such that

$$0 \leq k < 3D_a, \quad \phi^{-k}(x) \restriction_{[-10D_n, 10D_n]} \subseteq M_a,$$

provided that such a $k$ exists (here, $\phi^{-k}(x) \restriction_{[-10D_n, 10D_n]}$ denotes, with a slight abuse of notation, the basic open set in $2^{\mathbb{Z}}$ determined by $\phi^{-k}(x) \restriction_{[-10D_n, 10D_n]}$).

(2) Let $\hat{k}_2^a$ be the least integer $k$ such that

$$0 \leq k < 3D_a, \quad \phi^{k}(x) \restriction_{[-10D_n, 10D_n]} \subseteq M_a,$$

provided that such a $k$ exists.

(3) If both $\hat{k}_1^a$ and $\hat{k}_2^a$ exist and $(I_a, k_1^a, k_2^a)$ has not been defined at an earlier stage, let $k_1^a = \hat{k}_1^a$, $k_2^a = \hat{k}_2^a$, $I_a = x \restriction_{[-k_1^a, k_2^a]}$.

We define $f(x) \in \omega^\omega$ by letting $f(x)(n)$ be the integer that codes (in some reasonable manner) the following information:

Type I. For each interval $I_a$ defined at (exactly) the $n$th stage, code the following:
- (a) $I_a$.
- (b) $a$.
- (c) The translation of the right endpoint of $I_{a-1}$ to the right endpoint of $I_a$ (that is, $k_2^a - k_2^{a-1}$), if $I_{a-1}$ exists by the $n$th stage.
- (d) The translation of the right endpoint of $I_a$ to the right endpoint of $I_{a+1}$, if $I_{a+1}$ exists by the $n$th stage.

Type II. The string $s_n$ defined as follows. Let $l_n$ be the maximum of 0 and the integers $k_1^a$ determined by the $n$th stage, and $r_n$ be the maximum of 0 and the integers $k_2^a$ determined by the $n$th stage. Let $u_n = u_n(x) \doteq x \restriction_{[-(l_n+D_n), r_n+D_n]}$. Let $s_n$ be the least string which is a potential period of $u_n$, that is $u_n$ is a substring of $s_n^*$.

Type III. The pair of integers $\hat{a}_n, \hat{b}_n$ defined by the following:
- (a) $a_{-1} = b_{-1} = 0$
- (b) Let $a_n$ be the integer $0 \leq a_n < \mathrm{lh}(s_n)$, such that $u_n = \phi^{-a_n}(s_n^*) \restriction (\mathrm{lh}(u_n))$.
- (c) $b_n = l_n + D_n$.
- (d) $\hat{a}_n = a_n - a_{n-1}$, $\hat{b}_n = b_n - b_{n-1}$.

Notice that $a_n$ is the distance from the left boundary of $u_n(x)$ to the first start of a 'period' $s_n$. That is, if $u_n = t_n^0{}^\frown s_n{}^\frown \cdots {}^\frown s_n{}^\frown t_n^1$ where $t_n^0$ is a proper end segment of $s_n$ and $t_n^1$ is a proper initial segment of $s_n$, then $a_n = \mathrm{lh}(t_n^0)$. Also, $b_n$ is, roughly speaking, the distance from $x$ to the left boundary of $u_n(x)$.

To define $f(x)(n)$ we need to consider $x \upharpoonright_{[-14D_n, 14D_n]}$ only. Thus, this process will produce a continuous function. Also for $x \in F$ and $a \in \omega$, the triple $(I_a, k_1^a, k_2^a)$ will eventually be defined at some stage $n$ (so will be coded into $f(x)(n)$).

Claim 11. If $x \in F$ and $xE_Ty$ then $f(x)E_0f(y)$.

Proof. The disjointness property of the marker sets and their translations implies that for any $z \in 2^{\mathbb{Z}}$ and $k \in \mathbb{Z}$ there is at most one pair $(n, i) \in \omega \times \{0,1,2,3\}$ such that $\phi^{\pm i\cdot D_n + k}(z) \in M_n$. Suppose $x \in F$ and $xE_Ty$. Let $k_0$ be such that $\phi^{k_0}(x) = y$ and without loss of generality suppose $k_0 > 0$. Now for any $n \in \omega$,

$$\{\phi^t(x) \mid |t| \leq 3D_n\} - \{\phi^t(y) \mid |t| \leq 3D_n\} = \{\phi^{-3D_n+t}(x) \mid 0 \leq t < k_0\}.$$

Thus by the above property for some $N \in \omega$, for all $m \geq N$,

$$(\{\phi^k(x) : |k| \leq 3D_m\} \cap M_m) \subseteq (\{\phi^k(y) : |k| \leq 3D_m\} \cap M_m).$$

Similarly it follows that for some $M \in \omega$, for all $m \geq M$,

$$(\{\phi^k(y) : |k| \leq 3D_m\} \cap M_m) \subseteq (\{\phi^k(x) : |k| \leq 3D_m\} \cap M_m).$$

Thus there exists a $N_0$ such that for all $m \geq N_0$,

$$(\{\phi^k(x) : |k| \leq 3D_m\} \cap M_m) = (\{\phi^k(y) : |k| \leq 3D_m\} \cap M_m).$$

Another way to state this is for all $m \geq N_0$,

$$\{k : |k| \leq 3D_m \wedge \phi^k(x) \in M_m\} = \{k + k_0 : |k| \leq 3D_m \wedge \phi^k(y) \in M_m\}.$$

This implies for each $m \geq N_0$ that $f(x)$ and $f(y)$ are eventually at some stage coding the same interval for $I_m$. Let $N_1 \geq N_0$ be a stage so that for all $a \leq (N_0 + 1)$ the interval $I_a$ has been defined by the $N_1$th stage. Thus for all $m > N_1$ both $f(x)$ and $f(y)$ will be defining the same intervals $I_a$ for the $m$th stage with the same links to $I_{a-1}$ and $I_{a+1}$. Thus for all $m > N_1$, $f(x)(m)$ is coding the same Type I information as $f(y)(m)$.

Next we show that $f(x)$ and $f(y)$ will eventually code the same Type II information. Let $L$ be the maximum length of $u_{N_1}(x)$, $u_{N_1}(y)$. Choose $B \geq N_1$ such that for all $n \geq B$, $\mathrm{dist}(x, M_n) > L$ and $\mathrm{dist}(y, M_n) > L$. To see this can be done, note that since the marker sets are disjoint there can be only finitely many points of $\bigcup_n M_n$ within $L$ of either $x$ or $y$. Let $N_2 \geq N_1$ be a stage where $I_B$ has been defined. Note:

(1) Both $l_n$ and $r_n$ increase to infinity as $n$ goes to infinity.
(2) $l_{N_2} > l_{N_1}$ and $r_{N_2} > r_{N_1}$ for both $x$ and $y$.

(3) We have for all $m > N_1$, $f(x)(m)$ is coding the same Type I information as $f(y)(m)$ thus for all stages $m > N_1$ any $k_1^a$ defined for $y$ is $k_0$ larger than the $k_1^a$ defined for $x$ and any $k_2^a$ defined for $x$ is $k_0$ larger than the $k_2^a$ defined for $y$.

(4) Thus for all $m \geq N_2$ we have $l_m(y) = l_m(x)+k_0$ and $r_m(x) = r_m(y)+k_0$.

Thus for all stages $n \geq N_2$ it follows that $u_n(x) = u_n(y)$. This implies that for all $n \geq N_2$, $f(x)(n)$ and $f(y)(n)$ are coding the same Type II information.

We now show $f(x)$ and $f(y)$ will also eventually code the same Type III information. For all $n \geq N_2$ we have $u_n(x) = u_n(y)$. It follows that for all $n \geq N_2$ that the values of $s_n$, and hence $a_n$ defined for $x$ are the same as for $y$. We also clearly have that for $n \geq N_2$ that $b_n(x) = b_n(y) - k_0$. This implies that for all $n > N_2$ we have that $\hat{a}_n(x) = \hat{a}_n(y)$ and $\hat{b}_n(x) = \hat{b}_n(y)$. Thus for all $n > N_2$ we have that $f(x)(n) = f(y)(n)$. ⊣

Claim 12. If $x \in F$ and $f(x)E_0 f(y)$ then $xE_T y$.

Proof. Suppose $x \in F$ and $f(x)E_0 f(y)$. Since $f(x)(n)$ and $f(y)(n)$ eventually code the same type I information, there is some integer $N$ such that for $m \geq N$, $I_m(x) = I_m(y)$ and these intervals are linked to $I_{m-1}$, $I_{m+1}$ in the same way for both $x$ and $y$ (here $I_m(x)$ denotes the interval $I_m$ as defined for $x$). Now since the marker sets are disjoint, and on the free part we do define all $I_m$, by the disjointness property of the marker sets it follows that the right (left) endpoints of these intervals are of unbounded distances from $x$. This means that the intervals $\{I_m\}_{m\geq 1}$ and their links determine the equivalence class of $x$. In fact, any tail of $\{I_m\}_{m\geq 1}$ and their links will determine the equivalence class of $x$. Thus $xE_T y$. ⊣

We now turn to the periodic part of $2^{\mathbb{Z}}$. We use the following simple algebraic fact.

Claim 13. If $x \in 2^{\mathbb{Z}}$ is periodic with a period length $c$ and $N \geq c$ then for any string $s$ of length $d \leq c$ if $u \doteq x \restriction [-N, N]$ occurs as a substring of $s^*$ then $d = c$ and $s$ is a period of $x$.

Proof. First note the following simple fact. If $t, w, v \in 2^{<\omega}$ such that $t = w^\wedge v = v^\wedge w$ then there exists $z \in 2^{<\omega}$ and positive integers $l$, $k$ such that $w$ is $k$ copies of $z$ and $v$ is $l$ copies of $z$. This follows from an induction argument on the length of the longer of $w$ and $v$.

Suppose $d < c$ and $u \doteq x \restriction [-N, N]$ occurs as a substring of $t^*$ where $t$ has length $d < c$. Without loss of generality $N = c$, and so $u = s^\wedge s$ where $s$ is a period of $x$. We may also assume $t = x \restriction [-N, -N + d - 1]$. Let $m = c \bmod d$ and $w = (t_0, \dots, t_{m-1})$ and $v = (t_m, \dots, t_{d-1})$. Note $t = w^\wedge v$. Since $s^\wedge s$ is a substring of $t^*$ it follows by looking at the second copy of $s$ that $s$ extends $v^\wedge w$ and thus $v^\wedge w = w^\wedge v$. By the above simple fact it follows that $w$ and $v$ are both some number of copies of a string $z$, hence $s$ is also. This contradicts the period length of $x$ being $c$. ⊣

CLAIM 14. If $x \in 2^{\mathbb{Z}}$ is periodic and $xE_Ty$ then $f(x)E_0f(y)$.

PROOF. Suppose $x \in 2^{\mathbb{Z}}$ is periodic and $xE_Ty$. Choose $N_0$ such that $D_{N_0}$ is greater than three times the period of $x$. There can be no points from $M_{N_0}$ which are $E_T$ equivalent to $x$, since this would give points in $M_{N_0}$ within distance $D_{N_0}$ of each other. It follows that for both $x$ and $y$ and all large enough $n$ that there is no Type I information coded by $f(x)(n)$ or $f(y)(n)$. For large enough $n$ the Type II information coded by $f(x)(n)$ and $f(y)(n)$ also agrees; this follows immediately from Claim 13. Now we need to show the Type III information agrees for all stages past some point. As we mentioned above, for large enough $n$ there are no points of $M_n$ which are $E_T$ equivalent to $x$ or $y$. It follows that for large enough $n$ that $\hat{b}_n(x) = \hat{b}_n(y) = D_n - D_{n-1}$. For large enough $n$ we will also have that the period of $x$ divides $D_n$, and so $\hat{a}_n(x) = 0$. Likewise $\hat{a}_n(y) = 0$ for large enough $n$. Therefore for all large enough $n$ we have that $f(n)(x)$ codes the same Type III information as $f(n)(y)$. Thus $f(x)E_0f(y)$. ⊣

CLAIM 15. If $x \in 2^{\mathbb{Z}}$ is periodic and $f(x)E_0f(y)$ then $xE_Ty$.

PROOF. Since $x$ is periodic there is, for large enough $n$, no Type I information coded in $f(x)(n)$. Thus the same is true for $y$, and this in turn implies that $y$ is periodic. Considering now the Type II information coded by $f(x)(n)$ and $f(y)(n)$ we see that $x$ and $y$ have the same period $s$ (i.e., $xE_Ts^*$ and $yE_Ts^*$). Thus $xE_Ty$. ⊣

CLAIM 16. The function $f$ is one-to-one.

PROOF. Suppose $f(x) = f(y)$. We already know $xE_Ty$. Consider first the case where $x$ is periodic. Since for all $n$ the Type III information coded by $f(x)(n)$ and $f(y)(n)$ is the same, $\hat{a}_n(x) = \hat{a}_n(y)$ and $\hat{b}_n(x) = \hat{b}_n(y)$ for all $n$. This implies that $a_n(x) = a_n(y)$ and $b_n(x) = b_n(y)$ for all $n$. For large enough $n$, the $s_n$ coded in the type II information by $f(x)(n)$ will be a true period of $x$. From $a_n$ and $b_n$ we can then determine the offset of $x$ relative to $s_n^*$, more precisely, $x = \phi^d(s_n^*) = y$, where $d = -a_n + b_n \bmod p_n$, $p_n = \mathrm{lh}(s_n)$.

Now suppose $x \in F$. It has already been shown that for some $N_0$ for all $n \geq N_0$ we have that $u_n(x) = u_n(y)$. Furthermore $f(x) = f(y)$ implies $b_n(x) = b_n(y)$ for all $n$. From $u_n(x)$ and $b_n(x)$ we determine a basic neighborhood containing $x$. More precisely, for each $n \geq N_0$ we have that

$$y \in N_{x \restriction [-(l_n+D_n), r_n+D_n]} \qquad (\text{also } x \in N_{y \restriction [-(l_n+D_n), r_n+D_n]}).$$

This implies that $x = y$ since $D_n \to \infty$. Thus $f$ is one-to-one. ⊣

CLAIM 17. $f$ is a topological embedding, that is, a homeomorphism with its range.

PROOF. We have already observed that $f$ is continuous and one-to-one. Since $2^{\mathbb{Z}}$ is compact, it follows that $f^{-1}$ is also continuous. We can also see this directly as follows. Suppose $f(x) = y$ and fix $n_0 \in \omega$. We must show that there is an $n_1$ such that if $y' \in \mathrm{ran}(f)$ and $y' \restriction n_1 = y \restriction n_1$, then $x' \restriction -[n_0, n_0] = x \restriction [-n_0, n_0]$, where $x' = f^{-1}(y')$. Note that for any $n$, $y \restriction n$ determines $l_n$, $r_n$, $b_n$ (as in the definition of $f(x)$), and so determines $x \restriction [-(l_n + D_n), r_n + D_n]$. Since $l_n + D_n$, $r_n + D_n$ go to infinity with $n$, for large enough $n_1$ we have that $y \restriction n_1$ determines $x \restriction [-n_0, n_0]$. ⊣

At this point we have produced a continuous embedding from $(2^{\mathbb{Z}}, E_T)$ to $(\omega^\omega, E_0)$. The following simple lemma finishes the proof of Theorem 7.

LEMMA 18. *There is a continuous embedding* $\pi : (\omega^\omega, E_0) \to (2^\omega, E_0)$.

PROOF. Let $\theta : \omega \times \omega \to \omega$ be a bijection. For $x \in \omega^\omega$, let $\pi(x)(n) = i$ iff $\theta^{-1}(n) = (a, b)$ and if $x(a) = \sum_{j \in \omega} c_j 2^j$ is the 2-ary expansion of $x(a)$, then $c_b = i$. It is straightforward to check this works. ⊣

REMARK 19. We could avoid appealing to Lemma 18 by using the fact that there is a uniform bound $e_n$ on the size of $f(x)(n)$ for all $x$, and at stage $n$ output the binary expansion of $f(x)(n)$, padded to have length $e_n$.

We next consider the question of whether there is a topological embedding from $2^{\mathbb{Z}}$ with the shift to $2^\omega$ with the induced action of the odometer. Here we must restrict our attention to the free part $F$ of $2^{\mathbb{Z}}$ as $E_0$ has no finite classes. Recall $\phi$ is the shift action on $2^{\mathbb{Z}}$ and $\theta$ is the odometer action on $2^\omega$.

THEOREM 20. *There is a topological embedding* $f : F \to 2^\omega$ *such that for all* $x, y \in F$, $xE_Ty$ *iff* $f(x)E_0f(y)$ *and such that if* $y$ *is a positive shift of* $x$ *(i.e.,* $y = \phi^n(x)$ *for some* $n > 0$*) then* $f(y)$ *is a positive shift of* $f(x)$ *under the odometer action (i.e.,* $f(y) = \theta^m f(x)$ *for some* $m > 0$*).*

PROOF. Let $x \in F$. We define $f(x) \in 2^\omega$ in a manner similar to that of Theorem 7. We output bits of $f(x)$ in stages. At stage $n$, we first compute the integer $l_n \geq 0$ as in the proof of Theorem 7 except using the $\hat{k}_1^a$ instead of $k_1^a$. Thus $l_n$ is the maximum of 0 and the $\hat{k}_1^a$ for $0 \leq a \leq n$ which are defined. So $l_n$ is the distance to the "furthest marker to the left" that we have after the $n$th stage search. Let $I_n = x \restriction [-l_n, -l_n + 10D_n]$. Let $c_n = l_n - l_{n-1}$ (where $l_{-1} = 0$). For the $n$th stage output of $f(x)$ we output $I_n$ followed by the binary representation of $c_n$, padded by zeros at the end to have a fixed length depending only on $n$ (see Remark 19). This clearly defines a continuous map $f$ from $2^{\mathbb{Z}}$ to $2^\omega$.

First note that, as before, if $x, y \in F$ and $xE_Ty$ then for large enough $n$ we have that $l_n(x)$ picks the same marker point as $l_n(y)$. That is, if $y = \phi^k(x)$ then $l_n(y) = l_n(x) + k$. From this it follows that for large enough $n$ that $I_n(x) = I_n(y)$ and also that $c_n(x) = c_n(y)$ and so $f(x)E_0f(y)$. From the $E_0$ class of $f(x)$ we can reconstruct the $E_T$ class of $x$ since $I_n$ is of the form $x \restriction [-a_n, b_n]$ where both $a_n$, $b_n$ tend to infinity with $n$ and since $c_n$ describes how the interval $I_{n-1}$ sits inside $I_n$. $f$ is also one-to-one since from the $c_n$ we can compute the offset of $x$ relative to the $I_n$.

Finally, suppose that $y = \phi^k(x)$ where $k > 0$. If for all $n$ we have that $l_n(x)$ and $l_n(y)$ pick the same point (that is, $l_n(y) = l_n(x) + k$), then $I_n(x) = I_n(y)$ for all $n$, $c_0(x) < c_0(y)$, and $c_n(x) = c_n(y)$ for all $n > 0$. Thus, at the last point $l$ of disagreement between $f(x)$ and $f(y)$ we have that $f(x)(l) < f(y)(l)$. Hence $f(y) = \theta^m(f(x))$ for some $m > 0$. Otherwise, let $n$ be maximal such that $l_n(y) \neq l_n(x) + k$ (that is, $l_n(x), l_n(y)$ don't pick the same point). Consider the bits output at stage $n+1$ for $f(x)$ and $f(y)$. We claim that $c_{n+1}(x) < c_{n+1}(y)$. This suffices since the outputs for $f(x)$ and $f(y)$ clearly agree at all stages $> n+1$. To see the claim note that $l_{n+1}(x)$ picks the same point as $l_{n+1}(y)$, that is $l_{n+1}(y) = l_{n+1}(x) + k$, but $l_n(x)$ picks a point strictly to the left of $l_n(y)$, that is, $l_n(y) < l_n(x) + k$. This follows from the fact that for all $i$ we have $l_i(y) \leq l_i(x) + k$, which is easily checked from the definition of $l_i$. Thus, $c_{n+1}(x) < c_{n+1}(y)$. The continuity of $f^{-1}$ follows either from the compactness of $2^{\mathbb{Z}}$ or the argument of Claim 17. ⊣

Finally, it is possible to combine the arguments of Theorems 7 and 20 to obtain the following.

THEOREM 21. *There is a topological embedding $f : 2^{\mathbb{Z}} \to 2^{\omega}$ such that for all $x, y \in 2^{\mathbb{Z}}$, $xE_Ty$ iff $f(x)E_0f(y)$ and such that if $x, y \in F$ and $y$ is a positive shift of $x$ under $\phi$ then $f(y)$ is a positive shift of $f(x)$ under the odometer action $\theta$.*

PROOF (SKETCH). For $x \in 2^{\mathbb{Z}}$ we again define $f(x)$ in stages. At stage $n$, let $l_n$ again be as defined in Theorem 20 (and we set $l_{-1} = 0$). Our output at stage $n$ depends on whether $l_n > l_{n-1}$.

Case (I) $l_n > l_{n-1}$. As in Theorem 20, let $I_n = x \upharpoonright [-l_n, -l_n + 10D_n]$ and let $c_n = l_n - l_{n-1}$. Let $e_n, f_n$ be positive integers depending only on $n$ (and not $x$) such that the lengths of $I_n$ and the binary representation of $c_n$ will not exceed $e_n, f_n$ respectively. We output the binary representation for the length of $I_n$ padded by zeros at the end to have length $e_n$, followed by $I_n$ padded by zeros at the end to have length $e_n$, followed by the binary representation of $c_n$ padded at the end to have length $f_n$, followed by a single digit of 1.

Case (II) $l_n = l_{n-1}$. Let $u_n = x \upharpoonright [-l_n - 10D_n, -l_n + 10D_n]$. As in Theorem 7, let $s_n$ be the least string which is a potential period of $u_n$. As in Theorem 7, let $a_n$ be the distance from the left boundary of $I_n$ to the first start of an $s_n$ substring. Also, let $\hat{a}_n = a_n - a_{n-1}$ if $a_{n-1}$ is defined, and otherwise $\hat{a}_n = a_n$ (by $a_n - a_{n-1}$ we actually mean $a_n - a_{n-1} + g_n$ where $g_n$ depends only on $n$ and is large enough to guarantee the expression is positive). We may assume $e_n, f_n$ are larger than the output lengths of $s_n, \hat{a}_n$. We output the binary representation for the length of $s_n$ padded by zeros at the end to have length $e_n$, followed by $s_n$ padded at the end to have length $e_n$, followed by the binary representation of $\hat{a}_n$ padded at the end to have length $f_n$, followed by a single digit 0.

Clearly the map $f : 2^{\mathbb{Z}} \to 2^{\omega}$ we have defined is continuous. If $xE_Ty$, say $y = \phi^k(x)$, and $x, y \in F$, then for large enough $n$, $l_n(x)$ and $l_n(y)$ pick the same point, that is, $l_n(y) = l_n(x) + k$, and from this it follows easily that for large enough $n$ that $f(x)$ and $f(y)$ output the same bits at stage $n$. If $xE_Ty$ and both are periodic, then for large enough $n$ we are in case (II) for both $x$ and $y$. For large enough $n$, the value $s_n$ computed for both $x$ and $y$ will be the same, namely the true least string $s$ which is a period for $x$. Also, for large enough $n$ we will have $\hat{a}_n = 0$ using the fact that $\mathrm{lh}(s) \mid D_n$ for large enough $n$. The $E_0$ equivalence class of $f(x)$ determines the $E_T$ equivalence class of $x$ by an argument similar to that of Theorem 7. To see that $f$ is one-to-one note that $f(x)$ determines all of the $l_n(x)$. If $x \in F$, then from the $l_n$ and the $I_n$ we can reconstruct $x$. If $x$ is periodic, then from $f(x)$ we can recover the period $s$, and from the $\hat{a}_n$ we can recover the offset of $x$ relative to $s^*$. The continuity of $f^{-1}$ follows again either by the compactness of $2^{\mathbb{Z}}$ or a simple direct argument.

Finally, suppose $x, y \in F$ and $y = \phi^k(x)$ where $k > 0$. Let $n$ be least such that for all $m \geq n$, $l_n(y) = l_n(x) + k$. If $n = 0$, the argument is a minor variation of what follows, so assume $n > 0$. We claim that the bits output at stage $n$ for $f(y)$ are strictly greater (when viewed in the usual way as a binary representation for an integer) than those for $f(x)$. To see this, first note that $l_{n-1}(y) < l_{n-1}(x) + k$. If we are in case (I) for both $x$ and $y$ at stage $n$, then $c_n(y) > c_n(x)$ and the claim follows. If we are in case (II) for $x$ and case (I) for $y$, then the claim follows since the last non-padded bit added for this block was a 1 for $f(y)$ and a 0 for $f(x)$. The other cases cannot occur as $l_{n-1}(y) < l_{n-1}(x) + k$ (i.e., the point determined by $l_{n-1}(y)$ is to the right of that for $l_{n-1}(x)$). It is not difficult to check that for all $m \geq n + 1$ we will output the same bits for $f(x)$ and $f(y)$ at stage $m$ except perhaps in the case where $l_n(x) = l_{n-1}(x)$ and $l_n(y) > l_{n-1}(y)$. In this case we are in case (II) for $x$ at stage $n$, but case (I) for $y$. If $l_{n+1}(x) > l_n(x)$ (and hence the same for $y$), then it is easy to check that $f(x)$ and $f(y)$ output the same bits at all stages $m \geq n+1$. If $l_{n+1}(x) = l_n(x)$ (and so likewise for $y$), then we are in case (II) at stage $n + 1$ for both $x$ and $y$. For $x$ we have $\hat{a}_{n+1}(x) = a_{n+1}(x) - a_n(x)$ whereas $\hat{a}_{n+1}(y) = a_{n+1}(y)$. Also, $a_{n+1}(x) = a_{n+1}(y)$. So $\hat{a}_{n+1}(x) \leq \hat{a}_{n+1}(y)$. If strict inequality holds we are done as then $f(y)$ outputs a larger integer than $f(x)$ at stage $n + 1$ and it is easy to see that all later stages they agree. If equality holds then $f(x)$ and $f(y)$ have the same output at stage $n + 1$ since $s_{n+1}(x) = s_{n+1}(y)$ as well (since $u_{n+1}(x) = u_{n+1}(y)$). Since $f(y)$ had the larger output at stage $n$ we are done. ⊣

**§4. Concluding remarks.** Since $\mathbb{R}^\omega$ is topologically universal for Polish spaces (i.e., every Polish space embeds into $\mathbb{R}^\omega$), it follows immediately that the shift action of $\mathbb{Z}$ on $(\mathbb{R}^\omega)^{\mathbb{Z}}$ is universal for all equivalence relations generated by continuous actions of $\mathbb{Z}$ on Polish spaces. By this we mean that if $X$ is a Polish space and $\sim$ is an equivalence relation on $X$ corresponding to a continuous action of $\mathbb{Z}$ on $X$, then there is a continuous embedding

$f : (X, \sim) \to ((\mathbb{R}^\omega)^{\mathbb{Z}}, E_T)$, where $E_T$ again denotes the shift equivalence relation on $(R^\omega)^{\mathbb{Z}}$.

Of course, $(2^{\mathbb{Z}}, E_T)$ and $(2^\omega, E_0)$ cannot be universal for all continuous $\mathbb{Z}$ actions due to simple topological considerations, namely the spaces $2^{\mathbb{Z}}$ and $2^\omega$ are 0-dimensional so any continuous map from the connected space $\mathbb{R}^{\mathbb{Z}}$ to either of them is constant. The next result shows that in some sense this is the only obstruction to these spaces being universal.

THEOREM 22. *The equivalence relations $(2^{\mathbb{Z}}, E_T)$ and $(2^\omega, E_0)$ are both universal for the equivalence relations generated by continuous actions of $\mathbb{Z}$ on 0-dimensional Polish spaces.*

PROOF. Let $X$ be a 0-dimensional Polish space and $(n, x) \to n \cdot x \in X$ a continuous action of $\mathbb{Z}$ on $X$. Let $\sim$ be the corresponding equivalence relation on $X$. It suffices to show that $(X, \sim)$ continuously embeds into $(2^\omega, E_0)$. Define first $\pi : X \to X^{\mathbb{Z}}$ by $\pi(x)(n) = n \cdot x$. It is easy to check that $\pi$ is a continuous embedding from $(X, \sim)$ to $(X^{\mathbb{Z}}, E_T)$. Since $X$ is 0-dimensional it embeds, as a topological space, into $\omega^\omega$. This induces a continuous embedding from $(X^{\mathbb{Z}}, E_T)$ to $((\omega^\omega)^{\mathbb{Z}}, E_T)$. To complete the proof of Theorem 22 it suffices now to prove the following theorem, which is a generalization of Theorem 7. ⊣

THEOREM 23. *There is a continuous embedding from $((\omega^\omega)^{\mathbb{Z}}, E_T)$ into $(2^\omega, E_0)$.*

PROOF. Since the proof is just a slight generalization of that of Theorem 7, we just briefly mention the differences. We can identify the elements of $(\omega^\omega)^{\mathbb{Z}}$ with elements of $(\omega)^{\mathbb{Z}\times\omega}$. The shift map $\phi$ thus satisfies $\phi(x)(n, k) = x(n + 1, k)$. First note that the proof of Theorem 7 works as given for $\omega^{\mathbb{Z}}$ (with the shift equivalence $E_T$) instead of $2^{\mathbb{Z}}$. The main difference for the current proof is that the elements of our domain space are now functions from $\mathbb{Z} \times \omega$ to $\omega$ rather than $\mathbb{Z}$ to $\omega$. We again let $F$ denote the free part of $\omega^{\mathbb{Z}\times\omega}$. There is no problem in applying the method of Lemma 9 to construct marker sets which are relatively clopen in $F$ and which satisfy Lemma 9. As before we view these sets $M_n$ as extended to all of $\omega^{\mathbb{Z}\times\omega}$ as open sets. As before, $f(x)(n)$ will be an integer coding three types of information. To define the Type I information, we again define for some $a \le n$ triples $(I_a, k_1^a, k_2^a)$. We define $\hat{k}_1^a$ to be the least $0 \le k < 3D_a$, if one exists, such that the basic open set $\phi^{-k}(x) \restriction ([-10D_n, 10D_n] \times [0, D_n])$ in $\omega^{\mathbb{Z}\times\omega}$ is contained in $M_a$. If $k_1^a$, $k_2^a$ are both defined and $I_a$ has not been defined at an earlier stage, we set $I_a = x \restriction ([-k_1^a, k_2^a] \times [0, D_a])$. The rest of the Type I information is as before. For the Type II information, we define $l_n$, $r_n$ as before, and let now $u_n = x \restriction ([-(l_n + D_n), r_n + D_n] \times [0, D_n])$. The potential period $s_n$ now refers to a restriction of $u_n$ of the form $s_n = u_n \restriction ([a, b] \times [0, D_n])$ such that $s_n^*$ extends $u_n$, where $s_n^* : \mathbb{Z} \times [0, D_n] \to \omega$ is obtained by copying $s_n$ horizontally in both directions. The rest of the Type II and III information is as before. The proof that $f$ works is exactly as before.

The fact that $(2^\omega, E_0)$ continuously embeds into $(2^{\mathbb{Z}}, E_T)$ is a special case of the more general fact that $(2^\omega, E_0)$ continuously embeds into any "non-smooth" countable Borel equivalence relation (i.e., an equivalence relation having no Borel selector). This, in turn, is a special case of the result of [2] which extends this to all Borel equivalence relations. ⊣

We may also generalize Theorem 21 to $(\omega^\omega)^{\mathbb{Z}}$ as follows. The proof is entirely analogous to that of Theorem 21, we leave the details to the reader.

THEOREM 24. *There is a topological embedding $f : (\omega^\omega)^{\mathbb{Z}} \to 2^\omega$ such that for all $x, y \in (\omega^\omega)^{\mathbb{Z}}$, $x E_T y$ iff $f(x) E_0 f(y)$, and if $x, y$ are in the free part of $(\omega^\omega)^{\mathbb{Z}}$ and $y$ is a positive shift of $x$ then $f(y)$ is a positive shift under the odometer action of $f(x)$.*

Since any 0-dimensional Polish space topologically embeds in $\omega^\omega$, Theorem 24 gives immediately the following corollary. We say an auto-homeomorphism $\pi$ of a Polish space $X$ is free if for all $n \neq 0$, $\pi^n(x) \neq x$ for all $x \in X$.

COROLLARY 25. *Any free auto-homeomorphism of a 0-dimensional Polish space is topologically isomorphic to an induced action of the odometer on a subspace of $2^\omega$.*

We introduce the notion of dimension for countable Borel equivalence relations as follows.

DEFINITION 26. We say the countable Borel equivalence relation $E$ on the Polish space $X$ is $\leq n$ dimensional if there is a continuous embedding $f : (X, E) \to (Y^{\mathbb{Z}^n}, E_T)$ for some Polish space $Y$ (here $E_T$ again denotes the translation equivalence relation induced by the natural action of $\mathbb{Z}^n$ on $Y^{\mathbb{Z}^n}$). We say $(X, E)$ is exactly $n$-dimensional if it is $\leq n$ dimensional but not $\leq n-1$ dimensional.

Our results show that the equivalence relations $E_0(2)$, $E_0(\omega)$, $2^{\mathbb{Z}}$, $\omega^{\mathbb{Z}}$, and $(\omega^\omega)^{\mathbb{Z}}$ are all equivalent one-dimensional equivalence relations, that is, each of them continuously embeds into the others. This gives a fairly complete description of the one-dimensional equivalence relations. The situation is less clear for the higher dimensions. In particular, the following remains open.

QUESTION 27. Is $2^{\mathbb{Z}^n}$ (with translation equivalence $E_T$) exactly $n$-dimensional? Is $2^{\mathbb{Z}^{<\omega}}$ finite dimensional (i.e., $\leq n$ dimensional for some $n$)?

We do not know at the moment even if $2^{\mathbb{Z}\times\mathbb{Z}}$ is one-dimensional. We have presented here some new techniques for producing continuous embeddings between equivalence relations. Answers to the above questions could provide interesting limitations on these techniques.

## REFERENCES

[1] R. DOUGHERTY, S. JACKSON, and A. S. KECHRIS, *The structure of hyperfinite Borel equivalence relations*, ***Transactions of the American Mathematical Society***, vol. 341 (1994), pp. 193–225.

[2] L. Harrington, A. S. Kechris, and A. Louveau, *A Glimm-Effros dichotomy for Borel equivalence relations*, ***Journal of the American Mathematical Society***, vol. 3 (1990), pp. 903–927.

[3] S. Jackson, A. S. Kechris, and A. Louveau, *Countable Borel equivalence relations*, ***Journal of Mathematical Logic***, vol. 2 (2002), no. 1, pp. 1–80.

[4] A. S. Kechris, ***Classical descriptive set theory***, Graduate Texts in Mathematics, vol. 156, Springer-Verlag, 1995.

DEPARTMENT OF MATHEMATICS
PENNSYLVANIA STATE UNIVERSITY
DUBOIS, PA 15801, USA
*E-mail*: boykinc@psu.edu

DEPARTMENT OF MATHEMATICS
UNIVERSITY OF NORTH TEXAS
DENTON, TX 76203, USA
*E-mail*: jackson@unt.edu

# HAS THE CONTINUUM HYPOTHESIS BEEN SETTLED?

MATTHEW FOREMAN

**Abstract.** This paper states the Continuum Hypothesis, reviews its special status, and critically examines attempts to settle it.

**§1. Introduction.** Cantor's *Continuum Hypothesis* is the most celebrated open problem in Set Theory. Recent progress by Woodin has led to considerable optimism in some quarters that there either has been a solution or that there is likely to be a solution very soon. The main thesis of the paper is that the Continuum Hypothesis is an open problem.

This paper reviews elementary facts about the Continuum Hypothesis and discusses what it might mean to have a solution to it. It briefly describes the important and elegant work of Woodin and considers it in light of this discussion. An alternate solution is presented, and considered in the same light.

**§2. Background.** In 1873, Cantor proved that if $X$ is an infinite set, then the cardinality of the powerset $P(X)$ of $X$ (the collection of all subsets of $X$) is strictly larger than the cardinality of $X$. Immediately the natural question of the existence of intermediate cardinalities arose.

The most immediate and concrete version of this question is whether there is an intermediate cardinality between the cardinality of the natural numbers $\mathbb{N}$ and $P(\mathbb{N})$. Since $P(\mathbb{N})$ and $\mathbb{R}$ have the same cardinality, this is equivalent to asking whether there is an infinite subset of the real numbers that is not in one to one correspondence with either the natural numbers or the real numbers themselves.

The assertion that there is no such set is called the *Continuum Hypothesis* or *CH*. The assertion that there is no infinite set $X$ such that there is a set $Y \subset P(X)$ of cardinality intermediate between that of $X$ and of $P(X)$ is called the *Generalized Continuum Hypothesis* or *GCH*.

If $X$ is a set we will denote its cardinality by $|X|$. In analogy with finite sets, the cardinality of $P(X)$ is denoted $2^{|X|}$.

**Logic Colloquium '03**
Edited by V. Stoltenberg-Hansen and J. Väänänen
Lecture Notes in Logic, 24

With the Axiom of Choice, every cardinality has a canonical ordinal representative; namely the least ordinal of that cardinality. Moreover, since the ordinals are well-ordered, this implies that the infinite cardinalities are well-ordered. The cardinalities are usually enumerated as $\aleph_\alpha$ as $\alpha$ ranges over the ordinals.

With this observation the Continuum Hypothesis and the Generalized Continuum Hypothesis take on particularly elegant forms:

**Continuum Hypothesis:**

$$2^{\aleph_0} = \aleph_1$$

**Generalized Continuum Hypothesis:**

For all ordinals $\alpha$,

$$2^{\aleph_\alpha} = \aleph_{\alpha+1}.$$

These problems were recognized immediately as being of a fundamental nature in mathematics. Indeed in Hilbert's celebrated address to the 1900 International Congress of Mathematics [2], he gave a list of 10 open problems (later expanded to 23). The Continuum Hypothesis was the first problem on this list.

**§3. Why can't these problems be resolved in the "normal" way?** The Continuum Hypothesis and the Axiom of Choice were the first two statements of mathematics that were shown to be independent of ZFC that were not syntactically motivated.

Specifically, Gödel showed in 1938 that in the constructive universe $L$, the Axiom of Choice and the Generalized Continuum Hypothesis hold. Hence if ZFC is consistent then so are the Axiom of Choice and the GCH ([6]).

In 1963, Cohen [3] invented the method of Forcing. Using this method, Cohen gave examples of models of set theory where "ZF $+\neg$ AC" holds and models where "ZFC $+\neg$ CH" holds. Hence both the Axiom of Choice and the Continuum Hypothesis are independent of ZFC. The method turned out to be quite general and since this time many questions that arose in the course of ordinary mathematical investigation have been shown to be independent of ZFC.

Further work of Easton, Solovay and others showed that ZFC has very little influence on the behavior of the GCH at regular cardinals. In relatively recent work ([10]), Shelah showed that at many singular cardinals $\lambda$, the behavior of the GCH can be determined (or nearly so) by the behavior of the GCH at regular cardinals below $\lambda$ together with covering behavior.

In summary: the consensus view is that the axioms ZFC "capture" our clear knowledge of sets. In particular, they are adequate for all routine mathematical constructions and arguments. Assuming that ZFC is consistent (and therefore there are some examples of ZFC), there are examples of ZFC in which the Continuum Hypothesis holds and examples of ZFC where the Continuum Hypothesis fails. Hence, resolving the Continuum Hypothesis (or the GCH) requires some kind of argument or evidence that goes beyond the usual techniques of mathematics.

**§4. What to do?** The problem can be divided up in many different ways. The usual techniques for investigating propositions that are not resolvable by conventional mathematical methods can be summarized by the slogan:

(†) **Mathematics = 1st Order Logic + Axioms.**

This slogan allows one to "separate variables" in the following sense: First Order Logic is very well understood. Proofs in First Order Logic are finitary objects, the validities are enumerable, the proofs are decidable, and the Lowenheim-Skolem theorem provides a very useful tool for manipulating its models. As such, it presents itself as an ideal epistemic vehicle for investigating mathematics. By adopting First Order Logic, the evidential issues involved in settling independent mathematical statements are thrown entirely into the investigation of Axioms.

This choice appears to the author to be entirely a matter of convenience and efficacy: were investigations using some other logic (or another model of what mathematical reasoning is) to turn out to be fruitful and convincing it seems that there could be no objection to the use of those alternate methods. However, current alternatives to (†) that are known to the author "squeeze the balloon" to put the intrinsic foundational issues in forms where they are less clear and the methods of study are murky at best.

To illustrate an extreme case, Second Order Logic settles the Continuum Hypothesis. However, Second Order Logic yields no method of determining the truth value given to the Continuum Hypothesis.

At the other extreme is perhaps the mildest possible extension of First Order Logic, $\omega$-logic. Perhaps a new and unknown mathematical fact could be demonstrated from the axioms ZFC using $\omega$-logic. However, it is difficult to imagine such a *demonstration* that does not use some principle that extends the axioms of ZFC (such as an assumption that all integers are standard) and is formalizable that in First Order Logic over the universe of sets. Making this assumption $\phi$ explicit would yield a proof in First Order Logic in the axiom system ZFC $+ \{\phi\}$.

**§5. What Axioms to take?** Let us now take the project to be identifying collections of Axioms that are appropriate (in some vague sense for the moment) for the right hand side of the equation (†). It appears to the author that there are two types of reasons for adopting axioms:

- Intuitive Appeal
- Pragmatic or Utilitarian Reasons

In the first category, the author would put some of the axioms of ZFC such as the Union Axiom, which seem (even after mature reflection) to be inherent in the notion of set. Historically, the Axiom of Choice seems to be of this nature as well. (See [9] for anecdotes about opponents of the Axiom of Choice using it unconsciously in their work.)

In the second category one has axioms such as the Axiom of Foundation, which appears to the author to be restrictive and without clear motivation. Nonetheless, the Axiom of Foundation provides a very helpful organizational principle for the study of a universe $V$ and does not restrict in any way the phenomenon one intends to examine. (An ill-founded model can be studied by considering its well-founded parts.)

**5.1. The best of all possible worlds.** One hope would be the realization of simple and clearly stated axioms that settle all "interesting" problems and appeal directly to the intuition. The desire seems motivated by some ideal represented by Euclid's Axioms. However, Euclid's Geometry, as a transparent, poignant axiomatization, may be misleading. The phenomenon to be accounted for in mathematics (like predicting the weather) may simply be too complicated to admit such a description. Indeed, if there has been any progress at all in the last 40 years of Set Theory, it appears to point in this direction.

**5.2. The minimal adequate solution.** From a realist point of view, a minimally acceptable solution would be a recursive collection of Axioms that cohere with known facts and give an accurate *description* of the set theoretic universe.

In actual fact, our human limitations seem to dictate some kind of inter mediate answer (if an answer will be found at all). If we are able to find and identify a minimally acceptable solution, then it seems likely that it will be sufficiently conceptual and decipherable that it provides some kind of appealing insight.

The main point here is that the apparent complication of axioms or axiom systems is not prima facie evidence of their inadequacy.

**§6. How do we tell if we have good axioms?** The considerations that lead to the acceptance or rejection of Axioms seem to be of different levels of importance. There appears to be certain kinds of primary considerations that trump other kinds of evidence that is of a more circumstantial type.

In examples of situations where axioms are adopted, there is usually a combination of both primary evidence and secondary evidence, with the secondary evidence acting as confirmation and reinforcement of the primary evidence. When axioms are adopted for completely pragmatic reasons it appears that there are no alternative, contradictory competitors with the same level of plausibility.

**6.1. Primary considerations.** The most important consideration in evaluating an axiom or axiom system is what it *says*. Here the notion of *says* should be interpreted in the broadest possible sense. It should include not only the statement of the axiom but its mathematical consequences. Judgement of the consequences can be made in the ordinary way mathematicians make these judgements, which for lack of a better term I will call "intuition."

**6.1.1.** *Behavioral evidence.* In examining an axiom system for its intuitive appeal, one faces many of the obstacles that one might have for other personal intellectual or sensual judgements. For discussing the issue of the existence of clear intuitive content of an axiom system it appears more fruitful to examine the behavior of those who are familiar with the system than to attempt to directly discuss the experience of "intuition" itself.

Explicitly, if researchers in an area tend to agree on programs of investigation, make similar predictions and conjectures about open problems and are able to reach similar conclusions (perhaps without being in communication with each other) this provides behavioral evidence that the researchers have good intuitions about the content of the axiom systems.

If, in addition, these researchers tend to think about the axioms *as if* they were true, it provides evidence that they also have intuitions about the veridicality of the axioms. The classical example of such behavior comes with the controversy about the Axiom of Choice. Opponents of the adoption of the Axiom of Choice nonetheless (and apparently unconsciously) used the Axiom of Choice in their mathematics. (See [9] for a discussion of this.)

**6.1.2.** *Limitations on behavioral evidence.* All of these assertions need to be taken with appropriate qualifications and attributions of vagueness. For example, researchers may act as though various axioms are true but restrict their investigations to a *particular context.* This context acts like an unstated hypothesis in their intuitions. Their behavior then can be viewed as evidence about their mathematical perceptions of *that context.* Here the author has in mind (among others) mathematicians who work in constructive mathematics, or mathematicians who are interested primarily in $L(\mathbb{R})$ or larger determinacy models, and various other tribes.

This contextuality leads many to hold views accepting a plurality of contexts simultaneously. In the author's opinion, such views demand an explanation of this simultaneity, and an explanation of contextuality itself that does not presuppose a universal context of discourse. A discussion of this is well beyond the realm of this paper.

Another problem with the behavioral criteria is that mathematicians working with a particular axiom system tend to be self selecting. Those that do not find an axiom system intuitive or who reject the picture provided by those intuitions tend not to continue working with those axioms.

Thus behavioral evidence has all the problems of other psychological or sociological evidence: it is subject to personality, fashion, and even economic influences. Nonetheless, it appears that some valuable information can be gained by its consideration.

**6.1.3.** *Educated intuitions.* An important point about "mathematical intuition" is that it be *educated.* Many standard mathematical objects appear paradoxical on first inspection. Here one can enumerate items such as closed nowhere dense subsets of the unit interval that have positive measure, continuous nowhere differentiable functions and many other objects central to our understanding of mathematics.

One can imagine a scenario (perhaps in the late 19th century) where an *a fortiori* true axiom that precisely stipulated the existence of such an object might be naively and immediately rejected on the grounds that such objects are unintuitive. However, after having understood such objects we no longer find them unintuitive or paradoxical, indeed find them essential for modern analysis.

**6.1.4.** *Aesthetic judgements.* An interesting and amusing parallel might be drawn between educating intuitions and certain aesthetic judgements. Professional wine tasters surely have different preferences in wine, but seem to be able to agree on certain vocabulary that describes aspects of savor that are undetectable by novices. Moreover, to a large extent, professionals are able to agree on quality judgements of varieties of wines, even though they may not choose to drink a particular variety. These judgements may be obscure or impossible for novices.

**6.2. Secondary considerations.** Secondary (or circumstantial) considerations are easier to discuss, since many of them are less ephemeral and personal than "intuitions." We discuss here several secondary considerations that have been proposed:

**6.2.1.** *Predictions.* Evidence for an axiom system is given by statements $\phi$ that are proved in that axiom system and later *verified* by being proved in ZFC, or an even more concrete theory.

Gödel is said to have had the view that the use of infinite sets were verified by the existence of proofs (say in complex analysis) of number theoretic facts that used infinitary techniques. These number theoretic facts are then verified by (perhaps longer and less conceptual) proofs using only finitary assumptions.

**6.2.2.** *Coherence.* To be accepted a family of axioms should cohere both in consequence and basic conception and be compatible or cohere with "known" facts. Moreover a family of axioms may form a hierarchy with significant consequences following from stronger axioms in the hierarchy that don't follow from weaker axioms. We may call this latter property *gradations of consequence.*

**6.2.3.** *Hierarchies of consistency strength.* The existence of a linear hierarchy of consistency strength in a family of axioms is suggested as a reification of the content of the axioms. Moreover if a large number of disparate propositions (not included in the axiom family) can be similarly reified by this hierarchy, it is considered evidence for the truth of the axioms.

**6.2.4.** *Completeness.* If an axiom system is effective at settling interesting questions it is considered evidence for the truth of that axiom system. This effectiveness can be either of the form of directly giving answers to interesting questions or in showing that the questions "have answers" in that forcing (or other techniques) cannot be used to show them independent of the theory. A precise statement of a definition for such a completeness property can be given as follows:

> Suppose that $\Sigma$ is an axiom system extending ZFC. Let $\mathfrak{A}$ be a definable structure. Then the theory of $\mathfrak{A}$ is *forcing absolute* (or *generically absolute*) relative to $\Sigma$ iff whenever $V \models \Sigma$ and $V[G]$ is a forcing extension of $V$ and $V[G] \models \Sigma$,
>
> $$\mathfrak{A}^V \equiv \mathfrak{A}^{V[G]}.$$

Hence, while not giving an *actual* complete theory of $\mathfrak{A}$, the axioms *fix* the theory, in the sense that it cannot be changed by the known methods for producing independence results.

**6.2.5.** *Why are these* secondary *principles?* One can perform the following thought experiment to weigh the strength of each kind of evidence advanced here: Imagine an extremely distinguished and learned set theorist named "Kurt" who has in mind a theory $T$ proposed for the axioms of mathematics. Weigh the following examples as types of evidence:

E2.1 $T$ predicts number theoretic facts that, as far as we can verify, are true.
Is $T$ true?

E2.2 $T$ gives a coherent family of axioms.
Is $T$ true?

E2.3 $T$ belongs to a family of axioms that forms a hierarchy of consistency strengths.
Is $T$ true?

E2.4 $T$ gives answers to all interesting set theoretic questions.
Is $T$ true?

E1 Kurt believes $T$, describes direct intuitions that it is true, and has a clear ability to use it with understanding.
Is $T$ true?

It is clear that it is possible to give examples of axiom systems satisfying each of E2.1–2.4 that would not be seriously entertained as "true." We give examples below in our "case studies"; let it suffice here that the reader experiment with axiom systems such as:

$$\{\text{Con}\neg\text{Con(ZFC)}, \neg\text{ConCon}\neg\text{Con(ZFC)}, \text{Con}\neg\text{ConCon}\neg\text{Con(ZFC)}\ldots\}$$

to get examples satisfying E2.2–2.3 (and as far as we know E2.1) except for the fact that they do not seem to yield "gradations of consequence." As remarked below the axiom $V = L$ works very well in satisfying E2.4, in any form, including *generic absoluteness*.

**6.3. A remark on pluralism.** The lack of conclusion about choosing an axiom system to resolve the CH does not compel us to assert that all axiom systems are equally likely to be true or that the Continuum Problem is somehow indeterminate. Indeed, it appears to this author that this merely indicates that the problem is open. The fact that the questions we are asking are independent of a particular axiom system such as ZFC does not *a priori* confer on them any particular peculiarity in truth value (that their answers are somehow of a different *kind* than ordinary questions). To argue such a position one must also give persuasive evidence that ZFC has a special status, for example of capturing exactly (or at least including) those statements that are mathematically meaningful. That the statements we are interested in are independent of our working theory indicates merely that the techniques available for us to study them must be of a different kind than for statements that we are sure are settled by the theory.

**§7. Case studies.** We now consider several examples of well-known theories proposed as extensions or alternatives to ZFC as a foundation for mathematics. These examples are partly given to justify the categorization of the principles above as "primary" or "secondary" principles. In particular we give examples of theories that satisfy certain of the principles (such as generic absoluteness), while still being rejected for failing others (such as lack of intuitive content). This appears to the author as demonstrating the priority of one kind of evidence over another.

**7.1. New Foundations — a failed proposal.** Quine suggested a variant of type theory named *New Foundations* as a foundation for mathematics. While scattered individuals continue to do research on New Foundations, they are mostly concerned with establishing its consistency relative to a standard axiom system. The system has failed to attract mainstream adherents who develop standard mathematics inside the system.

The author proposes that the reason is primarily that the axiom system is extremely unintuitive.

**7.2. Large Cardinals — a successful axiom system.** The example of Large Cardinals is important because it does seem that there is a large group of set theorists who adopt large cardinals as axioms.

**7.2.1.** *Primary considerations.* Large cardinals have a large appealing and coherent body of consequences, especially in descriptive set theory. Essentially every classical question about effective sets that involves Lebesgue Measurability, the Property of Baire, and an effective *analogue* of the Axiom of Choice (uniformization) can be resolved on the basis of Large Cardinals.

Researchers who work with Large Cardinals around the world tend (on the whole) to make the same conjectures and have developed similar techniques for working with them. Moreover to those researchers, the consequences of large cardinals feel natural and fit well with other intuitions.

It is very rare to find a set theorist working with Large Cardinals because they *don't* believe them and are trying to refute them by considering their consequences.

It is a sociological fact that the dominant view among those actively searching for true axioms that extend ZFC is that Large Cardinal axioms are true.

**7.2.2.** *Secondary considerations.* Among the secondary evidence put forward by various authors we find:

**Predictions:** There is descriptive set theoretic structure that was discovered with the use of large cardinals that was later "verified" in second order arithmetic. (The most striking example of this kind is the Wadge hierarchy for Borel sets.)

**Coherence:** The statements of the axioms are conceptually uniform. (They are described below.) This uniformity allows the hierarchy to be understood with a small number of parameters and variation. Moreover the consequences vary uniformly with the axioms and fit well with similar variations that are provable in ZFC. Finally there are meaningful families of facts that follow from stronger axioms that do not follow from weaker axioms. (Measurability properties of definable sets for example.)

**Hierarchy of Consistency Strength:** Large cardinals form an (almost) linear hierarchy of consistency strengths. This hierarchy is very useful for calibrating the consistency strengths of virtually all interesting propositions that are independent of ZFC.

The study of this hierarchy has led to the discovery of canonical models for large cardinal axioms (*inner models* or *core models*). These models have a structure that is apparently transparent to those that understand them. These models are of great importance to the correspondence between descriptive set theory and large cardinals alluded to in the "gradations of consequences."

[While generally accepting the evidence cited above, the author is skeptical of some of the claims of the "inner model program." It is reported by those that understand inner models that the structure of the models is so clear that any (or many) questions about inner models can be answered by simply inspecting that understanding. This is apparently true for certain types of questions that are generated by the program itself, but not for other problems that remain open in even relatively simple inner models.

Moreover, this transparent structure supposedly verifies the consistency of axioms that have such a structure. Were this claim to be true, one would expect the inner model program to have uncovered inconsistencies in various axioms. However, in each instance of an inconsistent axiom, the inconsistency has been discovered by combinatorial means. The simple fact that very intelligent people have spent thousands of hours examining and proving consequences of the cardinals while studying inner models without uncovering an inconsistency *does* seem to be strong evidence for their consistency. However that this effort lead to the discovery of inner models for these cardinals does not alone seem to add weight to this evidence.]

**Completeness:** It is a consequence of Large Cardinal Axioms that the theory of $L(\mathbb{R})$ is invariant under forcings that preserve the large cardinal axioms. Hence Large Cardinal Axioms fix the theory of the hereditarily countable sets.

**7.2.3.** *Weaknesses.* The main weakness of the Large Cardinal axiom system is that it says (almost) *nothing* about questions in the vicinity of the $\aleph_n$'s or $2^{\aleph_0}$ that are not effective or that involve the Axiom of Choice in an essential way. In particular, Large Cardinals say nothing at all about the Continuum Hypothesis. (Theorems of Levy and Solovay make this rigorous by establishing that Large Cardinals are invariant under "small forcing.")

**7.2.4.** *The form of Large Cardinals.* For later comparison and to illustrate the conceptual coherence of Large Cardinal Axioms we give the basic form of a typical large cardinal:

There is an elementary embedding

$$j : V \to M$$

such that

1. $j$ moves certain ordinals certain places
2. $M$ is a sufficiently robust transitive subclass of the universe $V$.

Note that there are two parameters that determine the strength of the embedding: where ordinals go and the closure properties of $M$. Of particular prominence in discussing large cardinal embeddings is the *critical point of $j$* i.e. first ordinal moved by $j$. A typical example of a large cardinal axiom:

$\kappa$ is *$\lambda$-supercompact* iff there is an elementary embedding $j : V \to M$ with critical point $\kappa$ such that $M^\lambda \subset M$.

Variations on the basic form of large cardinal axioms are given by postulating classes of such elementary embeddings such as that of a *Woodin Cardinal*:

There is a cardinal $\lambda$ such that for all functions $f : \lambda \to \lambda$ there is an elementary embedding $j : V \to M$ such that the critical point of $j$ is closed under $f$ and $V_{j(f)(\kappa)} \subset M$.

Other variations include restricting the domain and/or range of $j$ to be of the form a $V_\lambda$ or some similar structure.

An obvious problem with this form of the statement of Large Cardinal axioms is that it is conspicuously *not* in first order logic. (For example, the assertion that an elementary embedding from $V$ to $M$ exists is an assertion that there is a proper class definable over $V$ that obeys some truth predicate ... ). However, the conceptual uniformity of the axiom system is clear enough that it gives a *sufficiently precise heuristic* for all of these statements to be shown by metamathematical means to have equivalent first order formulations. These formulations are typically in terms of ultrafilters on certain index sets that have various structure. So, for example, the axiom that $\kappa$ is $\lambda$-supercompact is equivalent to the statement:

There is a $\kappa$-complete, normal, fine ultrafilter on $[\lambda]^{<\kappa}$.

**7.3. The Axiom $V = L$.** This axiom is interesting because the working set theorists who accept this axiom are a distinct minority.

**7.3.1.** *Secondary considerations.* We begin by discussing the secondary considerations for acceptance of an axiom, where the axiom $V = L$ scores rather well:

**Predictions:** The study of $V = L$ has uncovered a large class of combinatorial properties such as $\diamondsuit$ and $\square$. These properties have become essential tools in the study of Set Theory. In this sense, the axiom $V = L$ has yielded *methodological predictions.*
Moreover, it is currently begin discovered that weakenings of the classical $\diamondsuit$ and $\square$ hold in ZFC. Examples of this include the existence of *club guessing sequences* and partitions of cardinals into stationary sets on which there are partial square sequences ([10]). This is an ongoing research program and there has been significant recent progress (e.g. Shelah's proofs of *middle diamond* in ZFC.)

**Coherence:** The axiom $V = L$ provides an adequate framework for all of mathematics and coheres with the verifiable data about mathematics. The constructible universe is very well understood and regular in its behavior. Indeed the "condensation" properties of $L$ are much desired in other models of set theory, as they imply useful coherence throughout the universe.
The axiom $V = L$ does not satisfy the demand of "gradations of consequence" but that is hardly relevant since it is only one axiom!

**Hierarchy of Consistency Strength:** Again, the axiom is but one axiom and therefore it is not meaningful to ask that it form a hierarchy of consistency strength.

**Completeness:** The axiom $V = L$ satisfies the very strongest kind of generic absoluteness: there are no non-trivial forcing notions that preserve the axiom $V = L$. In fact if $V$ and $W$ are two well-founded models of ZFC $+ V = L$ that have the same ordinals then $V = W$!

**7.3.2.** *Primary considerations.* It is the author's opinion that the axiom $V = L$ is largely rejected because it does not match the intuitions of most set theorists who are familiar with $L$ and forcing extensions of $L$.

The general intuitions reported are that $L$ is "too limiting" a notion of set. The expressions made by knowledgeable practitioners are to the effect that the requirement that every set be definable in a concrete way from earlier sets goes against the intended generality of the notion of set. The notion of set is supposed to be that of an *arbitrary* collection. The axiom $V = L$ says that even arbitrary collections of sets can be built very concretely (by Turing Machines (see [7])) from the thin spine of the ordinals. Moreover, even adherents to the axiom $V = L$ seem to have no problem finding a clear conception of collections that do not belong to $L$.

Other people have objections to specific consequences of the Axiom $V = L$. This is particularly so about facts in Descriptive Set Theory; there are easily definable "pathologies" in the Constructible Universe.

**7.3.3.** *Last word on L.* Despite receiving very high marks on most of the secondary criteria, the axiom $V = L$ was rejected because of what it says about the mathematical universe. Despite the fact that the axiom has as consequences total "generic absoluteness", and gives theory as complete and categorical as can possibly be hoped for, its consequences for the universe of set theory are considered highly unintuitive. In the face of a competing, more intuitive theory (Large Cardinals), it loses.

**§8. Woodin's work.** W.H. Woodin has recently proposed a negative solution to the problem of the Continuum Hypothesis. It is exposited in [11, 12] as well as [4] among other places. We will content ourselves with a very crude outline of the argument here.

Very roughly, Woodin has invented a new "logic" termed $\Omega$-logic. This logic is defined semantically and makes sense in the context of sufficiently many Large Cardinals.

While there is an analogue of "syntax", $\Omega$-logic is not a logic in the classical sense that the validities of the logic are recursively enumerable or that proofs are finitary objects. ("Proofs" are witnessed by weakly homogeneous trees, or structures with certain iteration properties, depending on how the logic is cast. Both of these are highly infinitary objects that depend on Large Cardinals for their existence.)

**8.1. A brief statement of the results.** The main results (as of August 2003) in Woodin's program are the following:

- There is a theory that is $\Omega$-complete for statements in $H(\aleph_2)$. Moreover this theory of $H(\aleph_2)$ cannot be changed by forcing.

  Woodin argues that this is the correct analogue of the invariance of the theory of $L(\mathbb{R})$ under forcing (in the presence of large cardinals).

Namely, it is a generalization of "generic absoluteness" from the theory of $H(\aleph_1)$ to $H(\aleph_2)$.

- The existence of such a theory implies that the Continuum Hypothesis fails.
- The theory (*) that is shown to exist implies that the continuum is $\aleph_2$.
- Woodin argues that the conditions required for the logic to make sense follow from the already accepted theory of Large Cardinals. Moreover, the generic absoluteness results are the correct generalization of the generic absoluteness provable from Large Cardinals alone. Hence, the theory (*) (or at least another theory giving the generic absoluteness) should be true.

In summary: (*) has a kind of generic absoluteness (or completeness) that generalizes the kind that can be shown for $L(\mathbb{R})$, and this is evidence for its truth. Moreover, any theory that implies this kind of absoluteness must imply that the CH fails. Hence the CH fails.

**8.2. How does one evaluate these results?** First of all, Woodin's mathematical results are theorems, and hence incontrovertible. The real question is the extra-mathematical question of the interpretation of the results. For this we need a digression on Forcing Axioms.

**8.2.1.** *Forcing Axioms.* When early independence results were being obtained by iterations of c.c.c. forcing, it was noticed that many of the proofs were remarkably similar. This led Martin to propose:

> **Martin's Axiom:** If $\mathbb{P}$ is a c.c.c. forcing $\gamma < 2^{\aleph_0}$ and $\langle D_\alpha : \alpha : \alpha < \gamma\rangle$ is a collection of dense subsets of $\mathbb{P}$, then there is a filter $\mathcal{F} \subset \mathbb{P}$ such that for all $\alpha < \gamma$, $\mathcal{F} \cap D_\alpha \neq \emptyset$.

In this form Martin's Axiom is a consequence of the CH and was viewed as a generalization. It was not originally viewed as playing the role of an "Axiom" in the sense of being an attempt at being a true axiom that extends the ZFC axioms.

The simplest form of Martin's Axiom that contradicts the CH is "$MA_{\aleph_1}$": This axiom has the same form, but demands that $\gamma = \aleph_1$. The only consequence for the value of the continuum that follows from $MA_{\aleph_1}$ is that the continuum is at least $\aleph_2$.

More recent research has found collections of partial orderings larger than the collection of c.c.c. partial orderings for which one can hope for filters that meet $\aleph_1$ dense sets. Among them is a maximal collection (the stationary set preserving partial orderings) which gives the axiom *Martin's Maximum* or MM. Another is the collection of ***Proper*** partial orderings which leads to the ***Proper Forcing Axiom*** or PFA. Another is a weakening of Martin's Maximum called the ***Bounded Martin's Maximum***, or BMM. In historical order the following results have been shown:

1. (Foreman, Magidor and Shelah) Assuming MM, for all regular $\kappa > \aleph_1$, $\kappa^{\aleph_1} = \kappa$. In particular, $2^{\aleph_0} = \aleph_2$, and the Singular Cardinals Hypothesis holds.
2. (Todorcevic/Velickovic) Assuming PFA, $2^{\aleph_0} = \aleph_2$.
3. (Todorcevic) Assuming BMM, $2^{\aleph_0} = \aleph_2$.

**8.2.2.** *What does Woodin's (allegedly true) theory say?* The theory (*) that Woodin proposes on the basis of generic absoluteness is produced abstractly and its contents are generally a mystery, even in the presence of a proper class of Woodin Cardinals. At the time of this writing essentially the only general consequence of the theory is BMM. Hence to evaluate Woodin's theory on the basis of primary considerations the only basis we have is BMM.

The author suggests then, that the main discussion of Woodin's Conclusion should be about the plausibility of BMM. Indeed it appears to the author that Woodin's results would provide circumstantial evidence that BMM is true, were BMM itself to be a convincing axiom. Since forcing axioms were originally proposed as convenient ways of achieving consistency results, in general a discussion of their "truth" has not been nearly as prominent as, say, Large Cardinals. (It should be noted that this discussion is beginning, see e.g. [1].)

An apparently *counterintuitive* fact about (*) is that its models have the property that $H(\aleph_2)$ is the $H(\aleph_2)$ of generic extensions of $L(\mathbb{R})$. This appears to this author as being an analogy to an assertion that $H(\aleph_1)$ should be the $H(\aleph_1)$ of a generic extension of $L$ (which is false assuming large cardinals). Without compelling further understanding this appears to require considerable explanation to be accepted. This may be a lack of educated intuitions on the author's part.

Another oddity of Woodin's solution is that it is immediately restricted to $P(\omega)$ and $H(\aleph_2)$. Without substantial modification and amendment it cannot be generalized to give a complete picture of cardinal arithmetic (or even the value of $2^{\aleph_2}$).

**8.2.3.** *Summary.* The kind of evidence that is cited by Woodin for the failure of CH is of the kind classified above as "secondary." The primary evidence would lie in the consequences of (*).

Lacking a clear understanding of the theory (*), it seems very hard to assent to Woodin's Solution. After all, the axiom $V = L$ has as many (and more) desirable secondary considerations as the theory (*). Could (*) be just as unintuitive and unsatisfying as $L$? Until we understand the theory (*) we cannot say.

**§9. A competing theory.** Were Woodin's solution the only viable candidate for a method of settling the Continuum Hypothesis, the case against its conclusiveness would be rather weak. However there are several other possibilities for a solution. We very briefly outline one here.

**9.1. Generic elementary embeddings.** Recall in Section 7.2.4, we gave the general form of a large cardinal axiom. We now give a general form that allows that the elementary embedding be defined, not necessarily as a definable class in $V$, but as a definable class in a generic extension of $V$:

There is an elementary embedding

$$j : V \to M$$

where:

1. $j$ moves certain ordinals certain places
2. $M$ is a sufficiently robust transitive subclass of a generic extension $V[G]$.

Note that because we have allowed the elementary embedding to be in a generic extension of $V$ we have introduced another parameter determining the strength of the axiom system. The three parameters of strength are:

1. Where ordinals go.
2. The closure properties of $M$.
3. The nature of the forcing $\mathbb{P}$ that produces $G$.

Because the closure properties of $M$ are often highly analogous to conventional large cardinals one often speaks of a cardinal $\kappa$ that is "generically supercompact" or "generically huge" by a forcing having certain properties.

As with conventional large cardinals, it is necessary to recast these axioms in terms of their first order equivalents. This is typically done in terms of ideals. Some common language for the equivalent ideal assumptions is:

There is a normal fine $\kappa$-complete ideal on $Z \subset P(X)$ which is:

- Precipitous
  or
- Saturated
  or
- has a small dense subset
- etc.

(Important names involved in discovering these ideal properties are Ulam, Jech, Solovay, Kunen and many others)

These axioms seem to fall into two incompatible classes:

**Generalized Large Cardinals:** This is a large and fairly coherent family of axioms including such axioms as generic huge embedding with critical point $\aleph_1$.

Generalized large cardinals vary from positing stationary set reflection (at the weaker end of the assumptions) to very strong Chang's Conjectures (such as $\aleph_\omega$ being Jonsson) at the stronger end.

**An apparently isolated example:** The statement that the non-stationary ideal on $\aleph_1$ is $\aleph_2$-saturated (and close variations).

As far as we know (in the summer of 2003) the latter class is related to results about sets of hereditary cardinality $\aleph_1$. It does not have analogues at accessible cardinals such as $\aleph_2$ or $\aleph_3$ or in conventional large cardinals.

**9.2. Generalized Large Cardinals and the GCH.** We begin by showing their relation to the Continuum Hypothesis.

THEOREM (Foreman, 1983). *If there is a* $j : V \to M \subset V[G]$ *where*:
a) $\mathrm{crit}(j) = \omega_1$
b) $M^{\omega_1} \cap V[G] \subset M$
c) $G \subset \mathrm{Col}(\omega, \omega_1)$ *is generic*
*Then the CH holds and* $2^{\aleph_1} = \aleph_2$.

NOTE. Jech showed that if there is a saturated ideal on $\aleph_1$ then the CH implies $2^{\aleph_1} = \aleph_2$.

Woodin improved this to show:

THEOREM. *If there is a countably complete, uniform* $\aleph_1$*-dense ideal on* $\aleph_2$ *then the CH holds and* $2^{\aleph_1} = \aleph_2$.

One can appeal to the following result to get the GCH:

THEOREM (Foreman). *Suppose that* $2^{\kappa} = \kappa^+$ *and there is a generic elementary embedding* $j : V \to M$ *such that*:
a) $\mathrm{crit}(j) = \kappa$, $j(\kappa) = \lambda$, $j(\kappa^+) = (\lambda^+)^V$
b) $j``\lambda^+ \in M$.
*Then* $2^{\lambda} = \lambda^+$.

**9.3. Generalized Large Cardinals as axioms.** We now discuss Generalized Large Cardinals in the terms we have considered other axiom systems.

**9.3.1.** *Primary criterion: content.* The author contends that "Generalized Large Cardinals" are straightforward generalizations of conventional large cardinals. Moreover, that the direct or indirect evidence for large cardinals, when suitably viewed, does not distinguish between conventional large cardinals and generic large cardinals and provides equally strong evidence for Generalized Large Cardinals. This assertion includes "educated intuitions."

On the other hand, the generic large cardinals are much less studied than conventional large cardinals, and so it is hard to supply the same kind of historical or sociological evidence for intuitive content of the large cardinals. (Note that arguments for *strong axioms of infinity* (cardinals such as inaccessible or Mahlo cardinals) that are based on the height or magnitude of the ordinals don't seem to apply to generic large cardinals. However they do not seem to suffice for even moderately strong large cardinals either.)

**9.3.2.** *Secondary criteria.*

**Predictions:** There are several *bona fide* examples of "predictions" made by Generalized Large Cardinals. These include normality properties from point-set topology that were first proved from ideal properties and later verified in ZFC. Probably the most convincing example of this kind is Silver's Theorem that if the GCH holds below a singular cardinal $\lambda$ of uncountable cofinality then it holds at $\lambda$.

As an historical fact, this theorem was proved by Magidor (e.g. at $\aleph_{\omega_1}$) under the assumption that there is a normal precipitous ideal on $\omega_1$. A matter of weeks later Silver improved this by eliminating the precipitousness assumption. (Moreover Silver's original proof used ill-founded generic ultraproducts.)

It remains true that Magidor's proof is much simpler (once the definition of "precipitous" is given) than the ZFC proof. Thus it appears to be a prediction in all ways: it was temporally prior and conceptually simpler, and "verified" in ZFC.

There are a fair number of examples of temporal predictions internal to the theory of generalized large cardinals. For example, in both the cases of regularity properties for $L(\mathbb{R})$ (measurability and property of Baire etc.) and with the CH, the author gave theorems that had generic huge cardinals as hypothesis, but Woodin improved these results to just ask for uniform ideals on regular cardinals. (See [5] for more discussion of these examples.)

**Coherence:** Large Cardinals form a 3-parameter family of axioms. Moreover on the axis of closure the family is directed under implications. There are a fair number of examples of the phenomenon "gradations of consequence." Chang's Conjectures are one example of this form. Another less well known example is given by the following:

- (Erdos-Rado) CH implies that
$$\omega_2 \to (\omega_1 + 1, \omega_2)$$
- (Laver, later Kanamori) CH + there is an $(\omega_2, \omega_2, \omega)$-saturated ideal on $\omega_1$ implies that
$$\omega_2 \to (\omega_1 \times 2 + 1, \omega_2)$$
- (Foreman-Hajnal) CH + there is an $\aleph_1$-dense ideal on $\omega_1$ implies:
$$\omega_2 \to (\omega_1^2 + 1, \omega_2)$$
- If $\kappa$ is a measurable cardinal and $\Omega$ is the first uniform indiscernible over bounded subsets of $\kappa$ that is above $\kappa$, then:
$$\kappa^+ \to (\Omega, \alpha) \quad \text{for all } \alpha < \kappa^+.$$

Yet another example is given by the following two theorems of Woodin:

- (Woodin) If there is an $\omega_2$-saturated uniform ideal on $\omega_2$ then $(\Theta)^{L(\mathbb{R})} < \omega_2$
- (Woodin) If there is an $\aleph_1$-dense, uniform ideal on $\omega_2$ then the CH holds.

Here it is easy to see that the first hypothesis is not sufficient for CH.

**Incoherence Phenomena:** Unfortunately, as was pointed out to the author by Woodin, there are examples of pairs of generalized large cardinals that can be individually shown to be consistent, but are mutually contradictory. For example, if one collapses a huge cardinal to be $\aleph_1$ with finite conditions then one gets an example of the axiom:

"There is a normal, fine, countably complete ideal on $[\lambda]^{\omega_1}$ with quotient algebra $\mathrm{Col}(\omega, <\lambda)$."

This axiom contradicts CH together with the existence of an "$(\aleph_2, \aleph_2, \aleph_1)$-saturated ideal on $\aleph_1$", an assumption shown to be consistent by Laver [8]. The reason for this is the fact that the first axiom implies the failure of the partition relation:

$$\begin{pmatrix}\omega_2\\ \omega_1\end{pmatrix} \to \begin{pmatrix}\omega_1\\ \omega_1\end{pmatrix}_{<\omega}$$

while the second implies that it holds.

**Hierarchies of Consistency Strength:** Generalized Large Cardinals contain various hierarchies of consistency strength, e.g.

THEOREM (Foreman). *The statements that "$\aleph_1$ is generically n-huge when forcing with* $\mathrm{Col}(\omega, \omega_1)$*" is a strict hierarchy of consistency strength (in the parameter n).*

Generalized large cardinals intertwine with conventional large cardinals in consistency strength; e.g.

THEOREM (Foreman). *The statement: "There is a cardinal $\kappa$ that is $\kappa^+$-supercompact" is equiconsistent with the statement that "$\omega_1$ is the critical point of a well determined generic $\omega_2$-supercompact embedding."*

Similar statements hold for your favorite large cardinal: generic huge cardinals, or ...

**Completeness:** There is an apparently coherent collection of Generalized Large Cardinals that imply:

1. (Foreman/improved by Woodin) The GCH.
2. (Foreman/improved by Woodin) There are Suslin trees on successors of regular cardinals.
3. There are no Kurepa trees on regular cardinals.
4. $\square_\kappa$ fails for all $\kappa$.
5. (Foreman) If $\alpha \leq \omega$, $\kappa' = \kappa^{+\alpha}$ and $\lambda' \leq \lambda^{+\alpha}$, then $(\kappa', \kappa) \twoheadrightarrow (\lambda', \lambda)$.
6. Stationary set reflection at regular cardinals (in the various possible forms).
7. (Foreman-Hajnal) If $\kappa^{<\kappa} = \kappa$ then
$$\kappa \to (\kappa^2 + 1, \alpha)$$
for all $\alpha < \kappa^+$.

[In the list above, if an item is not attributed, it was noticed by several people independently.]

Since the class of Generalized Large Cardinals include the class of conventional Large Cardinals, the desirable consequences of Large Cardinals in descriptive set theory can be said to follow also from Generalized Large Cardinals. However, these can often be proved directly. For example the following theorems were proved around 1980 (before Large Cardinals were known to have the same consequences):

THEOREM (Woodin). *Suppose that the CH holds and there is an $\aleph_1$-dense ideal on $\aleph_1$. Then every set of reals in $L(\mathbb{R})$ is Lebesgue Measurable, has the Property of Baire etc.*

A version of this theorem that did not use the CH was discovered independently by the author:

THEOREM (Foreman). *If there is a $j : V \to M \subset V[G]$ where*

a) $\text{crit}(j) = \aleph_1$, $j(\aleph_1) = |\mathbb{R}|$
b) $j``|\mathbb{R}| \in M$
c) $G \subset \text{Col}(\omega, \omega_1)$ *is generic.*

*Then every set of reals in $L(\mathbb{R})$ is Lebesgue Measurable, has the Property of Baire etc.*

**§10. Conclusion.** The case for Woodin's "solution" of the question of the Continuum Hypothesis is largely based on circumstantial evidence. Indeed, since so little is known about his proposed theory (*), it is hard to imagine how intuitions about its truth could have developed. The only primary evidence for the axiom (*) is that it implies BMM.

While it is possible that an accumulation of circumstantial evidence might be ultimately conclusive, it is the author's judgement that this is not yet the case for Woodin's solution. Moreover, viable alternatives to Woodin's solution exist and can be judged directly on their content. One of these alternatives is a simple generalization of Large Cardinals, and seems particularly effective at settling classical choice questions. For these reasons the Continuum Hypothesis should be viewed, at this time, as an open problem.

REFERENCES

[1] JOAN BAGARIA, *Bounded forcing axioms as principles of generic absoluteness*, ***Archive for Mathematical Logic***, vol. 39 (2000), no. 6, pp. 393–401.

[2] Felix E. Browder (editor), ***Mathematical developments arising from Hilbert problems***, American Mathematical Society, Providence, R.I., 1976.

[3] PAUL COHEN, *The independence of the continuum hypothesis*, ***Proceedings of the National Academy of Sciences of the United States of America***, vol. 50 (1963), pp. 1143–1148.

[4] PATRICK DEHORNOY, *Progres recents sur l'hypothese du continu (d'apres Woodin)*, Seminaire Bourbaki, expose 915, March 2003.

[5] MATTHEW FOREMAN, *Ideals and generic embeddings*, To appear in the ***Handbook of Set Theory***.

[6] KURT GÖDEL, ***The consistency of the continuum hypothesis***, Annals of Mathematics Studies, vol. 3, Princeton University Press, Princeton, N.J., 1940.

[7] PETER KOEPKE, *Seminar talk*, 2004.

[8] RICHARD LAVER, *An $(\aleph_2, \aleph_2, \aleph_0)$-saturated ideal on $\omega_1$*, ***Logic colloquium '80 (Prague, 1980)***, Stud. Logic Foundations Math., vol. 108, North-Holland, Amsterdam, 1982, pp. 173–180.

[9] GREGORY H. MOORE, ***Zermelo's axiom of choice***, Studies in the History of Mathematics and Physical Sciences, vol. 8, Springer-Verlag, New York, 1982.

[10] SAHARON SHELAH, ***Cardinal arithmetic***, Oxford Logic Guides, vol. 29, The Clarendon Press, Oxford University Press, New York, 1994.

[11] W. HUGH WOODIN, *The continuum hypothesis. I*, ***Notices of the American Mathematical Society***, vol. 48 (2001), no. 6, pp. 567–576.

[12] ———, *The continuum hypothesis. II*, ***Notices of the American Mathematical Society***, vol. 48 (2001), no. 7, pp. 681–690.

DEPARTMENT OF MATHEMATICS
UNIVERSITY OF CALIFORNIA, IRVINE
IRVINE, CA 92697, USA
*E-mail*: mforeman@math.uci.edu

# GEOMETRY OF INTERACTION IV: THE FEEDBACK EQUATION

JEAN-YVES GIRARD

**Abstract.** The first three papers on Geometry of Interaction [9, 10, 11] did establish the universality of the *feedback equation* as an explanation of logic; this equation corresponds to the fundamental operation of logic, namely cut-elimination, i.e., *logical consequence*; this is also the oldest approach to logic, *syllogistics*! But the equation was essentially studied for those Hilbert space operators coming from *actual* logical proofs.

In this paper, we take the opposite viewpoint, on the arguable basis that operator algebra is more primitive than logic: we study the *general* feedback equation of Geometry of Interaction, $\mathsf{h}(x \oplus y) = x' \oplus \sigma(y)$, where $\mathsf{h}, \sigma$ are hermitian, $\|\mathsf{h}\| \leq 1$, and $\sigma$ is a partial symmetry, $\sigma^3 = \sigma$. We show that the *normal form* which yields the solution $\sigma[\![\mathsf{h}]\!](x) = x'$ in the *invertible* case can be extended in a unique way to the general case, by various techniques, basically *order-continuity* and *associativity*.

From this we expect a definite break with *essentialism* à la Tarski: an interpretation of logic which does not presuppose logic!

**§1. Introduction.** We are essentially concerned with the technical contents of this precise paper in the series. For the general significance of Geometry of Interaction, see Appendix A.

**1.1. Non-commutativity in logic.** First of all, we are not using "non-commutative" in the sense of the non-commutative logic of Ruet and Abrusci [1], but rather in the sense of the non-commutative geometry of Connes [4]. The basic idea is to overcome the limitations of set theory[1] so as to give a sort of "quantum" interpretation of logic[2]. Set theory explains everything with *atoms*, a very useful *reduction*, even if criticised by various mathematicians. Category theory would rather focus on *morphisms*, but this approach does not quite expel the atoms, the *points*: it rather pretends not to see them. The non-commutative approach is more radical: the set-theoretic atoms appear as *eigenvectors*, they are related to interaction, observation. As far as the observer

[1] And category theory as well.

[2] Of course this is radically different from the project of the unfortunate quantum "logic", who tried to "tame" the quantum world.

**Logic Colloquium '03**
Edited by V. Stoltenberg-Hansen and J. Väänänen
Lecture Notes in Logic, 24

remains the same, "his" eigenvectors are used, and everything looks commutative, "set-theoretic"; technically speaking, we deal with diagonal matrices, i.e., stay in a *commutative* operator algebra. Non-commutativity is nothing but the *relativisation* of the subject, i.e., the oblivion of the distinguished "basis", or commutative algebra, associated with the observer.

We are concerned with the foundational part of logic, *proof-theory*. What follows is a short dictionary of the "non-commutative" analogues of familiar logical artifacts.

**Proofs, Functions, Programs:** Hermitians of norm at most 1. These hermitians need not be positive: the most basic *identity axiom* $A \vdash A$ is interpreted by the flip—a.k.a. extension cord—of $\mathcal{H}\oplus\mathcal{H}$ : $\mathsf{h}(x\oplus y) := y\oplus x$.

**Deduction, Composition:** Our hermitians usually come together with a *feedback* $\sigma$ which is a partial symmetry, i.e., a hermitian of spectrum within $\{-1, 0, +1\}$. The feedback corresponds to a logical deduction, in which the same formula occurs twice, both as a result (lemma) and as a hypothesis for the theorem: the feedback swaps the two copies. The basic artifact, corresponding to a proof with cuts, to a program before execution, is therefore a *cut-system* $(\mathcal{H}, \mathsf{h}, \sigma)$.

**Execution:** There is a dynamics, in the sense of a *performance*. Performing the cut, executing a program, amounts at actually "plugging" the feedback $\sigma$ with $\mathsf{h}$, i.e., at solving the *feedback equation*: $\mathsf{h}(x \oplus y) = x' \oplus \sigma(y)$ : $x$ is the input, $x'$ is the output and $y$ is the computation.

**1.2. Solving the feedback equation.** Section 4 is devoted to the most down-to-earth case, namely when the equation has always a solution, the *terminating case*. This solution is then shown to be unique as to $x'$, $y$ being unique up to a *deadlock*, i.e., something which is completely inaccessible, does not interfere with the system, and therefore can be ignored. A notion of "computational size", basically the norm of the operator yielding $x \oplus y$ as a function of $x$ is introduced, and is shown to enjoy a remarkable *associativity* inequality.

The most important (and natural) terminating case is *invertibility* (of $\mathrm{I}-\sigma\mathsf{h}$), in which case the feedback equation can be solved by inversion. The typical invertible cases arise from logic: if $\mathsf{h}$ interprets a logical proof, $\sigma\mathsf{h}$ is nilpotent, see [9], and the inverse is given by a finite power series. Logical rules[3] can be read as gimmicks enabling one to reduce complex feedback equations to simpler ones, basically by iterated substitutions, this is why the power series is finite.

Most of the paper is devoted to the study of the extension of the *normal form* $\sigma[\![\mathsf{h}]\!](x) = x'$ obtained in the invertible case to general systems.

**1.3. Continuity.** As usual, the first intuition is continuity. However, since the invertible case makes use of inversion, we cannot expect any reasonable

[3]Indeed, *cut-free rules*, i.e., the part of logic which doesn't deal with logical consequence.

topological continuity. Here there is a conflict between the mathematical tradition and some (recent) logical tradition: in the late sixties, following previous work by Kleene, Kreisel, Gandy ... Dana Scott introduced his non-Hausdorff *domains*, which are indeed ordered structures. The question whether or not this is topology is controversial, but an essential role is played anyway by a partial ordering of objects. The same phenomenon is observed here: there is a partial order (the usual pointwise ordering of hermitians) and the first important phenomenon is that:

*The normal form is monotonous (increasing).*

The suggestion coming from Scott domains is to try *order-continuity*, i.e., commutation to directed sups *and* infs. Encouraging point: the normal form is order-continuous in the invertible case. This is why we introduce *semi-invertibility*: lower-semi-invertibles (l.s.i.) are l.u.b. of invertible systems, upper-semi-invertibles (u.s.i.) are g.l.b. of invertibles, and the miracle is that invertibility is the same as the two semi-invertibilities. We can extend by means of l.u.b. (resp. g.l.b.) the normal form to l.s.i. (resp. u.s.i.) systems, and this consistently.

To come back to Scott domains and the logical "tradition": it was possible — using weird topologies — to style order-continuity "topological", only because the sole commutation was commutation to suprema. Here we have two conflicting commutations, one to suprema, one to infima, no global topology — weird or not — would account for order-continuity on both sides. Take the simplest example: on $\mathbb{R}$, upwards continuity corresponds to the open sets $]x, +\infty[$, whereas downwards continuity corresponds to the open sets $]-\infty, y[$; continuous functions from $X$ to $\mathbb{R}$ equipped with the "upwards" (resp. "downwards") topology are indeed l.s.c. (resp. u.s.c.) functions. These two $\mathcal{T}_0$-topologies have a supremum, the usual topology (since $]x, y[ = ]x, +\infty[ \cap ]-\infty, y[$), in other terms, full order-continuity does not correspond to continuity in any topological sense, natural or not.

But, even if order-continuity does not quite make sense topologically speaking, it can be used in conjunction with standard topologies (usually the weak and the strong topologies, both weaker than the norm topology). This is what makes our semi-invertible extension possible: order-continuity works like an *upgrader*, from weak to strong in the case of operators, from pointwise to uniform in Dini's theorem, see *infra*.

But order-continuity does not work beyond the semi-invertible case.

**1.4. Lebesgue integration.** The comparison with a classic, the Lebesgue integral, is illuminating. The problem at stake is the extension of the Riemann integral. We describe the main steps, pointing out the analogies/differences.

(i) The Riemann integral is a continuous linear map from the Banach space $\mathbb{R}([0,1])$ of continuous real-valued functions on $[0,1]$ into $\mathbb{R}$. Our case is similar: we start with a norm-continuous function $\sigma[\![\mathrm{h}]\!]$ (the normal form) defined on certain operators h (invertible case) of norm $\leq 1$. We can even assume the output space of dimension 1, see Proposition 9, hence belongs to $\mathbb{R}$. An essential difference: the normal form is not linear. However the output remains bounded: $\|\sigma[\![\cdot]\!]\| \leq 1$.

(ii) Coming back to integration, the next step is to remark that the Riemann integral is monotonous. The idea is to extend it to lower-semi-continuous functions, which are suprema of continuous functions. In the same way, we extend our normal form to l.s.i. systems by means of suprema.

(iii) Of course, only a monotonous function can be extended in that way. Both the Riemann integral and the normal form are monotonous. Moreover, the extension should be consistent with the starting point. In the case of integration, one uses Dini's theorem: if $f_n$ is an increasing sequence of continuous functions with a continuous supremum $f$, then the convergence $f_n \rightarrow f$ is uniform, i.e., a norm-convergence. The normal form is sup-continuous too: this relies on the strong convergence of bounded increasing nets, see Proposition 18, Appendix C.4, and the strong continuity of composition on balls.

(iv) This extension is, in both cases, consistent with the symmetric extension by infima. Because continuous = l.s.c. $\cap$ u.s.c. (resp. invertible = l.s.i. $\cap$ u.s.i.); moreover, the full semi-continuous (semi-invertible) case remains monotonous.

(v) Another step must be performed: after suprema, infima. This is the end of the story: w.r.t. Lebesgue integration, every measurable function is equivalent to an infimum $\inf_n f_n$ of l.s.c. functions. In our case, every system is the infimum of l.s.i. system.

(vi) But the answers are different: Lebesgue integration can perform this second step, basically because the extension to l.s.c. functions by suprema commutes to infima. Here we say goodbye to our model: the normal form, extended by suprema to l.s.i. systems, does not commute to infima, see Section 6.6. Something else must be found.

No doubt that this ultimate divergence is due to the non-commutativity of the normal form.

**1.5. Associativity.** To go beyond the semi-invertible case, one should introduce another idea, alien to the idea of approximation — topological, or order-theoretic. Here, we use one of the milestones of logic, the *Church-Rosser Property*, that we interpreted as *associativity* in ludics [13]. The question is the following: is an iterated normal form the same as a single normal form, in other terms, if the feedback splits as a direct sum $\sigma + \tau$, can we solve the equation in two steps, first with the sole feedback $\sigma$, then with the feedback $\tau$ applied to the solution?

The answer is positive in the invertible case; in the semi-invertible case, there is no obvious order-continuity argument, for we may have to relate semi-invertible systems which are on "different sides", l.s.i., versus u.s.i.; the situation seems desperate.

In fact this problem is the key to the full solution: every feedback is the difference of two projections, $\sigma = \sigma^+ - \sigma^-$. In case of a *positive* feedback, all systems are l.s.i., and the normal form is upwards continuous; in case of a *negative* feedback, all systems are u.s.i. and the normal form is downwards continuous. In order to get a full normal form, it is enough to prove associativity in case of *lopsided* feedbacks of opposite signs. This is not achieved by order-continuity, but by providing a sort of explicit formula. By the way, we heavily rely on the main property of positive hermitians: the existence of a square root.

**1.6. Stability.** I was surely one of the first persons to express strong doubts as to the topological nature of Scott domains. These doubts prompted me to introduce (rather to rediscover after Berry [3]) a competitor to the Scott ordering, the *stable* ordering, roughly corresponding to inclusion; this was the origin of *coherent spaces* and further developments such as *linear logic*.

The "inclusion" between hermitians is defined by $\mathsf{kh} = \mathsf{h}^2$, together with $\mathsf{k}\sigma\mathsf{h} = \mathsf{h}\sigma\mathsf{h}$, in case of a feedback. Contrarily to the standard ordering, inclusion has the structure of a (downwards) conditional lattice; moreover the solution $\sigma[\![\mathsf{h}]\!]$ of the feedback equation is monotonous and preserves conditional g.l.b., i.e., pull-backs. From this, it follows that the cut-system $(\mathcal{H}, \mathsf{h}, \sigma)$ has a unique *incarnation* $(\mathcal{H}, \mathsf{k}, \sigma)$, where $\mathsf{k} \sqsubset \mathsf{h}$ is the part of h "actually used" in the computation of $\sigma[\![\mathsf{h}]\!]$.

In a commutative — set-theoretic — setting, typically in ludics [13], inclusion is a refinement of the pointwise order. In a non-commutative world, the two orders are independent: typically $\mathsf{h} \sqsubset \mathsf{k} \Rightarrow -\mathsf{h} \sqsubset -\mathsf{k}$, whereas

$$\mathsf{h} \leq \mathsf{k} \Rightarrow -\mathsf{k} \leq -\mathsf{h}.$$

**1.7. Winning.** *Ludics* [13] mainly rests upon the notion of *polarity*, which roughly corresponds to the natural notion of *signature*, the most basic invariant of logic. Logic is then interpreted by sort of *bipartite graphs*, with positive/negative nodes (answers/questions, etc.).

What we try to mimic here by means of *bipartite hermitians*: assuming that our Hilbert space splits into a direct sum $\mathcal{H} = \mathcal{H}^+ \oplus \mathcal{H}^-$, we require that $\mathcal{H}^+\mathsf{h}\mathcal{H}^+ \geq 0$ and $\mathcal{H}^-\mathsf{h}\mathcal{H}^- \leq 0$. We must also restrict to feedbacks enjoying $\mathcal{H}^+\sigma\mathcal{H}^+ = 0$, $\mathcal{H}^-\sigma\mathcal{H}^- = 0$. Under such hypothesis, the surprising fact is that the solution of the feedback equation remains bipartite.

Indeed, the interpretation of logical proofs, $\lambda$-expressions, is not only bipartite, it enjoys $\mathcal{H}^+\mathsf{h}\mathcal{H}^+ = 0$, $\mathcal{H}^-\mathsf{h}\mathcal{H}^- = 0$. If we call this "winning", it is immediate that winning is preserved by composition, i.e., by normal forms. The importance of the preservation of winning lies in its relation to logical *consistency*, see [13].

**1.8. Immediate questions.** A few immediate questions that I had not the time to investigate:

(i) Assume that $(\mathcal{H}, \mathsf{h}, \sigma)$ is terminating. Does the solution correspond to the normal form?

(ii) More generally, if a normal form is achieved by various means, does this output corresponds to our normal form? The main example is *weak nilpotency*, $\langle (\mathsf{h}\sigma)^{\mathsf{n}}(x) \mid x \rangle \to 0$, which yields an *unbounded* execution, but a perfectly bounded output, see [10].

**1.9. Further work.** Further work should involve the definition of a notion of *polarity*[4] between hermitians, in such a way that, of two hermitians, at most one of them can be *winning*, and enjoying a *separation* property, see [13]. Then, last but not least, the remake of GoI, but not necessarily in a type I von Neumann algebra like $B(\mathcal{H})$: maybe a type II or a type III algebra[5] is more appropriate, especially in view of a subtler approach to logic.

## §2. Cut-systems.

### 2.1. The feedback equation.

DEFINITION 1 (Cut-Systems). A *cut-system* is a 3-tuple $(\mathcal{H}, \mathsf{h}, \sigma)$ such that:

- ⋆ $\mathcal{H}$ is a complex Hilbert space.
- ⋆ h is a hermitian — i.e., self-adjoint — operator on $\mathcal{H}$ of norm at most 1: $\mathsf{h}^* = \mathsf{h}$, $\|\mathsf{h}\| \leq 1$.
- ⋆ $\sigma$ (the *cut*, the *loop*, the *feedback*) is a *partial symmetry*, i.e., $\sigma = \sigma^* = \sigma^3$, see Appendix C.6.

Since $\sigma^3 = \sigma$, $\sigma^2$ is the orthogonal projection of a closed subspace $\mathcal{S} \subset \mathcal{H}$; we can therefore write $\mathcal{H} = \mathcal{R} \oplus \mathcal{S}$, with $\mathcal{R} := \mathcal{S}^\perp$. Accordingly to the standard abuse of notations, $\mathcal{S} = \sigma^2$, $\mathcal{R} = \mathrm{I} - \sigma^2$.

The cut-system $(\mathcal{H}, \mathsf{h}, \sigma)$ induces a *feedback equation*: given $x \in \mathcal{R}$ find $x' \in \mathcal{R}$, $y \in \mathcal{S}$, such that:

$$\mathsf{h}(x \oplus y) = x' \oplus \sigma(y). \tag{1}$$

Indeed, we are mostly interested in the "visible" part of the equation, i.e., the output, the *result* $x'$; the component $y$ is rather perceived as "internal", and quite corresponds to the *computation*.

To tell the truth, it is not quite true that this equation is always solvable, *stricto sensu*. But it is anyway our starting point.

REMARK 1. This definition covers everything done under the name "GoI" in [9, 10, 11], but for a small exception, namely the interpretation of *weakening* in [11], which introduces a non-hermitian operator, in contrast to what was

[4] Formerly called orthogonality.

[5] Algebras in which projections can be "halved", i.e., in which there is no minimal subspace, see [18].

previously done in [9]; this fancy variant was supposed to achieved effects of "connectedness", but this never worked. So let us get back to the original, hermitian, definition given in [9].

In fact, it turns out that the h constructed in [9, 10, 11] are not only hermitian, but also partial symmetries. But we cannot limit ourselves to partial symmetries, for the simple reason that, if we don't make any heavy additional hypothesis, the solution to the feedback equation need not be given by a partial symmetry.

REMARK 2. One may wonder why the loop is a partial symmetry, and not a projection. Obviously $(\mathcal{H}, \mathsf{h}, \sigma)$ behaves like $(\mathcal{H}, \mathsf{h}(\mathcal{R}+\sigma), \mathcal{S})$; but there is a deep conceptual difference, h is hermitian, $\mathsf{h}(\mathcal{R}+\sigma)$ is a product of hermitians, i.e., a nothing!

REMARK 3. The original equation is not between a hermitian and a feedback, it is between two hermitians, which are put in duality by the equation. The feedback equation is basically the remark that, w.l.o.g., we can assume that one of the two hermitians, the feedback, is of a very simple form. See Appendix B.2 for a discussion.

**2.2. Bipartism and winning.**

DEFINITION 2 (Bipartism). A *bipartite* cut-system is a cut-system $(\mathcal{H}, \mathsf{h}, \sigma)$, together with a decomposition $\mathcal{H} = \mathcal{H}^+ \oplus \mathcal{H}^-$, such that:
(i) $\mathcal{H}^+\mathsf{h}\mathcal{H}^+ \geq 0, \mathcal{H}^-\mathsf{h}\mathcal{H}^- \leq 0$.
(ii) $\mathcal{H}^+\sigma\mathcal{H}^+ = 0, \mathcal{H}^-\sigma\mathcal{H}^- = 0$.

DEFINITION 3 (Winning). A bipartite cut-system is *winning* when:

$$\mathcal{H}^+\mathsf{h}\mathcal{H}^+ = 0,\ \mathcal{H}^-\mathsf{h}\mathcal{H}^- = 0.$$

PROPOSITION 1. h *is winning iff* $-\mathsf{h}$ *is bipartite.*

PROOF. Trivial. Equivalently, h is winning if it remains bipartite when we swap $\mathcal{H}^+$ and $\mathcal{H}^-$. ⊣

REMARK 4. All cut-systems constructed in [9, 10, 11] are bipartite. Indeed, the notion of *signature* (positive, negative occurrences) induces a natural splitting of the Hilbert spaces at work.

Moreover, all these systems are winning. This should not surprise us, since logical rules are supposed to construct sort of "winning strategies", and GoI interprets logical rules. But, if we leave room for "losing" devices such as the *daimon* of ludics [13], then we may encounter bipartite cut-systems with $\mathcal{H}^+\mathsf{h}\mathcal{H}^+ \neq 0, \mathcal{H}^-\mathsf{h}\mathcal{H}^- \neq 0$.

Among the properties to be checked later, let us mention the fact that the solution of the feedback equation of a bipartite system is still bipartite; similarly, winning will be preserved.

### 2.3. Orderings.

DEFINITION 4 (Pointwise Order). We order cut-systems with the same underlying space and feedback in the obvious way:

$$(\mathcal{H},\mathsf{h},\sigma)\leq(\mathcal{H},\mathsf{k},\sigma)\Leftrightarrow \mathsf{k}-\mathsf{h}\geq 0. \tag{2}$$

The pointwise ordering[6] admits directed l.u.b. and g.l.b.; but it is not a lattice.

DEFINITION 5 (Order-continuity). A monotonous (increasing) map $\Phi$ from a subset of $Her(\mathcal{H})$ to $Her(\mathcal{R})$ is *order-continuous* when it preserves (directed) l.u.b. and g.l.b.

We shall apply this terminology to the *normal form* which associates to cut-systems $(\mathcal{H},\mathsf{h},\sigma)$ a hermitian of $\sigma[\![\mathsf{h}]\!]\in Her(\mathcal{R})$, to mean order-continuity w.r.t. the sole parameter h, consistently with Definition 4.

DEFINITION 6 (Stable order). The stable ordering is defined as follows:

$$(\mathcal{H},\mathsf{h},\sigma)\sqsubset(\mathcal{H},\mathsf{k},\sigma)$$

iff

(i) $\mathsf{kh}=\mathsf{h}^2$ and
(ii) $\mathsf{k}\sigma\mathsf{h}=\mathsf{h}\sigma\mathsf{h}$.

PROPOSITION 2. $(\mathcal{H},\mathsf{h},\sigma)\sqsubset(\mathcal{H},\mathsf{k},\sigma)$ *iff there exists a closed subspace* $\mathcal{E}$ *such that*:

(i) $\mathcal{E}\sigma=\sigma\mathcal{E}$ *and*
(ii) $\mathsf{h}=\mathsf{k}\mathcal{E}$.

PROOF. Assume that $(\mathcal{H},\mathsf{h},\sigma)\sqsubset(\mathcal{H},\mathsf{k},\sigma)$; then k and h coincide on the spaces $\mathrm{rg}(\mathsf{h})$ and $\mathrm{rg}(\sigma\mathsf{h})$, hence on the closure $\mathcal{E}$ of $\mathrm{rg}(\mathsf{h})+\mathrm{rg}(\sigma\mathsf{h})$. Obviously $\mathcal{E}\sigma=\sigma\mathcal{E}$ and $\mathsf{h}\mathcal{E}=\mathsf{k}\mathcal{E}$; moreover, $\mathcal{E}\mathsf{h}=\mathsf{h}$ hence, by taking adjoints, $\mathsf{h}\mathcal{E}=\mathsf{h}$ and $\mathsf{k}\mathcal{E}=\mathsf{h}$.

Conversely, if the condition of the proposition holds, then $\mathsf{h}=\mathcal{E}\mathsf{k}=\mathcal{E}\mathsf{h}$ and $\mathsf{kh}=\mathsf{k}\mathcal{E}\mathsf{h}=\mathsf{h}^2$; also, $\mathsf{k}\sigma\mathsf{h}=\mathsf{k}\sigma\mathcal{E}\mathsf{h}=\mathsf{k}\mathcal{E}\sigma\mathsf{h}=\mathsf{h}\sigma\mathsf{h}$. ⊣

COROLLARY 1. $\sqsubset$ *is an order relation. Moreover, any (non-empty) bounded family* $(\mathcal{H},\mathsf{h}_i,\sigma)$ *admits a g.l.b. with respect to* $\sqsubset$.

PROOF. Transitivity, antisymmetry are obvious. If $\mathcal{E}_i\sigma=\sigma\mathcal{E}_i$ and $\mathsf{h}_i=\mathsf{k}\mathcal{E}_i$, then $\prod_i \mathsf{h}_i:=\mathsf{k}\bigcap_i\mathcal{E}_i$ is the desired greatest lower bound. ⊣

Pointwise directed l.u.b. are like direct limits, whereas bounded stable g.l.b. are like pull-backs; some distributivity is therefore expected. We state it in the case of a binary pull-back, just for readability.

---

[6]Logicians would rather speak of "extensional" order; "pointwise", like in "pointwise convergence", seems however more standard.

PROPOSITION 3. *Assume that* $(\mathcal{H}, \mathsf{g}_i, \sigma)$, $(\mathcal{H}, \mathsf{h}_i, \sigma)$, *and* $(\mathcal{H}, \mathsf{k}_i, \sigma)$ *are directed increasing nets w.r.t. the pointwise order, with respective l.u.b.* $(\mathcal{H}, \mathsf{g}, \sigma)$, $(\mathcal{H}, \mathsf{h}, \sigma)$, *and* $(\mathcal{H}, \mathsf{k}, \sigma)$; *assume that* $(\mathcal{H}, \mathsf{g}_i, \sigma) \sqsubset (\mathcal{H}, \mathsf{k}_i, \sigma)$ *and* $(\mathcal{H}, \mathsf{h}_i, \sigma) \sqsubset (\mathcal{H}, \mathsf{k}_i, \sigma)$ *and let* $\mathsf{f}_i = \mathsf{g}_i \sqcap \mathsf{h}_i$. *Then* $(\mathcal{H}, \mathsf{g}, \sigma) \sqsubset (\mathcal{H}, \mathsf{k}, \sigma)$, *and* $(\mathcal{H}, \mathsf{h}, \sigma) \sqsubset (\mathcal{H}, \mathsf{k}, \sigma)$; *moreover* $(\mathcal{H}, \mathsf{h}_i, \sigma)$ *is an increasing directed net whose pointwise l.u.b.* $(\mathcal{H}, \mathsf{f}, \sigma)$ *is equal to* $(\mathcal{H}, \mathsf{g}, \sigma) \sqcap (\mathcal{H}, \mathsf{h}, \sigma)$.

PROOF. For instance, from $\mathsf{g}_i\mathsf{k}_i = \mathsf{g}_i^2$, we can, using a strong continuity argument (see Proposition 18, Appendix C.4) and the strong continuity of composition on balls, get $\mathsf{gk} = \mathsf{g}^2$. ⊣

REMARK 5. Since $(\mathcal{H}, \mathsf{h}, \sigma) \sqsubset (\mathcal{H}, \mathsf{k}, \sigma) \Rightarrow (\mathcal{H}, -\mathsf{h}, \sigma) \sqsubset (\mathcal{H}, -\mathsf{k}, \sigma)$, Proposition 3 works also for *decreasing* nets.

**§3. Invertible case.** The purpose of this section is to establish the existence of a "solution" to the feedback equation under a reasonable hypothesis, *invertibility*.

**3.1. Invertibility.**

PROPOSITION 4. *If* $\sigma$ *is a partial symmetry, then* $\sigma$ *can be written as the difference* $\sigma^+ - \sigma^-$ *of two projections such that* $\sigma^+ \cdot \sigma^- = 0$.

PROOF. This is a special case of a standard result mentioned in Appendix C.3. Indeed $\sigma^+$ and $\sigma^-$ are the orthoprojections of the eigenspaces corresponding to the values $+1$ and $-1$. Spectral calculus yields $\sigma^+ + \sigma^- = \sigma^2$ and $\sigma^+ = 1/2(\sigma^2 + \sigma)$, $\sigma^- = 1/2(\sigma^2 - \sigma)$. ⊣

DEFINITION 7. The cut-system $(\mathcal{H}, \mathsf{h}, \sigma)$ is *invertible* (resp. *upper-semi-invertible*: u.s.i., *lower-semi-invertible*: l.s.i.) when $\sigma - \sigma^2\mathsf{h}\sigma^2$ (resp. $\sigma^+ - \sigma^+\mathsf{h}\sigma^+$, $\sigma^- + \sigma^-\mathsf{h}\sigma^-$) is invertible as an endomorphism of $\mathcal{S} = \sigma^2$ (resp. of $\sigma^+, \sigma^-$).

By lemma 3.2.13. of [17] (approximate eigenvectors): a hermitian operator $\mathsf{f} \in B(\mathcal{H})$ is non-invertible iff there exists a sequence $(x_n)$, with $\|x_n\| = 1$ and $\mathsf{f}(x_n) \to 0$ as $n \to \infty$. Of course, $\|x_n\| = 1$ can be weakened into $\|x_n\| \geq a$ for some $a > 0$, replace $x_n$ with $1/\|x_n\| \cdot x_n$. Therefore, $(\mathcal{H}, \mathsf{h}, \sigma)$ is

**Invertible:** Iff there is no sequence $(x_n)$ with $\|x_n\| = 1$ such that $\sigma^2(x_n) = x_n$ and $\sigma^2\mathsf{h}(x_n) - \sigma(x_n) \to 0$.

**Upper-semi-invertible:** Iff there is no sequence $(x_n)$ with $\|x_n\| = 1$ such that $\sigma(x_n) = x_n$ and $\sigma^+\mathsf{h}(x_n) - x_n \to 0$.

**Lower-semi-invertible:** Iff there is no sequence $(x_n)$ with $\|x_n\| = 1$ such that $\sigma(x_n) = -x_n$ and $\sigma^-\mathsf{h}(x_n) + x_n \to 0$.

LEMMA 1. *If* $\mathsf{f}$ *is hermitian and* $\|\mathsf{f}\| \leq 1$, $x_n \in \mathcal{H}$, $\|x_n\| = 1$, *then* $\mathsf{f}(x_n) - x_n \to 0$ *iff* $\langle \mathsf{f}(x_n) \mid x_n \rangle \to 1$.

PROOF. If $\|\mathsf{f}(x_n) - x_n\| \to 0$, then $\langle \mathsf{f}(x_n) \mid x_n \rangle - 1 = \langle \mathsf{f}(x_n) - x_n \mid x_n \rangle \to 0$, hence $\langle \mathsf{f}(x_n) \mid x_n \rangle \to 1$.

Conversely, if $\langle f(x_n) \mid x_n \rangle \to 1$, since $\|f(x_n)\| \leq 1$, Cauchy-Schwarz implies that $\|f(x_n)\| \to 1$, and $\|f(x_n) - x_n\|^2 = \|f(x_n)\|^2 + \|x_n\|^2 - 2\langle f(x_n) \mid x_n \rangle \to 0$, hence $\|f(x_n) - x_n\| \to 0$. ⊣

Now, if $x_n \in \mathcal{H}$, $\|x_n\| = 1$, is such that $\sigma^2 h(x_n) - x_n \to 0$, we can apply the lemma to $f := \sigma^2 h \sigma^2$, and we conclude that $\langle \sigma^2 h(x_n) \mid x_n \rangle \to 1$, hence $\langle h(x_n) \mid x_n \rangle = \langle h(x_n) \mid \sigma^2(x_n) \rangle \to 1$. Applying the lemma in the other direction, we conclude that $h(x_n) - x_n \to 0$. The same can be done with $\sigma^+$ and $\sigma^-$, and we get the simplified characterisations: $(\mathcal{H}, h, \sigma)$ is

**Invertible:** Iff there is no sequence $(x_n)$ with $\|x_n\| = 1$ such that $\sigma^2(x_n) = x_n$ and $h(x_n) - \sigma(x_n) \to 0$.

**Upper-semi-invertible:** Iff there is no sequence $(x_n)$ with $\|x_n\| = 1$ such that $\sigma(x_n) = x_n$ and $h(x_n) - x_n \to 0$.

**Lower-semi-invertible:** Iff there is no sequence $(x_n)$ with $\|x_n\| = 1$ such that $\sigma(x_n) = -x_n$ and $h(x_n) + x_n \to 0$.

PROPOSITION 5. *$(\mathcal{H}, h, \sigma)$ is invertible iff it is both u.s.i. and l.s.i.*

PROOF. From the preliminary work just done, it is plain that invertibility implies upper- and lower-semi-invertibilities.

Conversely, assume $(\mathcal{H}, h, \sigma)$ not invertible, and let $(z_n)$ be s. t. $\|z_n\| = 1$, $\sigma^2(z_n) = z_n$ and $h(z_n) - \sigma(z_n) \to 0$. From $\|h(z_n)\| - \|\sigma(z_n)\| \to 0$, we get $\|h(z_n)\| \to 1$ and $\langle h^2(z_n) \mid z_n \rangle = \|h(z_n)\|^2 \to 1$; by Lemma 1, $h^2(z_n) - z_n \to 0$. With $x_n := z_n + h(z_n)$, $y_n := z_n - h(z_n)$, $h(x_n) - x_n \to 0$, $h(y_n) + y_n \to 0$, $\sigma(x_n) - x_n \to 0$, $\sigma(y_n) + y_n \to 0$. Let $X_n := \sigma(x_n) + x_n$, $Y_n := \sigma(y_n) - y_n$, then $\sigma(X_n) = X_n$ and $h(X_n) - X_n \to 0$, $\sigma(Y_n) = -Y_n$ and $h(Y_n) + Y_n \to 0$. Then one of the two sequences $(x_n), (y_n)$ has a subsequence of norm $\geq 1$, say $x_{n_k}$; then $\|X_{n_k}\| \geq 1$, and $(\mathcal{H}, h, \sigma)$ is not an u.s.i. system. ⊣

PROPOSITION 6. *If $(\mathcal{H}, h, \sigma) \leq (\mathcal{H}, k, \sigma)$, then:*

(i) *If $(\mathcal{H}, k, \sigma)$ is u.s.i., so is $(\mathcal{H}, h, \sigma)$.*

(ii) *If $(\mathcal{H}, k, \sigma)$ is l.s.i., then $(\mathcal{H}, \lambda h + (1 - \lambda)k, \sigma)$ is l.s.i. for $0 < \lambda < 1$.*

(iii) *If $(\mathcal{H}, k, \sigma)$ is l.s.i. and $(\mathcal{H}, h, \sigma)$ is u.s.i., then $(\mathcal{H}, \lambda h + (1 - \lambda)k, \sigma)$ is invertible for $0 < \lambda < 1$.*

(iv) *If $(\mathcal{H}, k, \sigma)$ is invertible, then $(\mathcal{H}, \lambda h + (1 - \lambda)k, \sigma)$ is invertible for $0 < \lambda < 1$.*

PROOF. (i) Take $(x_n)$ with $\|x_n\| = 1$ such that $\sigma(x_n) = x_n$; if $h(x_n) - x_n \to 0$, then Lemma 1 yields $\langle h(x_n) \mid x_n \rangle \to 1$, and since $h \leq k$,

$$\langle h(x_n) \mid x_n \rangle \leq \langle k(x_n) \mid x_n \rangle \leq 1$$

and $\langle k(x_n) \mid x_n \rangle \to 1$; the same lemma yields $k(x_n) - x_n \to 0$. So, the upper-semi-invertibility of k implies the upper-semi-invertibility of h.

(ii) Let $g := \lambda h + (1 - \lambda)k$, and take $(x_n)$ with $\|x_n\| = 1$ such that $\sigma(x_n) = -x_n$; if $g(x_n) + x_n \to 0$, then $\langle g(x_n) \mid x_n \rangle \to -1$, and necessarily

$$\langle k(x_n) \mid x_n \rangle \to -1.$$

(iii) Combination of (ii) with its dual version.

(iv) $(\mathcal{H}, \lambda h + (1-\lambda)k, \sigma)$ is l.s.i. because by (ii) and u.s.i. because $\lambda h + (1-\lambda)k \leq k$, and (i). ⊣

**3.2. The normal form.** In what follows, we shall deal with cut-systems, often called $(\mathcal{H}, h, \sigma)$, or $(\mathcal{H}, k, \sigma)$. It is convenient to adopt a matrix-like notations ("blocks", see Appendix C.7) corresponding to the direct sum decomposition $\mathcal{H} = \mathcal{R} \oplus \mathcal{S}$, with $\mathcal{S} = \sigma^2$; we shall implicitly assume that $h = \left[\begin{smallmatrix} A & B^* \\ B & C \end{smallmatrix}\right]$; in the concrete conditions we shall deal with, k will share the same coefficients A, B, i.e., $k = \left[\begin{smallmatrix} A & B^* \\ B & D \end{smallmatrix}\right]$.

Lemma 2. *If* f *is an invertible hermitian and* p *is positive, then the function* $(f - \lambda p)^{-1}$ *is defined and monotonous on an open neighbourhood of* 0 *in* $\mathbb{R}$.

Proof. $(f - \lambda p)^{-1} = f^{-1} + \lambda f^{-1} p f^{-1} + \lambda^2 f^{-1} p f^{-1} p f^{-1} + \cdots$. The power series converges for $|\lambda| < \|f^{-1}p\|^{-1}$. We can group the non constant terms: let $p_n := \lambda^{2n+1} . f^{-1} . (pf^{-1})^n \cdot \sqrt{p} \cdot (I + \lambda \sqrt{p}\, f^{-1} \sqrt{p}) \cdot \sqrt{p} \cdot (f^{-1}p)^n \cdot f^{-1}$, so that $(f - \lambda p)^{-1} = f^{-1} + \sum p_n$. For $|\lambda|$ small enough, $I + \lambda \sqrt{p}\, f^{-1} \sqrt{p}$ is positive and all $p_n$ are positive. From this, it is easy to conclude. ⊣

Theorem 3.1 (Monotonicity). *The map* $(\mathcal{H}, h, \sigma) \rightsquigarrow (\sigma - C)^{-1}$ *from invertible cut-systems to* $Her(\mathcal{S})$ *is monotonous (increasing) and order-continuous w.r.t. the parameter* $C$.

Proof. Keep in mind, although it does not quite matter here, that the parameters A, B of the block decomposition are kept constant. If the cut-systems $(\mathcal{H}, h, \sigma) \leq (\mathcal{H}, k, \sigma)$ are invertible, then, for $\lambda \in [0, 1]$, $(\mathcal{H}, \lambda h + (1-\lambda)k, \sigma)$ is invertible (Proposition 6). Let $p := D - C \in Her^+(\mathcal{S})$; the function

$$\varphi : \lambda \rightsquigarrow (\sigma - C + \lambda p)^{-1}$$

is a continuous map from [0, 1] to the Banach space $Her(\mathcal{S})$. By Lemma 2, this map is locally increasing, and it must be globally increasing. We conclude that $\varphi(0) \leq \varphi(1)$, i.e., $(\sigma - C)^{-1} \leq (\sigma - D)^{-1}$.

Assume now that $(\mathcal{H}, h_i, \sigma)$ $(i \in I)$ is an increasing *net* of invertible systems, with an invertible l.u.b. $(\mathcal{H}, h := \sup_{i \in I} h_i, \sigma)$. By Proposition 18 of Appendix C.4, $\sigma - C_i$ and $(\sigma - C_i)^{-1}$ respectively converge to $\sigma - C$ and some E in the strong-operator topology; by monotonicity, $E := \sup_{i \in I} (\sigma - C_i)^{-1} \leq (\sigma - C)^{-1}$. Now observe that the $\sigma - C_i$ have their norms bounded by 2; multiplication is strong operator-continuous, provided the left argument remains bounded: this implies $\sigma - C_i \cdot (\sigma - C_i)^{-1} \to (\sigma - C) \cdot E$. Hence $(\sigma - C) \cdot E = \mathcal{S}$, and, since $\sigma - C$ is invertible, $E = \sup_{i \in I} (\sigma - C_i)^{-1} = (\sigma - C)^{-1}$.

Downwards continuity is proved in the same way. ⊣

Definition 8 (Normal Form). If $(\mathcal{H}, h, \sigma)$ is invertible, its *normal form* is defined by the equation:

$$\sigma[\![h]\!] := A + B^*(\sigma - C)^{-1} B. \qquad (3)$$

The normal form corresponds to the "visible part" of the feedback equation:

THEOREM 3.2 (Normal Form). *If* $(\mathcal{H}, \mathsf{h}, \sigma)$ *is invertible, then the feedback equation admits the normal form* $\sigma[\![\mathsf{h}]\!]$ *as unique solution.* $\sigma[\![\mathsf{h}]\!]$ *is of norm at most* 1, *and it is norm-continuous, monotonous (increasing) and order-continuous w.r.t. the input* $\mathsf{h}$.

PROOF. We want to solve the equation: $\begin{bmatrix} \mathsf{A} & \mathsf{B}^* \\ \mathsf{B} & \mathsf{C}-\sigma \end{bmatrix} \begin{bmatrix} x \\ y \end{bmatrix} = \begin{bmatrix} x' \\ 0 \end{bmatrix}$. Invertibility quite means that $\sigma - \mathsf{C}$ is invertible. It is then obvious that

$$x' = (\mathsf{A} + \mathsf{B}^*(\sigma - \mathsf{C})^{-1}\mathsf{B})(x), \quad y = ((\sigma - \mathsf{C})^{-1}\mathsf{B})(x)$$

is a solution to the equation. This is indeed *the* solution: if

$$\begin{bmatrix} \mathsf{A} & \mathsf{B}^* \\ \mathsf{B} & \mathsf{C}-\sigma \end{bmatrix} \begin{bmatrix} 0 \\ y \end{bmatrix} = \begin{bmatrix} x' \\ 0 \end{bmatrix},$$

then $(\mathsf{C} - \sigma)(y) = 0$ and $x' = 0$. Hence, the (visible part of) the solution is given by $\sigma[\![\mathsf{h}]\!] := \mathsf{A} + \mathsf{B}^*(\sigma - \mathsf{C})^{-1}\mathsf{B}$.

It is plain that $\sigma[\![\mathsf{h}]\!]$ is hermitian and that the dependency $\mathsf{h} \rightsquigarrow \sigma[\![\mathsf{h}]\!]$ is norm-continuous.

Finally, from $\mathsf{h}(x \oplus y) = x' \oplus \sigma(y)$, we get $\|x'\|^2 + \|y\|^2 = \|x'\|^2 + \|\sigma(y)\|^2 \leq \|x\|^2 + \|y\|^2$, hence $\|x'\|^2 \leq \|x\|^2$: we just established that $\|\sigma[\![\mathsf{h}]\!]\| \leq 1$.

Let us now consider the behaviour w.r.t. the pointwise order. First observe that, if "coefficients" A, B in the "block" of h, k, ... are kept constant, then, by Theorem 3.1, we get order-monotonicity and order-continuity.

The general case is reduced to this case by means of the "Tortoise Principle" of Appendix B, and the reduction being extremely simple, the only problem is to avoid pedantism when saying something obvious: hence there might be some (slight) abuses of notations. The idea is to add two copies of $\mathcal{R}$, say $\mathcal{R}_2, \mathcal{R}_1$, hence $\mathcal{H}$ is replaced with $\mathcal{K} := \mathcal{R}_2 \oplus \mathcal{R}_1 \oplus \mathcal{H}$. If u is an isometry between $\mathcal{R}$ and $\mathcal{R}_1$, we can "extend" our feedback $\sigma$ into $\tau(x_2 \oplus x_1 \oplus x \oplus z) := 0 \oplus \mathsf{u}(x) \oplus \mathsf{u}^*(x_1) \oplus 0$. Similarly, if v is an isometry between $\mathcal{R}_1$ and $\mathcal{R}_2$, we can "extend" h into $\Psi(\mathsf{h})(x_2 \oplus x_1 \oplus y) := \mathsf{v}(x_1) \oplus \mathsf{v}^*(x_1) \oplus \mathsf{h}(y)$. There are a few obvious facts about this replacement:

(i) The map $\mathsf{h} \rightsquigarrow \Psi(\mathsf{h})$ is monotonous and order-continuous.

(ii) The system $(\mathcal{K}, \Psi(\mathsf{h}), \tau)$ is invertible; the best remains to solve the equation "manually", i.e., by "equality pushing". The normal form is explicitly given by:

$$\tau[\![\Psi(\mathsf{h})]\!] = \mathsf{vu}\sigma[\![\mathsf{h}]\!]\mathsf{u}^*\mathsf{v}^* \tag{4}$$

(iii) The normal form $\tau[\![\Psi(\mathsf{h})]\!]$ is monotonous and order-continuous w.r.t. the input $\Psi(\mathsf{h})$: this is because $\Psi(\mathsf{h})$ can be written (w.r.t. a decomposition $\mathcal{R}_2 \oplus \mathcal{R}_2{}^{\perp}$) as a block: $\begin{bmatrix} 0 & \mathsf{v} & 0 & 0 \\ \mathsf{v}^* & \cdot & \cdot & \cdot \\ 0 & \cdot & \cdot & \cdot \\ 0 & \cdot & \cdot & \cdot \end{bmatrix}$; we are therefore back to the case "A, B constant".

(iv) The normal form $\sigma[\![\mathsf{h}]\!]$ is monotonous and order-continuous w.r.t. the input $\mathsf{h}$: combination of the previous observations. ⊣

REMARK 6. Technically speaking, the Tortoise introduces a cut with an identity axiom, so that the whole *variable* part of the net is now "invisible".

### 3.3. Associativity.

DEFINITION 9 (Independence). Two feedbacks $\sigma$, $\tau$ are independent when $\sigma\tau = 0\ (= \tau\sigma)$.

In presence of two *independent* feedbacks $\sigma, \tau$, there are several possibilities to reach a "normal form" for $(\mathcal{H}, \mathsf{h}, \sigma + \tau)$. Either we directly form $(\sigma + \tau)[\![\mathsf{h}]\!]$, or first normalise $(\mathcal{H}, \mathsf{h}, \sigma)$, yielding the normal form $\sigma[\![\mathsf{h}]\!]$, and then normalise $(\mathcal{R}, \sigma[\![\mathsf{h}]\!], \tau)$, yielding $\tau[\![\sigma[\![\mathsf{h}]\!]]\!]$. We can also do it the other way around, leading to $\sigma[\![\tau[\![\mathsf{h}]\!]]\!]$. The question is whether or not these protocols yield the same output.

The presence of a double feedback occurs naturally when we deal with the static — category-theoretic — interpretation of logic: we must establish associativity of the composition[7] of morphisms, $f \circ (g \circ h) = (f \circ g) \circ h$. Now, composition corresponds to the cut-rule, i.e., to a feedback; here we are given two compositions, i.e., two independent feedbacks $\sigma, \tau$, together with a hermitian $\mathsf{k}$ corresponding to the three morphisms $f, g, h$ "put together". $f \circ (g \circ h)$ corresponds to $\sigma[\![\tau[\![\mathsf{k}]\!]]\!]$, whereas $(f \circ g) \circ h$ corresponds to $\tau[\![\sigma[\![\mathsf{k}]\!]]\!]$. To sum up, the question of equating the three protocols above is nothing but the soundness of the categorical approach to logic: this is why we speak of *associativity*. In traditional rewriting technology, associativity comes from the possibility of performing the rewritings in any order, i.e., from the familiar Church-Rosser property.

THEOREM 3.3 (Associativity). *Assume that $\sigma, \tau$ are independent and write $\mathcal{H} = \mathcal{R} \oplus \mathcal{S} \oplus \mathcal{T}$. Then $(\mathcal{H}, \mathsf{h}, \sigma + \tau)$ is invertible iff $(\mathcal{H}, \mathsf{h}, \tau)$ and $(\mathcal{R} \oplus \mathcal{S}, \tau[\![\mathsf{h}]\!], \sigma)$ are invertible. Moreover*

$$(\sigma + \tau)[\![\mathsf{h}]\!] = \sigma[\![\tau[\![\mathsf{h}]\!]]\!]. \tag{5}$$

PROOF. First assume that $(\mathcal{H}, \mathsf{h}, \tau)$ and $(\mathcal{R} \oplus \mathcal{S}, \tau[\![\mathsf{h}]\!], \sigma)$ are invertible; the left hand side corresponds to the equation, $x \in \mathcal{R}$ being given:

$$\mathsf{h}(x \oplus y \oplus z) = x' \oplus \sigma(y) \oplus \tau(z) \tag{6}$$

whose "visible part" is $x' = (\sigma + \tau)[\![\mathsf{h}]\!](x)$. This equation can be solved in two steps: first, given $x \in \mathcal{R}, y' \in \mathcal{S}$, solve:

$$\mathsf{h}(x \oplus y' \oplus z) = x' \oplus y'' \oplus \tau(z) \tag{7}$$

[7] See also Appendix B.2.

whose visible part is $x' \oplus y'' = \tau[\![\mathsf{h}]\!](x \oplus y)$. Then, add the constraint $y'' = \sigma(y')$, which amounts at solving:

$$\tau[\![\mathsf{h}]\!](x \oplus y) = x' \oplus \sigma(y) \tag{8}$$

and whose visible part is given by $x' = \sigma[\![\tau[\![\mathsf{h}]\!]]\!](x)$. We just established equation (5) "manually". We need a little more care concerning invertibility matters.

(i) If $(\mathcal{H}, \mathsf{h}, \tau)$ is not invertible, take a sequence $(x_n) \in \mathcal{T}$ of norm 1 such that $\mathsf{h}(x_n) - \tau(x_n) \to 0$; since $\sigma(x_n) = 0$, we just found approximate eigenvectors for $\mathsf{h} - (\sigma + \tau)$.

(ii) If $(\mathcal{H}, \mathsf{h}, \tau)$ is invertible but $(\mathcal{R} \oplus \mathcal{S}, \tau[\![\mathsf{h}]\!], \sigma)$ is not invertible, take a sequence $(y_n) \in \mathcal{S}$ of norm 1 such that $\tau[\![\mathsf{h}]\!](y_n) - \sigma(y_n) \to 0$. There exists $z_n$ such that $\mathsf{h}(y_n \oplus z_n) = \tau[\![\mathsf{h}]\!](y_n) \oplus \tau(z_n)$. Then $(y_n \oplus z_n)$ is such that $\mathsf{h}(y_n \oplus z_n) - (\sigma + \tau)(y_n \oplus z_n) \to 0$, and, once more, $(\mathcal{H}, \mathsf{h}, \sigma + \tau)$ is not invertible. Summing up, we just proved that the invertibility of the full system implies the invertibility of the partial systems corresponding to a two-step "normalisation".

(iii) W.r.t. the decomposition $\mathcal{H} = \mathcal{R} \oplus \mathcal{S} \oplus \mathcal{T}$, write $\mathsf{h} = \begin{bmatrix} \cdot & \cdot & \cdot \\ \cdot & \mathsf{A} & \mathsf{B}^* \\ \cdot & \mathsf{B} & \mathsf{C} \end{bmatrix}$ Assuming that $\tau - \mathsf{C}$ and $\sigma - \tau[\![\mathsf{h}]\!]$ are invertible, we want to show that

$$\mathsf{k} := \begin{bmatrix} \sigma - \mathsf{A} & -\mathsf{B}^* \\ -\mathsf{B} & \tau - \mathsf{C} \end{bmatrix}$$

is invertible. Define:

$$\mathsf{g} := \begin{bmatrix} (\sigma - \tau[\![\mathsf{h}]\!])^{-1} & (\sigma - \tau[\![\mathsf{h}]\!])^{-1}\mathsf{B}^*(\tau - \mathsf{C})^{-1} \\ (\tau - \mathsf{C})^{-1}\mathsf{B}(\sigma - \tau[\![\mathsf{h}]\!])^{-1} & (\tau - \mathsf{C})^{-1}\mathsf{B}(\sigma - \tau[\![\mathsf{h}]\!])^{-1}\mathsf{B}^*(\tau - \mathsf{C})^{-1} + (\tau - \mathsf{C})^{-1} \end{bmatrix}$$

A straightforward (but painful) computation yields $\mathsf{gk} = \begin{bmatrix} \mathcal{S} & 0 \\ 0 & \mathcal{T} \end{bmatrix}$. This proves the left-invertibility of $\mathsf{k}$; since $\mathsf{k}$ is hermitian, it is invertible: from $\mathsf{gk} = \mathrm{I}$, we get $\mathsf{k}^*\mathsf{g}^* = \mathsf{kg} = \mathrm{I}$. ⊣

REMARK 7. Putting things together, i.e., using the theorem, the inverse can be expressed in a more symmetrical way, typically:

$$\mathsf{k}^{-1} = \begin{bmatrix} (\sigma - \tau[\![\mathsf{h}]\!])^{-1} & (\sigma - \tau[\![\mathsf{h}]\!])^{-1}\mathsf{B}^*(\tau - \mathsf{C})^{-1} \\ (\tau - \sigma[\![\mathsf{h}]\!])^{-1}\mathsf{B}(\sigma - \mathsf{A})^{-1} & (\tau - \sigma[\![\mathsf{h}]\!])^{-1} \end{bmatrix}$$

or the adjoint expression

$$\mathsf{k}^{-1} = \begin{bmatrix} (\sigma - \tau[\![\mathsf{h}]\!])^{-1} & (\sigma - \mathsf{A})^{-1}\mathsf{B}^*(\tau - \sigma[\![\mathsf{h}]\!])^{-1} \\ (\tau - \mathsf{C})^{-1}\mathsf{B}(\sigma - \tau[\![\mathsf{h}]\!])^{-1} & (\tau - \sigma[\![\mathsf{h}]\!])^{-1} \end{bmatrix}$$

**3.4. Stability and incarnation.**

PROPOSITION 7. *If* $(\mathcal{H}, \mathsf{h}, \sigma) \sqsubset (\mathcal{H}, \mathsf{k}, \sigma)$ *and* $(\mathcal{H}, \mathsf{k}, \sigma)$ *is invertible, then* $(\mathcal{H}, \mathsf{h}, \sigma)$ *is invertible.*

PROOF. Completely immediate. ⊣

In the next theorem, all systems are supposed to be invertible.

THEOREM 3.4 (Stability). *The normal form is compatible with stability, more precisely*:

(i) *If* $(\mathcal{H}, \mathsf{h}, \sigma) \sqsubset (\mathcal{H}, \mathsf{k}, \sigma)$, *then* $\sigma[\![\mathsf{h}]\!] \sqsubset \sigma[\![\mathsf{k}]\!]$.

(ii) *If* $(\mathcal{H}, \mathsf{h}_i, \sigma) \sqsubset (\mathcal{H}, \mathsf{k}, \sigma)$, *then* $\sigma[\![\bigsqcap_i \mathsf{h}_i]\!] = \bigsqcap_i \sigma[\![\mathsf{h}_i]\!]$.

PROOF. (i) Assume that $\mathsf{h} = \mathsf{k}\mathcal{E}, \sigma\mathcal{E} = \mathcal{E}\sigma$; then $\mathsf{h} = \mathcal{E}\mathsf{k}$ and $\mathcal{E}$ commutes to both of $\mathcal{R}, \mathcal{S}$, and let $\mathcal{F} := \mathcal{R}\mathcal{E}, \mathcal{G} := \mathcal{S}\mathcal{E}$. If $\mathsf{h}(x \oplus y) = x' \oplus \sigma(y)$, then using $\mathsf{h} = \mathsf{k}\mathcal{E}$, we get $\mathsf{k}(\mathcal{F}(x) \oplus \mathcal{G}(y)) = x' \oplus \sigma(y)$; using $\mathsf{h} = \mathcal{E}\mathsf{k}$, we get $x' \oplus \sigma(y) = \mathcal{F}(x') \oplus \mathcal{G}(\sigma(y)) = \mathcal{F}(x') \oplus \sigma(\mathcal{G}(y)))$. Summing up, we get $\mathsf{k}(\mathcal{F}(x) \oplus \mathcal{G}(y)) = x' \oplus \sigma(\mathcal{G}(y))$, i.e., $\sigma[\![\mathsf{k}]\!](\mathcal{F}(x)) = \sigma[\![\mathsf{h}]\!](x)$.

(ii) Immediate. ⊣

This justifies the following definition:

DEFINITION 10 (Incarnation). If $(\mathcal{H}, \mathsf{h}, \sigma)$ is an invertible cut-system, its *incarnation* is the smallest $(\mathcal{H}, \mathsf{k}, \sigma) \sqsubset (\mathcal{H}, \mathsf{h}, \sigma)$ such that $\sigma[\![\mathsf{k}]\!] = \sigma[\![\mathsf{h}]\!]$.

Incarnation, namely the "useful part" of a system, plays a central role in ludics. Remember that for instance the *mystery of incarnation*[8] reduces the Cartesian product to an intersection!

**3.5. Normal forms and winning.** Let us investigate the normal form in the *bipartite* case. The next proposition is quite surprising:

PROPOSITION 8. *If a bipartite hermitian is invertible, its inverse is still bipartite.*

PROOF. Let $x = \mathsf{h}(x' \oplus y')$, with $x, x' \in \mathcal{H}^+, y' \in \mathcal{H}^-$; then

$$\begin{aligned}\langle \mathsf{h}^{-1}(x) \mid x\rangle &= \langle x' \oplus y' \mid x\rangle = \langle x' \mid x\rangle = \langle x' \mid \mathsf{h}(x' \oplus y')\rangle \\ &= \langle x' \mid \mathsf{h}(x')\rangle + \langle x' \mid \mathsf{h}(y')\rangle\end{aligned}$$

and $\langle x' \mid \mathsf{h}(y')\rangle = \langle \mathsf{h}(x') \mid y'\rangle = \langle x \mid y'\rangle - \langle \mathsf{h}(y') \mid y'\rangle = -\langle \mathsf{h}(y') \mid y'\rangle$. Summing up, we find $\langle \mathsf{h}^{-1}(x) \mid x\rangle = \langle \mathsf{h}(x') \mid x'\rangle - \langle \mathsf{h}(y') \mid y'\rangle \geq 0$. In the same way one proves that $\langle \mathsf{h}^{-1}(y) \mid y\rangle \leq 0$ for $y \in \mathcal{H}^-$. ⊣

Let us now introduce an inessential — but useful — tool: assume that, w.r.t. the decomposition $\mathcal{H} = \mathcal{R} \oplus \mathcal{S}$, $\mathsf{h} = \begin{bmatrix} \mathsf{A} & \mathsf{B}^* \\ \mathsf{B} & \mathsf{C} \end{bmatrix}$ and given $x \in \mathcal{R} = \mathcal{S}^\perp$, with $\|x\| \leq 1$, we can define $\mathsf{h}_x$ as a hermitian of $\mathbb{C} \oplus \mathcal{S}$ : $\mathsf{h}_x := \begin{bmatrix} \langle \mathsf{A}(x) \mid x\rangle & \mathsf{B}(x)^* \\ \mathsf{B}(x) & \mathsf{C} \end{bmatrix}$. Obviously, $\|\mathsf{h}_x\| \leq \|\mathsf{h}\|$.

PROPOSITION 9. *If* $\mathsf{h}$ *is invertible and* $x \in \mathcal{R}$, $\|x\| \leq 1$, *then*:

$$\sigma[\![\mathsf{h}_x]\!] = \langle \sigma[\![\mathsf{h}]\!](x) \mid x\rangle.$$

---

[8] See [13].

PROOF. Completely obvious, e.g., from the explicit formula for $\sigma[\![\mathsf{h}]\!]$:

$$\begin{aligned}\langle(\mathsf{A}+\mathsf{B}^*(\sigma-\mathsf{C})^{-1}\mathsf{B})(x)\mid x\rangle\\ &= \langle\mathsf{A}(x)\mid x\rangle + \langle((\sigma-\mathsf{C})^{-1}\mathsf{B})(x)\mid \mathsf{B}(x)\rangle\\ &= \langle\mathsf{A}(x)\mid x\rangle + \mathsf{B}(x)^*\cdot(\sigma-\mathsf{C})^{-1}\cdot\mathsf{B}(x).\end{aligned}$$ ⊣

THEOREM 3.5 (Winning). *The normal form of a bipartite invertible closed system is bipartite. Furthermore, if the system is winning, so is its normal form.*

PROOF. Assume that $\mathcal{R}=\mathcal{R}^+\oplus\mathcal{R}^-$; we must show that, for $x\in\mathcal{R}^+$ (resp. $x\in\mathcal{R}^-$) $\langle\sigma[\![\mathsf{h}]\!](x)\mid x\rangle\geq 0$ (resp. $\langle\sigma[\![\mathsf{h}]\!](x)\mid x\rangle\leq 0$). Let us treat the case where $x\in\mathcal{R}^+$, and let us assume that $\|x\|\leq 1$. By Proposition 9, we can reduce the problem to the case of $\mathsf{h}_x$, i.e., to the case where $\mathcal{H}=\mathbb{C}\oplus\mathcal{S}$, the component $\mathbb{C}$ being declared positive. Let us enlarge[9] $\mathcal{H}$ into $\mathcal{K}:=\mathbb{C}\oplus\mathbb{C}\oplus\mathcal{H}$, the first $\mathbb{C}$ being positive, the second one being negative. Then, starting with the bipartite $\mathsf{h}=\begin{bmatrix} a & y^* \\ y & \mathsf{C}\end{bmatrix}$ (in particular $a\geq 0$), consider

$$\mathsf{k}=\begin{bmatrix} 0 & 1 & 0 & 0\\ 1 & 0 & 0 & 0\\ 0 & 0 & a & y^*\\ 0 & 0 & y & \mathsf{C}\end{bmatrix},$$

still bipartite. If $\tau=\begin{bmatrix} 0&0&0&0\\0&0&1&0\\0&1&0&0\\0&0&0&0\end{bmatrix}$, then $\sigma,\tau$ are independent and $(\mathcal{K},\mathsf{k},\sigma+\tau)$ is invertible; indeed $\tau[\![\sigma[\![\mathsf{k}]\!]]\!]=\sigma[\![\mathsf{h}]\!]$. Let

$$\mathsf{g}=\begin{bmatrix} 0 & 1 & 0\\ 1 & a & y^*\\ 0 & -y & \sigma-\mathsf{C}\end{bmatrix},\qquad\text{and}\qquad \mathsf{g}^{-1}=\begin{bmatrix} b & \cdot & \cdot\\ \cdot & \cdot & \cdot\\ \cdot & \cdot & \cdot\end{bmatrix},$$

then $\sigma[\![\mathsf{h}]\!]=\tau[\![\sigma[\![\mathsf{k}]\!]]\!]=(\sigma+\tau)[\![\mathsf{k}]\!]=b$; $-\mathsf{g}$ (and its inverse $-\mathsf{g}^{-1}$, by Proposition 8) is bipartite. $b$ is located in the second copy of $\mathbb{C}$, declared negative, so $-b\leq 0$ and $\sigma[\![\mathsf{h}]\!]\geq 0$. ⊣

COROLLARY 2. *If a bipartite system is winning, so is its normal form.*

PROOF. We observed (Proposition 1) that winning is the same as remaining bipartite when $\mathcal{H}^+$ and $\mathcal{H}^-$ are swapped. If $(\mathcal{H},\mathsf{h},\sigma)$ is winning, then its normal form is bipartite w.r.t. the decomposition $\mathcal{R}^+,\mathcal{R}^-$, but also w.r.t. the opposite decomposition, hence it is winning. ⊣

**§4. Termination.** This section is concerned with the "natural" solution of the feedback equation (1). The subsection on deadlocks is devoted to the unicity of the "invisible" component $y$: essentially, we can assume unicity, i.e., "remove deadlocks". The subsection on computational size makes use of

[9] Another application of the Tortoise Principle, see Appendix B.

the norm of the *execution operator* and proves an inequality corresponding to associativity.

**4.1. Deadlocks.** We use the notations of Section 3.2: w.r.t. the decomposition $\mathcal{H} = \mathcal{R} \oplus \mathcal{S}$, $\mathsf{h} = \left[\begin{smallmatrix} \mathsf{A} & \mathsf{B}^* \\ \mathsf{B} & \mathsf{C} \end{smallmatrix}\right]$.

DEFINITION 11 (Deadlocks). The cut-system $(\mathcal{H}, \mathsf{h}, \sigma)$ is *deadlock-free* when $\sigma - \mathsf{C}$ is *injective* as an endomorphism of $\mathcal{S} = \sigma^2$.

REMARK 8. The notion of *deadlock-free* algorithm of [10] ($\sigma\mathsf{C}$ weakly nilpotent) is less general.

Invertible systems are deadlock-free, but the converse is not true: in infinite dimension, injectivity does not imply invertibility. But there is a relation between the two notions:

**Invertible:** Means the absence of a *approximate* eigenvectors (for the value 0), $x_n \in \mathcal{S}$, with $\|x_n\| = 1$, and $\sigma(x_n) - \mathsf{h}(x_n) \to 0$; we know that $\mathsf{h}$ can be replaced with $\mathsf{C} : \sigma(x_n) - \mathsf{C}(x_n) \to 0$.

**Deadlock-free:** No longer the approximate notion, the exact one: $x_n$ constant. In particular, $\sigma - \mathsf{C}$ is injective on $\mathcal{S}$ iff $\sigma - \mathsf{h}$ is injective on the same $\mathcal{S}$.

DEFINITION 12 (Deadlocks). The *deadlock* space $\mathcal{Z} \subset \mathcal{S}$ is defined by $\mathcal{Z} := \ker(\sigma - \mathsf{C})$, and its (non-zero) elements are called *deadlocks*. To be deadlock-free therefore means that there is no deadlock.

PROPOSITION 10. *The deadlock space $\mathcal{Z}$ (or rather the associated orthoprojection) commutes to $\sigma$, $\mathsf{C}$, $\mathsf{h}$.*

PROOF. The proof is a sort of simplified version of Proposition 5. If $z \in \mathcal{Z}$, then $\mathsf{C}(z) = \sigma(z)$ and $\|\mathsf{C}(z)\| = \|\sigma(z)\| = \|z\|$; since $\|\mathsf{C}\| \leq 1$, $\mathsf{C}^2(z) = z$ (use Lemma 1 of Section 3.1) and let us consider $x := z + \mathsf{C}(z)$, $y := z - \mathsf{C}(z)$. Then $\mathsf{C}(x) = x$, but also $\sigma(x) = \sigma(z) + \sigma\mathsf{C}(z) = \mathsf{C}(z) + \sigma^2(z) = x$; in the same way, $\mathsf{C}(y) = \sigma(y) = -y$. $\mathcal{Z}$ therefore appears as the sum of two orthogonal subspaces, $\mathcal{Z}^+ = \{x; \mathsf{C}(x) = \sigma(x) = x\}$ and $\mathcal{Z}^- = \{y; \mathsf{C}(y) = \sigma(y) = -y\}$. From this $\mathcal{Z}$ commutes to both of $\sigma$ and $\mathsf{C}$. Finally, if $z \in \mathcal{Z}$, $\mathsf{h}(z) = \mathsf{B}^*(z) \oplus \mathsf{C}(z)$ and $\|\mathsf{C}(z)\| = \|z\|$ together with $\|\mathsf{h}(z)\| \leq \|z\|$ force $\mathsf{B}^*(z) = 0$. This shows that $\mathsf{B}^*\mathcal{Z} = 0$ and $\mathcal{Z}$ commutes with $\mathsf{h}$. ⊣

REMARK 9. The previous proposition explains why the kernel of $\sigma - \mathsf{C}$ (indeed, of $\sigma - \mathsf{h}$) is styled the *deadlock space*: nothing enters, nothing exits. Deadlocks are excluded from logical systems and even from $\lambda$-calculi[10]. But they are rather friendly: don't bother them, they don't bother you. In particular, we can always "remove" a deadlock by replacing $\mathsf{h}$ with $(\mathsf{I} - \mathcal{Z})\mathsf{h}$ without any essential prejudice to the feedback equation.

[10] In proof-nets they correspond to *short trips*, or cycles in the Danos-Regnier criterion [5].

If $0 \in \mathrm{sp}(\sigma - \mathsf{C})$[11] we must be cautious: in case 0 is isolated, this corresponds to a deadlock that we can easily ignore. The situation is quite different when 0 is an accumulation point of $\mathrm{sp}(\sigma - \mathsf{C})$[12]. We are no longer dealing with deadlocks, but with infinite computations, and things are not that easy!

**4.2. Computational size.** Let $(\mathcal{H}, \mathsf{h}, \sigma)$ be a cut-system and let $\mathcal{Z}$ be its deadlock space. Two different solutions of the feedback equation (1) $x' \oplus y$, $x'' \oplus y'$ w.r.t. the same input $x$ are such that $x' = x''$: from

$$\mathsf{h}(0 \oplus y - y') = x' - x'' \oplus \sigma(y - y') \tag{9}$$

and $\|h\| \leq 1$, $\|\sigma(y - y')\| = \|y - y'\|$, we get $\|x' - x''\| = 0$, hence $x' = x''$[13]. But obviously two different solutions $x' \oplus y$, $x', y'$ can still differ on their invisible parts $y, y'$, in which case $y - y'$ is a deadlock.

Using Proposition 10, we easily obtain a sort of "unicity":

(i) If $x' \oplus y$ is a solution to the feedback equation w.r.t. the input $x$, so is $x' \oplus \mathcal{Z}(y)$.

(ii) $x' \oplus \mathcal{Z}(y)$ is — among all solutions $x' \oplus y'$ corresponding to the input $x$ — the one of smallest norm.

(iii) This smallest choice amounts at "removing deadlocks", i.e., at replacing the system $(\mathcal{H}, \mathsf{h}, \sigma)$ with $(\mathcal{H}, (\mathsf{I} - \mathcal{Z})\mathsf{h}, \sigma)$.

DEFINITION 13 (Termination). The cut-system $(\mathcal{H}, \mathsf{h}, \sigma)$ is *terminating* when the feedback equation (1) has a solution $x' \oplus y$ for every input $x \in \mathcal{R}$. In which case we define the *execution* operator $\mathrm{ex}(\mathsf{h}, \sigma)$ as the bounded operator from $\mathcal{R}$ to $\mathcal{H}$ assigning to $x \in \mathcal{R}$ the vector $x \oplus \mathcal{Z}(y)$[14].

REMARK 10. The typical terminating case is invertibility; in which case (with the notations of Definition 8), $\mathrm{ex}(\mathsf{h}, \sigma) = \mathcal{R} \oplus (\sigma - \mathsf{C})^{-1}\mathsf{B}$. But there are many other cases of termination, for instance when $\mathsf{B} = 0$.

DEFINITION 14 (Computational Size). If $(\mathcal{H}, \mathsf{h}, \sigma)$ is terminating, its *computational size* is defined by

$$\mathrm{size}(\mathsf{h}, \sigma) := \|\mathrm{ex}(\mathsf{h}, \sigma)\|. \tag{10}$$

Termination may be a rather accidental property; this is why the following theorem is less powerful than its original model, Theorem 3.3. In what follows, we use the (undefined) notation $\tau[\![\mathsf{h}]\!]$ in the obvious sense; but it might be inconsistent with the general definition given later.

---

[11] Seen as an operator on $\mathcal{S}$.

[12] Upper-semi-invertibility means that 0 is not limit of strictly positive points of the spectrum.

[13] This was first noticed during the proof of the normal form Theorem 3.2, when showing that the normal form is of norm at most 1.

[14] This is not a misprint, I didn't mean $x' \oplus \mathcal{Z}(y)$, see the proof of Theorem 4.1 below.

**Theorem 4.1** (Associativity of Size). *Assume that $\sigma, \tau$ are independent, and that $(\mathcal{H}, \mathsf{h}, \tau)$ and $(\mathcal{R} \oplus \mathcal{S}, \tau[\![\mathsf{h}]\!], \sigma)$ are terminating; then $(\mathcal{H}, \mathsf{h}, \sigma + \tau)$ is terminating too. Moreover*:

$$\text{(11)} \qquad \mathrm{size}(\mathsf{h}, \sigma + \tau) \leq \mathrm{size}(\tau[\![\mathsf{h}]\!], \sigma) \cdot \mathrm{size}(\mathsf{h}, \tau).$$

Proof. First assume all systems deadlock-free. We reproduce the argument in the beginning of the proof of Theorem 3.3. Rewrite equation (7) as:

$$\text{(12)} \qquad \mathsf{h}(x \oplus y' \oplus \psi(x \oplus y')) = x' \oplus y'' \oplus \tau(\psi(x \oplus y'))$$

and equation (8) as:

$$\text{(13)} \qquad \tau[\![\mathsf{h}]\!](x \oplus \varphi(x)) = x' \oplus \sigma(\varphi(x)).$$

Both can be combined to yield:

$$\text{(14)} \qquad \mathsf{h}(x \oplus \varphi(x) \oplus \psi(x \oplus \varphi(x))) = x' \oplus \sigma(\varphi(x)) \oplus \tau(\psi(x \oplus \varphi(x))).$$

Now observe that: $x \oplus \varphi(x) = \mathrm{ex}(\tau[\![\mathsf{h}]\!], \sigma)(x)$, $x \oplus y \oplus \psi(x \oplus y) = \mathrm{ex}(\mathsf{h}, \tau)(x)$, $x \oplus \varphi(x) \oplus \psi(x \oplus \varphi(x)) = \mathrm{ex}(\mathsf{h}, \sigma + \tau)(x)$. Hence:

$$\text{(15)} \qquad \mathrm{ex}(\mathsf{h}, \sigma + \tau) = \mathrm{ex}(\mathsf{h}, \tau) \cdot \mathrm{ex}(\tau[\![\mathsf{h}]\!], \sigma)$$

and from this (11) follows.

In general, "remove the deadlocks" in $(\mathcal{H}, \mathsf{h}, \tau)$ and $(\mathcal{R} \oplus \mathcal{S}, \tau[\![\mathsf{h}]\!], \sigma)$. Then we get an equality close to (15), but for the point that we are not sure that the left hand side corresponds to the smallest solution. Anyway the inequality (11) still holds. ⊣

Remark 11. Our definition of size comes from the naive power series expansion of the solution of (1):

$$\text{(16)} \qquad \sigma[\![\mathsf{h}]\!] = \mathcal{R}(\mathsf{h} + \mathsf{h}\sigma\mathsf{h} + \mathsf{h}\sigma\mathsf{h}\sigma\mathsf{h} + \cdots)\mathcal{R}$$

a formula which is for instance correct when $\sigma\mathsf{h}$ is nilpotent:

$$\text{(17)} \quad \mathrm{ex}(\mathsf{h}, \sigma) = \mathcal{R} + \mathsf{B}^*\sigma\mathsf{B} + \mathsf{B}^*\sigma\mathsf{C}\sigma\mathsf{B} + \mathsf{B}^*\sigma\mathsf{C}\sigma\mathsf{C}\sigma\mathsf{B} + \cdots + \mathsf{B}^*\sigma(\mathsf{C}\sigma)^{n-2}\mathsf{B}$$

where $n$ is the greatest integer such that $(\sigma\mathsf{C})^n \neq 0$ (the "order of nilpotency" of $\sigma\mathsf{C}$). Then $\mathrm{size}(\mathsf{h}, \sigma) \leq n$; in practice, e.g., for a converging normalisation in — say — system $\mathbb{F}$, $\mathrm{size}(\mathsf{h}, \sigma) \sim \sqrt{n}$.

It might be more natural to replace the size with its logarithm: Danos observed in his Thesis (see, e.g., [6]) that, in $\lambda$-calculus, this "order of nilpotency" is indeed exponential in the number of reduction steps needed to normalise the term interpreted by $(\mathcal{H}, \mathsf{h}, \sigma)$. With such an alternative definition, Theorem 4.1 would involve a sum instead of a product.

## §5. Semi-invertible case.

**5.1. Order approximations.** When we speak of "the l.u.b. of a net", we implicitly assume that we are speaking of directed increasing net (with non-empty index set); symmetrically for "the g.l.b. of a net". Observe that l.s.i. are closed under l.u.b., u.s.i. are closed under g.l.b.: this is immediate from Proposition 6(i).

PROPOSITION 11. *Invertible systems are dense w.r.t. the pointwise order. More precisely:*

(i) *Every system is the g.l.b. of a net of l.s.i. systems.*
(ii) *Every system is the l.u.b. of a net of u.s.i. systems.*
(iii) *Every u.s.i. system is the g.l.b. of a net of invertible systems.*
(iv) *Every l.s.i. system is the l.u.b. of a net of invertible systems.*

PROOF. (i) For $0 < \mu < 1$, let $h^\mu := \mu h + (1-\mu)I$; since $h \leq I$, and $(\mathcal{H}, I, \sigma)$ is l.s.i., the convex combinations $h^\mu$ is l.s.i., by Proposition 6(ii). The $h^\mu$ form a decreasing net, with g.l.b. h.
(ii) Symmetrical: use the increasing net $h_\lambda := \lambda h + (\lambda - 1)I$.
(iii) If h is assumed to be u.s.i., then the $h^\mu$ are invertible, by Proposition 6(i) or (iii); and h is the g.l.b. of a net of invertible systems.
(iv) Symmetrical. ⊣

**5.2. Lower-semi-invertible case.**

THEOREM 5.1 (Normal Form). *There exists a unique extension of the normal form to l.s.i. systems which commutes with least upper bounds. This extension is order-monotonous; furthermore, the theorems established in the invertible case, and styled "associativity", "winning", "stability", still hold.*

PROOF. With the notations of Proposition 11, define:

$$\sigma[\![h]\!] := \sup_\lambda \sigma[\![h_\lambda]\!]. \tag{18}$$

Since the normal form is order-continuous in the invertible case, equation (18) holds in this case, hence our definition extends the original one. Moreover, this extension is monotonous: if $h \leq k$, then $h_\lambda \leq k_\lambda$, and $\sup_\lambda \sigma[\![h_\lambda]\!] \leq \sup_\lambda \sigma[\![k_\lambda]\!]$. The next point is that this definition commutes with suprema: if $h = \sup_i h[i]$, then $\sigma[\![h]\!] = \sup_\lambda \sigma[\![h_\lambda]\!] = \sup_i \sup_\lambda \sigma[\![h[i]_\lambda]\!] = \sup_\lambda \sup_i \sigma[\![h[i]_\lambda]\!] = \sup_i \sigma[\![h[i]]\!]$: besides a triviality on double suprema, one uses the order continuity of the normal form in the invertible case, i.e., that $\sup_i \sigma[\![h[i]_\lambda]\!] = \sigma[\![h_\lambda]\!]$. Unicity is a trivial consequence of commutation to suprema. Let us now check the extension of the main theorems:

**Associativity:** If $(\mathcal{H}, h, \sigma + \tau)$ is l.s.i., then the $(\mathcal{H}, h_\lambda, \sigma + \tau)$ are invertible. From this, $\tau[\![h]\!] = \sup_\lambda \tau[\![h_\lambda]\!]$ is l.s.i., and $(\mathcal{R} \oplus \mathcal{S}, \tau[\![h]\!], \sigma)$ is l.s.i., as supremum of l.s.i. systems. Finally,

$$\sigma[\![\tau[\![h]\!]]\!] = \sup_\lambda \sigma[\![\tau[\![h_\lambda]\!]]\!] = \sup_\lambda (\sigma + \tau)[\![h_\lambda]\!] = (\sigma + \tau)[\![h]\!].$$

Conversely, assume that $\tau[\![h]\!]$ is l.s.i., but $(\mathcal{H}, h, \sigma + \tau)$ is not l.s.i.; then some $(\mathcal{H}, h_\lambda, \sigma + \tau)$ is not invertible, but, since $\tau[\![h_\lambda]\!]$ is invertible, Theorem 3.3 shows that $(\mathcal{R} \oplus \mathcal{S}, \tau[\![h_\lambda]\!], \sigma)$ is not invertible. Since

$h_\lambda \leq \lambda(\mathcal{R}+\mathcal{S})$, we get $\tau[\![h_\lambda]\!] \leq \tau[\![\lambda I]\!] \leq \lambda I$, hence $(\mathcal{R}\oplus\mathcal{S}, \tau[\![h_\lambda]\!], \sigma)$ is u.s.i. by Proposition 6(i), and therefore not l.s.i.

**Winning:** If h is bipartite (resp. winning), so are the $(h_\lambda)$, and so are the $\sigma[\![h_\lambda]\!]$, and their supremum $\sigma[\![h]\!]$.

**Stability:** Proved in the same tautological way, relying on a property that I like to state independently, see next proposition. ⊣

PROPOSITION 12. *Assume that $(\mathcal{H}, h, \sigma) \sqsubset (\mathcal{H}, k, \sigma)$, and that $(\mathcal{H}, k, \sigma)$ is l.s.i.; then $(\mathcal{H}, h, \sigma)$ is l.s.i. too.*

PROOF. Assume that $h = k\mathcal{E}$ and $\mathcal{E}\sigma = \sigma\mathcal{E}$; then $\mathcal{E}h = k\mathcal{E}$. If $h(x_n)+x_n \to 0$, with $\sigma(x_n) = -x_n$, and $\|x_n\| = 1$, then $k(\mathcal{E}(x_n)) + \mathcal{E}(x_n) = \mathcal{E}(h(x_n)+x_n) \to 0$, and $\sigma(\mathcal{E}(x_n)) = \mathcal{E}(\sigma(x_n)) = -\mathcal{E}(x_n)$. Moreover, since $\|k\mathcal{E}(x_n)\| = \|h(x_n)\| \to 1$ and $\|k\| \leq 1$, we conclude that $\|\mathcal{E}(x_n)\| \to 1$. The sequence $\mathcal{E}(x_n)$ is an approximate eigenvector sequence for $-\sigma^- - k$. ⊣

**5.3. Semi-invertible case.** Of course, the normal form can be symmetrically extended to u.s.i. systems.

THEOREM 5.2 (Normal Form). *There exists a unique extension of the normal form to u.s.i. systems which commutes with greatest lower bounds. This extension is order-monotonous; furthermore, the theorems established in the invertible case, and styled "associativity", "stability", "winning", still hold.*

PROOF. With the notations of Proposition 11, we define:

$$\sigma[\![h]\!] := \inf_{\mu} \sigma[\![h^{\mu}]\!] \tag{19}$$

etc. ⊣

We have therefore two extensions of the normal form; these extensions are consistent, since there is no conflict (the intersection of the two domains, l.s.i. and u.s.i., consists of invertibles).

PROPOSITION 13. *The normal form is monotonous.*

PROOF. $h \leq k$, we must show that $\sigma[\![h]\!] \leq \sigma[\![k]\!]$. This is already taken care of when h, k are "on the same side". In view of Proposition 6(i), it remains to consider the case where h is u.s.i. and k is l.s.i. But then, the interpolant $g := 1/2(h+k)$ is invertible by Proposition 6(iii): hence

$$\sigma[\![h]\!] \leq \sigma[\![g]\!] \leq \sigma[\![h]\!].$$ ⊣

No clear order-continuity can be stated, this is due to the fact that one side commutes to sups, the other to infs[15]. But order order-continuity can still be used as a tool. For instance, one can prove the following:

PROPOSITION 14.

$$(-\sigma)[\![-h]\!] = -\sigma[\![h]\!]. \tag{20}$$

[15]And this result is optimal, see Section 6.6.

PROOF. Immediate: first checked in the invertible case, then extended to the — say — l.s.i. case, using an (increasing) order-continuity argument. ⊣

Stability remains, essentially because there is no conflict l.s.i./u.s.i., the same for winning. But *associativity* is problematic. This is the central problem and also the key to the general case, hence let us keep this point for the ultimate section.

## §6. General case.

### 6.1. Lopsided feedbacks.

DEFINITION 15 (Lopsided Feedbacks). A feedback $\sigma$ is *lopsided* iff it is either a projection, i.e., $\sigma^2 = \sigma$ (*positive* feedback), or the opposite of a projection, i.e., $\sigma^2 = -\sigma$ (*negative* feedback).

Roughly speaking, the three following words socialise: *positive* feedbacks, *l.s.i.* nets, and *l.u.b.*. If $\sigma$ is positive, then all cut-systems $(\mathcal{H}, \mathsf{h}, \sigma)$ are l.s.i., in particular, we have a very satisfactory notion of *normal form*, expressed by Theorem 5.1; therefore any system is the l.u.b. of a (directed, increasing) net of invertibles. The socialisation of the cocktail "positive + l.s.i. + l.u.b." is expressed by the equation:

$$\sigma[\![\sup_i \mathsf{h}_i]\!] = \sup_i \sigma[\![\mathsf{h}_i]\!] \tag{21}$$

which holds for any (increasing, directed) net.

Symmetrically, the words *negative*, u.s.i., g.l.b. associate well. In case of a negative feedback, all systems are u.s.i., etc. and

$$\sigma[\![\inf_i \mathsf{h}_i]\!] = \inf_i \sigma[\![\mathsf{h}_i]\!]. \tag{22}$$

**6.2. Associativity: a first attempt.** Let us come back to the semi-invertible case, Section 5.3. If $(\mathcal{H}, \mathsf{h}, \sigma + \tau)$ is semi-invertible, it is — say —, l.s.i.; then both of $(\mathcal{H}, \mathsf{h}, \tau)$ and $(\mathcal{R} + \mathcal{S}, \tau[\![\mathsf{h}]\!], \sigma)$ are l.s.i., and equation (5) holds. The converse is less obvious, since $(\mathcal{H}, \mathsf{h}, \sigma + \tau)$ and $(\mathcal{R} + \mathcal{S}, \tau[\![\mathsf{h}]\!], \sigma)$ may be respectively u.s.i. and l.s.i.

This case occurs naturally when $\sigma = \pi$ ($\pi$ for "positive", "projection"), $\tau = \nu$ ($\nu$ for "negative"). Every system $(\mathcal{H}, \mathsf{h}, \nu)$ is u.s.i., every system $(\mathcal{R} \oplus \mathcal{S}, \mathsf{k}, \pi)$ is l.s.i., hence one can form $\pi[\![\nu[\![\mathsf{h}]\!]]\!]$, and for symmetrical reasons, $\nu[\![\pi[\![\mathsf{h}]\!]]\!]$. Without any hypothesis on h, $(\pi + \nu)[\![\mathsf{h}]\!]$ does not make sense, but, with $\pi := \sigma^+$, $\nu := -\sigma^-$, $\pi[\![\nu[\![\mathsf{h}]\!]]\!]$ and $\nu[\![\pi[\![\mathsf{h}]\!]]\!]$ yield two candidates for the normal form $\sigma[\![\mathsf{h}]\!]$. To sum up:

(i) Associativity is the key to the general case.
(ii) It is enough to investigate associativity in the case of *lopsided* feedbacks.

Obviously:

(i) When the feedbacks are positive, everybody is l.c.i. and associativity holds: $\pi'[\![\pi[\![\mathsf{h}]\!]]\!] = \pi[\![\pi'[\![\mathsf{h}]\!]]\!]$ without any hypothesis on h.

(ii) Symmetrically when the feedbacks are negative: $\nu'[\![\nu[\![\mathsf{h}]\!]]\!] = \nu[\![\nu'[\![\mathsf{h}]\!]]\!]$.
(iii) When the feedbacks are of different sign, we know that associativity works under the strong hypothesis that $(\mathcal{H}, \mathsf{h}, \pi + \nu)$ is semi-invertible. Without hypothesis on $\mathsf{h}$, we can prove an inequality.

LEMMA 3. *If $\pi$ is positive, then $\pi[\![\mathsf{h}]\!] = \inf_\mu \pi[\![\mathsf{h}^\mu]\!]$.*

PROOF. W.r.t. the decomposition $\mathcal{H} = \mathcal{R} \oplus \pi$, let $\mathsf{h} = \begin{bmatrix} \mathsf{A} & \mathsf{B}^* \\ \mathsf{B} & \mathsf{C} \end{bmatrix}$. Assuming $(\mathcal{H}, \mathsf{h}, \pi)$ invertible, $\pi[\![\mathsf{h}^\mu]\!] = \mu\mathsf{A} + \mu^2\mathsf{B}^*(\pi - \mu\mathsf{C} - (1-\mu)\pi)^{-1}\mathsf{B} + (1-\mu)\mathcal{R}$. Observing that $\pi - \mu\mathsf{C} - (1-\mu)\pi = \mu(\pi - C)$, we eventually get:

$$\pi[\![\mathsf{h}^\mu]\!] = \mu\pi[\![\mathsf{h}]\!] + (1-\mu)\mathcal{R} = \pi[\![\mathsf{h}]\!]^\mu \tag{23}$$

Since both sides of equation (23) commute to l.u.b., we conclude that the equation holds for arbitrary $\mathsf{h}$. From this one easily concludes. ⊣

LEMMA 4. *If $\nu$ is negative, then $\nu[\![\mathsf{h}]\!] = \sup_\lambda \nu[\![\mathsf{h}_\lambda]\!]$.*

PROOF. Symmetrical. ⊣

PROPOSITION 15. *If the independent feedbacks $\pi, \nu$ are respectively positive and negative and $(\mathcal{H}, \mathsf{h}, \pi + \nu)$ is a cut-system, then:*

$$\pi[\![\nu[\![\mathsf{h}]\!]]\!] \leq \nu[\![\pi[\![\mathsf{h}]\!]]\!]. \tag{24}$$

PROOF. For the semi-invertible $\mathsf{h}_\lambda, \mathsf{h}^\nu$, associativity holds. Then

$$\pi[\![\nu[\![\mathsf{h}]\!]]\!] = \sup_\lambda(\pi + \nu)[\![\mathsf{h}_\lambda]\!],$$

essentially by Lemma 4. Symmetrically, Lemma 3 yields $\nu[\![\pi[\![\mathsf{h}^\mu]\!]]\!] = \nu[\![\pi[\![\mathsf{h}]\!]]\!]$. Using $\mathsf{h}_\lambda \leq \mathsf{h} \leq \mathsf{h}^\mu$, we eventually get (24). ⊣

But there is no hope to prove equality following this pattern. This the ultimate point one can reach by "continuity" techniques.

**6.3. Positive feedbacks.** In what follows, $\pi$ is a positive feedback, and

$$\mathsf{h} = \begin{bmatrix} \mathsf{A} & \mathsf{B}^* \\ \mathsf{B} & \mathsf{C} \end{bmatrix}$$

w.r.t. the direct sum decomposition $\mathcal{H} = \mathcal{R} \oplus \pi$.

THEOREM 6.1 (Minimality). *$\pi[\![\mathsf{h}]\!] - \mathsf{A}$ is the smallest operator $\mathsf{A}'$ of $B(\mathcal{R})$ such that*

$$\begin{bmatrix} \mathsf{A}' & -\mathsf{B}^* \\ -\mathsf{B} & \pi - \mathsf{C} \end{bmatrix} \geq 0. \tag{25}$$

PROOF. We first assume $(\mathcal{H}, \mathsf{h}, \pi)$ invertible; let $\mathsf{k}$ be the block corresponding to the choice $\mathsf{A}' = \mathsf{B}^*(\pi - \mathsf{C})^{-1}\mathsf{B}$ in equation (25). Then

$$\begin{aligned} \langle \mathsf{k}(x \oplus y) \mid x \oplus y \rangle = \langle (\mathsf{B}^*(\pi - \mathsf{C})^{-1}\mathsf{B})(x) \mid x \rangle - \langle \mathsf{B}(x) \mid y \rangle \\ - \langle \mathsf{B}^*(y) \mid x \rangle + \langle (\pi - \mathsf{C})(y) \mid y \rangle. \end{aligned} \tag{26}$$

Let $x' := (\pi - \mathsf{C})^{-1/2}\mathsf{B}(x)$, $y' := \sqrt{\pi - C}(y)$. Then the right-hand side of (26) rewrites as $\langle x' \mid x' \rangle - \langle y' \mid x' \rangle - \langle x' \mid y' \rangle + \langle y' \mid y' \rangle = \langle x' - y' \mid x' - y' \rangle$, and is therefore positive. This expression vanishes for $x' = y'$, and this shows that the first term $\langle x' \mid x' \rangle = \langle \mathsf{B}^*(\pi - \mathsf{C})^{-1}\mathsf{B}(x) \mid x \rangle$ actually takes *the* minimum possible value making (26) positive.

Let us quickly conclude: if $\pi - \mathsf{C}$ is not invertible, then $\pi - \mathsf{C} = \inf_\lambda(\pi - \mathsf{C}_\lambda)$. In (26), change C into $\mathsf{C}_\lambda$, then $\sigma[\![\mathsf{h}]\!]$ remains equation; solution of the since $\sigma[\![\mathsf{h}]\!] = \sup_\lambda \sigma[\![\mathsf{h}_\lambda]\!]$, it is indeed the smallest solution working for *all* $\lambda$, i.e., working for C instead of the $\mathsf{C}_\lambda$. ⊣

REMARK 12. Since $\mathsf{h} \leq \mathrm{I}$, $\mathsf{A}' = \mathcal{R} - \mathsf{A}$ enjoys (25); from this we get $\pi[\![\mathsf{h}]\!] \leq \mathcal{R}$, consistently with $\|\pi[\![\mathsf{h}]\!]\| \leq 1$.

The next result relies on the folklore of operators see Appendix C.8.

THEOREM 6.2 (Resolution). $\pi[\![\mathsf{h}]\!] - \mathsf{A} = \psi^*\psi$, *where* $\psi$ *is uniquely determined by the conditions*:

$$\sqrt{\pi - \mathsf{C}} \cdot \psi = \mathsf{B} \tag{27}$$
$$\mathrm{dom}(\pi - \mathsf{C}) \cdot \psi = \psi.$$

PROOF. Assume that $\sqrt{\pi - \mathsf{C}} \cdot \psi = \mathsf{B}$, and let $\mathsf{A}' := \psi^*\psi$; then equation (25) holds: in the computation of (26), replace $(\pi - \mathsf{C})^{-1/2}\mathsf{B}$ with $\psi$. The expression vanishes for $x' = y'$, i.e., $\psi(x) = \sqrt{\pi - \mathsf{C}}(y)$. This is possible only if $\psi$ has its range included in $\mathrm{dom}(\sqrt{\pi - \mathsf{C}})$, equivalently, see remark 17 in Appendix C.8, only if $\mathrm{dom}(\pi - \mathsf{C}) \cdot \psi = \psi$.

It remains to show the existence $\psi$; in view of Proposition 19 (and remark 17), it is enough to show that $\mathsf{BB}^* \leq \mathsf{k} \cdot (\pi - \mathsf{C})$, for a certain real $k$. Let $x \oplus y \in \mathcal{H}$; since $\mathsf{h} \leq \mathrm{I}$:

$$\begin{aligned}\langle (\mathrm{I} - \mathsf{h})(x \oplus y) \mid x \oplus y \rangle = {} & \langle (\mathcal{R} - \mathsf{A})(x) \mid x \rangle + \langle \mathsf{B}(x) \mid y \rangle \\ & + \langle \mathsf{B}^*(x) \mid y \rangle + \langle (\pi - \mathsf{C})(y) \mid y \rangle \geq 0\end{aligned}$$

Now, the same high-school technique used in the proof of the Cauchy-Schwarz inequality yields $\|\langle \mathsf{B}(x) \mid y \rangle\|^2 \leq \langle (\mathcal{R} - \mathsf{A})(x) \mid x \rangle \cdot \langle (\pi - \mathsf{C})(y) \mid y \rangle$. Taking $x := \mathsf{B}^*(y)$, we get $\langle \mathsf{BB}^*(\mathrm{y}) \mid y \rangle^2 \leq \|\pi - \mathsf{C}\| \cdot \|\mathsf{B}^*(y)\|^2 \cdot \langle (\pi - \mathsf{C})(y) \mid y \rangle$. Using the familiar $\langle \mathsf{BB}^*(\mathrm{y}) \mid y \rangle = \|\mathsf{B}^*(y)\|^2$, we eventually get:

$$\mathsf{BB}^* \leq \|\mathcal{R} - \mathsf{A}\| \cdot (\pi - \mathsf{C}). \tag{28}$$

We got our inequality, with $k = \|\mathcal{R} - \mathsf{A}\|$. ⊣

DEFINITION 16 (Resolvant). Let $(\mathcal{H}, \mathsf{h}, \pi)$ be a cut-system with a positive feedback, and let $\psi$ be as in Theorem 6.2. We define the *resolvant*

$$\mathrm{res}(\mathsf{h}, \pi) := \begin{bmatrix} \mathsf{A} & \psi^* \\ \psi & 0 \end{bmatrix}.$$

REMARK 13. It is easily shown that $\|\,\mathrm{res}(\mathsf{h}, \pi)\| = \|\mathsf{A} + \psi^*\psi\| \leq 1$.

COROLLARY 3. *The cut-system* $(\mathcal{H}, \mathsf{h}, \pi)$ *is equivalent to its resolvant*:

$$\pi[\![\mathsf{h}]\!] = \pi[\![\mathrm{res}(\mathsf{h}, \pi)]\!] \tag{29}$$

PROOF. $\pi[\![\mathrm{res}(\mathsf{h},\pi)]\!] = \mathsf{A} + \psi^*\psi$. ⊣

**6.4. Negative feedbacks.** The case of a negative feedback $\nu$ is symmetrical. One introduces the resolvant $\mathrm{res}(\mathsf{h},\nu) := -\mathrm{res}(-\mathsf{h},-\nu)$ and one checks, using Proposition 14, that:

$$\nu[\![\mathsf{h}]\!] = \nu[\![\mathrm{res}(\mathsf{h},\nu)]\!] \tag{30}$$

PROPOSITION 16. *If $\pi$ positive and $\nu$ negative are independent, then*

$$\pi[\![\nu[\![\mathsf{h}]\!]]\!] = \nu[\![\pi[\![\mathsf{h}]\!]]\!].$$

PROOF. W.r.t. the decomposition $\mathcal{H} = \mathcal{R} \oplus \pi \oplus (-\nu)$, let $\mathsf{h} = \begin{bmatrix} \mathsf{A} & \mathsf{B}^* & \mathsf{D}^* \\ \mathsf{B} & \mathsf{C} & \mathsf{E}^* \\ \mathsf{D} & \mathsf{E} & \mathsf{F} \end{bmatrix}$. Let

$$\mathsf{k}_1 := \pi[\![\nu[\![\mathsf{h}]\!]]\!] \quad \text{and} \quad \mathsf{k}_2 := \nu[\![\pi[\![\mathsf{h}]\!]]\!] : \mathsf{k}_i = \begin{bmatrix} \mathsf{A} & \mathsf{b}^*{}_i & \mathsf{d}_i{}^* \\ \mathsf{b}_i & 0 & \mathsf{e}_i{}^* \\ \mathsf{d}_i & \mathsf{e}_i & 0 \end{bmatrix}.$$

We must show that $\mathsf{k}_1 = \mathsf{k}_2$, what we do coefficientwise. All these coefficients are unique in some sense, the unicity being characterised by their range and/or domain see Definition 18 in Appendix C.8. For instance $\mathsf{b}_i$ and $\mathsf{d}_i$ are with respective ranges included in $\mathrm{dom}(\pi - \mathsf{C})$ and $\mathrm{dom}(\mathsf{F} - \nu)$. The case of $\mathsf{e}_i$ is more complex: its range is included in $\mathrm{dom}(\mathsf{F} - \nu)$, and its domain is included in $\mathrm{dom}(\pi - \mathsf{C})$. So far so good, the coefficients satisfy the same range/domain constraints.

Now let us look at the precise equalities defining our coefficients:

(i) $\sqrt{\pi - \mathsf{C}} \cdot \mathsf{b}_i = \mathsf{B}$, hence $\mathsf{b}_1 = \mathsf{b}_2$.
(ii) $\sqrt{\mathsf{F} - \nu} \cdot \mathsf{d}_i = \mathsf{D}$, hence $\mathsf{d}_1 = \mathsf{d}_2$.
(iii) $(\sqrt{\mathsf{F} - \nu} \cdot \mathsf{e}_1) \cdot \sqrt{\pi - \mathsf{C}} = \mathsf{E}$, $\sqrt{\mathsf{F} - \nu} \cdot (\mathsf{e}_2 \cdot \sqrt{\pi - \mathsf{C}}) = \mathsf{E}$; again, $\mathsf{e}_1 = \mathsf{e}_2$.

⊣

**6.5. The solution.** It is plain from the discussion of Section 6.2 that, if we can prove that lopsided feedbacks of opposite sides associate, we can define:

DEFINITION 17 (Normal Form). Let $(\mathcal{H}, \mathsf{h}, \sigma)$ be a cut-system; we define its *normal form* by:

$$\sigma[\![\mathsf{h}]\!] := (-\sigma^-)[\![\sigma^+[\![\mathsf{h}]\!]]\!] = \sigma^+[\![(-\sigma^-)[\![\mathsf{h}]\!]]\!]. \tag{31}$$

THEOREM 6.3 (Lopsided Associativity). *Assuming that $\pi$ (positive) and $\nu$ (negative) are independent*:

$$\pi[\![\nu[\![\mathsf{h}]\!]]\!] = \nu[\![\pi[\![\mathsf{h}]\!]]\!] \tag{32}$$

PROOF. $\pi[\![\nu[\![\mathsf{h}]\!]]\!] = \pi[\![\nu[\![\mathrm{res}(\mathsf{h},\nu)]\!]]\!]$; since $(\mathcal{H}, \mathrm{res}(\mathsf{h},\nu), \sigma)$ is l.s.i.,

$$\begin{aligned} \pi[\![\nu[\![\mathrm{res}(\mathsf{h},\nu)]\!]]\!] &= \nu[\![\pi[\![\mathrm{res}(\mathsf{h},\nu)]\!]]\!] \\ &= \pi[\![\nu[\![\mathrm{res}(\mathrm{res}(\mathsf{h},\nu)),\pi]\!]]\!] \\ &= \sigma[\![\mathrm{res}(\mathrm{res}(\mathsf{h},\nu)),\pi]\!]. \end{aligned}$$

In the same way, $\nu[\![\pi[\![h]\!]]\!] = \sigma[\![\mathrm{res}(\mathrm{res}(h,\pi)),\nu]\!]$. The theorem is therefore a consequence of Proposition 16. ⊣

COROLLARY 4. *With notations coming from the proof of Proposition* 16, *the normal form is given by*:

$$\sigma[\![h]\!] = A + b^*(e^*e+\pi)b - b^*e^*(ee^* - \nu)^{-1}d - d^*(ee^* - \nu)^{-1}eb - d^*(ee^* - \nu)^{-1}d. \tag{33}$$

PROOF. Basically one must inverse $\begin{bmatrix} \pi & -e^* \\ -e & \nu \end{bmatrix}$, whose square is $\begin{bmatrix} \pi+e^*e & 0 \\ 0 & -\nu+ee^* \end{bmatrix}$. From this one easily gets (33). ⊣

REMARK 14. Since $(ee^* - \nu)^{-1}e = e(e^*e+\pi)^{-1}$, we can also express the normal form by:

$$\sigma[\![h]\!] = A + b^*(e^*e+\pi)b - b^*(e^*e+\pi)^{-1}e^*d - d^*e(e^*e+\pi)^{-1}b - d^*(ee^* - \nu)^{-1}d. \tag{34}$$

THEOREM 6.4 (Full Normal Form). *The normal form* $\cdot[\![\cdot]\!]$ *is the only associative and order-monotonous extension of Definition* 8 (*the invertible case*) *commuting to l.u.b.* (*resp. g.l.b.*) *of l.s.i.* (*resp. u.s.i.*) *systems. It enjoys the analogues of the stability Theorem* 3.4 *and the winning Theorem* 3.5.

PROOF. There are obviously enough constraints to make this extension unique. As to monotonicity, and — say — commutation to l.u.b. of l.s.i. systems, use the definition $\sigma[\![h]\!] := \sigma^+[\![(-\sigma^-)[\![h]\!]]\!]$: if $(\mathcal{H}, h, \sigma)$ is l.s.i., then $(\mathcal{H}, h, -\sigma^-)$ is invertible (this is the definition). This shows that our definition extends the l.s.i. case, and we are back to Section 5.2. Associativity is almost obvious:

$$\begin{aligned}(\sigma+\tau)[\![h]\!] &= (\sigma^- - \tau^-)[\![(\sigma^+ + \tau^+)[\![h]\!]]\!] = \tau^-[\![\sigma^-[\![\tau^+[\![\sigma^+[\![h]\!]]\!]]\!]]\!] \\ &= \tau^-[\![\tau^+[\![\sigma^-[\![\sigma^+[\![h]\!]]\!]]\!]]\!] = \tau[\![\sigma[\![h]\!]]\!].\end{aligned}$$

As to stability, if $(\mathcal{H}, h, \sigma) \sqsubset (\mathcal{H}, k, \sigma)$, then there exists $\mathcal{E}$ such that $\sigma\mathcal{E} = \mathcal{E}\sigma$ and $h = k\mathcal{E}$. But then $(\sigma + \sigma^2)\mathcal{E} = \mathcal{E}(\sigma + \sigma^2)$, i.e., $\sigma^+\mathcal{E} = \mathcal{E}\sigma^+$, hence $(\mathcal{H}, h, \sigma^+) \sqsubset (\mathcal{H}, k, \sigma^+)$. From this $\sigma^+[\![h]\!] \sqsubset \sigma^+[\![k]\!]$. Since $\sigma^-\mathcal{E} = \mathcal{E}\sigma^-$, we get $(\mathcal{H}, \sigma^+[\![h]\!], \sigma^-) \sqsubset (\mathcal{H}, \sigma^+[\![k]\!], \sigma^-)$, i.e., $\sigma[\![h]\!] \sqsubset \sigma[\![k]\!]$. We skip the part on conditional infima.

Finally, we already know that bipartism is preserved by the normal form in case of a l.s.i. system. In order to conclude, it is therefore enough to remark that the normal form commutes to g.l.b. of decreasing nets of the form $h^\mu$, see Proposition 17 below. ⊣

We like to state independently the last fact mentioned in the proof: this limited amount of general order-continuity is useful!

**Proposition** 17. $\sigma[\![\mathsf{h}]\!] = \sup_\lambda \sigma[\![\mathsf{h}_\lambda]\!] = \inf_\mu \sigma[\![\mathsf{h}^\mu]\!]$.

**Proof.** While proving Proposition 15, we established the inequality

$$\sigma^+[\![\sigma^-[\![\mathsf{h}]\!]]\!] \leq \sup_\lambda \sigma[\![\mathsf{h}_\lambda]\!] \leq \inf_\mu \sigma[\![\mathsf{h}^\mu]\!] \leq \sigma^-[\![\sigma^+[\![\mathsf{h}]\!]]\!],$$

but

$$\sigma^+[\![\sigma^-[\![\mathsf{h}]\!]]\!] = \sigma[\![\mathsf{h}]\!] = \sigma^-[\![\sigma^+[\![\mathsf{h}]\!]]\!]. \qquad \dashv$$

**Remark** 15. The previous commutations *do not* ensure general order-continuity, see next section.

**6.6. Order-continuity: a counter-example.** The counterexample is obtained by successive simplifications: commutation to g.l.b. in the case of a positive feedback, indeed a hyperplane.

(i) Let $\mathcal{H} := \mathbb{C}\oplus\ell^2$; if $y = (y_i) \in \ell^2$ is of norm $1/4$, then $\|1/2\oplus -2y\|^2 = 1/2$, and $(1/2 \oplus -2y)(1/2 \oplus -2y)^* = \begin{bmatrix} 1/4 & -y^* \\ -y & 4yy^* \end{bmatrix}$ is of norm $1/2$. Then $\begin{bmatrix} 0 & y^* \\ y & -4yy^* \end{bmatrix}$ is of norm $1/2$ as well.

(ii) Let $\pi_n \subset \ell^2$, $\pi_n := \{(x_i); \forall j \geq n\ x_j = 0\}$ and let $\pi = \ell^2$. Then $\pi/2 - \pi_n/4$ and $\pi/4$ are of norm $\leq 1/2$.

(iii) If $\mathsf{C}_n := \pi/2 - \pi_n/4 - 4yy^*$, $\mathsf{C} := \pi/4 - 4yy^*$, and $\mathsf{h}_n := \begin{bmatrix} 0 & y^* \\ y & \mathsf{C}_n \end{bmatrix}$, then $(\mathsf{h}_n)$ is a decreasing net of hermitians of norm $\leq 1$, whose g.l.b. is $\mathsf{h} := \begin{bmatrix} 0 & y^* \\ y & \mathsf{C} \end{bmatrix}$.

(iv) Assume that all coefficients $y_i$ are non-zero. Then $y \notin \mathrm{rg}(\pi_n)$. From this $\sup\{a; ayy^* \leq \pi_n/4\} = 0$. Since $\|yy^*\| = 1/16$,

$$\sup\{a; ayy^* \leq \pi/2 + \pi_n/4 + 4yy^*\} = 12.$$

But $\pi/2+\pi_n/4+4yy^* = \pi - \mathsf{C}_n$, and $ayy^* \leq \pi - \mathsf{C}_n$ iff $\mathsf{h}_n := \begin{bmatrix} 1/a & y^* \\ y & \pi-\mathsf{C}_n \end{bmatrix}$ is positive. From this we get $\pi[\![\mathsf{h}_n]\!] = 1/12$.

(v) The same computation, but done for h, yields:

$$\sup\{a; ayy^* \leq 3\pi/4 + 4yy^*\} = 16.$$

From this $\pi[\![\mathsf{h}_n]\!] = 1/16$.

(vi) $\pi[\![\inf_n \mathsf{h}_n]\!] < \inf_n \pi[\![\mathsf{h}_n]\!]$, that's our counter-example.

Summing up:

**Theorem** 6.5 (Order-Discontinuity). *The normal form $\sigma[\![\cdot]\!]$ is not order-continuous.*

**Remark** 16. Remark that the $\mathsf{C}_n$ do not commute. If they were commuting, the commutation to infima would reduce to pointwise computation w.r.t. a basis — may be "continuous", in which all the $\mathsf{C}_n$ are "diagonal".

## Appendix A. Geometry of interaction.

**A.1. Cut-elimination.** The essential[16] principle of reasoning is the use of *lemmas*: in order to prove $B$, first prove it under the hypothesis $A$, then prove (the lemma) $A$; this is expressed by *Modus Ponens*:

$$\frac{A \quad A \Rightarrow B}{B} \tag{35}$$

reformulated by Gentzen in his *sequent calculus*; a *sequent* $\Gamma \vdash \Delta$ consists of two finite sequences $\Gamma = A_1, \ldots, A_m$ and $\Delta = B_1, \ldots, B_n$ of formulas separated by the "turnstile" $\vdash$, with the intended meaning that the conjunction of the $A_i$ implies the disjunction of the $B_j$. Sequent calculus is organised along the "rule you love to hate", the *cut-rule*:

$$\frac{\Gamma \vdash \Delta, A \quad A, \Gamma' \vdash \Delta'}{\Gamma, \Gamma' \vdash \Delta, \Delta'} \tag{36}$$

which contains *Modus Ponens* as the particular case $\Gamma = \Delta = \Gamma' = \emptyset$, $\Delta' = B$. The cut-rule emphasises the fact that $A$ occurs twice, once negatively (a hypothesis) once positively (a conclusion). These two occurrences "cancel" each other; later on, this cancellation will be rendered by a partial symmetry $\sigma$ swapping the two occurrences of $A$.

This cancellation is virtual; however there is a process known as *cut-elimination*[17] which allows one to replace — under certain hypotheses — a proof with cuts with one without cuts, a *cut-free* proof. This process is long and tedious, and uses many "cooking recipes". A proof without cuts, without lemmas, is usually unreadable, but more explicit; and, by the way, cut-elimination is related to the problem of finding "elementary proofs" in number-theory[18]. The structure of cut-elimination is better understood through *natural deduction*, or through (isomorphic) functional calculi, such as *$\lambda$-calculus*, in which *Modus Ponens* is nothing but the application of a function $f$ which maps $A$ into $B$ to an argument $a$ in $A$, yielding $f(a) \in B$, compare equation (35) with:

$$\frac{a \in A \quad f \in A \Rightarrow B}{f(a) \in B} \tag{37}$$

The various versions of $\lambda$-calculus are governed by the equation:

$$(\lambda x t)u = t[u/x]. \tag{38}$$

$\lambda x t$ is the function associating $t$, an expression containing $x$, to the argument represented by a variable $x$; the equation looks therefore like a mere triviality

[16] A basic reference for this section is [15].

[17] With variants, such as *normalisation* in natural deduction or $\lambda$-calculi, see below.

[18] And has sometimes been successfully used to this effect.

... but a very powerful triviality, the extant $\lambda$-calculi being able to represent computable functions:

**Typed $\lambda$-calculi:** Typically system $\mathbb{F}$ which contains all computable *terminating* functions whose termination can be proved within second-order arithmetic. There is no hope to find a terminating computable function not representable in $\mathbb{F}$, the only exceptions being obtained through *ad hoc* diagonalisations, i.e., by *cheating*.

**Pure $\lambda$-calculus:** In typed $\lambda$-calculi, there is a "super-ego", *typing*, which forbids certain "non-logical" combinations — in the same way the choice of names can be used to avoid incest[19]. In pure $\lambda$-calculus, a function can be applied to anything, including itself. This calculus contains *all* computable functions, most of them partial, i.e., non-terminating. In the absence of typing, there is no way to tell the wheat from the tares, i.e., to individuate the *total* functions[20].

**A.2. Categorical interpretations.** The Church-Rosser theorem[21] states that cut-elimination is *associative*, i.e., that we can apply equation (38) in any order. Concretely, this means that, in $\lambda$-calculi, the composition of functions is associative; the functional intuition is therefore correct. But where to find such functions?

Set theory is inadequate: the interpretation of self-application would lead to Russell's paradox, and, by the way, one can see pure $\lambda$-calculus, which is a "naive" function theory, as the "correct" version of naive set-theory. In both cases, every operation has a fixpoint: in set-theory the fixpoint of negation — constructed by Russell's paradox — is a contradiction; pure $\lambda$-calculus avoids the pitfall by allowing *undefined* objects, i.e., non-terminating computations: for instance, if we try to mimic the fixpoint of negation we obtain a never ending process.

Set-theory being too brutal, people turned their attention towards category theory. Is it possible to replace "functions" with *morphisms*? The answer is positive, and at work in *Scott domains*. The idea is to make a category out of certain topological spaces, so that our functions are continuous morphisms. The main difficulty is to have the function space $\hom(X, Y)$ to be in turn a topological space of the same nature, not to speak of the *continuity* of the canonical operations, e.g., the composition of morphisms. To make the long story short, the operation succeeded, the equation

$$\mathbb{D} \simeq \hom(\mathbb{D}, \mathbb{D}) \tag{39}$$

which produces a *domain* isomorphic with its function space, has a solution: what was exactly needed for $\lambda$-calculus.

---

[19] In GoI, typing is responsible for nilpotency.

[20] The problem of making total a partial algorithm is of the same nature as the problem of extending an unbounded operator to the full Hilbert space, a pure nonsense!

[21] A good reference for this section is [2].

... But the patient was dead; the challenge of making continuous too many canonical morphisms was too heavy, and Scott domains are far astray from standard topology: for instance, they are never Hausdorff.

More recent investigation with Banach spaces [12] help us to understand the problem in standard topological terms: the function space $\hom(X, Y)$ can be seen as the space of *analytical* maps from the *open* unit ball of $X$ to the *closed* unit ball of $Y$. Composition is problematic, since it would involve the extension of a bounded analytic map defined on $\{x; \|x\| < 1\}$ to $\{x; \|x\| \leq 1\}$, an operation already desperate when $X = \mathbb{C}$. So there is no *real*[22] continuity in cut-elimination.

If we look carefully, Scott domains are indeed ordered sets, and continuity is just commutation to least upper bounds. This aspect of Scott's contribution is beyond criticism: more, it is extremely important. But this does not justify the building of a sort of "counter-topology". In this paper, cut-elimination is explained by the inversion of a hermitian operator on Hilbert space. The solution found in the invertible case is extended by various methods, including l.u.b. and g.l.b., to the general case, but is not even order-continuous.

**A.3. Stability.** Instead of explaining commutation to l.u.b. by topology, I prefered to use commutation to *direct limits*; direct limits usually socialise well with *pull-backs*, and this apparently minor change of viewpoint introduced *stability* as commutation to pull-backs, a notion with no topological interpretation. This led to *coherent spaces*[23], partly rediscovering earlier work of Berry [3]. Anticipating on a further discussion, there are two ways of presenting coherent spaces:

**Essentialist:** A coherent space $X$ is the pair $(|X|, \frown\!\!\!\!\smile)$ of a *carrier* (a set) $|X|$ and a *coherence* $\frown\!\!\!\!\smile$ on $|X|$, i.e., a binary and reflexive relation. A *clique* $a \sqsubset X$ is any subset of the carrier made of pairwise coherent points. The main operations on cliques are directed unions (i.e., direct limits) and conditional intersections (i.e., pull-backs) $a \cap b$, provided $a \cup b$ is a clique . The main theorems basically rest on the (linear) negation $\sim X := (|X|, \asymp)$ — $\asymp$ meaning "incoherent or equal" — and on the basic remark that a clique and an *anti-clique* intersect on at most one point:

$$a \sqsubset X,\ b \sqsubset {\sim}X \Rightarrow \sharp(a \cap b) \leq 1. \tag{40}$$

**Existentialist:** Instead of working with coherence (cliques), we admit arbitrary subsets of the carrier $|X|$, and we say that two such subsets $a, b \subset |X|$ are *polar* when $\sharp(a \cap b) \leq 1$. We can define a coherent space as a set of subsets of the carrier equal to its bipolar ${\sim\sim}X$. This alternative definition is shown to be equivalent to the "official" one:

[22] Neither "actual", nor "compatible with $\mathbb{R}$."
[23] See for instance [15].

observe that, if $X = \sim\sim X$ and $x, y \in |X|$ then either $\{x, y\} \in X$ or $\{x, y\} \in \sim X$, the disjunction being exclusive when $x \neq y$; from this one easily recovers the coherence of $X : x \bigcirc y \Leftrightarrow \{x, y\} \in X$.

This discussion may seem extremely philosophical, i.e., for the common sense, a gilding of the lily. It takes all its significance when one tries to get rid of commutativity, i.e., when one considers the carrier as the basis — among others — of a complex vector space. There is no way of speaking of a coherence relation on a vector space, but there is still the existentialist version, for instance, we could replace subsets with positive hermitians, and the cardinal (the dimension) with the *trace*, so as to define polarity by $\mathrm{tr}(\mathsf{hk}) \leq 1$. This is what we did in *quantum coherent spaces*, see [14]. By the way there are still major variants of coherent spaces, typically the *hypercoherences* of Ehrhard [7], which lack an existentialist approach; and this is not a meaningless "philosophical" digression.

**A.4. Geometry of interaction.** Equation (38) defines a universal algorithmics[24], but there is something strange in this equality, *one side is more equal than the other*. The equation is indeed treated as a *rewriting*:[25]

$$(\lambda xt)u \rightsquigarrow t[u/x] \tag{41}$$

which *a priori* makes no sense in category theory. The question is therefore to decide whether this rewriting is pure engineering, or — in the same way the original "cooking recipes" of cut-elimination admitted a categorical (static) interpretation — if it admits a decent mathematical interpretation.

*Geometry of interaction* (GoI) is a *dynamic* explanation of logic, based on operator algebras. A (cut-free) proof (of a sequent) is represented by a square matrix whose entries are bounded operators on a given Hilbert space $\mathcal{H}$; the dimension of the matrix corresponds to the number of formulas in the *sequent*.

Here, beware of possible misunderstandings: operators are functions on Hilbert space, and they are in turn used to interpret functions (those coming from $\lambda$-calculus); but the composition of functions has nothing to do with the composition of the associated operators. For instance, the *identity axiom* of logic $A \vdash A$, (which roughly corresponds to the identity function $\lambda xx$) is interpreted by the anti-diagonal matrix: $\begin{bmatrix} 0 & \mathrm{I} \\ \mathrm{I} & 0 \end{bmatrix}$ where I is the identity operator on $\mathcal{H}$. The matrix is $2 \times 2$, the two rows/columns corresponding to the two occurrences of $A$. The anti-diagonal matrix should be viewed as a common electronic device, the *extension cord*: the two $A$ (the two copies of $\mathcal{H}$) correspond to the two plugs through which some alternative current (an element of $\mathcal{H}$) may enter/exit. The matrix says that everything coming from the left exits on the right without change, and similarly from right to left: in real life,

---

[24]For this section, the reference is the first four papers on Geometry of interaction [8, 9, 10, 11].

[25]See for instance [15].

an extension cord works in this way, in both directions, even if our choice of plugs male/female tends to make them unidirectional; but this is only a "super-ego" designed to avoid accidents[26].

In presence of cuts, the pattern is slightly modified; typically, if cut-free proofs of $\Gamma \vdash \Delta, A$ and $A, \Gamma' \vdash \Delta'$ are interpreted by matrices $M$ and $M'$ (respectively indexed by $\Gamma, \Delta, A$ and $A, \Gamma', \Delta'$) then (renaming the second $A$ as $A'$), the proof obtained by applying the cut-rule (36) is the matrix $\mathsf{h} = \begin{bmatrix} M & 0 \\ 0 & M' \end{bmatrix}$ indexed by $\Gamma, \Delta, A, A', \Gamma', \Delta'$; to this matrix is added another matrix, the *feedback* $\sigma = \begin{bmatrix} 0&0&0&0 \\ 0&0&I&0 \\ 0&I&0&0 \\ 0&0&0&0 \end{bmatrix}$ with only two non-zero entries, $\sigma_{AA'} = \sigma_{A'A} = I$. The name *feedback* suggests that some output of $\mathsf{h}$ is given back to $\mathsf{h}$ trough $\sigma$. Indeed, Geometry of Interaction explains cut-elimination as the I/O diagram of this system.

We shall have plenty of space to conceptualise this, so let us solve the equation in the simplest case, namely that of a *Modus Ponens* with an identity axiom:

$$\frac{\vdash A \quad A \vdash A}{\vdash A} \tag{42}$$

Obviously, if the original proof of $A$ corresponds to a $1 \times 1$ matrix $[\mathsf{a}]$ then $\mathsf{h} = \begin{bmatrix} \mathsf{a}&0&0 \\ 0&0&I \\ 0&I&0 \end{bmatrix}$ and $\sigma = \begin{bmatrix} 0&I&0 \\ I&0&0 \\ 0&0&0 \end{bmatrix}$. The I/O equation corresponding to the feedback consists in, given $z \in \mathcal{H}$, finding $x, x', y, y', z' \in \mathcal{H}$ such that

$$\begin{aligned} \mathsf{h}(x \oplus y \oplus z) &= x' \oplus y' \oplus z' \\ \sigma(x' \oplus y' \oplus z') &= x \oplus y \oplus 0. \end{aligned} \tag{43}$$

The solution is obvious: $x' = \mathsf{a}(x)$, $y' = z$, $z' = y$, $y = x'$, $x = y'$, hence $x = y' = z$, $z' = y = x' = \mathsf{a}(x) = \mathsf{a}(z)$. Viewed from outside, the system yields, given the input $z$, the output $z' = \mathsf{a}(z)$. If we rename the three occurrences of $A$, from left to right, as $A, A', A''$, what we actually did was, given some device $\mathsf{a}$ communicating through plug $A$ and some extension cord exchanging plugs $A', A''$, to physically plug $A$ with $A'$. The resulting system behaves like $\mathsf{a}$, but now trough the only "bachelor" plug, $A''$.

**A.5. GoI: main results.** First the good news, it works. The interpretation of logic, or $\lambda$-calculus, involves the identity axiom and the cut-rule, that we both explained; it also involves the interpretation of various logical operations (conjunction, disjunction). It turns out that all these operations are indeed $*$-isomorphisms between matrix algebras; typically binary connectives make use of an isomorphism replacing a $n+1 \times n+1$ matrix with a $n \times n$ matrix with coefficients in the same $B(\mathcal{H})$, which can be done by means of an

[26]If the cord were not bi-directional, there would be no need for heavy precautions, a different colour would suffice.

isometry $\mathcal{H} \oplus \mathcal{H} \simeq \mathcal{H}$, equivalently two partial isometries[27] $\mathsf{p}, \mathsf{q}$ of $\mathcal{H}$ such that

$$\begin{aligned} \mathsf{p}^*\mathsf{p} = \mathsf{q}^*\mathsf{q} &= \mathrm{I} \\ \mathsf{p}\mathsf{p}^* + \mathsf{q}\mathsf{q}^* &= \mathrm{I}. \end{aligned} \tag{44}$$

The main result of GoI is that cut-elimination, normalisation, correspond to the solution of the feedback equation between $\mathsf{h}$ and $\sigma$. In fact the correspondence is not quite exact, the two coincide only in certain cases, but these cases are the "important" ones. In fact GoI corresponds to sophisticated reduction techniques, typically Lamping's *optimal reduction* for pure $\lambda$-calculus, see [16].

A last idea came from GoI: it clearly distinguishes between "before" and "after", i.e., the two sides of the rewriting (41), in this respect, this is a progress over categories. But more, it gives a meaning to the computation itself: remember, when we solved the basic, case of the feedback equation (43), we had to give values, in function of the input $z$, to the output $z'$, but also to "internal values", $x, x', y, y'$. The operator associating $x' \oplus y' \oplus z'$ to the input $z$ plays the role of the computation process. In this respect, the typed (logical) cases treated in [9, 11] differ from the (non-logical) case of pure $\lambda$-calculus treated in [10]: in the typed case, the execution operator is bounded, whereas in the pure, untyped case, the execution is generally unbounded.

**A.6. Augustinian considerations.** The opposition between *essentialism* and *existentialism*, Thomas and Augustine, is central in logic. Essentialism explains things as "coming from a hat", existentialism is more interactive, and, maybe, more modern. An essentialist version of ethics is "follow the rule, because it is the rule", an existentialist version would be "try to defeat the rule". Logically speaking, essentialism refuses *untyped* notions, i.e., objects that are not born with a pedigree; existentialism accepts them all, and later puts some labels on them, depending on their behaviour, see [14] for a discussion. "Proofs as functions" and "Proofs as actions", can be seen under the light of this opposition:

- ⋆ Logic is surely born essentialist. If you don't understand what this means, just remember Tarski's definition of truth "$A \wedge B$ is true iff $A$ is true *and* $B$ is true": behind the conjunction $\wedge$ stands a "meta-conjunction". Essentialism, the primality of essence, is the claim that everything preexists as a meta ... , and the meta as a metameta, of course!
- ⋆ The functional paradigms, typically untyped $\lambda$-calculus, would rather present functions as primitive, let us say that they are given by programs, and formulas as comments, *specifications*. It is plain that the essence (the

[27] See C.6.

specification) is posterior to existence (the program)[28]; bad programs do exist, they breakdown, but if only good programs were in use, certain companies wouldn't sell that much!

Even the idea of a category-theoretic interpretation can be read in a Thomist way: equality is the application of diagrammatic essences (e.g., limits, colimits).

**A.7. GoI: limitations.** The first truly existentialist explanation of logic came with the *proof-nets* of linear logic, which inverted the tradition: starting with diagrams as sort of "wild graphs", compose edges by shortening of *paths*. The logico-categorical (essentialist) description of vertices as formulas/objects is now posterior to existence (here, the physical drawing of paths in a graph); the formula written on a vertex of the graph is a comment on the topological status of this precise vertex inside the graph. This is the meaning of the *correctness criterion* of proof-nets, seen as a topological property of graphs, *acyclicity*, see [5].

Geometry of interaction is nothing but an infinite-dimensional generalisation of proof-nets, but it no longer meets our implicit Augustinian standards.

**Operators:** To any proof/$\lambda$-expression, GoI associates a pair $(\mathsf{h}, \sigma)$. The feedback equation is solved by proving that $\sigma\mathsf{h}$ is nilpotent, i.e., $(\sigma\mathsf{h})^n = 0$ for some $n$ [9, 11], with a weaker version for pure $\lambda$-calculus [10], *weak nipotency*: $\langle(\sigma\mathsf{h})^n(x) \mid x\rangle \to 0$ for all $x \in \mathcal{H}$.

**Associativity:** It was necessary to establish some structural properties of the solution, typically *associativity*, which deals with iterated feedbacks. Here we started to work *against* the spirit of operator algebras; in order to ensure that certain compositions of partial isometries are still partial isometries, we were led to very artificial restrictions.

To sum up: when we follow logic, h and $\sigma$ are always partial symmetries. But one can represent them in such a way that, w.r.t. a basis given in advance, they correspond to partial bijections. In other terms, all this work eventually amounts as a calculus of partial involutions of $\mathbb{N}$! Again the work was nontrivial, but didn't really go into the very heart of operator algebras.

**A.8. Quantum coherent spaces.** Under the influence of *Quantum Computing*, esp. the recent work of Selinger [20], I was able to revisit the static interpretation (coherent spaces) in the spirit of *quantum* interaction. The result is *Quantum Coherent Spaces* [14], QCS for short. To sum up, QCS are rather satisfactory, as an Augustinian approach to logic, even if they are limited to finite dimension. The basic idea is that the points in Scott domains, coherent spaces, ... are like the distinguished basis of a Hilbert space, and

[28] The functional explanation of logic, due to Kolmogorov has been violently attacked by another figure of Thomism, Kreisel: in [19] he claimed that everything should be relativised to a *given* formal system. Logic should presuppose logic ...

that everything coming from logic, $\lambda$-calculus is well-behaved, e.g., diagonal w.r.t. this basis. If we forget this distinguished basis, it turns out that everything still makes sense, but it now looks "quantum". That's an existentialist twist: when a function meets an argument which was not designed for it, nothing happens in the essentialist world (forbidden!); in the real world, the interaction takes place anyway (measurement, reduction of the wave packet).

The limitation of QCS to finite dimension seems to be absolute. This is why the idea of using GoI (much more flexible) is extremely natural[29]. But, in this new round of GoI, we shall get rid of any artificial (essentialist) restriction, in particular, of any commitment to a particular basis.

**Appendix B. The Tortoise Principle.** The Tortoise Principle, at work in the proofs of theorems 3.2 and 3.5, exchanges a structural simplification of the feedback equation, against an enlargement of the Hilbert space. By the way, the feedback equation itself is an application of the Tortoise Principle, to a simpler, but less manageable, equation: see Section B.2 below.

**B.1. An unexpected contributor.** Although he was Professor of Logic in Oxford, Lewis Carroll didn't make any significant contribution to his official field. He is mainly remembered for his photographs of young girls like Alice Liddell, and also for nonsense books like the two *Alice*. His short story "What the Tortoise said to Achilles" is a sort of endless chasing of *Modus Ponens*, our *cut*, our *feedback*. Basically, a *Modus Ponens* between $A$ and $A \Rightarrow B$ is replaced with another one between $A \wedge (A \Rightarrow B)$ and $(A \wedge (A \Rightarrow B)) \Rightarrow B$; this replacement — or rather simplified versions of it — is what I call the "Tortoise Principle". To understand the technical interest for a modern reader, let us follow the convention of using subscripts, to avoid confusions. One starts with a *Modus Ponens* between $A_1$ and $A_2 \Rightarrow B_1$; for us, it means a hermitian $\mathsf{f}$ of $\mathcal{F} := \mathcal{H}_1 \oplus \mathcal{H}_2 \oplus \mathcal{G}_1$, together with a feedback exchanging $\mathcal{H}_1$ and $\mathcal{H}_2$; the output (eventual solution of the feedback equation) is a hermitian of $\mathcal{G}_1$. The second *Modus Ponens* between $A_1 \wedge (A_2 \Rightarrow B_1)$ and $(A_3 \wedge (A_4 \Rightarrow B_2)) \Rightarrow B_3$, is a hermitian $\mathsf{k}$ of $\mathcal{K} := (\mathcal{H}_1 \oplus \mathcal{H}_2 \oplus \mathcal{G}_1) \oplus (\mathcal{H}_3 \oplus \mathcal{H}_4 \oplus \mathcal{G}_2 \oplus \mathcal{G}_3)$, together with a feedback exchanging $\mathcal{H}_1$ and $\mathcal{H}_3$, $\mathcal{H}_2$ and $\mathcal{H}_4$, $\mathcal{G}_1$ and $\mathcal{G}_2$. The output of the feedback equation is a hermitian of $\mathcal{G}_3$, strictly isomorphic with the output found in the original case.

So what did we gain in this complication? Look at $\mathsf{f}$: we know very little about it. Now look at $\mathsf{k}$; if we adopt the block decomposition suggested by our use of parentheses when we defined $\mathcal{K}$, it can be written: $\begin{bmatrix} \mathsf{f} & 0 \\ 0 & \mathsf{h} \end{bmatrix}$. The point is that $\mathsf{h}$ is perfectly well-known: it is the interpretation of the standard tautology $(A \wedge (A \Rightarrow B)) \Rightarrow B$. What is unknown, "variable", i.e., $\mathsf{f}$, is now wholly located in the support $\mathcal{G}_3{}^{\perp}$ of the feedback.

[29] In both cases, the identity axiom flips two copies of the same $\mathcal{H}$; in QCS, it flips $\mathcal{H} \otimes \mathcal{H}$, in GoI, it flips $\mathcal{H} \oplus \mathcal{H}$, which is incredibly better!

If you are not convinced of the interest of this Tortoise principle, try to prove directly that the normal form of a bipartite system is still bipartite! No doubt, you can make it, by a tedious equality chasing, after correcting a few errors of signs. But, with an appropriate use of the Tortoise, it reduces to showing that the inverse of a bipartite hermitian is bipartite, and the length of the equality chasing in Proposition 8 remains decent.

If we count the number of atoms in the original *Modus Ponens*, they are four (we have to count the conclusion); Tortoise doubles it to eight, the following step to sixteen, etc. But all the conceptual simplification, and the relevance to logic, is located in the first step. The infinite iteration performed by Carroll — in order to get a sort of mock Zeno's paradox — is pointless, as expected from the master of nonsense: a pleasant — but somewhat superficial author — who stumbled — accident or intuition, who knows? — on ideas that would only take shape in the mid 1930's — with the work of a genuine logician, Gehrard Gentzen.

**B.2. Example: cut vs. composition.** Indeed the Tortoise Principle has implicitly been used in our very basic formulation of what is a cut-system. If we come back to the original idea of *composition*, we must deal with a sort of general *Modus Ponens*, essentially the syllogism *Barbara* "all *R* are *S*, all *S* are *T*, hence all *R* are *T*":

$$\frac{R \Rightarrow S \quad S \Rightarrow T}{R \Rightarrow T} \tag{45}$$

If the (given proofs of) the two premises are expressed by means of operators $\mathsf{f} \in \mathcal{B}(\mathcal{R} \oplus \mathcal{S})$, $\mathsf{g} \in \mathcal{B}(\mathcal{S} \oplus \mathcal{T})$, then composition amounts at solving, $x \in \mathcal{R}$, $z \in \mathcal{T}$ being given:

$$\begin{aligned} \mathsf{f}(x \oplus y) &= x' \oplus y' \\ \mathsf{g}(y' \oplus z) &= y \oplus z' \end{aligned} \tag{46}$$

and the output is the operator which yields $x' \oplus z'$ as a function $\mathsf{k}(x \oplus z)$.

The main advantage of this formulation is that it is close to what we have actually in mind: a duality. But, technically speaking, it is awfully complex: besides the fact that $\mathsf{f}, \mathsf{g}$ are hermitians of norm at most 1, we know strictly nothing.

Here comes the Tortoise: introduce the space $\mathcal{H} = \mathcal{R} \oplus \mathcal{S} \oplus \mathcal{S} \oplus \mathcal{T}$, and let $\sigma$ be the partial symmetry swapping the two $\mathcal{S}$, $\sigma(x \oplus y \oplus y' \oplus z) := 0 \oplus y \oplus y' \oplus 0$. It is immediate that solving (46) is the same as solving the feedback equation for $(\mathcal{H}, \mathsf{f} \oplus \mathsf{g}, \sigma)$. Moreover, the feedback equation is just a particular instance of (46), where $\mathcal{R}, \mathcal{S}, \mathcal{T}, \mathsf{f}, \mathsf{g}$ have been respectively replaced with $\mathcal{R} \oplus \mathcal{T}, \mathcal{S} \oplus \mathcal{S}, 0, \mathsf{f} \oplus \mathsf{g}, \tau$, where $\tau$ is the total symmtry swapping the two copies of $\mathcal{S}$.

We obvious lost the symmetric character, for instance we seldom allow changes of feedback[30]. But we gained the fact that the feedback is a partial symmetry, i.e., almost the simplest case of a hermitian operator.

Associativity is the question arising when we add a third equation to (46) (with $\mathsf{h} \in B(\mathcal{T} \oplus \mathcal{U})$):

$$\mathsf{h}(z' \oplus w) = z \oplus w' \tag{47}$$

we can either "solve" (46) + (47) in a single step, or do it in two steps, e.g., first solve (46), which yields $\mathsf{k} \in B(\mathcal{R} \oplus \mathcal{T})$, then solve (47)+ (48)

$$\mathsf{k}(x \oplus z) = z' \oplus w' \tag{48}$$

The Tortoise expresses the system of three equations (46) + (47) as a cut-system $(\mathcal{R}\oplus\mathcal{S}\oplus\mathcal{S}\oplus\mathcal{T}\oplus\mathcal{T}\oplus\mathcal{U}, \mathsf{f} + \mathsf{g} + \mathsf{h}, \sigma+\tau)$, where $\tau$ swaps the two copies of $\mathcal{T}$, and associativity really translates as $(\sigma + \tau)[\![\mathsf{f} + \mathsf{g} + \mathsf{h}]\!] = \tau[\![\sigma[\![\mathsf{f} + \mathsf{g} + \mathsf{h}]\!]]\!]$.

**Appendix C. Operator-theoretic basics.** These materials are covered by many textbooks, my favourite one being [17].

**C.1. Bounded operators.** We are working on a complex Hilbert space $\mathcal{H}$, equipped with the sesquilinear form $\langle x \mid y\rangle$, linear in $x$, anti-linear in $y$. We are mostly interested in the space $B(\mathcal{H})$ of *bounded*, i.e., continuous, operators (i.e., linear endomorphisms) on $\mathcal{H}$. As usual, $\mathsf{u}^*$ denotes the *adjoint* of u, i.e., the unique operator satisfying $\langle \mathsf{u}(x) \mid y\rangle = \langle x \mid \mathsf{u}^*(y)\rangle$ for all $x, y \in \mathcal{H}$. The identity operator of $\mathcal{H}$ is noted I, or even $\mathcal{H}$ in case of ambiguity as to the underlying Hilbert space; the null operator is noted 0.

**C.2. Topologies.** Several topologies are of interest on the complex vectors space $B(\mathcal{H})$, we list most of them below, in decreasing *strength*; remember that to be stronger means to have more open (more closed) sets, i.e., less converging nets[31]. The three topologies below make sum and scalar multiplication continuous:

**Norm:** The norm $\|\mathsf{u}\| = \sup\{\|\mathsf{u}\|(x); \|x\| \leq 1\}$ makes both product and adjunction continuous, indeed $\|\lambda\mathsf{u}\| = |\lambda| \cdot \|\mathsf{u}\|$, $\|\mathsf{uv}\| \leq \|\mathsf{u}\| \cdot \|\mathsf{v}\|$, $\|\mathsf{u}^*\| = \|\mathsf{u}\|$; $B(\mathcal{H})$ is complete, i.e., it is an involutive *Banach algebra*. But, last but not least, as a consequence of the Cauchy-Schwarz inequality, $\|\mathsf{uu}^*\| = \|\mathsf{u}\|^2$, making $B(\mathcal{H})$ a $C^*$*-algebra*.

**Strong:** The net $(\mathsf{u}_i)$ *strongly* converges to u when for all $x \in \mathcal{H}$ $\|\mathsf{u}_i(x) - \mathsf{u}(x)\|$ converges to 0. The strong topology seems a bit weird, since adjunction is not strongly continuous. The main point is that composition $\mathsf{u}, \mathsf{v} \leadsto \mathsf{uv}$ is strongly continuous, provided the argument u remains bounded in norm.

---

[30]Only exception: associativity.

[31]Generalisation of sequences: a net is a family indexed by a non-empty directed ordered set.

**Weak:** The net $(\mathsf{u}_i)$ *weakly* converges to u when for all $x, y \in \mathcal{H}$ $\langle \mathsf{u}_i(x) \mid y \rangle$ converges to $\langle \mathsf{u}(x) \mid y \rangle$. Good news, adjunction is weakly continuous, but composition is only separately continuous, which is not enough in practice. But, as a compensation, the unit ball $B_1(\mathcal{H}) := \{\mathsf{u}; \|\mathsf{u}\| \leq 1\}$ is compact.

For strong convergence, $\|x\| \leq 1$ is enough, and for weak convergence, $\|x\|, \|y\| \leq 1$, and even $x = y$ is enough. The inequality: $|\langle \mathsf{u}(x) \mid y \rangle| \leq \|\mathsf{u}(x)\| \leq \|\mathsf{u}\|$ is responsible for the relative strength of the topologies.

**C.3. Normal operators.** An operator is *normal* when it commutes to its adjoint: $\mathsf{u}\mathsf{u}^* = \mathsf{u}^*\mathsf{u}$. A normal operator generates a commutative $C^*$-algebra, which is isomorphic — through the *spectral calculus* — with the space of continuous complex-valued functions $\mathbb{C}(\mathrm{sp}(\mathsf{u}))$ on the *spectrum* $\mathrm{sp}(\mathsf{u})$. The *spectral calculus* maps u to the inclusion map $\iota_{\mathsf{u}} : \mathrm{sp}(\mathsf{u}) \subset \mathbb{C}$.

The most common operators are normal, among them:

**Unitaries:** $\mathsf{u}\mathsf{u}^* = \mathsf{u}^*\mathsf{u} = \mathrm{I}$. They correspond to *isometries of Hilbert space*; by the spectral calculus, they are exactly those normal operators with spectrum in the unit circle $\mathbb{T} := \{z; |z| = 1\}$.

**Hermitian:** They are such that $\mathsf{h} = \mathsf{h}^*$, they are also called *self-adjoint*. By the spectral calculus, they are exactly those normal operators with spectrum in $\mathbb{R}$. *Positive* hermitians are exactly those normal operators with spectrum in $\mathbb{R}^+$; the typical positive hermitian is any operator $\mathsf{u}\mathsf{u}^*$. Indeed any positive hermitian h is of this form, and u can even be chosen positive, just take $\mathsf{u} = \sqrt{\mathsf{h}}$, which makes sense, since the function $\sqrt{\cdot}$ is defined and continuous on $\mathrm{sp}(\mathsf{h}) \subset \mathbb{R}^+$. Among the standards of the spectral calculus, the decomposition $\mathsf{h} = \mathsf{h}^+ - \mathsf{h}^-$ of a hermitian as the difference of two positive hermitians: apply the spectral calculus to the real functions $x^+ := \sup(x, 0)$ and $x^- := \sup(-x, 0)$, and observe that $\mathsf{h}^+\mathsf{h}^- = \mathsf{h}^-\mathsf{h}^+ = 0$. We use the notation $Her(\mathcal{H})$ for the set of bounded hermitians operating on $\mathcal{H}$; more generally, we can indicate the spectrum, e.g., $Her_{]-1,+1]}(\mathcal{H})$ will denote hermitians such that $\mathrm{sp}(\mathsf{h}) \subset ]-1, +1]$, or the norm, e.g., $Her_{\leq 1}(\mathcal{H})$ consists of hermitians of norm $\leq 1$, and $Her_{<1}(\mathcal{H})$ of hermitians of norm $< 1$. Finally, $Her^+(\mathcal{H})$ will stand for *positive* hermitians.

Normality is not that interesting beyond these two cases: for normal operators don't socialise. This is neither the case for unitaries (closed under product, inversion, multiplication by a scalar of modulus 1), nor hermitians (closed under addition, multiplication by a real scalar; positive hermitians are closed under addition and multiplication by a positive scalar).

**C.4. The pointwise order.** The standard definition of positivity is pointwise:

$$\langle \mathsf{h}(x) \mid x \rangle \geq 0 \quad (x \in \mathcal{H}) \tag{49}$$

In fact the quadratic form $Q(x) := \langle \mathsf{h}(x) \mid x \rangle$ determines h, and conversely, any bounded and positive quadratic form

$$
\begin{array}{lll}
 & 0 \leq Q(x) \leq M\|x\|^2 & (x \in \mathcal{H}) \\
(50) & Q(\lambda x) = |\lambda|^2 Q(x) & (x \in \mathcal{H},\ \lambda \in \mathbb{C}) \\
 & Q(x+y) + Q(x-y) = 2(Q(x) + Q(y)) & (x, y \in \mathcal{H})
\end{array}
$$

can uniquely be written $Q(x) := \langle \mathsf{h}(x) \mid x \rangle$, with $\mathsf{h} \in Her^{+}_{\leq M}(\mathcal{H})$.

Positive hermitians induce a partial ordering of $Her(\mathcal{H})$, which is defined pointwise by:

$$
(51) \qquad \mathsf{h} \leq \mathsf{k} \Leftrightarrow \forall x \in \mathcal{H} \quad \langle \mathsf{h}(x) \mid x \rangle \leq \langle \mathsf{k}(x) \mid x \rangle
$$

A (monotone increasing) *net* of hermitians of $Her(\mathcal{H})$ is a family $(\mathsf{h}_i)$ $(i \in I)$ indexed by a (non-empty) directed ordered set $I$ ($I$ is not supposed to be denumerable), and such that

$$
(52) \qquad i \leq j \Rightarrow \mathsf{h}_i \leq \mathsf{h}_j
$$

A bounded net admits a l.u.b. $\mathsf{h} = \sup_{i \in I} \mathsf{h}_i$ defined by

$$
(53) \qquad \langle \mathsf{h}(x) \mid x \rangle := \sup_{i \in I} \langle \mathsf{h}_i(x) \mid x \rangle
$$

The equation makes sense since the $\langle \mathsf{h}_i(x) \mid x \rangle$ are bounded, and it defines a hermitian, i.e., a positive quadratic form, because of the directedness of $I$. Although h is defined as a *weak* limit, it appears to be a strong limit:

PROPOSITION 18. *If* $\mathsf{h} = \sup_{i \in I} \mathsf{h}_i$, *then* $\mathsf{h}_i \to \mathsf{h}$ *in the* strong-operator *topology.*

PROOF. See [17, Lemma 5.1.4]. ⊣

For obvious reasons, symmetric results hold for *monotone decreasing* nets (existence of g.l.b., strong convergence).

**C.5. Projections and symmetries.** An operator which is both hermitian and unitary enjoys $\mathsf{u}^2 = \mathsf{u}$, let us call it a *symmetry*; symmetries are exactly those normal operators with spectrum in $\{-1, +1\} = \mathbb{R} \cap \mathbb{T}$, i.e., in $Her_{\{-1,+1\}}(\mathcal{H})$. An idempotent hermitian is called a *projection*, and among normal operators, projections are those with spectrum in $\{0, 1\}$, i.e., in $Her_{\{0,+1\}}(\mathcal{H})$. On the Hilbert space $\mathcal{H}$, a projection can be identified with its range $\mathcal{R} = \mathrm{rg}(\mathsf{h})$, which is a closed subspace: using the orthogonal decomposition $\mathcal{H} = \mathcal{R} \oplus \mathcal{R}^{\perp}$, $\mathsf{h}(x \oplus y) = x$, i.e., h acts as the orthoprojection on its range $\mathcal{R}$. This justifies the abusive notational identification between h and its range.

Every symmetry $\sigma$ is induced by an orthogonal decomposition $\mathcal{H} = \mathcal{R} \oplus \mathcal{S}$: with the abuse of notations just introduced, let $\mathcal{R} := (\mathrm{I} + \sigma)/2$, $\mathcal{S} := (\mathrm{I} - \sigma)/2$, so that $\sigma(x \oplus y) = x - y$. Conversely, a projection $\mathcal{R}$ induces the symmetry $2\mathcal{R} - \mathrm{I}$.

**C.6. Partial isometries.** A *partial isometry* is any u such that uu* is a projection; partial isometries must be handled with care, since they need not be normal. Since uu* is normal, a spectral characterisation is $\mathrm{sp}(uu^*) \subset \{0,1\}$, which yields[32] $\mathrm{sp}(u^*u) \subset \mathrm{sp}(uu^*) \cup \{0\} \subset \{0,1\}$: hence u*u is also a projection.

A partial isometry establishes an isomorphism between the spaces (i.e., projections) u*u and uu*. But one can hardly compose them: if u, v are partial isometries, then uv is a partial isometry exactly when the projections vv* and u*u commute.

A *partial symmetry* is a hermitian partial isometry. Partial symmetries are those normal operators $\sigma$ such that $\mathrm{sp}(\sigma) \subset \{-1,0,1\}$, i.e., belong to $Her_{\{-1,0,+1\}}(\mathcal{H})$; in other terms, among the normal operators, those enjoying $\sigma^3 = \sigma$. Then $\sigma^2$ is a projection $\mathcal{R}$, and $\sigma$ restricted to $\mathcal{R}$ is a symmetry. Any partial symmetry $\sigma$ can be uniquely written as the difference of two projections $\sigma^+ - \sigma^-$, with $\sigma^+ \cdot \sigma^- = 0$.

**C.7. Blocks and matrices.** We explain our conventions about matrices.

**Blocks:** In case of a (Hilbert) direct sum decomposition $\mathcal{H} = \bigoplus_1^n \mathcal{H}_i$, we can write any operator f on $\mathcal{H}$, bounded or unbounded, as the sum $\sum_{ij} \mathcal{H}_i f \mathcal{H}_j$. What one can write, for instance, when $n = 2$ as

$$f = \begin{pmatrix} \mathcal{H}_1 f \mathcal{H}_1 & \mathcal{H}_1 f \mathcal{H}_2 \\ \mathcal{H}_2 f \mathcal{H}_1 & \mathcal{H}_2 f \mathcal{H}_2 \end{pmatrix}.$$

Such blocks compose in the usual way, i.e., $h_{ik} = \sum_j f_{ij} g_{jk}$. However, they are not the real thing, we did hardly more than provide a useful, readable, notation.

**Matrices:** They correspond to an isomorphism $B(\mathcal{H}) \sim B(\mathcal{K}) \otimes M_n(\mathbb{C})$. In other terms, in $\begin{bmatrix} f_{11} & f_{12} \\ f_{21} & f_{22} \end{bmatrix}$ (observe the different style of brackets), all coefficients belong to $B(\mathcal{K})$, not to $B(\mathcal{H})$. This is the real thing.

The two notions can be related in a particular case, namely when we are given partial isometries $\alpha_{ij}$, such that $\alpha_{ii} = \mathcal{H}_i$, $\alpha_{ji} = \alpha_{ij}{}^*$, $\alpha_{ik} = \alpha_{ij}\alpha_{jk}$. Then f can be represented by the actual matrix (with coefficients in $B(\mathcal{H}_1)$), $f_{ij} = \alpha_{1i} f \alpha_{j1}$, e.g., when $n = 2$,

$$f = \begin{bmatrix} \alpha_{11} f \alpha_{11} & \alpha_{11} f \alpha_{21} \\ \alpha_{12} f \alpha_{11} & \alpha_{12} f \alpha_{21} \end{bmatrix}.$$

**C.8. Inclusion of ranges.** What follows is basically folklore, a variation on the *polar decomposition* of operators, see [18, p. 401].

DEFINITION 18 (Domain). Let $u \in B(\mathcal{H})$; we define $\mathrm{dom}(u) := (\ker(u))^\perp$, and clrg(u) as dom(u*), so that the two notions coincide in the hermitian case.

clrg(u) is the closure of rg(u), i.e., $\mathrm{rg}(u)^{\perp\perp}$.

---

[32] See [17], proposition 3.2.8.

PROPOSITION 19. *Assume that* $0 \leq \mathsf{f} \leq \mathsf{h}$; *then* $\mathrm{rg}(\sqrt{\mathsf{f}}) \subset \mathrm{rg}(\sqrt{\mathsf{h}})$; *indeed* $\sqrt{\mathsf{f}} = \sqrt{\mathsf{h}} \cdot \varphi$ *for an appropriate* $\varphi$ *of norm at most* 1.

PROOF. Using $\langle \mathsf{f}(x) \mid x \rangle = \|\sqrt{\mathsf{f}}(x)\|^2$, etc., we easily obtain

$$\|\sqrt{\mathsf{f}}(x)\| \leq \|\sqrt{\mathsf{h}}(x)\|.$$

Therefore we can define the linear map $\psi$ from $\mathrm{rg}(\sqrt{\mathsf{h}})$ to $\mathrm{rg}(\sqrt{\mathsf{f}})$, by

$$\psi(\sqrt{\mathsf{h}}(x)) := \sqrt{\mathsf{f}}(x).$$

It is immediate that:

(i) $\psi$ is well-defined;
(ii) $\|\psi(y)\| \leq \|y\|$.

In two steps, we can:

(i) First extend $\psi$ by norm-continuity to the closed subspace $\mathrm{dom}(\sqrt{\mathsf{h}})$;
(ii) Next extend it to the full $\mathcal{H}$ by making it null on the orthocomplement $\ker(\sqrt{\mathsf{h}})$ $(= \ker(\mathsf{h}))$ of $\mathrm{dom}(\sqrt{\mathsf{h}})$ $(= \mathrm{dom}(\mathsf{h}))$.

If this extension is called $\varphi^*$, it is plain that $\|\varphi^*\| \leq 1$ and $\sqrt{\mathsf{f}} = \varphi^* \cdot \sqrt{\mathsf{h}}$. From this, $\sqrt{\mathsf{f}} = \sqrt{\mathsf{h}} \cdot \varphi$ and we are done. ⊣

REMARK 17. More generally, if $\mathsf{u}\mathsf{u}^* \leq \mathsf{h}$, there exists a $\psi$ of norm at most 1 such that $\mathsf{u} = \sqrt{\mathsf{h}} \cdot \psi$. This $\psi$ is made unique by the requirement $\mathrm{rg}(\psi) \subset \mathrm{dom}(\mathsf{h})$. By the way, if $\mathsf{u} = \sqrt{\mathsf{h}} \cdot \theta$; then $\psi = \mathrm{dom}(\mathsf{h}) \cdot \theta$.

NON SI NON LA

## REFERENCES


[1] V. M. ABRUSCI and P. RUET, *Non-commutative logic I: the multiplicative fragment*, ***Annals of Pure and Applied Logic***, vol. 101 (2000), pp. 29–64.

[2] R. AMADIO and P.-L. CURIEN, ***Domains and lambda-calculi***, Cambridge Tracts in Theoretical Computer Science, vol. 46, Cambridge University Press, 1998.

[3] G. BERRY, *Stable models of typed lambda-calculi*, ***Proceedings of the 5th international colloquium on automata, languages and programming***, Lecture Notes in Computer Science, vol. 62, Springer Verlag, 1978.

[4] A. CONNES, ***Non-commutative geometry***, Academic Press, San Diego, CA, 1994.

[5] V. DANOS and L. REGNIER, *The structure of multiplicatives*, ***Archive for Mathematical Logic***, vol. 28 (1989), pp. 181–203.

[6] ———, *Proof-nets and the Hilbert space*, ***Advances in linear logic*** (Girard, Lafont, and Regnier, editors), London Mathematical Society Lecture Note Series, vol. 222, Cambridge University Press, 1995.

[7] T. EHRHARD, *Hypercoherences: a strongly stable model of linear logic*, ***Advances in linear logic*** (Girard, Lafont, and Regnier, editors), Cambridge University Press, Cambridge, 1995, pp. 83–108.

[8] J.-Y. GIRARD, *Towards a geometry of interaction*, ***Categories in computer science and logic***, Proceedings of Symposia in Pure Mathematics, no. 92, American Mathematical Society, Providence, 1988, pp. 69–108.

[9] ———, *Geometry of interaction I : interpretation of system F*, ***Logic colloquium '88*** (Ferro, Bonotto, Valentini, and Zanardo, editors), North-Holland, Amsterdam, 1989, pp. 221–260.

[10] ———, *Geometry of interaction II : deadlock-free algorithms*, ***Proceedings of COLOG 88*** (Martin-Löf and Mints, editors), Lecture Notes in Computer Science, vol. 417, Springer-Verlag, Heidelberg, 1990, pp. 76–93.

[11] ———, *Geometry of interaction III : accommodating the additives*, ***Advances in linear logic*** (Girard, Lafont, and Regnier, editors), Cambridge University Press, Cambridge, 1995, pp. 329–389.

[12] ———, *Coherent Banach spaces : a continuous denotational semantics*, ***Theoretical Computer Science***, vol. 227 (1999), pp. 275–297.

[13] ———, *Locus Solum*, ***Mathematical Structures in Computer Science***, vol. 11 (2001), pp. 301–506.

[14] ———, *Between logic and quantic : a tract*, Technical report, Institut de Mathématiques de Luminy, October 2003.

[15] J.-Y. Girard, Y. Lafont, and P. Taylor, ***Proofs and types***, Cambridge tracts in theoretical computer science, vol. 7, Cambridge University Press, Cambridge, 1990.

[16] G. Gonthier, M. Abadi, and J.-J. Levy, *The geometry of optimal lambda-reduction*, ***POPL'92*** (ACM Press, editor), Birkhäuser, Boston, 1992.

[17] R. V. Kadison and J. R. Ringrose, ***Fundamentals of the theory of operator algebras, vol. I***, Pure and Applied Mathematics, Academic Press, 1983, (vol. III contains the solutions of exercises of vol. I).

[18] ———, ***Fundamentals of the theory of operator algebras, vol. II***, Pure and Applied Mathematics, Academic Press, 1986.

[19] G. Kreisel, *Mathematical logic*, ***Lectures in modern mathematics, vol. III*** (T. L. Saaty, editor), Wiley & Sons, New York, 1965, pp. 99–105.

[20] P. Selinger, *Towards a quantum computing language*, ***Mathematical Structures in Computer Science***, vol. 14 (2004), pp. 527–586.

INSTITUT DE MATHÉMATIQUES DE LUMINY
UPR 9016 – CNRS 163
AVENUE DE LUMINY, CASE 930
F-13288 MARSEILLE CEDEX 09, FRANCE
*E-mail*: girard@iml.univ-mrs.fr

# ON LOCAL MODULARITY IN HOMOGENEOUS STRUCTURES

TAPANI HYTTINEN†

**Abstract.** As an attempt to generalize classical results from geometric stability theory to homogeneous model theory, the following two theorems are proved:

THEOREM 1. *If $M$ is an $\omega$-stable (large) homogeneous structure and $D \subseteq M$ is unimodular strongly minimal set with weak elimination of canonical bases, then $D$ is 2-pseudolinear.*

THEOREM 2. *If $M$ is an $\omega$-stable (large) homogeneous structure and $D \subseteq M$ is a 2-pseudolinear strongly minimal set, then either $M$ interprets a non-classical group or $D$ is locally modular.*

In the work described in [Zi1], B. Zilber showed that the geometry of a strongly minimal set in a model of an $\omega$-categorical $\omega$-stable theory is locally modular (and by $\omega$-categoricity and compactness it is locally finite and thus either trivial or projective or affine over a finite field). In [Hr], E. Hrushovski generalized this result by replacing the $\omega$-categoricity assumption by unimodularity of the set, see Definition 0.5(ii) below. In [Zi2], Zilber studied these questions in the context of non-elementary classes. Reformulating in the context of homogeneous classes (Zilber's context was a bit more general), he showed that a strongly minimal (large) homogeneous structure has locally modular geometry (in fact he showed that the geometry is, essentially, either trivial or projective or affine over a field), if $dcl = bcl$ i.e. if $t(a/A)$ is bounded, then $a$ is the only realization of $t(a/A)$. Notice that $dcl = bcl$ is a very strong form of unimodularity.

In this paper we make an attempt to generalize Hrushovski's result to the context of homogeneous classes. See [HS2] for a discussion on the justification of the studies of non-elementary classes.

Hrushovski's theorem is proved in two steps: one shows first that unimodularity implies that the set is 2-pseudolinear and then that 2-pseudolinearity implies local modularity. When we tried to generalize the theorem, a problem arises in each step, a problem that we were not able to solve.

The argument behind the first step is based on clever calculations on multiplicities. It is essential that all the multiplicities we come across are finite. In the elementary case, compactness guarantees this trivially. In the homoge-

---

2000 *Mathematics Subject Classification.* Primary 03C52; Secondary 03C45.
*Key words and phrases.* Homogeneous model theory, locally modular.
†Partially supported by the Academy of Finland, grant 40734.

**Logic Colloquium '03**
Edited by V. Stoltenberg-Hansen and J. Väänänen
Lecture Notes in Logic, 24

neous case, this is not so clear. Unimodularity guarantees this as long as we work in $M$ (= the monster model of the class), but in the argument we need objects that are not necessarily in $M$ i.e. we must work in $M^{eq}$ and in $M^{eq}$ the multiplicities may, as far as we can see, be infinite. So we have added an assumption we call weak elimination of canonical bases, which takes care of our problem. We want to point out that the assumption is stronger than is really needed (e.g. we do not need that the automorphisms $f$ from Definition 0.3(ii) fix $b$, it is enough that the number of images of $b$ under such automorphisms is finite). It is a compromise between the weakness and the aesthetic values.

The argument behind the second step is based on a construction of an algebraically closed field. Here the problem is our old friend: the problem of the existence of a non-classical group. See [HLS] for more on these groups (or e.g. [Hy1], where these groups were called bad). However, we want to point out that this problem is open.

Throughout this paper, $M$ is an $\omega$-stable large homogeneous structure. We assume that the reader is familiar with [Hy3] and use results, concepts and conventions from [Hy3] freely. However, the main result from [Hy3] needed here is recalled below.

The basic stability theory for homogeneous classes of models is developed in [HS1]. In the paper, notions like independence ($\downarrow$) and Lascar strong types ($Lstp$) are defined and their basic properties are proved.

DEFINITION 0.1. (i) We say that $t(a/A)$ is bounded if the number of realizations of the type in $M$ is $< |M|$. If $t(a/A)$ is not bounded, we say that it is unbounded.

(ii) By $bcl(A)$ we denote the set of all $a$ such that $t(a/A)$ is bounded.

(iii) We write $a \in dcl(A)$ if $a$ is the only realization of $t(a/A)$.

(iv) We say that $Lstp(a/A)$ is parallel to $Lstp(b/B)$ if there is $c$ such that $Lstp(c/A) = Lstp(a/A)$, $Lstp(c/B) = Lstp(b/B)$, $c \downarrow_A B$ and $c \downarrow_B A$.

Next we define what we mean by a strongly minimal set in homogeneous model theory (term quasiminimal is also used for these sets). These are the objects that arise naturally e.g. in the proof of Morley's theorem for homogeneous classes.

DEFINITION 0.2. (i) We say that unbounded $p = t(a/A)$ is strongly minimal if the following holds

(a) $a \downarrow_A A$,

(b) if $a \not\downarrow_A B$, then $a \in bcl(A \cup B)$,

(c) if $b$ and $c$ realize $p$, $b \downarrow_A B$ and $c \downarrow_A B$, then $t(b/B) = t(c/B)$ (i.e. $t(a/A)$ is stationary).

(ii) We say that $D \subseteq M^n$ is strongly minimal (over $A$) if there is a strongly minimal type $p$ over a finite set (over $A$) such that $D$ is the set of realizations of $p$.

Next we explain what canonical bases are in homogeneous model theory.

DEFINITION 0.3. (i) We write $b = cb(Lstr(a/A))$ $(= cb(a/A))$ if the following holds:

(a) $b \in dcl(a \cup A)$,

(b) $b \in bcl(A)$,

(c) $a \downarrow_b A$,

(d) if $f$ is an automorphism of $M$, then $f(b) = b$ iff $Lstp(a/A)$ is parallel to $Lstr(f(a)/f(A))$.

(ii) Suppose $D$ is strongly minimal over finite $A$. We say that $D$ weakly eliminates canonical bases if the following holds: if $B \subseteq D$ is finite, $b_1, b_2 \notin bcl(B \cup A)$ are elements of $D$ and $(b_1, b_2) \not\downarrow_A B$, then there is a finite sequence $c \in D$ such that the following holds:

(a) $(b_1, b_2) \downarrow_{A \cup c} B$,

(b) for all automorphisms $f$ of $M$, which fix $A$ pointwise, if $Lstp((b_1, b_2)/A \cup B)$ is parallel to $Lstp((f(b_1), f(b_2))/A \cup f(B))$, then $f(c) = c$.

Notice the following: If $T$ is a complete first-order theory, $T$ is strongly minimal and $T$ has the elimination of imaginaries, then the monster model of $T$ weakly eliminates canonical bases. So e.g. algebraically closed fields weakly eliminate canonical bases. Also if $M$ is modular, strongly minimal and $bcl = dcl$, then $M$ weakly eliminates canonical bases. However, strong minimality and $bcl = dcl$ alone do not imply this (e.g. affine spaces over the field of rationals).

The basic properties of the independence allow us to define $U$-rank for types over sufficiently saturated models, see [Hy2]. However, if $M$ is of finite $U$-rank in the sense of this $U$-rank, the independence has all the usual properties of non-forking and we can define $U$-rank for all complete types over any set, again see [Hy2].

LEMMA 0.4 ([Hy3]). *Suppose $M$ is of finite $U$-rank. Then there is $M^{eq}$ such that the following holds*:

(i) *$M^{eq}$ is $\omega$-homogeneous.*

(ii) *For all finite $A \subseteq M^{eq}$ and $a \in M^{eq}$, there is an element $b$ of $M^{eq}$ such that $b = cb(a/A)$ (and $t(a/b)$ is stationary).*

(iii) *In $M^{eq}$, $\cup\{dcl(A) \mid A \subseteq M$ finite$\} = M^{eq}$.*

(iv) *In $M^{eq}$, there is an independence notion $\downarrow$ which has all the usual properties of non-forking (and the restriction of $\downarrow$ to $M$ is the usual independence notion in $M$).*

(v) *In $M^{eq}$, for all $a$ and $A$, $t(a/bcl(A))$ is stationary.*

PROOF. See [Hy3] (the fact that '$a \downarrow_{cb(a/A)} A$' follows from the proof of [Hy3, Theorem 1.17]). ⊣

DEFINITION 0.5. (i) For all $a$ and $A$, mult$(a/A)$ denotes the cardinality of the set

$$\{Lstp(b/A) \mid t(b/A) = t(a/A)\}.$$

(ii) We say that a strongly minimal $D$ over $A$ is unimodular if for all elements $a_0, \dots, a_n$ and $b_0, \dots, b_n$ of $D$

$$\text{mult}((a_0, \dots, a_n)/\{b_0, \dots, b_n\}) = \text{mult}((b_0, \dots, b_n)/\{a_0, \dots, a_n\}) < \omega$$

if the following holds:

(a) $U((a_0, \dots, a_n)/A) = U((b_0, \dots, b_n)/A) = n + 1$,

(b) $bcl(\{a_0, \dots, a_n\}) = bcl(\{b_0, \dots, b_n\})$.

(iii) We say that a strongly minimal $D$ over finite $A$ is $k$-pseudolinear if the following holds: Suppose $B \subseteq D$ is finite, $b_1, b_2 \in D - bcl(B)$ are elements of $D$ and $(b_1, b_2) \not\downarrow_A B$, then $U(c/A) \leq k$, where $c \in D^{eq}$ (notice that $D$ is a homogeneous structure itself) is such that $c = cb((b_1, b_2)/B \cup A)$.

Definition 0.6. We say that a structure $(N, R_0, \dots, R_n)$ is interpretable in $M$ ($M$-interpretable) if there are a finite $A \subseteq M$, $N^* \subseteq M^{<\omega}$, an equivalence relation $E \subseteq (N^*)^2$ and $R_i^* \subseteq (N^*/E)^{\#R_i}$, $i \leq n$, such that all these sets are preserved under all automorphisms of $M$ which fix $A$ pointwise and $(N^*/E, R_0^*, \dots, R_n^*) \cong (N, R_0, \dots, R_n)$.

## §1. Pseudolinearity.

We recall from the abstract that in this paper the following two theorems are proved.

Theorem 1. *If $M$ is $\omega$-stable and $D \subseteq M$ is unimodular strongly minimal set with weak elimination of canonical bases, then $D$ is 2-pseudolinear.*

Theorem 2. *If $M$ is $\omega$-stable and $D \subseteq M$ is a 2-pseudolinear strongly minimal set, then either $M$ interprets a non-classical group or $D$ is locally modular.*

Notice that from the proof of Theorem 2, it follows easily that in the theorem, 2 can be replaced by any $k < \omega$ and the theorem still holds.

In this section we prove Theorem 1. This section is, modulo few lemmas, a verbatim copy of [Pi, Chapter 2, Section 4] (of course, in this paper, the meaning of words like independent, $\omega$-stable, strongly minimal, canonical bases etc. is different from their meaning in [Pi]).

Throughout this section we assume that $M$ is strongly minimal, unimodular and that it weakly eliminates canonical bases. We work in $M^{eq}$.

Definition 1.1. (i) By $bcl^*(A)$ we denote the set $bcl(A) \cap M$.

(ii) Let $c \in M^{eq}$ and $d_1, \dots, d_n \in M$ be independent elements of $M$. Suppose $c \in dcl(bcl^*(d))$, where $d = (d_1, \dots, d_n)$ and for some $k < n$, $(d_1, \dots, d_k) \downarrow_\emptyset c$ and for all $k < i \leq n$, $d_i \in bcl(c, d_1, \dots, d_k)$. Then

$$Z_d(c) = \text{mult}(c/d)/\,\text{mult}(d/(c, d_1, \dots, d_k))$$

if the multiplicities are finite, and otherwise $Z_d(c)$ is undefined.

(iii) We say that $c \in M^{eq}$ is good, if for all $d = (d_0, \dots, d_n)$ and $k$ as in (ii) above, both $\text{mult}(c/d)$ and $\text{mult}(d/(c, d_1, \dots, d_k))$ are finite.

(iv) We let $C^{cb} \subseteq M^{eq}$ be the set of all $c$ such that for some finite $A \subseteq M$ and elements $a_1, a_2 \in M - bcl(A)$ such that $(a_1, a_2) \not\perp_{\emptyset} A$, $c = cb((a_1, a_2)/A)$.

LEMMA 1.2. *If $c$ is a finite sequence of elements of $M \cup C^{cb}$, then $c$ is good.*

PROOF. Let $d = (d_1, \ldots, d_n)$ and $k$ be as in Definition 1.1(ii) for $c$.

We prove that $\text{mult}(d/(c, d_0, \ldots, d_k))$ is finite by finding a sequence $c^*$ such that $\text{mult}(d/(c, d_0, \ldots, d_k))$ is less than or equal to $\text{mult}(d/(c^*, d_0, \ldots, d_k))$ and $\text{mult}(d/(c^*, d_0, \ldots, d_k))$ is finite.

If $e \in M$ and $a \in bcl^*(e)$, then by the definition of unimodularity, $\text{mult}(a/e)$ is finite. Also if $c' \in c \cap C^{cb}$, then $c' = cb((b_1, b_2)/d)$ for some $b_1, b_2 \in M - bcl(d)$ with $(b_1, b_2) \not\perp_{\emptyset} d$. Let $c^*$ be as in the definition of weak elimination of canonical bases for $b_1, b_2$ and $d$. Then $c^* \in dcl(c')$ and $c' \in bcl(c^*)$. Thus $\text{mult}(d/c \cup \{d_0, \ldots, d_k\}) \leq \text{mult}(d/c^* \cup \{d_0, \ldots, d_k\})$ is finite.

To see that $\text{mult}(c/d)$ is finite, it suffices to show that if $c'$ is as above, then $\text{mult}(c'/d)$ is finite. But this is clear, since $\text{mult}(c'/d)$ is at most $\text{mult}(b_2/d \cup \{b_1\})$. ⊣

LEMMA 1.3. *If $c$ is good, then $Z_d(c)$ does not depend on $d$.*

PROOF. See (the proof of) [Pi, Chapter 2, Lemma 4.6]. ⊣

So by $Z(c)$ we mean $Z_d(c)$, where $d$ is any sequence as in the definition of $Z_d(c)$. By $Z(a/b)$ we mean $Z(ab)/Z(b)$. Also clearly $Z(a/b)$ depends on $t(ab/\emptyset)$ only. Thus if $p$ is a (complete) type over $b$, by $Z(p)$ we mean $Z(a/b)$, where $a$ is any sequence that realizes $p$.

Suppose $\Phi(x)$ is a collection of formulas. We say that $\Phi$ is good if the following holds:

(i) If $a$ realizes $\Phi$, then $a$ is a sequence of elements of $M \cup C^{cb}$.

(ii) The number of (complete) types $p$ over $\emptyset$ such that $\Phi \subseteq p$ is finite.

If $\Phi(x, y)$ is good, then by $\exists x\Phi(x, y)$ we mean the set of all formulas $\phi(y)$ such that if $(a, b)$ realizes $\Phi$, then $\models \phi(b)$.

LEMMA 1.4. *If $\Phi(x, y)$ is good, then also $\exists x\Phi(x, y)$ is good.*

PROOF. Clearly, there are only finitely many complete types $p$ such that if $b$ realizes $p$, then there is $a$ such that $(a, b)$ realizes $\Phi$. Let $p_i$, $i < n$, be a complete list of these. For all $i < n$, choose $\phi_i \in p_i$ such that $\phi_i \notin p_j$ for any $j \neq i$. Then $\exists x\Phi(x, y)$ contains the following set:

$$\{\vee_{i<n}\phi_i\} \cup \{\phi_i \rightarrow \psi \mid i < n, \psi \in p_i\}.$$

Clearly it follows that $\exists x\Phi(x, y)$ is good. ⊣

Suppose $\Phi(x, y)$ is good and $b$ realizes $\exists x\Phi(x, y)$. By $Z(\Phi(x, b))$ we mean $\sum_{i<n} Z(p_i)$, where $p_i$, $i < n$, is a complete list (without repetition) of complete types $p \supseteq \Phi(x, b)$ over $b$.

LEMMA 1.5. *Suppose $A \subseteq M$ is finite, $a$ and $b$ are elements of $M$, $c$ and $d$ are elements of $C^{cb}$, $a, b \notin bcl(A \cup \{c\})$, $a, b \notin bcl(A \cup \{d\})$, $c =$*

$cb((a,b)/A \cup \{c\}), d = cb((a,b)/A \cup \{d\}), U((a,b)/c) = U((a,b)/d) = 1$, $U(c/A)$ *and* $U(d/A)$ *are* $\geq 2$ *and* $c \downarrow_A d$. *Let* $p(x,y,u,w) = t((a,b,c,A)/\emptyset)$, $q(x,y,v,w) = t((a,b,d,A)/\emptyset)$ *and* $r(u,v,w) = t((c,d,A)/\emptyset)$. *Then* $\Phi = p \cup q \cup r$ *is good.*

PROOF. An easy calculation with $U$-rank shows that if $c'$ realizes

$$p(a,b,u,A) \cup r(u,d,A),$$

then $c' \downarrow_{A \cup \{a,b\}} d$. Thus, since even in homogeneous model theory, Lascar strong types are stationary, the number of completions of $\Phi$ is at most $\text{mult}(c/A \cup \{a,b\})$. So it suffices to prove the following claim. ⊣

CLAIM 1.6. Suppose $A \subseteq M$ is finite and $b$ is an element of $C^{cb}$. Then $\text{mult}(b/A)$ is finite.

PROOF. Suppose not. Let $b_i$, $i < \omega$, witness this. For all $i < \omega$, choose a sequence $a_i \in M$ such that $b_i \in dcl(a_i)$. Clearly, we may choose these so that for all $i < j < \omega$, $t(b_i \cup a_i/A) = t(b_j \cup a_j/A)$ $(t(b_i/A) = t(b_j/A))$. But then for all $i < j < \omega$, $Lstp(a_i/A) \neq Lstp(a_j/A)$ (this is because if $(c_i)_{i<\omega} \subseteq M$ is an infinite indiscernible sequence over $A$ and $d_i$, $i < \omega$, are such that $d_0 \in dcl(A \cup c_0)$ and for all $i < \omega$, $t(d_i \cup c_i/A) = t(d_0 \cup c_0/A)$, then $(d_i)_{i<\omega}$ is an indiscernible sequence over $A$ and either infinite or contains just one element). Thus $\text{mult}(a_0/A)$ is infinite. This contradicts our assumption that $M$ is (strongly minimal and) unimodular. ⊣

With the lemmas above, the proof of Theorem 1 is a verbatim copy of the proof of [Pi, Chapter 2, Theorem 4.15]. So we leave the rest of the proof to the reader.

**§2. Finding a group.** From now on, in this paper, we assume that $M$ is strongly minimal $\omega$-stable, 2-pseudolinear and does not interpret a non-classical group and start to work towards the proof of Theorem 2.

In this section we use ideas from the proof of [Pi, Chapter 5, Proposition 3.2]. The main difference is that we do not have as smooth theory of groups interpretable in $M$ as in the elementary case. So to find the algebraically closed field $F$ that gives the final contradiction we must work much harder than in [Pi] ($F$ is defined in Section 4). The assumption that $M$ does not interpret a non-classical group is used to show that the multiplication of the field is commutative. Without this assumption the author of this paper can not even prove that $F$ is a division ring.

Recall that $S \subseteq M^{eq}$ is $M$-definable over $A$ if every automorphism of $M^{eq}$ fixes $S$ setwise if it fixes $A$ pointwise. For finite $A$, we say also that $S$ is $A$-definable if $S$ is $M$-definable over $A$. If $S \subseteq M^{eq}$ is a set $M$-definable over $A$ and $B \subseteq M^{eq}$, then we say that $a \in S$ is generic over $B$ if $U(a/A \cup B) = \max\{U(c/A) \mid c \in S\}$.

We define an equivalence relation $E^*$ on $M$ so that $aE^*b$ if $a \in bcl(b)$. We usually denote $a/E^* \in M^{eq}$ as $a$. It is always clear from the context which one we mean.

In [Hy3] it is shown that if $M$ is linear (= 1-pseudolinear), then the pregeometry is locally modular. So for a contradiction we assume that $M$ (and so $M/E^*$) is not linear, i.e. that there are a finite sequence $e \in M$ and elements $a_0, a_1 \in M/E^* - bcl(e)$ such that $U(a/e) = 1$ and $U(cb(t(a/bcl(e)))/\emptyset) = 2$, where $a = (a_0, a_1)$.

We let $D$ be the set of all pairs $b$ such that $t(b/bcl(e)) = t(a/bcl(e))$. Then $D$ is strongly minimal.

DEFINITION 2.1. We let $G'$ be the set of triples $p = (e', b, c)$ such that

(i) $e'$ is a sequence of elements of $M$ of (any) finite length,
(ii) $b, c \in D$,
(iii) for all $b' \in D - bcl(e \cup e')$ there is exactly one $c' \in D - bcl(e \cup e')$ such that $t((b', c')/bcl(e \cup e')) = t((b, c)/bcl(e \cup e'))$ and for all $c' \in D - bcl(e \cup e')$ there is exactly one $b' \in D - bcl(e \cup e')$ such that $t((b', c')/bcl(e \cup e')) = t((b, c)/bcl(e \cup e'))$.

For all $b' \in D - bcl(e \cup e')$ and $p = (e', b, c) \in G'$, we denote by $p(b')$ the unique $c' \in D - bcl(e \cup e')$ such that $t((b', c')/bcl(e \cup e')) = t((b, c)/bcl(e \cup e'))$.

Notice that since $M$ is 2-pseudolinear, for all $(e', b, c) \in G'$,

$$U(cb(t((b, c)/bcl(e \cup e'))/e) \leq 2.$$

We define an equivalence relation $E$ on $G'$ as follows: $pE\sigma$ holds, $p = (e', b, c)$ and $\sigma = (e'', b', c')$, if there is $b'' \in D - bcl(e \cup e' \cup e'')$ such that $p(b'') = \sigma(b'')$.

We let $G^* = G'/E \subseteq M^{eq}$ and denote $p = p/E$. It is easy to see that $G^*$ is a group and that for all $p \in G^*$, $U(p/e) \leq 2$ ($p\sigma = \tau$ if for some (all) $b \in D - bcl(e \cup \{p, \sigma, \tau\})$, $\tau(b) = p(\sigma(b))$).

LEMMA 2.2. *There is $p \in G^*$ such that $U(p/e) = 2$. In particular, $G^* \neq \{1\}$.*

PROOF. Let $e'$ be such that $t(e'/bcl(\emptyset)) = t(e/bcl(\emptyset))$ and $e' \downarrow_\emptyset e$. Let $b = (b_0, b_1) \in D$ be such that $b \downarrow_e e'$. Choose $c_0 \in M$ be such that $t(e'b_1c_0/bcl(\emptyset)) = t(ea_0a_1/bcl(\emptyset))$. Clearly $c_0 \notin bcl(e)$ and so there is $c_1$ such that $c = (c_0, c_1) \in D$. As in [Pi, Chapter 5, Lemma 3.3], we can see the following:

(*) If $t((b, c')/bcl(e \cup e')) = t((b, c)/bcl(e \cup e'))$, then $c'_0 = c_0$ (i.e. $c'_0 \in bcl(c_0)$) and $c'_1 = c_1$ i.e. $c' = c$.

Similarly, one can see:

(**) If $t((b', c)/bcl(e \cup e')) = t((b, c)/bcl(e \cup e'))$, then $b' = b$.

By (*) and (**), $p = (e', b, c) = (e', (b_0/E^*, b_1/E^*), (c_0/E^*, c_1/E^*)) \in G'$.

It is also easy to see that

$$U((\rho/E)/e) = U(cb(t((b,c)/bcl(e \cup e')))/e) = U(cb(t((b_1,c_0)/bcl(e')))/e) = 2.$$

Since $\rho/E \in G^*$, we are done. ⊣

We now name by new constants the elements of $e$ making $bcl(\emptyset) = bcl(e)$.

We say that an $A$-definable subgroup $H$ of $G^*$ has unique generics if for all finite $A \subseteq B \subseteq M^{eq}$ and generic $g, f \in H$ over $B$, there is an automorphism $F$ of $M^{eq}$ such that $F \restriction B = id$ and $F(g) = f$.

LEMMA 2.3. *Suppose $H$ is an $A$-definable subgroup of $G^*$ and*

$$U(H/A) = \max\{U(h/A) \mid h \in H\} = n > 0.$$

*Then there is $A$-definable $H' \subseteq H$ such that $H'$ has unique generics and*

$$U(H'/A) = n.$$

*In particular, there is $\emptyset$-definable $G^\circ \subseteq G^*$ such that $U(G^\circ/\emptyset) = 2$ and $G^\circ$ has unique generics.*

PROOF. Let $S = \{Lstp(g/A) \mid g \in H \text{ is generic over } A\}$. For $p \in S$ and $g \in H$, let $g(p) = Lstp(gf/A)$, where $f \in H$ is generic over $\{g\} \cup A$ and $Lstp(f/A) = p$. It is easy to see that this defines an action of $H$ on $S$. Let $H'$ be the center of this action. Then $H'$ is $A$-definable. Since $H/H'$ is bounded, it is easy to see that $U(H'/A) = n$. Finally, $H'$ has unique generics: For a contradiction, suppose $g, f \in H'$ are generic over $A$ and $Lstp(g/A) \neq Lstp(f/A)$. Since $H'$ is $A$-definable, we can choose these so that $g$ is generic over $\{f\} \cup A$. Then $fg^{-1}(Lstp(g/A)) = Lstp(f/A)$ (since $g$ is generic over $\{fg^{-1}\} \cup A$). But then $fg^{-1} \notin H'$, a contradiction. ⊣

LEMMA 2.4. *Suppose $a_0, \ldots, a_3 \in D$ be such that $U((a_0, \ldots, a_3)/\emptyset) = 4$. Then there is $\rho \in G^\circ$ such that $U((a_0, a_1)/\rho) = 2$ and $\rho(a_0) = a_2$ and $\rho(a_1) = a_3$.*

PROOF. Choose $\rho = (e', b, c) \in G^\circ$ so that $U(\rho/\emptyset) = 2$ and $\rho \downarrow_\emptyset (a_0, a_1)$. It suffices to show that $U((a_0, a_1, \rho(a_0), \rho(a_1))/\emptyset) = 4$ (use suitable strong automorphism of $M^{eq}$).

For this, it suffices to show that $\rho \in bcl(\{a_0, a_1, \rho(a_0), \rho(a_1)\})$. For a contradiction, suppose this is not the case. Then $(a_1, \rho(a_1)) \downarrow_{\{a_0, \rho(a_0)\}} \rho$ (or it is the case that $(\alpha_0, \rho(a_0)) \downarrow_\emptyset \rho$, which is easily seen to be impossible). But then there is $e^* \in bcl(\{a_0, \rho(a_0)\}) \cap bcl(\rho)$ such that $e^* = cb(t((b,c)/bcl(e')))$. Since $U(e^*/\emptyset) \leq 1$, $U(\rho/\emptyset) \leq 1$, a contradiction. ⊣

**§3. Getting a total action.** Our group $G^\circ$ acts on the pregeometry $(D, bcl)$ only generically. On the other hand, we have a really good theory on such groups only in the case when the action is total. Using the idea from [Pi, Chapter 5, Remark 1.10], we show that $(D, bcl)$ can be extended to a minimal set in $M^{eq}$ in such a way that $G^\circ$ acts on it totally and the action extends the action of $G^\circ$ on $D$ (and in well-controlled way). This does not make it possible for us to apply existing theory (the extended action does not preserve the closure operation), but it is most useful when we develop a part of the theory in our setting. In the end of this section we show that we can further change the group and the minimal set a bit so that the group acts on the set strictly 2-transitively. This section contains the main body of the work done in this paper.

Let $\mathcal{D}$ be the set of all pairs $(g, a)$, where $g \in G^\circ$ and $a \in D$. We define an equivalence relation $\sim$ on $\mathcal{D}$ so that $(g, a) \sim (f, b)$ if for some (= all, by Lemma 2.3) $h \in G^\circ$ generic over $\{g, f, a, b\}$, $(hg)(a) = (hf)(b)$. It is easy to see that $\sim$ is an equivalence relation. We let $P'$ be the set of all equivalence classes $(g, a)/\sim$, $(g, a) \in \mathcal{D}$. We define the action of $G^\circ$ on $P'$ by $f((g, a)/\sim) = (fg, a)/\sim$.

Lemma 3.1. *The action of $G^\circ$ on $P'$ is well-defined.*

Proof. Suppose $(g, a) \sim (g', a')$ and $f \in G^\circ$. We need to show that $(fg, a) \sim (fg', a')$. For this let $h \in G^\circ$ be generic over $\{f, g, g', a, a'\}$. We need to show that $hfg(a) = hfg'(a')$. Since $h$ is generic over $\{f, g, g', a, a'\}$, $hf$ is generic over $\{g, g', a, a'\}$ ($h \in bcl(\{f, hf\})$). But then $(h(fg))(a) = ((hf)g)(a) = ((hf)g')(a') = (h(fg'))(a')$. ⊣

Lemma 3.2. *If $a \notin bcl(\{f\})$, then $(g, a) \sim (gf^{-1}, f(a))$. In particular, if $a \notin bcl(\{g\})$, then $(g, a) \sim (1, g(a))$.*

Proof. Let $h \in G^\circ$ be generic over $\{g, f, a\}$. Then $hg$ is generic over $\{f, a\}$ and so also $hgf^{-1}$ is generic over $\{f, a\}$. Thus since $a \notin bcl(\{f\})$, we get $a \notin bcl(\{f, hgf^{-1}\})$. But then $hg(a) = ((hg)(f^{-1}f))(a) = ((hg)f^{-1})(f(a)) = h(gf^{-1})(f(a))$. ⊣

In order to simplify the notation, we write $(g, a)$ also for $(g, a)/\sim$ (unless it is important to make a distinction). Since $a \in D$ and $(1, a)$ are 'interdefinable', if wanted, we can identify them. Also by Lemma 3.2, if $a \notin bcl(\{g\})$, then $g((1, a)) = (1, g(a))$. So our action extends the generic action of $G^\circ$ on $D$. Finally, if $(1, a), (1, b) \notin bcl(X)$ for finite $X$, then there is an automorphism $F$ of $M^{eq}$ such that $F \restriction X = id$ and $F((1, a)) = (1, b)$. So to show that $P' - bcl(\emptyset)$ is strongly minimal, it is enough to prove (i) from the following lemma.

Lemma 3.3.

(i) *If $(g, a) \not\sim (1, b)$ for all $b \in D$, then $(g, a)/\sim \in bcl(\emptyset)$.*

(ii) *If $x \in P'$ is generic over $\{f, g\}$ and $f(x) = g(x)$, then $f = g$.*

Proof. (i): Let $f$ be generic over $\{g, a\}$. By Lemma 3.2 (and the definition of $\sim$), it suffices to show that $b = (fg)(a) \in bcl(\{f, a\})$ (if $(g_i, a_i)$, $i < \lambda$, is a collection of such pairs, w.l.o.g., we may assume that for all $i < j < \lambda$, $a_i = a_j$). But even $b \in bcl(\{f\})$, since otherwise by Lemma 3.2 (notice that $fg$ is generic over $\{g, a\}$), $(g, a) \sim (f^{-1}, b) \sim (1, f^{-1}(b))$.

(ii): Since $x$ is generic over $\{f, g\}$, $x = (1, a)$ for some $a \in D$ and $a$ is generic over $\{f, g\}$. By Lemma 3.2, $(1, g(a)) = g(x) = f(x) = (1, f(a))$. Thus $f(a) = g(a)$ and so $f = g$. ⊣

We want that $G^\circ$ acts 2-transitively on a minimal set. For this we make small changes to both $P'$ and $G^\circ$.

We define an equivalence relation $\approx$ on $P'$ by $x \approx y$ if for all $\rho \in G^\circ$, $U((\rho(x), \rho(y))/\emptyset) \leq 1$.

LEMMA 3.4. (i) *$x \approx y$ iff for some (all) $\rho \in G^\circ$ generic over $\{x, y\}$, $U((\rho(x), \rho(y))/\emptyset) \leq 1$*

(ii) *$\approx$ is an equivalence relation.*

(iii) *If $U((x, y)/\emptyset) = 2$, then $x \not\approx y$.*

(iv) *If $x \approx y$, then $\rho(x) \approx \rho(y)$ for all $\rho \in G^\circ$.*

PROOF. (i): Clearly it suffices to show that if $x \not\approx y$, then for some $\rho \in G^\circ$ generic over $\{x, y\}$, $U((\rho(x), \rho(y))/\emptyset) \geq 2$. Since $x \not\approx y$, there is some $\sigma \in G^\circ$ such that $U((\sigma(x), \sigma(y))/\emptyset) \geq 2$. Let $\rho \in G^\circ$ be generic over $\{x, y, \sigma\}$. Then $\rho$ is generic over $\{\sigma(x), \sigma(y)\}$ and thus $U((\rho(\sigma(x)), \rho(\sigma(y)))/\emptyset) \geq 2$. Since $\rho\sigma$ is generic over $\{x, y\}$, the claim follows.

(ii): Suppose $x \approx y$ and $y \approx z$ but $x \not\approx z$. Then there is $\rho \in G^\circ$ such that $(x, y, z) \downarrow_\emptyset \rho$ and $\rho(x) \notin bcl(\rho(z))$. Then $bcl(\rho(x)) \cap bcl(\rho(z)) = bcl(\emptyset)$. On the other hand, $\rho(y) \in bcl(\rho(x)) \cap bcl(\rho(z))$. Since $\rho$ is generic over $\{y\}$, $\rho(y) \notin bcl(\emptyset)$ and we have a contradiction.

(iii): Let $\rho = id$.

(iv): Suppose not. Then there is $\sigma \in G^\circ$ such that $U((\sigma(\rho(x)), \sigma(\rho(y))/\emptyset) = 2$. But then $\sigma\rho \in G^\circ$ contradicts the assumption. ⊣

So by Lemma 3.4, $G^\circ$ acts on $P = P'/\approx$. We write $\rho \approx \sigma$ if for some (all) $a \in P - bcl(\{\rho, \sigma\})$, $\rho(a) = \sigma(a)$. Then clearly $\approx$ is a congruence on $G^\circ$ and we let $G = G^\circ/\approx$.

LEMMA 3.5. *Suppose $f, g \in G^\circ$. If $f \approx g$, then $f \in bcl(\{g\})$.*

PROOF. Let $a \in P'$ be generic over $\{f, g\}$. By Lemma 3.4(ii), $f(a) \in bcl(\{g(a)\})$. Thus $U((a, f(a))/\{g\}) = 1$. It follows that $(a, f(a)) \downarrow_{\{g\}} f$. Then $cb(stp((a, f(a))/\{f\}) \in bcl(\{g\})$. Since

$$f \in bcl(cb(Lstp((a, f(a))/\{f\}))),$$

the claim follows. ⊣

LEMMA 3.6. *Suppose $g, f \in G^\circ$ are such that $g \approx f$ and $x = (h, a)/\sim \in P'$. Then $g(x) \approx f(x)$.*

PROOF. Suppose not. Then there is $h' \in G^\circ$ generic over $\{f, g, h, a\}$ such that

$$(*) \qquad U((h'fh)(a), (h'gh)(a))/\emptyset) = 2.$$

By Lemma 3.4(iii), it is easy to see that if $f_0 \approx f_1$ and $g_0 \approx g_1$, then $f_0 g_0 \approx f_1 g_1$. Thus $h'fh \approx h'gh$. Thus by Lemma 3.5, $h'fh \in bcl(\{h'gh\})$. But then $(1, a)$ is generic over $\{h'fh, h'gh\}$. Since

$$h'fh \approx h'gh,\ U(((h'fh)(a), (h'gh)(a))/\emptyset) \leq 1.$$

This contradicts (*). ⊣

So $G$ acts on $P$.

For $x, y \in \mathcal{D}$, we write $x \approx y$ if $(x/\sim) \approx (y/\sim)$. Notice that $(g_0, a_0) \approx (g_1, a_1)$, if for all $f \in G^\circ$ generic over $\{g_0, a_0, g_1, a_1\}$,

$$U((fg_0)(a_0), (fg_1)(a_1))/\emptyset) \leq 1.$$

Again to simplify the notation, we write $(g, a)$ also for $(g, a)/\approx$.

LEMMA 3.7. *For all $(g_i, a_i) \in P$, $i < 4$, if $(g_0, a_0) \neq (g_1, a_1)$ and $(g_2, a_2) \neq (g_3, a_3)$, then there is $f \in G$ such that $f((g_0, a_0)) = (g_2, a_2)$ and $f((g_1, a_1)) = (g_3, a_3)$.*

PROOF. Let $h' \in G^\circ$ be generic over $\{g_i, a_i \mid i < 4\}$ and $h$ generic over $\{h'\} \cup \{g_1, a_i \mid i < 4\}$. Then $U((h'g_0)(a_0), (h'g_1)(a_1), (hg_2)(a_2)), (hg_3)(a_3))/\emptyset) = 4$. By Lemma 2.4, there is $g \in G^\circ$ such that for $i < 2$, $g((h'g_i)(a_i)) = hg_{i+2}(a_{i+2})$. We claim that $h^{-1}gh'/\approx$ is as wanted: for $i < 2$,

$$\begin{aligned}
&(h^{-1}gh')((g_i, a_i)/\approx) \\
&= (h^{-1}g)((h'g_i, a_i)/\approx) = (h^{-1}g)((1, h'g_i(a_i))/\approx) \\
&= h^{-1}((1, (hg_{i+2})(a_{i+2}))/\approx) = h^{-1}((hg_{i+2}, a_{i+2})/\approx) \\
&= (g_{i+2}, a_{i+2})/\approx.
\end{aligned}$$

⊣

LEMMA 3.8. *Suppose $H \subseteq G$ is an $A$-definable subgroup of $G$, has unique generics and $U(H/A) = 1$. Then $H$ is Abelian and $H$ is 1-determined i.e. for all $f, g \in H$ and $x \in P$, if $f(x) = g(x)$, then $f = g$.*

PROOF. We define a pregeometry $cl$ on $H$ as follows: $g \in cl(\{g_0, \ldots, g_n\})$ if for some (all) $x \in P$ generic over

$$\{g, g_0, \ldots, g_n\} \cup A,\ g(x) \in bcl(\{g_0(x), \ldots, g_n(x)\} \cup A).$$

It is easy to see that this is a pregeometry and that $(H, cl)$ is a group carrying an $\omega$-homogeneous pregeometry (see [HLS]). So either $H$ is Abelian or non-classical. By our assumptions on $M$, $H$ is not non-classical.

We are left to show that $H$ is 1-determined. Let $f, g$ and $x$ be as in the claim. Let $h \in H$ be generic over $\{f, g, x\}$. Then $h^{-1}(x)$ is generic over $\{f, g\}$. Since $H$ is Abelian, $g(h^{-1}(x)) = h^{-1}(g(x)) = h^{-1}(f(x)) = f(h^{-1}(x))$. Thus $g = f$ by the definition of $G$. ⊣

We say that $G$ has hereditarily unique generics, if it has unique generics and for all $x \in P$, there is a group $H' \subseteq \{g \in G \mid g(x) = x\}$ such that $U(H'/\{x\}) = 1$ and $H'$ has unique generics.

LEMMA 3.9. *G has hereditarily unique generics.*

PROOF. By the proof of Lemma 2.3, it suffices to show that $G$ has unique generics. This is clear by Lemma 3.5 and the choice of $G^\circ$. ⊣

LEMMA 3.10. *$G$ is weakly 2-determined i.e. if $f, g \in G$, $x, y \in P$ are generic over $\emptyset$, $x \neq y$ and $g(x) = f(x)$ and $g(y) = f(y)$, then $f = g$.*

PROOF. We show first that we may assume that $u((x, y)/\emptyset) = 2$: Let $h \in G$ be generic. Then also $h^{-1}$ is generic and thus letting $x' = h^{-1}(x)$ and $y = h^{-1}(y)$, $U((x', y')/\emptyset) = 2$. Also $fh(x') = gh(x')$ and $fh(y') = gh(y')$. Since it suffices to show that $fh = gh$, the claim follows.

For a contradiction, suppose $f \neq g$. Then there is $h \in G - \{1\}$ such that $U(\{z \in P \mid h(z) = z\}/\emptyset) = 2$. We claim that $h$ is a generic element of $G$. For this, choose $z_0, z_1$ so that $z_0$ is generic over $\{z \in P \mid h(z) = z\}$ and $z_1$ is generic over $\{z \in P \mid h(z) = z\} \cup \{z_0, h(z_0)\}$. It suffices to show that $U((z_0, z_1, h(z_0), h(z_1))/\emptyset) = 4$. But this is clear since otherwise there is $z \notin bcl(\{z_0, z_1, h(z_0), h(z_1)\})$ such that $h(z) = z$, which would imply that $h = 1$ (since $h \in bcl(\{z_0, z_1, h(z_0), h(z_1)\})$).

Let $z \in P$ be generic and let $H'$ be as in the definition of hereditarily unique generics for $x = z$. Let $h' \in H'$ be a generic element of $H'$. By Lemma 3.8, $z$ is the only element in $P$ such that $h'(z) = z$. Thus by what was shown above and the uniqueness of generics, $h'$ is not a generic element of $H$. We get the required contradiction by showing that $h'$ is in fact a generic element of $H$.

Again, choose $z_0, z_1 \in P$ so that $z_0$ is generic over $\{z, h'\}$ and $z_1$ is generic over $\{z, h'\} \cup \{z_0, h'(z_0)\}$. It suffices to show that $U((z_0, z_1, h'(z_0), h'(z_1))/\emptyset) = 4$. If not then since $z, h' \in bcl(\{z_0, z_1, h'(z_0), h'(z_1)\})$, $U((z, h', z_0, z_1)/\emptyset) < 4$. But then $U(h'/\{z\}) = 0$ i.e. $h'$ is not generic. ⊣

COROLLARY 3.11. *$G$ is 2-determined i.e. if $x \neq y$ and $f(x) = h(x)$ and $f(y) = h(y)$, then $f = h$.*

PROOF. Let $x = (g_0, a_0)/\approx$ and $y = (g_1, a_1)/\approx$ Let $g' \in G$ be generic over $\{g_0, g_1, a_0, a_1\}$. Then by Lemma 3.2, $(g_0, a_0) \sim (g_0 g', (g')^{-1}(a_0))$. Also $(g_0 g')^{-1}$ is generic over $\{g_1, a_1\}$. But then $(g_0 g')^{-1} g_1$ is generic over $\{a_1\}$ i.e. $a_1 \notin bcl(\{(g_0 g')^{-1} g_1\})$. Thus by Lemma 3.2,

$$(g_1, a_1) \sim (g_0 g', ((g_0 g')^{-1} g_1)(a_1));$$

i.e., we may assume that $g_0 = g_1 = g$. Let $h'$ be generic over $\{g, h, f, a_0, a_1\}$. Then $a_i \notin bcl(\{h' f g\})$ and $a_i \notin bcl(\{h' h g\})$ for $i < 2$. So since $f((g, a_i)/\approx) = h((g, a_i)/\approx)$ for $i < 2$, $(h' f g)((1, a_i)/\approx) = (h' h g)((1, a_i)/\approx)$ for $i < 2$. By Lemma 3.10, $h' h g = h' f g$ and so $h = f$. ⊣

§4. **Finding a field.** In this section we show that the action of $G$ on $P$ gives rise to an $M$-interpretable algebraically closed field. We follow the construction from [HLS], which is based on a construction due to Hrushovski.

We let $I = \{g \in G \mid g^2 = 1\}$ and we fix generic $a = (1, a^*)/\approx \in P$ and let

$$N = \{g \in G \mid \{g'(a) \mid g' \in I \wedge gg' \notin I\} \text{ is of bounded dimension}\}$$

and $G_a = \{g \in G \mid g(a) = a\}$.

LEMMA 4.1.

(i) *If* $g, g' \in I, b \in P$ $g(b) = g'(b)$ *and* $g(b) \neq b$, *then* $g = g'$.
(ii) *If* $g, g' \in I$, $g(a) \downarrow_{\emptyset} a \cup g'(a)$ *and* $g'(a) \downarrow_{\emptyset} a \cup g(a)$, *then* $gg' \in N$.
(iii) *If* $g, g' \in N$ *and* $g(a) = g(a')$, *then* $g = g'$.
(iv) $N$ *is a subgroup of* $G$.

PROOF. (i): Since $g(g(b)) = b$ and similarly for $g'$, the claim follows from Corollary 3.11.

(ii): It is easy to see that $g'(a) \downarrow_A a \cup g(g'(a))$ and since $gg'g' = g \in I$, $gg'f \in I$ holds for all $f \in I$ generic over $\{gg', a\}$.

(iii): Assume first that $g(a) \neq a$. Choose $f \in I$ and $b \in P$ such that $f$ is generic over $\{a, g, g'\}$ and $f(b) = a$. Then $gf, g'f \in I$ and since $gf(b) = g'f(b)$, by (i), $gf = g'f$ and thus $g = g'$.

Assume then that $g(a) = a$. We show that $g = 1$. Assume not. By Corollary 3.11, if $b \in P - \{a\}$, then $g(b) \neq b$. Let $f \in I$ be generic over $\{a, g\}$ and let $b = f(a) \neq a$. Since $g \in N$, $gfgf(a) = a$ which can easily be seen to be impossible.

(iv): Let $g, g' \in N$. We show first that $gg' \in N$. Choose $f \in I$ generic over $\{a, g, g'\}$. Then it is easy to see that $g'(f(a)) \downarrow_{\emptyset} a \cup g(a)$ and since $g' \in N$ $g'f \in I$. But then since $g \in N$, $g(g'f) \in I$.

Let $g \in N$. We show that $g^{-1} \in N$. If $g^2 = 1$, the claim is clear. So we assume that $g^2 \neq 1$. By (iii), $g(a) \neq a$. Let $f \in I$, be generic over $\{a, g\}$. Then $gf \in I$ and thus $gfgf = 1$ i.e. $g^{-1} = fgf$. But by (ii), $fgf \in N$. ⊣

LEMMA 4.2. *$N$ is Abelian.*

PROOF. From Lemma 4.1(iii) it follows easily that $N$ carries an $\omega$-homogeneous pregeometry. Also clearly $N$ is $\boldsymbol{M}$-interpretable, and so the claim follows since $\boldsymbol{M}$ does not interpret non-classical groups. ⊣

We want to point out that it is possible to show that $N$ is Abelian also without using the assumption that there are no $\boldsymbol{M}$-interpretable non-classical groups, see [HLS].

LEMMA 4.3. *$G_a$ is a subgroup of $G$ and Abelian.*

PROOF. The claim that $G_a$ is a subgroup of $G$ is trivial. The other claim follows from Corollary 3.11 as in the proof of Lemma 4.2. ⊣

THEOREM 4.4.

(i) *$G_a$ acts on $N$ via conjugation and the action is sharply transitive.*
(ii) *There is an $\boldsymbol{M}$-interpretable algebraically closed field $F = (F, +, \cdot)$. Moreover, $(F, +)$ is $N$.*

PROOF. (i): First we show that $G_a$ acts on $N - \{1\}$ via conjugation, i.e. that if $g \in N$ and $f \in G_a$, then $g^f \in N$. For this, let $h \in I$ be generic over $\{a, f, g\}$. Since conjugation is a permutation of $I - \{1\}$ $h^f \in I$ and $h^f$ is generic over $\{a, g^f\}$. Thus it is enough to show that $g^f h^f \in I$ i.e. that $fghf^{-1} \in I$. This clear since $gh \in I$.

By Lemma 3.7, it is easy to see that the action of $G_a$ on $N$ is transitive. By Corollary 3.11, the action is sharply transitive.

(ii): In this proof we denote 1 by 0 for the obvious reason. We let $N$ be the additive group of the field and we pick some $g \in N - \{0\}$, generic over $\{a\}$ and let it be the 1 in the field and from now on, this $g$ is denoted by 1. For all $g \in N - \{0\}$, we let $f_g \in G_a$ be the unique element such that $1^{f_g} = g$. Then for all $g, g' \in N$, we define the multiplication by $g \cdot g' = (g')^{f_g}$. This makes $N$ a field. We show that the multiplication is commutative, the rest is obvious:

$$\begin{aligned} g \cdot g' &= f_g g' f_g^{-1} = (f_g f_{g'}) 1 (f_{g'}^{-1} f_g^{-1}) \\ &= (f_{g'} f_g) 1 (f_g^{-1} f_{g'}^{-1}) = f_{g'} g f_{g'}^{-1} = g' \cdot g. \end{aligned}$$

Also it is clear that this field is $\boldsymbol{M}$-interpretable. Finally, as before, it is easy to see that $F$ carries an $\omega$-homogeneous pregeometry. Thus by [HLS], $F$ is algebraically closed. ⊣

We can now prove the second of our main theorems.

PROOF OF THEOREM 2. Suppose not. Then $\boldsymbol{M}$ does not interpret a non-classical group. So as we have seen above, there is a $U$-rank 1 set $N \subseteq \boldsymbol{M}^{eq}$ with unique generic type (over some finite set) such that on $N$ there is a structure of an algebraically closed field and the structure is invariant (over the finite set) under the automorphisms of $\boldsymbol{M}^{eq}$. As in the elementary case, it follows that the unique generic type of $N$ is not 2-pseudolinear (see [Pi]). On the other hand, for all $a \in N$, there is $b \in \boldsymbol{M}$ such that $a \in bcl(b)$. So since $\boldsymbol{M}$ is 2-pseudolinear, it follows that $N$ is 2-pseudolinear (if $c \in bcl(d)$, $d \in bcl(c)$, $C \subseteq bcl(D)$ and $D \subseteq bcl(C)$, then $c \downarrow_{cb(d/D)} C$ and thus $cb(c/C) \in bcl(cb(d/D))$), a contradiction. ⊣

## REFERENCES

[Hr] E. HRUSHOVSKI, *Locally modular regular types*, ***Classification theory*** (J. Baldwin, editor), Springer-Verlag, Berlin, 1985.

[Hy1] T. HYTTINEN, *Groups acting on geometries*, ***Logic and algebra, Proceedings of conferences on logic and algebra 2000–2001*** (Y. Zhang, editor), Contemporary Mathematics, vol. 302, 2002, pp. 221–233.

[Hy2] ———, *Finiteness of U-rank implies simplicity in homogeneous structures*, ***Mathematical Logic Quarterly***, vol. 49 (2003), pp. 576–578.

[Hy3] ———, *Finitely generated substructures of a homogeneous structure*, ***Mathematical Logic Quarterly***, vol. 50 (2004), pp. 77–98.

[HLS] T. Hyttinen, O. Lessmann, and S. Shelah, *Interpreting groups and fields in some nonelementary classes*, Journal of Mathematical Logic, to appear.

[HS1] T. Hyttinen and S. Shelah, *Strong splitting in stable homogeneous models*, ***Annals of Pure and Applied Logic***, vol. 103 (2000), pp. 201–228.

[HS2] ———, *Main gap for locally saturated elementary submodels of a homogeneous structure*, ***The Journal of Symbolic Logic***, vol. 66 (2001), pp. 1286–1302.

[Pi] A. Pillay, ***Geometric stability theory***, Oxford Logic Guides, vol. 32, Clarendon Press, Oxford, 1996.

[Zi1] B. Zilber, *Structural properties of models of $\omega_1$-categorical theories*, ***Logic, methodology and philosophy of science VII***, North-Holland, Amsterdam, 1986, pp. 115–128.

[Zi2] ———, *Hereditarily transitive groups and quasi-urbanic structures*, ***American Mathematical Society Translations***, vol. 195 (1999), pp. 165–186.

DEPARTMENT OF MATHEMATICS
P.O. BOX 4
00014 UNIVERSITY OF HELSINKI
FINLAND
*E-mail*: Tapani.Hyttinen@helsinki.fi

# DESCRIPTIVE SET THEORY AND UNCOUNTABLE MODEL THEORY

MICHAEL C. LASKOWSKI[†]

**Abstract.** We survey arguments in which methods of classical descriptive set theory are used to obtain information about uncountable models of theories in a countable language. Silver's theorem about Borel equivalence relations is used in the computation of the uncountable spectrum of certain theories and the fact that analytic sets have the property of Baire is useful in the analysis of stable, unsuperstable theories.

In the early days of the development of model theory it was considered natural and was certainly beneficial to assume that the theories under investigation were in a *countable* language. The primary advantage of this assumption was the presence of the Omitting Types Theorem of Grzegorczyk, Mostowski, and Ryll-Nardzewski [1], which generalized arguments of Henkin [3] and Orey [8]. Following this, Vaught [13] gave a very pleasing analysis of the class of countable models of such a theory. This led to Morley's categoricity theorem [7] for certain classes of uncountable models of theories in a countable language.

The landscape was completely altered by the subsequent work of Shelah (see e.g. [11]). He saw that the salient features of Morley's proof did not require the assumption of the language being countable. Indeed, many of notions that were central to Shelah's work, including unstability, the fcp, the independence property and the strict order property, are *local*. That is, a theory possesses such a property if and only if some formula has the property. Consequently, the total number of formulas in the language is not relevant. Still other notions, such as superstability, are not local but can be described in terms of *countable fragments* of the theory. That is, a theory of any cardinality is superstable if and only if all of its reducts to countable fragments of the theory are superstable. Using a vast collection of machinery, Shelah was able to answer literally hundreds of questions about the class of *uncountable* models of certain theories. Most[1] of his arguments do not depend on the cardinality of the underlying language. In particular, he gave a proof of Łoś' conjecture,

[†] Partially supported by NSF Research Grant DMS 0300080.

[1] One place where countability of the theory is useful is his analysis of theories with NOTOP, the negation of the omitting types order property.

Logic Colloquium '03
Edited by V. Stoltenberg-Hansen and J. Väänänen
Lecture Notes in Logic, 24

that the analogue of Morley's theorem[2] holds for theories in languages of any size. Somewhat curiously, whereas Shelah's methods were very good in classifying uncountable models of a theory, they had considerably less to say about the countable models of a theory.

The 'party line' changed as a result of this — a general belief was that if one were concerned with uncountable models of a theory, then the countability of the language should not be relevant. In what follows we give two examples where this is not the case. In both examples the countability of the language allows us to apply classical methods of Descriptive Set Theory in order to prove theorems about the uncountable models of a theory. The first section is by now folklore. It is included to set notation and to indicate that rudimentary descriptive set theory plays a role in the basic underpinnings of model theory. The second section highlights the overall argument of [2] wherein new dividing lines are given to complete the classification of the uncountable spectra of complete theories in countable languages. The third section highlights the main result of [4], which proves a structure theorem for saturated models of a stable theory in a countable language. One consequence of the main theorem is that one dividing line in Shelah's attempt at describing the class of $\aleph_1$-saturated models of a countable theory is redundant (Corollary 3.6(2)).

This article is definitely intended to be a survey. Many definitions, although standard, are not given and proofs of theorems are merely sketched. The reader who is interested in more rigor is referred to [2] for the material in Section 2 and to [4] for Section 3. Throughout the article, classical descriptive set-theoretic results are referred to as 'Facts.' It is noteworthy that all of the facts needed here are relatively soft.

**§1. Stone spaces and $T^{eq}$.** Fix a complete theory $T$ in a language $L$. For a fixed choice of variables $\bar{x} = \langle x_1, \ldots, x_n \rangle$, the quotient of all $L$-formulas $\varphi(\bar{x})$ whose free variables are among $\bar{x}$ modulo $T$-equivalence is a Boolean algebra. Its associated Stone space $S_{\bar{x}}$ consists of the complete types in the variables $\bar{x}$. The Stone space is naturally topologized by declaring the set of subsets $U_\varphi$ to be a basis, where each $U_\varphi = \{p \in S_{\bar{x}} : \varphi \in p\}$.

It is routine to verify that each of the spaces $S_{\bar{x}}$ is compact, Hausdorff, and totally disconnected. Many model theoretic concepts translate easily into this setting. As examples, for any model $M$ of $T$, the set of types in $S_{\bar{x}}$ realized in $M$ is dense, and a type $p$ is isolated in $S_{\bar{x}}$ if and only if it contains a complete formula (i.e., is a principal type).

If, in addition, we assume that the language $L$ is countable, then each of the spaces $S_{\bar{x}}$ can be endowed with a complete metric, hence it is a Polish space.[3]

[2]Specifically, if a theory in a language of size $\kappa$ is categorical in some cardinality greater than $\kappa$, then it is categorical in all cardinals greater than $\kappa$.

[3]Unlike some texts, we allow Polish spaces to have isolated points.

FACT 1.1. Every Polish space is either *countable* (in which case the isolated types are dense) or it contains a perfect subset.

THEOREM 1.2. *For $T$ a complete theory in a countable language,*

1. *$T$ has a countable saturated model if and only if $S_{\bar{x}}$ is countable for all $\bar{x}$; and*
2. *$T$ has a prime (atomic) model if and only if the isolated types of $S_{\bar{x}}$ are dense for all $\bar{x}$.*

In order to 'beautify' these statements and to provide a natural setting for arguments such as Shelah's existence theorem for semiregular types, one commonly passes from the theory $T$ to its expansion $T^{eq}$. Specifically, let $\mathcal{E}$ denote the set of all $L$-formulas $E(\bar{x}, \bar{y})$ that are equivalence relations in models of $T$. We form the language $L^{eq}$ by adjoining to $L$ a new unary predicate symbol $U_E$ and a new function symbol $f_E$ for each $E \in \mathcal{E}$. Then each model $M$ of $T$ has a canonical expansion and extension to an $L^{eq}$-structure $M^{eq}$ which can be described as follows:

The universe of $M^{eq}$ is the disjoint union of the interpretations of the $U_E$'s, where for each $2n$-ary relation $E$, $U_E(M^{eq}) = \{\bar{a}/E : \bar{a} \in M^n\}$. We identify the original structure $M$ with $U_=(M^{eq})$. For each $2n$-ary $E \in \mathcal{E}$, $f_E$ is interpreted as the canonical mapping from $U_=(M^{eq})^n$ onto $U_E(M^{eq})$ given by $\bar{a} \mapsto \bar{a}/E$.

We let $T^{eq}$ denote the theory of any $M^{eq}$ and denote *the* Stone space of $T^{eq}$ to consist of all complete $L^{eq}$ 1-types that contain exactly one formula of the form $U_E(x)$. Since powers of $M$ embed naturally into $M^{eq}$ (for each $n$, $M^n$ is identified with $U_n(M^{eq})$, where $U_n$ corresponds to the $2n$-ary equivalence relation of componentwise equality) $S$ can be thought of as being a 'disjoint union' of the spaces $S_{\bar{x}}$, together with Stone spaces of other, nontrivial quotients.

This space $S$ is not compact, but it is locally compact and Hausdorff. Furthermore, if the original language $L$ is countable, then so is $L^{eq}$ and $S$ is a Polish space. So, the 'beautification' of Theorem 1.2 above is immediate:

THEOREM 1.3. *For $T$ a complete theory in a countable language,*

1. *$T$ has a countable saturated model if and only if $S$ is countable; and*
2. *$T$ has a prime (atomic) model if and only if the isolated types of $S$ are dense.*

We note in passing that the classical theorem of Vaught asserting that for a complete theory $T$ in a countable language, if $T$ has a countable saturated model then $T$ has an atomic model follows immediately from this and Fact 1.1.

FACT 1.4 (Baire Category Theorem). Polish spaces are not meagre (i.e., they are not the countable union of nowhere dense sets).

From this and the characterizations given above, one can easily obtain the following strengthenings of the omitting types theorem:

THEOREM 1.5. *Let $T$ be a complete theory in a countable language.*

1. *If $\Gamma \subseteq S$ is meagre and no $p \in \Gamma$ is isolated, then there is a model $M$ of $T$ omitting each $p \in \Gamma$;*
2. *If, in addition, $T$ does not have a prime model, then there are $2^{\aleph_0}$ nonisomorphic countable models omitting every $p \in \Gamma$.*

The proof of (1) is merely a Henkin construction with conditions added to ensure that no type in $\Gamma$ gets realized. For (2) one performs a Henkin construction on a tree of models indexed by $2^{<\omega}$. The continuum models are described by branches through the tree. At any stage of the construction, one has assigned 'formula-much' information to finitely many nodes in the tree with the understanding that any model that is coded by a branch passing through the node will satisfy the formula. Conditions are included to ensure that each branch will produce a model of the theory that omits every type in $\Gamma$. Additional conditions ensure that distinct branches give rise to nonisomorphic models. In spirit, the proof is like arguments employing Sacks forcing. Variations of this technique are used in the verification of Proposition 2.4 in the next section.

**§2. The uncountable spectrum of a theory.** Given a complete theory $T$ in a countable language, it is natural to ask how many nonisomorphic models of $T$ are present in any cardinality $\kappa$. Whereas the case of $\kappa = \aleph_0$ is still problematic, a full answer is known when $\kappa$ is uncountable. As notation, let $I(T, \kappa)$ denote the number of pairwise nonisomorphic models of $T$ of size $\kappa$. For a theory $T$, its *uncountable spectrum* is the mapping $\kappa \mapsto I(T, \kappa)$ for $\kappa > \aleph_0$.

The following theorem appears in [2]:

THEOREM 2.1. 1. *Among all theories in countable languages, there are exactly twelve 'species' of uncountable spectra (some of which involve a parameter);*

2. *Fix a countable language $L$. The equivalence relation on complete $L$-theories of 'having the same uncountable spectrum' is $\Pi^1_1$ in the standard topology of complete $L$-theories.*

In other words, the determination of the uncountable spectrum of $T$ can be made by analyzing which *countable* configurations of elements embed into models of $T$. Thus, despite the fact that the uncountable spectrum involves classes of uncountable models, the determination of which species the map $\kappa \mapsto I(T, \kappa)$ belongs to reduces to questions about the class of *countable* models of $T$. It is because of this that methods of Descriptive Set Theory proved useful in the proof of Theorem 2.1. In addition to the facts mentioned in the previous section, two more classical results, respectively due to Mazurkiewicz (see e.g., [10]) and Silver [12] are relevant:

FACT 2.2. 1. A $G_\delta$ subspace of a Polish space is Polish.
2. A Borel equivalence relation on a Borel subset of a Polish space has either countably many or $2^{\aleph_0}$ equivalence classes.

Much of the work in computing uncountable spectra was done by Shelah. His analysis proceeds in a 'top-down' fashion. First, he described the theories having the maximal spectrum (i.e., $I(T,\kappa) = 2^\kappa$ for all uncountable $\kappa$). It turns out that the uncountable spectrum is maximal unless $T$ is superstable and, for any triple of models[4] $M_0, M_1, M_2$, if $\{M_1, M_2\}$ are independent over $M_0$, then there is a prime and minimal model over their union, which we denote by $M_1 \bigoplus_{M_0} M_2$. Using these algebraic facts Shelah provided a structure theorem for the class of uncountable models of theories whose spectrum is not maximal: Any uncountable model of such a theory is prime and minimal over a well-founded, independent tree of *countable* submodels. As there are at most $2^{\aleph_0}$ nonisomorphic countable models, an upper bound on the number of nonisomorphic models of size $\kappa$ can be computed from a bound on the depth of the trees that can occur. If the theory admits trees of infinite depth, then Shelah is able to use coding tricks (using certain levels of the tree as 'markers') to obtain a matching lower bound. So we concentrate on countable theories in which every uncountable model is prime and minimal over an independent tree of countable models in which each branch has length at most some fixed finite number $d$. For such theories, the naive upper bound on the number of nonisomorphic models of size $\aleph_\alpha$ alluded to above is $\beth_{d-1}(|\omega + \alpha|^{2^{\aleph_0}})$.

In order to find lower bounds on $I(T,\aleph_\alpha)$ we consider the following scenario: Fix $1 \le n < d$ and an increasing chain

$$\overline{\mathcal{M}} := M_0 \subset_{na} M_1 \subset_{na} \cdots \subset_{na} M_{n-1}$$

of countable models satisfying $wt(M_{i+1}/M_i) = 1$ and (when $i > 0$) $M_{i+1}/M_i$ is orthogonal to $M_{i-1}$. We wish to describe the set of chains $\overline{\mathcal{N}}$ of length $n+1$ extending $\overline{\mathcal{M}}$ and count the number models of a fixed uncountable cardinality that are prime and minimal over a tree of countable models, every branch of which is isomorphic to some $\overline{\mathcal{N}}$.

Accordingly, we call a countable model $N$ a *leaf* of $\overline{\mathcal{M}}$ if $wt(N/M_{n-1}) = 1$ and $N/M_{n-1}$ is orthogonal to $M_{n-2}$. Leaves of $\overline{\mathcal{M}}$ are connected with the space of types

$$R(\overline{\mathcal{M}}) : \{p \in S(M_{n-1}) : p \text{ regular and } p \perp M_{n-2}\}$$

(when $n = 1$ we delete the final condition). Indeed, if $a$ realizes a type in $R(\overline{\mathcal{M}})$ and $N$ is dominated by $M_{n-1} \cup \{a\}$, then $N$ is a leaf of $\overline{\mathcal{M}}$. Conversely, every leaf of $\overline{\mathcal{M}}$ realizes at leastone type in $R(\overline{\mathcal{M}})$. Unfortunately, the

[4] It is equivalent to require that $M_0, M_1, M_2$ be *countable* models.

correspondence between leaves of $\overline{\mathcal{M}}$ and $R(\overline{\mathcal{M}})$ is not tight. It is possible for a single leaf to contain realizations of several types in $R(\overline{\mathcal{M}})$, and it is possible that the same type in $R(\overline{\mathcal{M}})$ can be realized in several nonisomorphic leaves.

For a subset $Y$ of leaves, a *$Y$-tree* is an independent tree in which every branch is isomorphic to $\overline{\mathcal{M}}\frown\langle N\rangle$ for some $N \in Y$, and a model of $T$ is a *$Y$-model* if it is prime and minimal over a $Y$-tree. In order to get a lower bound on the number of $Y$-trees of a certain cardinality we examine subsets of leaves with certain 'separation properties.' Specifically, we call a family $Y$ of leaves of $\overline{\mathcal{M}}$ *diffuse* if

$$N \oplus_{M_{n-1}} V \not\cong_V N' \oplus_{M_{n-1}} V$$

for all distinct $N, N' \in Y$ and all $Y$-models $V$. Similarly, a set $Y$ of leaves is *diverse* if $N \bigoplus_{M_{n-2}} V \not\cong_V N' \bigoplus_{M_{n-2}} V$, again for all distinct $N, N' \in Y$ and all $Y$-models $V$. (It is not hard to show that a diffuse family is diverse.) The following is the content of Propositions 5.6 and 5.8 of [2].

Proposition 2.3. *If there is a diffuse set $Y$ of leaves of cardinality $2^{\aleph_0}$, then for any $\alpha > 0$ there are* $\min\{2^{\aleph_\alpha}, \beth_{n-1}(|\omega+\alpha|^{2^{\aleph_0}})\}$ *nonisomorphic $Y$-models of size $\aleph_\alpha$. If such a $Y$ is diverse then there are at least* $\min\{2^{\aleph_\alpha}, \beth_{n+1}\}$ *nonisomorphic $Y$-models of size $\aleph_\alpha$.*

Our method for computing the uncountable spectra will be to translate dichotomies occurring from descriptive set theory into dichotomies among theories, which will be expressed in terms of the existence or nonexistence of large diffuse or diverse sets of leaves. As an example, if $\{p_i : i \in \kappa\} \subseteq R(\overline{\mathcal{M}})$ are pairwise orthogonal and $Y = \{N_i : i \in \kappa\}$ is a set of leaves of $\overline{\mathcal{M}}$ such that each $N_i$ realizes $p_i$, then $Y$ is diffuse (see e.g., Lemma 3.6 of [2]). We will see below that $R(\overline{\mathcal{M}})$ is a Borel subset of $S(M_{n-1})$ and that nonorthogonality is a Borel equivalence relation. Hence, by Silver's theorem the number of nonorthogonality classes (i.e., the size of a maximal diffuse family obtained in this fashion) is either countable or has size $2^{\aleph_0}$.

In the case when the theory $T$ is totally transcendental (equivalently $\aleph_0$-stable) choosing types from different nonorthogonality classes is the only way of obtaining diffuse sets of leaves. When $T$ is t.t. any leaf $N$ realizes a *strongly regular* type $p \in R(\overline{\mathcal{M}})$. In addition, this type $p$ is realized in any leaf that realizes a type that is nonorthogonal to $p$. Moreover, for any $a$ realizing such a type $p$, there is a prime model over $M_{n-1} \cup \{a\}$. Thus, there is a canonical choice of a leaf $N$ among all possible leaves that realize a certain nonorthogonality class. Hence the size of a maximal diffuse family of leaves is precisely the number of nonorthogonality classes realized in $R(\overline{\mathcal{M}})$. As noted above this number is finite, countably infinite, or $2^{\aleph_0}$.

When $T$ is superstable but not t.t. there are other mechanisms for producing diffuse or diverse sets of leaves. It might be that a leaf $N$ does not realize any strongly regular types. Furthermore, even if $N$ does realize a strongly regular type there need not be a prime model over $M_{n-1}$ and a realization of the type. The effect of one of these 'failures' depends on whether the nonorthogonality class consists of *trivial* or *nontrivial* regular types. The following Proposition, which is the content of Propositions 3.21, 5.6 and 5.8 of [2], illustrates the effect of such failures. The constructions of the large families of leaves are similar to the construction of a large family of countable models in Theorem 1.5(2) in this paper.

PROPOSITION 2.4. *Fix a type $q \in R(\overline{\mathcal{M}})$, a realization a of q, and an uncountable cardinal $\aleph_\alpha$. Suppose that* either *there is a perfect set of types in $R(\overline{\mathcal{M}})$ nonorthogonal to q* or *there is no prime model over $M_{n-1} \cup \{a\}$. Then:*

1. *If q is trivial then there is a diffuse set Y of leaves of size $2^{\aleph_0}$ (hence there are $\min\{2^{\aleph_\alpha}, \beth_{n-1}(|\omega+\alpha|^{2^{\aleph_0}})\}$ nonisomorphic Y-models of size $\aleph_\alpha$); and*
2. *If q is nontrivial then there is a diverse set Y of leaves of size $2^{\aleph_0}$ (hence there are at least $\min\{2^{\aleph_\alpha}, \beth_{n+1}\}$ nonisomorphic Y-models of size $\aleph_\alpha$).*

Sorting all of this out (i.e., computing $I(T, \aleph_\alpha)$ in all of the scenarios) is where Descriptive Set Theory comes into play. We begin with some rather crude computations.

LEMMA 2.5. *Let T be a stable theory in a countable language and let $M_0 \subseteq M$ be countable models of T. Then:*

1. *$\{p \in S(M) : p$ is trivial, weight 1$\}$ is a $G_\delta$ (i.e., $\Pi_2$);*
2. *$\{p \in S(M) : p$ is almost orthogonal to $M_0$ over $M\}$ is a $G_\delta$;*
3. *$\{p \in S(M) : p$ is almost orthogonal to $q\}$ is a $G_\delta$ for any fixed $q \in S(M)$;*
4. *$\{p \in S(M) : p$ is regular and orthogonal to $M_0\}$ is $\Pi_4$;*
5. *$\{p \in S(M) : p$ is not orthogonal to $q\}$ is $\Sigma_4$ for any fixed $q \in S(M)$.*

The verifications of all of these are routine. One should note the disparity in the complexity of determining 'almost orthogonality' versus 'orthogonality' in the statements above. The good news is that these bounds suffice to show that $R(\overline{\mathcal{M}})$ is a Borel subset of $S(M_{n-1})$ and that nonorthogonality is a Borel equivalence relation on $R(\overline{\mathcal{M}})$. The bad news is that $\Pi_4$ and $\Sigma_4$ sets of types do not have very good closure properties. Thus, we attempt to improve these bounds by showing that these sets have simpler descriptions in some restricted settings.

Our analysis begins as in the t.t. case. If $R(\overline{\mathcal{M}})$ has $2^{\aleph_0}$ nonorthogonality classes then there is a diffuse family $Y$ of leaves of size $2^{\aleph_0}$, so the number of $Y$-models equals the naive upper bound mentioned above. Consequently, we can ignore this case and henceforth *assume that $R(\overline{\mathcal{M}})$ has only countably many nonorthogonality classes.*

Next, we concentrate on the *trivial* regular types in $R(\overline{\mathcal{M}})$. Note that any trivial, weight 1 type is necessarily regular. Also, a trivial, regular type is orthogonal to $M_0$ if and only if it is almost orthogonal to $M_0$. Thus, by

Lemma 2.5(1) and (2),

$$R_{tr}(\overline{\mathcal{M}}) = \{p \in R(\overline{\mathcal{M}}) : p \text{ is trivial, regular, } p \perp M_{n-2}\}$$

is a $G_\delta$ subset of $S(M_{n-1})$ hence is a Polish space itself by Fact 2.2(1).

Let $\{r_i : i < j \leq \omega\}$ be a set of representatives of the nonorthogonality classes of $R_{tr}(\overline{\mathcal{M}})$. For each $i < j$, let

$$X_i = \{p \in R_{tr}(\overline{\mathcal{M}}) : p \text{ is not orthogonal to } r_i\}.$$

Whereas this is typically a complicated set (*cf.* Lemma 2.5(5)), note that a type $p \in R_{tr}(\overline{\mathcal{M}})$ is an element of $X_i$ if and only if it is orthogonal to some (equivalently to every) type in $X_k$ for all $k < j$, $k \neq i$. But, since orthogonality is equivalent to almost orthogonality for trivial types, it follows from Lemma 2.5(3) and the countability of $j$ that each $X_i$ is a $G_\delta$ subset of $R_{tr}(\overline{\mathcal{M}})$, hence is a Polish space. But now, by Fact 1.1, either each $X_i$ is countable, or for some $i < j$, $X_i$ contains a perfect subset. In the latter case Proposition 2.4 yields a diffuse set of leaves of size $2^{\aleph_0}$. As we have dispensed that case, we may now *assume that each $X_i$ is a countable Polish space.* Hence by Fact 1.1, each $X_i$ has an isolated point. In this context, an isolated point must be strongly regular (see e.g., [6, D.15]). So, to finish the analysis of the trivial types, we ask whether there is any trivial, strongly regular type $p$ such that there is no prime model over $M_{n-1}$ and a realization of $p$. If a prime model fails to exist, then by Proposition 2.4 there is a diffuse family of size $2^{\aleph_0}$. If the requisite prime models do exist, then the set of trivial types in $R(\overline{\mathcal{M}})$ acts as in the t.t. case.

Now, if we are not done already (i.e., produced a diffuse family of leaves of size continuum) then we have argued that only countably many nonorthogonality classes are represented in $R(\overline{\mathcal{M}})$ and the set $R_{tr}(\overline{\mathcal{M}})$ is countable. Let $Q$ be a complete set of representatives of regular types represented in $R(\overline{\mathcal{M}})$. *Since we are free to work in $T^{eq}$* it follows from Lemma 8.2.20 of [9] that for each $q \in Q$ there is a regular type $q'$ nonorthogonal to $q$ and a formula $\theta \in q'$ such that $\theta$ is $q'$-simple, has $q'$-weight 1 and moreover $q'$-weight is definable inside $\theta$. Without loss, we may assume that $q' = q$. Then the set

$$Z_q = \{p \in S(M_{n-1}) : \theta \in p,\ w_q(p) = 1\}$$

is a closed subset of $S(M_{n-1})$. Note that if $p \in Z_q$, then $p \not\perp q$, hence $p \perp M_{n-2}$, so $p \in R(\overline{\mathcal{M}})$. Conversely, if $p \in R(\overline{\mathcal{M}})$, $p \not\perp q$, and $\theta \in p$, then $p \in Z_q$. The analysis is now similar to the above. If $Z_q$ is uncountable for some $q$ then by Lemma 1.1 it contains a perfect subset, so there is a diverse subset of size continuum by Proposition 2.4. On the other hand, if each $Z_q$ is countable, then it contains an isolated point. That is, every nonorthogonality class represented in $R(\overline{\mathcal{M}})$ has a strongly regular representative. Next, if there is no prime model over $M_{n-1}$ and a realizationof some strongly regular type in $R(\overline{\mathcal{M}})$, then there is a diverse family of leaves of size $2^{\aleph_0}$. Finally, if there is always a prime model over such sets, then as in the t.t. case, we have a canonical choice of a leaf corresponding to each nonorthogonality class of $R(\overline{\mathcal{M}})$.

These considerations allow us to compute the spectrum of a theory in almost all cases.

**§3. Stable theories and the death of DIDIP.** In this section we consider strictly stable theories (i.e., stable but not superstable) in a countable language. The important distinction is that in a superstable theory, any type over a model is based on a *finite* subset of the model, hence by a single element if we pass to $T^{eq}$. It is because of this fact that almost all of the common stability-theoretic adjectives (see e.g., Lemma 2.4) give rise to Borel subsets of the Stone space. However, when $T$ is countable and strictly stable then types over models are based on *countable* subsets of the model. Thus, from the point of view of Descriptive Set Theory, we are forced upward into the projective hierarchy. Whereas the application of DST in this section only requires consideration of $\Sigma^1_1$ sets, it is my belief that further theorems about strictly stable theories in countable languages may require one to consider more complicated projective sets. The key to the proof of Theorem 3.3, which is the main theorem of this section, is the following classical theorem of Lusin and Sierpiński [5].

FACT 3.1. Any $\Sigma^1_1$-subset of a Polish space has the property of Baire (i.e., if $A$ is $\Sigma^1_1$ then there is an open set $U$ such that the symmetric difference $A \triangle U$ is meagre).

Fix a strictly stable theory $T$ in a countable language. Since $T$ is not superstable, results of Shelah (see e.g., [11]) demonstrate that one cannot reasonably classify the class of all uncountable models of $T$, so we pass to the subclass of $\aleph_1$-saturated models of $T$.[5] In this context, call a model $M$ *$\aleph_1$-prime over $A$* if $A \subseteq M$, $M$ is $\aleph_1$-saturated, and $M$ embeds elementarily over $A$ into any $\aleph_1$-saturated model containing $A$. In [11] Shelah proves that $\aleph_1$-prime models exist over any sets $A$.

DEFINITION 3.2. Let $T$ be a countable theory. $T$ has NDOP if for all triples $\{M, M_0, M_1\}$ of $\aleph_1$-saturated models with $M = M_0 \cap M_1$ and $\{M_0, M_1\}$ independent over $M$, if $N$ is $\aleph_1$-prime over $M_0 \cup M_1$ and $p \in S(N)$ is nonalgebraic, then $p$ is not orthogonal to some $M_i$.

More generally, for any infinite cardinal $\mu$, $T$ has $\mu$-NDOP if for all $\alpha < \mu$, all sets $\{M, N\} \cup \{M_i : i < \alpha\}$ of $\aleph_1$-saturated models such that $M \subseteq M_i$ for all $i < \alpha$, $\{M_i : i < \alpha\}$ are independent over $M$ and $N$ is $\aleph_1$-prime over $\bigcup\{M_i : i < \alpha\}$, every nonalgebraic type over $N$ is not orthogonal to some $M_i$.

[5] In this setting a model is $\aleph_1$-saturated if and only if it is $F^a_{\aleph_1}$-saturated in the terminology of Shelah [11] if and only if it is an 'a-model' in the terminology of Pillay [9].

Theorem 1.3 of [4] shows that it is equivalent to require that $N$ be $\aleph_1$-minimal over $\bigcup\{M_i : i < \alpha\}$, so the definition of NDOP given here coincides with the definition given in [11]. In [11] Shelah shows that if $T$ does not have NDOP then $T$ has $2^\kappa$ nonisomorphic $\aleph_1$-saturated models of size $\kappa$ for all $\kappa \geq 2^{\aleph_0}$.

It is easily proved by induction on $\alpha < \omega$ that if $T$ has NDOP then $T$ has $\omega$-NDOP. Furthermore, if $T$ happened to be superstable and had $\omega$-NDOP, then as every type over $N$ is based and stationary over a finite subset of $N$ (hence over an $\aleph_1$-prime model over $\bigcup\{M_i : i \in X\}$ for some finite set $X \subseteq \alpha$) $T$ would have $\mu$-NDOP for all infinite cardinals $\mu$. Arguing similarly, since our $T$ is strictly stable in a countable language, all types over $N$ are based and stationary over a countable subset. Thus, $\aleph_1$-NDOP implies $\mu$-NDOP for all cardinals $\mu$. However, at least at first glance there seems to be a gap between the notions of NDOP and $\aleph_1$-NDOP for such theories. Somewhat surprisingly (as demonstrated by Corollary 3.6(2)) this gap does not exist.

THEOREM 3.3. *For stable theories in a countable language, if $T$ has NDOP then $T$ has $\aleph_1$-NDOP (hence $\mu$-NDOP for all $\mu$).*

The full proof of this theorem is given in Section 5 of [4], but it is instructive to see how it follows from the Lusin-Sierpiński theorem mentioned above. Fundamentally the proof of Theorem 3.3 is very much like the argument that there is no $\Sigma^1_1$-definable nonprincipal ultrafilter on $\mathcal{P}(\omega)$, so we first review that argument. First of all, we endow $\mathcal{P}(\omega)$ with its standard topology by declaring the set $\mathcal{B} = \{U_{A,B} : A, B \text{ are finite subsets of } \omega\}$, where

$$U_{A,B} = \{X \in \mathcal{P}(\omega) : A \subseteq X, B \cap X = \emptyset\}$$

to be a basis of open sets. It is easily checked that $\mathcal{P}(\omega)$ is a Polish space. Note that a subset $Y \subseteq \mathcal{P}(\omega)$ is a nowhere dense if and only if for all pairs of disjoint finite sets $(E, F)$, there is a pair of disjoint finite sets $(E', F')$ with $E \subseteq E'$, $F \subseteq F'$ and $U_{E',F'} \cap Y = \emptyset$.

Now suppose that $V$ is a $\Sigma^1_1$-definable ultrafilter on $\mathcal{P}(\omega)$. We will show that it is principal. By Fact 3.1 there is an open subset $U$ such that $V \triangle U$ is meagre. That is, *either* $V$ is meagre *or* there is a disjoint pair $(A, B)$ of finite subsets of $\omega$ such that $U_{A,B} \setminus V$ is meagre. However, if $V$ is meagre, say $V = \bigcup\{Y_n : n \in \omega\}$ where each $Y_n$ is nowhere dense, then by iteratively employing the characterization of nowhere denseness mentioned above, we could find a sequence $\langle (E_n, F_n) : n \in \omega\rangle$ of pairs of disjoint finite subsets of $\omega$ such that $E_0 = F_0 = \emptyset$, $E_n \subseteq E_{n+1}$, $F_n \subseteq F_{n+1}$, $U_{E_n,F_n} \cap Y_n = \emptyset$ and $U_{F_n,E_n} \cap Y_n = \emptyset$ for all $n \in \omega$. But then, if we let $X = \bigcup\{E_n : n \in \omega\}$, neither $X$ nor $(\omega \setminus X)$ would be elements of any $Y_n$. In particular, neither $X$ nor $(\omega \setminus X)$ would be elements of $V$, contradicting the assumption that $V$ is an ultrafilter.

We conclude that $U_{A,B} \setminus V = \{Y_n : n \in \omega\}$ for some finite, disjoint sets $A, B \subseteq \omega$ and some nowhere dense sets $Y_n$. For each $n$ let

$$Y_n^* = \{X : (X \cup A \setminus B) \in Y_n\}.$$

Since any two nonempty basic open subsets of $\mathcal{P}(\omega)$ are homeomorphic, each $Y_n^*$ is nowhere dense. Now construct a sequence $\langle (E_n, F_n) : n \in \omega \rangle$ of pairs of disjoint finite subsets of $\omega$ such that $E_0 = A$, $F_0 = B$, $E_n \subseteq E_{n+1}$, $F_n \subseteq F_{n+1}$, $U_{E_n,F_n} \cap Y_n = \emptyset$ and $U_{F_n,E_n} \cap Y_n^* = \emptyset$ for all $n \in \omega$. Let $X = \bigcup\{E_n : n \in \omega\}$ and $Z = (\omega \setminus X) \cup A \setminus B$. Clearly both $X$ and $Z$ are in $U_{A,B}$. Since $X \in U_{E_n,F_n}$ for each $n$, $X \notin Y_n$ for every $n$, hence $X \in V$. Similarly, $(\omega \setminus X) \notin Y_n^*$ for all $n$, so $Z \in V$. Since $V$ is a filter, $X \cap Z = A \in V$. Since $A$ is finite, $V$ is principal.

Towards the proof of Theorem 3.3, assume that $T$ has NDOP and choose $\aleph_1$-saturated models $M, N$ and $\{M_i : i \in \omega\}$ such that $M \subseteq M_i$ for all $i \in \omega$, $\{M_i : i \in \omega\}$ are independent over $M$, and $N$ is $\aleph_1$-prime over their union. As well, fix a nonalgebraic type $p \in S(N)$. We will show that $p$ is nonorthogonal to some $M_i$. To accomplish this let $J$ denote the finite subsets of $\omega$. We inductively build a 'stable system' $\{M_A : A \in J\}$ of submodels of $N$ such that each $M_A$ is $\aleph_1$-prime over $\bigcup\{M_B : B \subsetneq A\}$ and $N$ is $\aleph_1$-prime over $\bigcup\{M_A : A \in J\}$. As notation, for any $X \subseteq \omega$ let $M_X = \bigcup\{M_A : A \in J \cap \mathcal{P}(X)\}$. For any finite $\Delta \subseteq L$, let

> $W_\Delta = \{X \subseteq \omega : p \not\perp N'$ for some $N' \preceq N$ such that $N'$ is $\aleph_1$-prime over $M_X$, $N$ is $\aleph_1$-prime over $N' \cup M_\omega$, and the nonorthogonality is witnessed by some $\varphi(x, y) \in \Delta\}$.

Much of the argument is devoted to showing that each $W_\Delta$ is a $\Sigma_1^1$ subset of $\mathcal{P}(\omega)$ (see [4, Claim 5.18]). So the set $W = \bigcup\{W_\Delta : \Delta \text{ finite}\}$ is a $\Sigma_1^1$-subset of $\mathcal{P}(\omega)$ as well. The set $W$ is not precisely an ultrafilter, but the following lemma shows that it is close to being one:

LEMMA 3.4. 1. *For all $\Delta$ and all $X \subseteq Y \subseteq \omega$, if $X \in W_\Delta$ then $Y \in W_\Delta$;*
2. *For all $X \subseteq \omega$, either $X \in W$ or $(\omega \setminus X) \in W$;*
3. (*Weak intersection*) *If, for some $\Delta$ and $A \subseteq \omega$, there are sets $\{X_j : j \in \omega\} \subseteq W_\Delta$ such that $X_j \cap X_k = A$ for all $j < k < \omega$, then $A \in W_\Delta$.*

Note that Condition (1) is trivial and (2) follows from NDOP. The verification of (3) follows easily from Proposition 2.16 of [4].

The argument now splits into cases as in the ultrafilter argument given above. On one hand, if every $W_\Delta$ is meagre then $W$ is meagre. As in the ultrafilter argument we construct a set $X \in \mathcal{P}(\omega)$ such that $X \notin W$ and $(\omega \setminus X) \notin W$, contradicting Condition (2). On the other hand, if some $W_\Delta$ is nonmeagre then there is a pair $(A, B)$ of disjoint finite sets such that $U_{A,B} \setminus W_\Delta$ is meagre. Now, by diagonalizing across the nowhere dense sets spanning this difference it is easy to produce an infinite family $\{X_j : j \in \omega\} \subseteq U_{A,B} \cap W_\Delta$ as in Condition (3). But this implies that $A \in W_\Delta$, hence $p$ is nonorthogonal to

$M_A$ with $A$ finite. Since NDOP implies $\omega$-NDOP, Theorem 1.3 of [4] implies that $p$ is nonorthogonal to $M_i$ for some $i \in A$ and we finish.

In [4] we obtain the following corollaries to this theorem. The first is reminiscent of Shelah's 'Main Gap' for the class of models of a classifiable theory and the second is rather unexpected.

DEFINITION 3.5. Let $T$ be a stable theory in a countable language.

1. $T$ is *deep* if there is an infinite elementary chain $\langle M_n : n \in \omega \rangle$ of models of $T$ such that $M_{n+1}/M_n$ is orthogonal to $M_{n-1}$ for all $n \geq 1$.
2. $T$ is *shallow* if it is not deep.
3. $T$ has *DIDIP* if there is an elementary chain $\langle M_n : n \in \omega \rangle$ of $\aleph_1$-saturated models of $T$ and a type $p \in S(N)$, where $N$ is $\aleph_1$ prime over $\bigcup_{n \in \omega} M_n$, that is orthogonal to every $M_n$.

Proofs of the following Corollaries appear in [4].

COROLLARY 3.6. *Let T be a stable theory in a countable language.*

1. *If T has NDOP and is shallow then every saturated model of size at least* $2^{\aleph_0}$ *is* $\aleph_1$*-prime and minimal over an independent tree of* $\aleph_1$*-saturated models of size* $2^{\aleph_0}$.
2. *If T has NDOP and is shallow then T does not have DIDIP.*

REFERENCES

[1] A. GRZEGORCZYK, A. MOSTOWSKI, and C. RYLL-NARDZEWSKI, *Definability of sets of models in axiomatic theories*, ***Bulletin de l'Academie Polonaise de Sciences. Serie des Sciences Mathematiques Astronomiques et Physiques***, vol. 9 (1961), pp. 163–167.

[2] B. HART, E. HRUSHOVSKI, and M.C. LASKOWSKI, *The uncountable spectra of countable theories*, ***Annals of Mathematics. Second Series***, vol. 152 (2000), no. 1, pp. 207–257.

[3] L. HENKIN, *A generalization of the concept of ω-consistency*, ***The Journal of Symbolic Logic***, vol. 19 (1954), pp. 183–196.

[4] M.C. LASKOWSKI and S. SHELAH, *Decompositions of saturated models of stable theories*, submitted to *Fundamenta Mathematicae*.

[5] N. LUSIN and W. SIERPIŃSKI, *Sur quelques propriétés des ensembles (A)*, ***Bulletin de l'Academie des Sciences de Cracovie***, (1918), pp. 35–48.

[6] M. MAKKAI, *A survey of basic stability theory with particular emphasis on orthogonality and regular types*, ***Israel Journal of Mathematics***, vol. 49 (1984), pp. 181–238.

[7] M. MORLEY, *Categoricity in power*, ***Transactions of the American Mathematical Society***, vol. 114 (1965), pp. 514–538.

[8] S. OREY, *On ω-consistency and related properties*, ***The Journal of Symbolic Logic***, vol. 21 (1956), pp. 246–252.

[9] A. PILLAY, ***Geometric stability theory***, Oxford University Press, 1996.

[10] R. POL, *The works of Stefan Mazurkiewicz in topology*, ***Handbook of the history of general topology, vol. 2 (San Antonio, TX, 1993)***, Kluwer Acad. Publ., Dordrecht, 1998, pp. 415–430.

[11] S. SHELAH, ***Classification theory***, North-Holland, 1990.

[12] J. SILVER, *Counting the number of equivalence classes of borel and coanalytic equivalence relations*, ***Annals of Mathematical Logic***, vol. 18 (1980), pp. 1–28.

[13] R. Vaught, *Denumerable models of complete theories*, ***Infinitistic methods***, Pergamon, London, 1961, pp. 303–321.

DEPARTMENT OF MATHEMATICS
UNIVERSITY OF MARYLAND
COLLEGE PARK
MD 20742, USA
*E-mail*: mcl@math.umd.edu

# DECIDABLE PROPERTIES OF LOGICAL CALCULI AND OF VARIETIES OF ALGEBRAS

LARISA MAKSIMOVA

**Abstract.** We give an overview of decidable and strongly decidable properties over the propositional modal logics K, GL, S4, S5 and Grz, and also over the intuitionistic logic Int and the positive logic Int$^+$.

We consider a number of important properties of logical calculi: consistency, tabularity, pretabularity, local tabularity, various forms of interpolation and of the Beth property. For instance, consistency is decidable over K and strongly decidable over S4 and Int; tabularity and pretabularity are decidable over S4, Int and Pos; interpolation is decidable over S4 and Int$^+$ and strongly decidable over S5, Grz and Int; the projective Beth property is decidable over Int, Int$^+$ and Grz, etc. Some complexity bounds are found.

In addition, we state that tabularity and many variants of amalgamation and of surjectivity of epimorphisms are base-decidable in varieties of closure algebras, of Heyting algebras and of relatively pseudocomplemented lattices.

**§1. Introduction.** Propositional calculi are usually defined by systems of axioms schemes and rules of inference. Natural problems arising in general study of logical calculi, for example, the problem of equivalence or the problem of determining for arbitrary calculus whether it is consistent or not, are, in general, undecidable. The first undecidability results for propositional calculi were found by S. Linial and E. Post in 1949 [16]. In particular, it was proved that the property "to be an axiomatization of the classical propositional logic" is undecidable. In 1963 A. V. Kuznetsov [13] found an essential extension of this result. He proved that for every superintuitionistic logic L the property "to be an axiomatization of L" is undecidable. In particular, consistency and many other properties of calculi are, in general, undecidable. In [3] A. V. Chagrov has given a survey of results on undecidable properties of propositional calculi. Here we concentrate on decidable properties.

When we restrict ourself by considering particular families of calculi, for instance, propositional calculi extending intuitionistic or some modal logic, many important properties of calculi appear to be decidable (see a survey

The work was supported by Russian Foundation for Basic Research, project no. 03-06-80178.

**Logic Colloquium '03**
Edited by V. Stoltenberg-Hansen and J. Väänänen
Lecture Notes in Logic, 24

in [4]). When the rules of inference are fixed, for any given finite system of additional axioms, one can effectively decide consistency problem for normal modal logics, tabularity and interpolation problems for extensions of the intuitionistic logic or of the modal system S4 and some other problems. Also we can take rules of inference into consideration and study more general question: which properties are decidable when not only axioms but also rules of inference may vary. Sets of new postulates must necessarily be finite because of Kuznetsov's statement: No non-trivial property of logics is decidable under recursive axiomatization (see [4]).

Let a logical propositional calculus $L_0$ be given. We consider arbitrary extensions of $L_0$ by adding finitely many new axiom schemes and rules of inference. We say that a property P of logical calculi is *decidable over* $L_0$ if there is an algorithm which for any finite set $Ax$ of axiom schemes decides whether the system $L_0 + Ax$ has the property P or not; P is *strongly decidable over* $L_0$ if there is an algorithm which for any finite system *Rul* of axiom schemes and rules of inference decides whether the system $L_0 + Rul$ has the property P or not. We consider only so-called structural rules of inference which are invariant under subsitution.

We take as $L_0$ some standard calculus for intuitionistic propositional logic Int, its positive fragment $\text{Int}^+$ or modal logic S4 and present a number of properties decidable over $L_0$, in particular, consistency, equivalence to some known logic, tabularity, interpolation property over Int, $\text{Int}^+$ and S4, local tabularity over S4, projective Beth's property over Int and $\text{Int}^+$ and so on. Some of these results were already presented in [4]. Here we bring complexity bounds for recognizing the above-mentioned properties of calculi. Also we state that many of these properties are not only decidable but also strongly decidable.

Our results work for both Gentzen-style and Hilbert-style systems. Since we use algebraic methods for our proofs, Hilbert-style systems are more convenient. As for sequent calculi, we may use well-known reduction of sequent calculi to Hilbert-style calculi and replace any sequent of a form $A_1, \ldots, A_n \Rightarrow B_1, \ldots, B_m$ by the formula $A_1 \& \cdots \& A_n \to B_1 \vee \cdots \vee B_m$ etc.

In Section 2 we give the main definitions. In Section 3 we show that some known classifications of calculi over Int and S4 are strongly decidable. The problem of equivalence of propositional calculi is discussed in Section 4. In Section 5 a catalog of decidable properties is presented. Some complexity bounds are given.

Most of our results are proved by algebraic methods and have an algebraic interpretation. There is a duality between normal modal logics and varieties of Heyting algebras, between superintuitionistic logics and varieties of Heyting algebras and between positive logics and varieties of pseudo-complemented lattices. We rewrite our decidability results in the algebraic language in the last section.

Most of the algorithms found in the papers [18]–[34] were given in algebraic form. We note that they can be rewritten in terms of Kripke semantics, and we represent them in that form in our paper.

**§2. Basic definitions.** We take as a basic calculus L some standard formulation for the intuitionistic propositional logic Int (with modus ponens or cut among its postulates), or its positive fragment $\mathrm{Int}^+$, or one of modal systems K, K4, S4, S5, GL, Grz (with modus ponens and necessity rule $A/\Box A$ among the postulates).

Each calculus determines its *logic*, i.e. the set of its theorems. We write $L_1 \leq L_2$ if all theorems of $L_1$ are theorems of $L_2$. Two calculi are *equivalent* if they determine the same logic.

**2.1. Positive and superintuitionistic calculi.** The language of positive calculi contains $\&, \vee, \rightarrow$ and $\top$ as primitive connectives, for superintuitionistic calculi we add the constant $\bot$ ("absurdity") and consider negation as an abbreviation: $\neg A \rightleftharpoons A \rightarrow \bot$.

A *superintuitionistic logic* is a set of formulas containing the set Int of all intuitionistically valid formulas and closed under substitution and modus ponens. A *positive logic* is a set of positive formulas containing $\mathrm{Int}^+$ and closed under the same rules. A positive logic is determined by some set of axiom schemes added to $\mathrm{Int}^+$. It is clear that one can replace a finite set of axiom schemes with their conjunction. We denote by $L + A$ the extension of a logic $L$ by an extra axiom scheme $A$.

For a given logic $L$, the set of all logics containing $L$ is denoted by $E(L)$.

Let us bring some standard denotations for positive and superintuitionistic calculi.

$$\mathrm{Cl} = \mathrm{Int} + (p \vee \neg p),\ \mathrm{Cl}^+ = \mathrm{Int}^+ + (p \vee (p \rightarrow q)),$$
$$\mathrm{LC} = \mathrm{Int} + (p \rightarrow q) \vee (q \rightarrow p),$$
$$\mathrm{LC}^+ = \mathrm{Int}^+ + (p \rightarrow q) \vee (q \rightarrow p),$$
$$For = \mathrm{Int} + \bot,\ For^+ = \mathrm{Int}^+ + p,$$
$$\mathrm{KC} = \mathrm{Int} + (\neg p \vee \neg\neg p).$$

For a superintuitionistic or positive calculus $L$, $\Gamma \vdash_L A$ denotes that $A$ is derivable from $\Gamma$ and axioms of L by modus ponens.

**2.2. Normal modal calculi.** The language of modal logic contains all the connectives of the intuitionstic logic and, in addition, the modal operators $\Box$ and $\Diamond$, where $\Diamond A \rightleftharpoons \neg\Box\neg A$. Here we consider only normal modal calculi that contain modus ponens and the necessity rule $A/\Box A$ among their postulates. The minimal normal modal calculus K is defined by adding one axiom schema $\Box(A \rightarrow B) \rightarrow (\Box A \rightarrow \Box B)$ and the necessity rule to the postulates of the classical propositional logic. Further,

$$\begin{aligned}
\mathrm{K4} &= \mathrm{K} + (\Box p \to \Box\Box p),\\
\mathrm{GL} &= \mathrm{K4} + \Box(\Box p \to p) \to \Box p,\\
\mathrm{S4} &= \mathrm{K4} + (\Box p \to p),\\
\mathrm{S4.1} &= \mathrm{S4} + (\Box\Diamond p \to \Diamond\Box p),\\
\mathrm{S4.2} &= \mathrm{S4} + (\Diamond\Box p \to \Box\Diamond p),\\
\mathrm{S4.1.2} &= \mathrm{S4} + (\Box\Diamond p \leftrightarrow \Diamond\Box p),\\
\mathrm{S4.3} &= \mathrm{S4} + \Box(\Box p \to q) \vee \Box(\Box q \to p);\\
\mathrm{Grz} &= \mathrm{S4} + (\Box(\Box(p \to \Box p) \to p) \to p);\\
\mathrm{Grz} &= \mathrm{S4} + (\Box(\Box(p \to \Box p) \to p) \to p),\\
\mathrm{Grz.2} &= \mathrm{Grz} + \mathrm{S4.2};\\
\mathrm{Grz.3} &= \mathrm{Grz} + \mathrm{S4.3};\\
\mathrm{S5} &= \mathrm{S4} + (p \to \Box\Diamond p).
\end{aligned}$$

For modal $L$ and any set of formulas $\Gamma$ we write $\Gamma \vdash_L A$ if there is a derivation of $A$ from $\Gamma$ and axioms of $L$ by modus ponens and necessity rules.

A *normal modal logic* is any set of modal formulas containing all the axioms of K and closed under modus ponens, necessitation and substitution rules. By $NE(L)$ we denote the set of all normal modal logics containing $L$.

**2.3. Admissible rules.** We denote by $A_1, \dots, A_n/B$ the rule of inference which for every substitution $s$ derives $s(B)$ from $s(A_1), \dots, s(A_n)$.

Let $L$ be a propositional Hilbert style calculus. We say that a rule

$$A_1(p_1, \dots, p_k), \dots, A_n(p_1, \dots, p_k)/B(p_1, \dots, p_k)$$

is *admissible in L* if it preserves provability in $L$, i.e. for any formulas $C_1, \dots, C_k$ such that $A_i(C_1, \dots, C_k)$ are theorems of $L$ for $i = 1, \dots, n$, the formula $B(C_1, \dots, C_k)$ is a theorem of $L$.

We say that the rule

$$A_1, \dots, A_n/B$$

is *derivable in L* if $A_1, \dots, A_n \vdash_L B$. It is easy to see that any rule derivable in $L$ is admissible in $L$. The converse holds only for certain calculi, in particular, for those axiomatizing the classical propositional logic.

It is evident that for superintuitionistic and modal Hilbert style systems any additional rule $A_1, \dots, A_n/B$ is equivalent w.r.t admissibility to the rule $(A_1 \& \cdots \& A_n)/B$ with only one premise, and any additional axiom scheme $A$ is equivalent to the rule $\top/A$. So we can consider additional rules of inference with only one premise, and identify additional axiom schemes with corresponding rules.

Obviously, two calculi are equivalent if and only if they have the same admissible rules.

V. Rybakov [40] solved a difficult problem of admissibility of inference rules for the intuitionistic logic Int, modal S4 and many other logics. He found efficient criteria of admissibility, and proved that most of known superintuitionistic and modal systems are decidable with respect to admissibility of inference rules. We used Rybakov's results to prove strong decidability of some problems.

**2.4. Kripke semantics.** Recall that the logic $\mathrm{Int}^+$ is complete with respect to intuitionistic Kripke models $\mathbf{M} = (W, \leq, \models)$, where $W$ is a non-empty set and $\leq$ is a partial order on $W$, and $\models$ satisfies the conditions:

(I1) if $x \models p$ and $x \leq y$ then $y \models p$ for any variable $p$ (monotonicity),
(I2) $x \models (A \to B) \Leftrightarrow \forall y((x \leq y \text{ and } y \models A) \Rightarrow y \models B)$,
(I3) $x \models (A\&B) \Leftrightarrow (x \models A \text{ and } x \models B)$,
(I4) $x \models (A \vee B) \Leftrightarrow (x \models A \text{ or } x \models B)$.

For the intuitionistic logic Int one more condition is necessary:

(I5) $x \not\models \bot$.

We say that a formula $A$ is *true in a model* $\mathbf{M} = (W, \leq, \models)$ if $x \models A$ for all $x \in W$, and *refutable in* $\mathbf{M}$ otherwise. A formula $A$ is *intuitionistically valid in a frame* $(W, \leq)$ if it is true in any intuitionistic model $\mathbf{M} = (W, \leq, \models)$ based on $(W, \leq)$. A logic or calculus $L$ is said to be *complete* with respect to a class $C$ of Kripke frames if for any formula $A$, $L \vdash A$ if and only if $A$ is valid in all frames in $C$.

The logics LC and $\mathrm{LC}^+$ are characterized by linearly ordered frames, and Cl and $\mathrm{Cl}^+$ by frames with one-element set $W$. Further, the logics

$$\mathrm{LP}_2 = \mathrm{Int} + p \vee (p \to q \vee \neg q) \quad \text{and} \quad \mathrm{LP}_2^+ = \mathrm{Int}^+ + (p \to q \vee (q \to r))$$

are determined by partially ordered frames without three-element chains, and the logics

$$\mathrm{Z}_3 = \mathrm{LC} + \mathrm{LP}_2 \quad \text{and} \quad \mathrm{Z}_3^+ = \mathrm{LC}^+ + \mathrm{LP}_2^+$$

are characterized by one two-element chain.

The well known definition of a Kripke model for modal logics differs from intuitionistic models. A Kripke model is a triple $\mathbf{M} = (W, R, \models)$, where $W$ is a non-empty set, $R$ is any binary relation on $W$ and $\models$ satisfies the following conditions:

(M1) $x \models \neg A \Leftrightarrow x \not\models A$,
(M2) $x \models (A \to B) \Leftrightarrow (x \not\models A \text{ or } x \models B)$,
(M3) $x \models \Box A \Leftrightarrow \forall y(xRy \Rightarrow y \models A)$.

We say that a formula $A$ is *true in a model* $\mathbf{M} = (W, R, \models)$ if $x \models A$ for all $x \in W$, and *refutable in* $\mathbf{M}$ otherwise. A formula $A$ is *valid in a frame* $(W, R)$ if it is true in any model $\mathbf{M} = (W, R, \models)$ based on $(W, R)$. A logic or calculus

$L$ is said to be *complete* with respect to a class $C$ of Kripke frames if for any formula $A$, $L \vdash A$ if and only if $A$ is valid in all frames in $C$.

Recall that the logic K is complete under the class of all frames, K4 under transitive frames, GL under transitive and irreflexive frames without infinitely increasing chains, S4 under transitive and reflexive frames, Grz under partially ordered frames without infinitely increasing chains, and S5 under frames with a total relation $R$.

For the sequel, we will need some sequences of frames. For $n \geq 1$ we denote by

- $S_n$ the set $\{1, \ldots, n\}$ with natural ordering relation;
- $X_n$ the same set, where $xRy$ for all $x$, $y$;
- $Y_{n+1}$ the set $\{1, \ldots, n+1\}$, where $xRy$ for all $x, y \leq n$ and $xR(n+1)$ for all $x$;
- $U_{n+1}$ the set $\{0, 1, \ldots, n+1\}$, where $0RxR(n+1)$ for all $x$ and $\neg xRy$ for $1 \leq x, y \leq n, x \neq y$;
- $V_n$ is the subframe of $U_{n+1}$ obtained by deleting $(n+1)$.

Also we define a frame $X_n''$, where $X_n'' = X_n \cup \{0, n+1\}$, and

$$xR''y \iff (x = 0 \text{ or } 1 \leq x,\ y \leq n \text{ or } x = y = n+1),$$

and its subframe $X_n'$ obtained by deleting 0.

A rule $A/B$ is *satisfiable* in a (modal or intuitionistic) model $\mathbf{M} = (W, R, \models)$ provided

$$\forall x(x \models A) \Rightarrow \forall x(x \models B),$$

and *refutable in* $\mathbf{M}$ otherwise. A rule $A/B$ is *valid* in a frame $(W, R)$ if it is satisfiable in any model $\mathbf{M} = (W, R, \models)$ based on this frame.

A mapping $\theta$ of a frame $\mathbf{W} = (W, R)$ onto $\mathbf{W}' = (W', R')$ is a *p-morphism* if it satisfies the conditions (a) $xRy \Rightarrow \theta(x)R'\theta(y)$ and (b) $\theta(x)R'y \Rightarrow \exists z(xRz \text{ and } \theta(z) = y)$. It is known that $\theta$ is a p-morphism if and only if $\varphi = \theta^{-1}$ is a monomorphism of $\mathbf{W}'^{+}$ into $\mathbf{W}^{+}$. We note that validity of rules in frames is preserved by p-morphisms but is not preserved by transfer to subframes.

Let $L$ be any calculus. We say that a frame $\mathbf{W}$ *validates* (or *satisfies*) $L$ if all the axioms and rules of $L$ are valid in $\mathbf{W}$.

**2.5. Complexity.** We recall some notions of Complexity Theory [11, 36].

With any set $X$ of formulas one can associate a decision problem: for arbitrary formula $A$ to determine whether $A$ is in $X$ or not. Complexity classes P, NP, EXP, NEXP consist of sets $X$ such that the problem of membership in $X$ can be decided on Turing machines in polynomial time (with respect to the size of the formula $A$), non-deterministic polynomial time, exponential time and non-deterministic exponential time respectively. For sets in PSPACE this problem can be decided in polynomial space. A set $X$ is in coNP or in coNEXP if and only if its complement is in NP or in NEXP respectively. The class DP consists of all intersections $X \cap Y$ such that $X \in$ NP and $Y \in$ coNP. The class $\Delta_2^p$ contains all sets decidable by polynomial-time oracle machines with an oracle in NP. It is known that

$$\mathsf{P} \neq \mathsf{EXP},\ \mathsf{P} \subseteq \mathsf{NP} \cap \mathsf{coNP},$$
$$\mathsf{NP} \cup \mathsf{coNP} \subseteq \mathsf{DP} \subseteq \Delta_2^p \subseteq \mathsf{PSPACE} \subseteq \mathsf{EXP} \subseteq \mathsf{NEXP} \cap \mathsf{coNEXP}.$$

Each of the classes NP, coNP, $\Delta_2^p$, PSPACE, EXP, NEXP, coNEXP is closed under finite unions and intersections of sets, and $\Delta_2^p$ contains all boolean combinations of sets in NP. In addition, the classes P, $\Delta_2^p$ and PSPACE are closed under complements.

Let $\mathcal{C}$ be a complexity class. A decision problem of $X$ is *$\mathcal{C}$-hard* if any set $Y$ in $\mathcal{C}$ is polynomially reducible to $X$. The problem is *$\mathcal{C}$-complete* if it is in $\mathcal{C}$ and $\mathcal{C}$-hard. To prove that a decision problem of $X$ is $\mathcal{C}$-hard, it is sufficient to show that some $\mathcal{C}$-hard problem is polynomially reducible to $X$. To prove that $X$ is in $\mathcal{C}$, it is sufficient to reduce $X$ by a polynomial to some $\mathcal{C}$-complete problem.

One can find lists of $\mathcal{C}$-complete problems for known complexity classes in [11, 36]. Satisfiability problem of the classical propositional logic is a standard example of NP-complete problem, and validity and non-satisfiability in *Cl* are typical examples of coNP-complete problems. The most known example of DP-complete problem is SAT-UNSAT: Given two boolean formulas $\varphi$ and $\psi$, to determine whether it is true that $\varphi$ is satisfiable and $\psi$ is not [36].

For various problems concerning propositional calculi, we find some complexity bounds with respect to the size $|Ax|$ or $|Rul|$ of the sets of additional postulates. A *size* of such a set is the general number of occurrences of variables and logical connectives in formulas of this set. From [35] (Propositions 3.1 and 3.3) and [33] we can obtain some lower bounds of complexity:

PROPOSITION 2.1. *Let $L_0$ be any consistent and finitely axiomatizable logic in $E(\mathrm{Int})$, $E(\mathrm{Int}^+)$ or in $NE(\mathrm{S4})$, $S$ some family of extensions of $L_0$. Then the problem "$(L_0 + Ax) \in S$?" is*

- *NP-hard if there are finitely axiomatizable extensions $L'$ and $L$ of $L_0$ such that $L' < L$, $L \in S$ and $L' \notin S$,*
- *coNP-hard if there are finitely axiomatizable extensions $L$ and $L''$ of $L_0$ such that $L < L''$, $L \in S$ and $L'' \notin S$,*
- *DP-hard if there are finitely axiomatizable extensions $L$, $L'$ and $L''$ of $L_0$ such that $L' < L < L''$, $L \in S$ and $L', L'' \notin S$.*

A logic $L$ is called *Kripke-complete* if provability in $L$ is equivalent to $L$-validity. A logic $L$ is said to have the *finite model property* (*FMP*) if provability in $L$ is equivalent to validity in all finite $L$-frames. It is clear that any logic with FMP is Kripke-complete. Each tabular logic $L$ can be characterized by a suitable finite $L$-frame. A logic $L$ is *polynomially* (*exponentially*) *approximable* if any formula $A$ non-provable in $L$ is refutable in some finite $L$-frame whose cardinality is bounded by a polynomial (respectively, exponential) function of the size of $A$.

REMARK. In our consideration, the notion of refutability is more important than satisfiability. In the case of modal logics one can easily see that a formula $A$ is satisfiable in some model if and only if $\neg A$ is refutable in the same model. There is no such evident direct reduction for intuitionistic models.

In our calculation we use a lemma whose proof is in fact given in [8, Proposition 3.1]:

LEMMA 2.2. *Given a model $M$ and a formula $A$, there is an algorithm for calculating the value of $A$ in $M$ that runs in time $O(\|M\| \times |A|)$, where $\|M\|$ is the sum of the number of elements in the frame and the number of pairs in $R$.*

As a consequence, we easily get

LEMMA 2.3. *For each fixed finite (intuitionistic or modal) frame* **W**, *refutability in* **W** *is NP-complete and validity in* **W** *is coNP-complete.*

It is known that any property of frames expressible in first order language with one binary relation is recognizable on a finite frame in polynomial time w.r.t. the size of the frame ([36, Theorem 5.1]). By Lemma 2.2 we get

LEMMA 2.4. *If a logic $L$ in $E(\mathrm{Int})$, $E(\mathrm{Int}^+)$ or in $NE(\mathrm{K})$ is polynomially (or exponentially) approximable by a class of frames definable by finitely many first order formulas then $L$-refutability problem is in NP (resp. in NEXP) and $L$-provability problem is in coNP (resp. in coNEXP).*

By Proposition 2.1 we conclude

LEMMA 2.5. *If a consistent logic $L$ in $E(\mathrm{Int})$, $E(\mathrm{Int}^+)$ or in $NE(\mathrm{K})$ is polynomially approximable by a class of frames definable by finitely many first order formulas then $L$-refutability problem is NP-complete and $L$-provability problem is coNP-complete.*

Complexity of provability and satisfiability problems in many non-classical logics, for instance, in intuitionistic logic, various systems of modal logic, temporal and dynamic logics was studied in [7, 8, 15, 37, 42, 43]. R. Ladner [15] proved that provability problem is PSPACE-complete for modal logics K, T and S4 and coNP-complete for S5. R. Statman [43] proved that the problem of determining if an arbitrary implicational formula is intuitionistically valid is PSPACE-complete. PSPACE-completeness of provability in Grz and GL is shown in [4]. In the same book NP-completeness of refutability problem for many superintuitionistic and modal logics is proved.

In the next sections we consider the problem of recognizing properties of logical calculi and bring complexity bounds for some properties which are known to be decidable.

**§3. Strongly decidable classifications.** In this section we consider some known classifications of superintuitionistic and positive logics, and also of normal extensions of the modal logic S4. The results of this section were proved for extensions of Int and S4 in [31] in terms of the algebraic semantics. The proofs for positive logics are similar to those for superintuitionistic logics. We rewrite the statements of [31] in terms of Kripke frames. We consider some formulas which are used for axomatization of well-known logics, and bring out their characterization via special Kripke frames.

**3.1. Classification by slices.** Classification of superintuitionistic logics by slices was suggested by T. Hosoi [10]. We denote

$$\pi_0 = p_0, \quad \pi_{n+1} = p_{n+1} \vee (p_{n+1} \to \pi_n).$$

We say that a superintuitionistic or positive calculus $L$ is *of the nth slice* for $0 < n < \omega$ if $L \vdash \pi_n$ and $L \nvdash \pi_{n-1}$; $L$ is *of finite slice* if $L \vdash \pi_n$ for some $n \geq 0$, and *of infinite slice* otherwise.

It is known that for any partially ordered frame **W**, the formula $\pi_n$ is intuitionistically valid in **W** if and only if there is no $(n+1)$-element chain in **W**. It follows that for any calculus $L$ of the $n$th slice, where $n > 0$, the length of chains in any frame validating $L$ is bounded by $n$.

For $n \geq 0$ we define

$$\sigma_0 = \bot, \quad \sigma_{n+1} = \Box p_{n+1} \vee \Box(\Box p_{n+1} \to \sigma_n).$$

For any transitive and reflexive frame **W**, the formula $\sigma_n$ is valid in **W** if and only if **W** contains no $(n+1)$-element chain, where

$$x < y \Leftrightarrow (xRy \text{ and not } yRx).$$

We say that a normal extension $L$ of S4 is *of the nth slice* for $0 < n < \omega$ if $L \vdash \sigma_n$ and $L \nvdash \sigma_{n-1}$; $L$ is *of finite slice* if $L \vdash \sigma_n$ for some $n \geq 0$, and *of infinite slice* otherwise.

Recall from [10, 41] that LC is the greatest among superintuitionistic logics of infinite slice, and Grz.3 is the greatest logic of infinite slice in $NE$(S4). For $0 < n < \omega$, the $n$th slice in $E$(Int) contains the least element Int $+ \pi_n$ and the greatest element LC $+ \pi_n$, and the $n$th slice of $NE$(S4) has the least element S4 $+ \sigma_n$ and the greatest element Grz $+ \sigma_n$.

All logics of finite slices are *locally tabular*, i.e. for any finite set of variables there exist only finitely many pairwise non-equivalent formulas [14, 41].

Recall that $S_n$ denotes the set $\{1, \ldots, n\}$ with the natural ordering relation. From [31, Proposition 3.4.1], we obtain

PROPOSITION 3.1. *Let* $L =$ Int $+$ *Rul or* $L =$ Int$^+$ $+$ *Rul. Then*

*$L \vdash \pi_n$ iff there is a rule in Rul refutable in the frame* $S_{n+1}$;

It means that $L =$ Int$+$*Rul* or $L =$ Int$^+$ $+$*Rul* is of the $n$th slice $(0 < n < \omega)$ if and only if all the rules in *Rul* are intuitionistically valid in the frame $S_n$ but there is a rule in *Rul* refutable in $S_{n+1}$. Also we have (cf. [31, Proposition 4.4.1]):

PROPOSITION 3.2. *Let* $L = \mathrm{S4} + Rul$. *Then*

$L \vdash \sigma_n$ *iff there is a rule in Rul refutable in the frame* $S_{n+1}$;

From [31, Theorem 5.2], we obtain

PROPOSITION 3.3.

(a) *Let* $L = \mathrm{Int} + Rul$ *or* $L = \mathrm{Int}^+ + Rul$, *where Rul is finite, and k be the maximal number of variables in rules of Rul. Then L is of infinite slice if and only if all the rules in Rul are intuitionistically valid in the frame* $S_{k+1}$.

(b) *Let* $L = \mathrm{S4} + Rul$, *where Rul is finite, and k be the maximal number of occurrences of modalities in rules of Rul. Then L is of infinite slice if and only if all the rules in Rul are valid in the frame* $S_{k+2}$.

So we get (cf. [31, Theorem 5.3])

THEOREM 3.4.

(a) *There exists an algorithm calculating for any finite system Rul the number of a slice containing the logic* $L = \mathrm{Int} + Rul$.

(b) *There exists an algorithm calculating, for any finite system Rul of positive rules, the number of a slice containing the logic* $L = \mathrm{Int}^+ + Rul$.

(c) *There exists an algorithm calculating, for any finite system Rul of rules in modal language, the number of a slice containing the logic* $L = \mathrm{S4} + Rul$.

(d) *The properties "to be a logic of a finite slice" and "to be a logic of the nth slice" for* $0 \leq n \leq \omega$ *are strongly decidable over* Int, $\mathrm{Int}^+$ *and* S4.

By Propositions 2.1, 3.1–3.3 and Lemma 2.2 one can show that for any fixed $n$, the problem whether the system Int + *Rul*, $\mathrm{Int}^+$ + *Rul* or S4 + *Rul* axiomatises a logic of $n$th slice is

- DP-complete if $0 < n < \omega$,
- NP-complete if $n = 0$,
- coNP-complete if $n = \omega$.

**3.2. Classification by clusters.** In [24] we introduced two characteristics $\mu_1$ and $\mu_2$ of S4-frames (i.e., reflexive and transitive frames) and logics in $NE(\mathrm{S4})$ as follows. With any element $a$ of a reflexive and transitive frame $\mathbf{W}$ we associate its *cluster*, i.e. $\{x \in \mathbf{W} \mid aRxRa\}$. A cluster of $a$ is *internal* if there is $y$ such that $aRy$ and $\neg yRa$, and *external* otherwise. We define $\mu_1(\mathbf{W})$ and $\mu_2(\mathbf{W})$ as the suprema of cardinalities of external and of internal clusters in $\mathbf{W}$ respectively. If $L$ is a calculus, we define

$$\mu_i(L) = \sup\{\mu_i(\mathbf{W}) \mid \mathbf{W} \text{ is a finite frame validating all theorems of } L\}.$$

These characteristics can be expressed in modal language in the following way.

With each finite frame $\mathbf{W}$ we associate a formula $\delta(\mathbf{W})$ as follows. Take a variable $p_a$ for each $a \in \mathbf{W}$. Then $\delta(\mathbf{W})$ is defined as the conjunction of the following formulas:

(1) $\Box(p_x \to \neg p_y)$ for all $x \neq y$,
(2) $\Box(p_x \to \Diamond p_y)$ for all $x, y$ such that $xRy$,
(3) $\Box(p_x \to \neg\Diamond p_y)$ for all $x, y$ such that $\neg xRy$,
(4) $\Box(\bigvee\{p_a \mid a \in \mathbf{W}\})$.

In Section 2 we defined some special frames. For $n \geq 1$ we denoted by

$X_n$ the set $\{1, \ldots, n\}$, where $xRy$ for all $x, y$;

$Y_{n+1}$ the set $\{1, \ldots, n+1\}$, where $xRy$ for all $x, y \leq n$ and $xR(n+1)$ for all $x$.

Also we defined a frame $X_n''$, where $X_n'' = X_n \cup \{0, n+1\}$, and

$$xR''y \iff (x = 0 \text{ or } x, y \in \{1, \ldots, n\} \text{ or } x = y = n+1),$$

and its subframe $X_n'$ obtained by deleting 0.

We denote

$$\xi_n = \delta(X_n) \to \neg p_1;$$
$$\eta_n = \delta(Y_n) \to \neg p_1.$$

The following Lemma was proved in [24].

LEMMA 3.5. *For each $L \in NE(\mathrm{S4})$ and for any $n > 0$*

(a) *$\mu_1(L) \leq n$ iff $\xi_{n+1}$ is a theorem of $L$;*
(b) *$\mu_2(L) \leq n$ iff $\eta_{n+2}$ is a theorem of $L$.*

It was proved in [20] that a logic

$$PM4 = \mathrm{S4} + \sigma_2 + (\Box\Diamond p \leftrightarrow \Diamond\Box p)$$

is one of the five pretabular logics in $NE(\mathrm{S4})$, and for arbitrary $L \in NE(\mathrm{S4})$

$$\mu_2(L) = \omega \quad \text{iff } L \leq PM4.$$

This logic is characterized by all frames $Y_n$ for $n \geq 2$. Another pretabular logic S5 is characterized by all frames $X_n$ for $n > 0$, and for each $L \in NE(\mathrm{S4})$

$$\mu_1(L) = \omega \quad \text{iff } L \leq \mathrm{S5}.$$

We rewrite the statements of [31] concerning $\mu_i$, $\eta_n$ and $\xi_n$ in terms of Kripke frames.

PROPOSITION 3.6. *Let $L = \mathrm{S4} + Rul$. Then*

(a) *$L \vdash \eta_{n+1}$ iff there is a rule in Rul refutable in the frame $Y_{n+1}$;*
(b) *for $n > 1$, $L \vdash \xi_n$ iff none of the frames $X_n$, $X_n'$, $X_n''$ satisfies $L$.*

PROPOSITION 3.7. *Let $L = \mathrm{S4}+Rul$, $k$ be the maximal number of occurrences of modalities in rules of Rul. Then the following are equivalent:*

(a) *$\mu_1(L)$ is finite;*
(b) *$\mu_1(L) \leq k$;*
(c) *none of the frames $X_{k+1}$, $X_{k+1}'$, $X_{k+1}''$ satisfies $L$.*

PROPOSITION 3.8. *Let $L = \mathrm{S4} + Rul$, $k$ be the maximal number of modalities in rules of Rul. Then the following are equivalent:*

(a) $\mu_2(L)$ *is finite*;
(b) $\mu_2(L) \leq k$;
(c) *some rule in Rul is refutable in the frame* $Y_{k+2}$.

From Lemma 3.5 and Proposition 3.6 we get

PROPOSITION 3.9. *Let* $L = \mathrm{S4} + Rul$. *Then*
(a) $\mu_1(L) \leq n$ *iff none of the frames* $X_n$, $X'_n$, $X''_n$ *satisfies* $L$;
(b) $\mu_2(L) \leq n$ *iff some rule in Rul is refutable in the frame* $Y_{n+2}$.

The last three propositions give

THEOREM 3.10. *There exist algorithms calculating* $\mu_1(L)$ *and* $\mu_2(L)$ *for any* $L = \mathrm{S4} + Rul$.

What happens in the case where the calculus contains no additional rules but has only new axiom schemes? Then we can simplify Propositions 3.9(a) and 3.7 as follows:

PROPOSITION 3.11. *Let* $L = \mathrm{S4} + Ax$. *Then*
(a) $\mu_1(L) \leq n$ *iff* $Ax$ *is refutable in the frame* $X_{n+1}$;
(b) $\mu_1(L) = \omega$ *iff* $Ax$ *is refutable in the frame* $X_{k+2}$, *where* $k$ *is the number of occurrences of modalities in* $Ax$.

Note that for calculating $\mu_2(\mathrm{S4} + Ax)$ we use the algorithms given by Propositions 3.8 and 3.9(b).

By Propositions 3.7–3.11 we can prove that for every fixed $n$, each of the problems "$\mu_1(\mathrm{S4} + Rul) = n$?", "$\mu_2(\mathrm{S4} + Rul) = n$?", "$\mu_1(\mathrm{S4} + Ax) = n$?" and "$\mu_2(\mathrm{S4} + Ax) = n$?" is DP-complete over S4 for $0 < n < \omega$ and coNP-complete for $n = \omega$.

**§4. Equivalence and inclusion.** The most of the problems considered in the next section can be reformulated in terms of equivalence to some particular calculus. For instance, a superintuitionistic calculus is consistent if and only if it is not equivalent to the logic *For* consisting of all formulas. Similarly, a normal modal calculus is consistent iff it is not equivalent to the logic consisting of all modal formulas.

Two logical calculi are *equivalent* if they have the same set of theorems. We write $L_1 \leq L_2$ if all the theorems of $L_1$ are theorems of $L_2$. It is clear that $L_1$ is equivalent to $L_2$ if and only if $L_1 \leq L_2$ and $L_2 \leq L_1$.

It is well known that, in general, the problem of recognizing whether two calculi are equivalent or not is undecidable even if rules of inference are fixed. Nevertheless, we know that in some particular cases the problem is decidable. In this section we note that the problem of equivalence to $L_1$ over Int is strongly decidable if $L_1$ is in some list of well-known superintuitionistic calculi. Also we prove strong decidability of the problem of equivalence for many modal calculi extending the modal system S4.

Let $L_0 \leq L_1$. Obviously, the inclusion $L_0 + Ax \leq L_1$ is the same as provability of all axioms of $Ax$ in $L_1$. At the same time, admissibility of all rules of *Rul* in $L_1$ implies $L_0 + Rul \leq L_1$ but the converse, in general, does not hold.

Nevertheless, we have an evident criterion of equivalence:

PROPOSITION 4.1. *Let a calculus $L_1$ be axiomatized by adding some set of axiom schemes to $L_0$. Then $L_0 + Rul$ is equivalent to $L_1$ if and only if all rules in Rul are admissible in $L_1$ and, moreover, all axioms of $L_1$ are theorems of $L_0 + Rul$.*

Indeed, if all rules in *Rul* are admissible in $L_1$ then all theorems of $L = L_0 + Rul$ are derivable in $L_1$. If, in addition, all axioms of $L_1$ are derivable in $L$ then $L_0 + Rul$ is equivalent to $L_1$. The converse is evident.

Using this proposition we can prove

PROPOSITION 4.2. *Let a calculus $L_1$ be axiomatized by adding finitely many axiom schemes to $L_0$.*

(a) *If the admissibility problem is decidable in $L_1$ and, moreover, the inclusion $L_0 + Rul \geq L_1$ is decidable, then the problem of equivalence to $L_1$ is strongly decidable over $L_0$.*
(b) *If the problem of equivalence to $L_1$ is strongly decidable over $L_0$ then the admissibility problem is decidable in $L_1$ and the inclusion $L_0 + Ax \geq L_1$ is decidable.*

V. Rybakov proved that the admissibility problem is decidable for many superintuitionistic and modal logics. We collect his results in

STATEMENT 4.3 (Rybakov [40]). (a) All finitely axiomatizable logics of finite slices in $E$(Int) and in $NE$(S4) are decidable w.r.t. admissibility.

(b) Superintuitionistic logics Int, KC and LC are decidable w.r.t. admissibility.

(c) Modal logics S4, S4.1, S4.2, S4.3, Grz, Grz.2, Grz.3 are decidable w.r.t. admissiubility.

Using this result, we proved in [31]

THEOREM 4.4. *Let $L_1$ be a finitely axiomatizable logic of finite slice in $E$*(Int) *(or in $NE$*(S4)*). Then the problem of equivalence to $L_1$ is strongly decidable over* Int *(resp. over* S4*).*

Also Rybakov's results make possible to state the following

THEOREM 4.5 ([31]).

(a) *Let $L_1$ be one of the superintuitionistic logics* Int, LC, KC. *Then the problem of equivalence to $L_1$ is strongly decidable over* Int.
(b) *Let $L_1$ be one of the modal logics* S4, S4.1, S4.2, S4.3, Grz, Grz.2, Grz.3, S5. *Then the problem of equivalence to $L_1$ is strongly decidable over* S4.

For example, Int + *Rul* = Int iff all rules in *Rul* are admissible in Int; Int + *Rul* = LC iff all rules in *Rul* are admissible in LC and, moreover, none of the frames $V_2$ and $U_3$ (intuitionistically) validates *Rul* (see Section 2.4). Further, S4 + *Rul* = S4.1 iff all rules in *Rul* are admissible in S4.1 and none of the frames $X_2, X_2', X_2''$ validates *Rul*; S4 + *Rul* = S5 iff all rules in *Rul* are admissible in S5 and some rule in *Rul* is refutable in the frame $S_2$.

We state a theorem on inclusions.

THEOREM 4.6.

(a) *The problem of inclusion* $L \le$ LC *is strongly decidable over* Int.

(b) *Let* $L_1$ *be one of* Grz.3, S5, *PM*4. *Then the problem of inclusion* $L \le L_1$ *is strongly decidable over* S4.

PROOF. It is known that LC is the greatest logic of infinite slice in $E(\text{Int})$, and Grz.3 is the greatest logic of infinite slice in $NE(\text{Grz})$. So the statement about LC and Grz.3 follows immediately from Proposition 3.3. For S5 and *PM*4 we use Propositions 3.7 and 3.8, since these logics are the greatest among the logics with infinite $\mu_1$ and, respectively, $\mu_2$. ⊣

We note that Proposition 3.3 gives easy algorithms for recognizing admissibility in LC and Grz.3. For instance, a rule is admissible in Grz.3 if and only if it is valid in the frame $S_{k+2}$, where $k$ is the number of occurrences of modalities in the rule.

Also we see that if a rule is admissible in some logic contained in LC then it is admissible in LC. The analogous statement holds for Grz.3 and also for *PM*4. In general, it may happen that a rule is admissible in some system but not admissible in some of its extension.

## §5. Decidable properties of calculi and complexity bounds.

In this section a catalog of decidable properties is presented. All complexity bounds given in this section concern extensions of basic calculi by axiom schemes but not rules of inference. The algorithms and proofs are presented in the corresponding papers. Some of them can be found also in [4].

### 5.1. Consistency.

A calculus is *consistent* if it differs from the set of all formulas.

It is known that for any formula $A$ without modalities, Int + $A$ is consistent iff $A$ is a classical two-valued tautology. The same criterion holds for $\text{Int}^+$ (where $A$ is positive). Further, D. Makinson [17] proved in 1966 that for any formula $A$ of modal language, the calculus K + $A$ is consistent iff $A$ is valid in at least one of two one-element frames $\langle W, R\rangle$, where $R = \emptyset$ or $R = W^2$. This implies coNP-completeness of consistency problem [35]. So we have

STATEMENT 5.1. Consistency is decidable over Int, $\text{Int}^+$ and K.

STATEMENT 5.2. Consistency problem over Int, $\text{Int}^+$ and K is *coNP*-complete.

It was proved in [31] that

STATEMENT 5.3. Consistency is strongly decidable over Int and $D = K + \Diamond\top$.

In order to prove that a calculus Int + *Rul* is consistent, one must show that all rules in *Rul* are valid in the two-valued Boolean algebra. The calculus D + *Rul* is consistent if and only if all rules in *Rul* are valid in the one-element reflexive frame [31].

PROBLEM. Is Consistency strongly decidable over K, K4 or GL?

**5.2. Tabularity and related problems.** A logic is called *tabular* if it can be characterized by finitely many finite models; and *pretabular* if it is maximal among non-tabular logics.

A logic $L$ is called *locally tabular* if for any finite set $P$ of propositional variables there exist only finitely many formulas of $P$ non-equivalent in $L$.

STATEMENT 5.4. Tabularity and Pretabularity are decidable over

- Int (Kuznetsov, Maksimova 1972 [18]),
- Int$^+$ (Verhozina 1978 [44]),
- S4 (Maksimova 1975 [20], Esakia, Meskhi 1977 [6], Rautenberg 1977 [38]),
- GL (Chagrov 1989 [2]).

There are exactly three pretabular logics over Int, two over Int$^+$ and five over S4. So the pretabularity problem is just the problem of equivalence to some of pretabular logics.

One can show that Int + $Ax$ is tabular iff $Ax$ is intuitionistically refutable in each of the frames $S_{n+1}$, $V_{n+1}$ and $U_{n+2}$, where $n = |Ax|$. A calculus Int$^+$ + $Ax$ is tabular iff $Ax$ is intuitionistically refutable in each of the frames $S_{n+1}$ and $V_{n+1}$, where $n = |Ax|$. A calculus S4 + $Ax$ is tabular iff $Ax$ is refutable in each of the frames $S_{n+1}$, $V_{n+1}$, $U_{n+2}$, $X_{n+1}$ and $Y_{n+2}$, where $n$ is the number of occurrences of modalities in $Ax$.

STATEMENT 5.5. Local Tabularity is decidable over

- S4 (Maksimova 1975 [19]),
- GL (Rybakov 1982 [39]).

OPEN PROBLEM. Is Local Tabularity decidable over Int or Int$^+$?

The following complexity bounds were found by Maksimova, Voronkov 2003 [35] for Int and S4, and by L. Maksimova 2002 [33] for Int$^+$.

STATEMENT 5.6.

- The tabularity problems over Int, Int$^+$ and S4 are *NP*-complete.
- The pretabularity problems over Int, Int$^+$ and S4 are *DP*-hard and in $\Delta_2^p$.
- The local tabularity problems over S4 and GL are *NP*-complete.

The assertions on strong decidability were proved by Maksimova 2000 [31]:

STATEMENT 5.7. The following properties are strongly decidable:

- Pretabularity over Int and S4,
- Tabularity over LC and S4.3,
- Local tabularity over S4.

To prove that pretabularity is strongly decidable, we essentially used V. Rybakov's results [40]. He proved that the problem of admissibility of inference rules is decidable in many important logical systems, in particular, in all pretabular logics over Int or S4.

OPEN PROBLEM. Is Tabularity strongly decidable over Int, $\mathrm{Int}^+$ or S4?

**5.3. Interpolation and Beth's properties.** A logic $L$ is said to have *Craig's interpolation property (CIP)*, if for every formula $(A \to B) \in L$ there exists $C$ such that

(i) both $A \to C$ and $C \to B$ belong to $L$,
(ii) any variable of $C$ occurs in both $A$ and $B$.

A logic $L$ is said to have *interpolation property (IPD)*, if $A \vdash_L B$ implies that exists a formula $C$ such that

(i) $A \vdash_L C$, $C \vdash_L B$, and
(ii) any variable of $C$ occurs in both $A$ and $B$.

*Beth's definability properties* have as their source the theorem on implicit definability proved by E. Beth in 1953 [1] for the classical first order logic:

Any predicate implicitly definable in a first order theory is explicitly definable.

We formulate some analogs of Beth's property for propositional logics.

Let $\mathbf{x}$, $\mathbf{q}$, $\mathbf{q}'$ be disjoint lists of variables not containing $y$ and $z$, $A(\mathbf{x}, \mathbf{q}, y)$ a formula.

PB1. If $\vdash_L A(\mathbf{x},\mathbf{q},y) \& A(\mathbf{x},\mathbf{q}',z) \to (y \leftrightarrow z)$, then $\vdash_L A(\mathbf{x},\mathbf{q},y) \to (y \leftrightarrow B(\mathbf{x}))$ for some formula $B(\mathbf{x})$.

PB2. If $A(\mathbf{x},\mathbf{q},y), A(\mathbf{x},\mathbf{q}',z) \vdash_L (y \leftrightarrow z)$, then $A(\mathbf{x},\mathbf{q},y) \vdash_L (y \leftrightarrow B(\mathbf{x}))$ for some formula $B(\mathbf{x})$.

We get weaker versions B1 and B2 of Beth's property by deleting $\mathbf{q}$ in PB1 and PB2 respectively.

We note that:

On the family NE(K) of normal modal logics:

$$\text{PB1} \iff \text{B1} \iff \text{CIP},\ \text{PB1} \Rightarrow \text{PB2} \Rightarrow \text{B2},\ \text{CIP} \Rightarrow \text{IPD},$$

B2 and IPD incomparable;

$$\text{IPD} \iff \text{CIP},\ \text{PB1} \iff \text{PB2 and B1} \iff \text{B2 over Int and over Int}^+.$$

We collect L. Maksimova's decidability results concerning above-mentioned properties in

STATEMENT 5.8. The following properties are decidable:

- CIP over Int and $\text{Int}^+$ (1977 [22], [21])
- CIP and IPD over S4 (1979 [23])
- PB2 over Int (2001 [32])
- PB2 over Grz and over $\text{Int}^+$ (2003 [34]).

The property B2 is decidable over Int, $\text{Int}^+$ and over K4 due to

STATEMENT 5.9. B2 is valid:

- for all extensions of Int and $\text{Int}^+$ (Kreisel 1960 [12])
- for all extensions of K4 (Maksimova 1992 [28]).

OPEN PROBLEM. Decidability of PB2 over S4.

Recall that CIP is undecidable over GL [3, 4].

Complexity bounds were found by Maksimova and Voronkov [35] for extensions of S4 and of Int and by Maksimova 2002 [33] for extensions of $\text{Int}^+$:

STATEMENT 5.10. • CIP over Int and over Grz is *PSPACE*-complete
• CIP and PB2 over $\text{Int}^+$ are *PSPACE*-complete
• CIP and IPD over S4, PB2 over Int and over Grz are in *coNEXP* and *PSPACE*-hard.

Strong decidability results were found by Maksimova 2000 [31]:

STATEMENT 5.11. The following properties are strongly decidable:

- CIP and IPD over Int and over Grz
- CIP, IPD and PB2 over S5.

CONJECTURE. CIP and IPD over S4, PB2 over Int and over Grz are strongly decidable.

It was proved that there are finitely many superintuitionistic and positive logics with CIP or PB2 [22, 30, 33]. All these logics are fully described, and the algorithms recognizing CIP and PB2 are found [32, 33].

It should be noted that CIP and IPD problems over S4 are not yet completely solved even for axiomatic extensions of S4. In [23]–[26] we have found a list of 49 logics containing all logics with IPD in $NE(\text{S4})$. All these logics are well described, and the problem of equivalence to each of the 49 logics is decidable. But we do not know if all of those logics really have CIP or IPD. We only know that 31 of the 49 logics have CIP, in particular, S4, S4.1, S4.2, S4.1.2. S5, Grz, Grz.2 and many other known logics. In addition, there are 12 logics in $NE(\text{S4})$ that have IPD but do not possess CIP. There are 6 logics in the list, for which CIP and IPD problems are not yet solved. These logics are $\text{S4}+\eta_4$, $\text{S4}+\eta_4+\xi_3$, $\text{S4}+\eta_4+\xi_2$, $\text{S4.2}+\eta_4$, $\text{S4.2}+\eta_4+\xi_3$, $\text{S4.2}+\eta_4+\xi_2$ (formulas $\xi_n$ and $\eta_n$ were defined in Section 3.2).

**§6. Algebraic equivalents.** It is well known that there is a duality between normal modal logics and varieties of modal algebras. The least normal modal logic K is defined by the variety of all modal algebras, K4 by the variety of so-called transitive modal algebras, S4 by topoboolean algebras (or closure algebras), GL by diagonalizable algebras, Grz by Grzegorczyk algebras, S5 by epistemic algebras and so on. Also there is a duality between superintuitionistic logics and varieties of Heyting algebras, between positive logics and varieties of relatively pseudocomplemented lattices. For a given logic $L$, its associated variety $V(L)$ is determined by identities $A = \top$ for $A$ provable in $L$.

The properties of logics considered in this paper have natural algebraic equivalents.

PROPOSITION 6.1. *For any logic L in* $E(\mathrm{Int})$, $E(\mathrm{Int}^+)$ *or in* $NE(\mathrm{K})$:

- *L has CIP iff* $V(L)$ *has the super-amalgamation property SAP* (Maksimova 1977, 1991 [22], [27]),
- *L has IPD iff* $V(L)$ *has the amalgamation property AP* (Maksimova 1979 [23], Czelakowski 1982 [5]),
- *L has B2 iff* $V(L)$ *has epimorphisms surjectivity ES** (Nemeti 1985 [9]),
- *L has PB2 iff* $V(L)$ *has strong epimorphisms surjectivity SES* (Maksimova 1998 [29]).

We recall the definitions of amalgamation, super-amalgamation, strong amalgamation StrAP, etc.

AP: If $\mathbf{A}$ is a common subalgebra of algebras $\mathbf{B}$ and $\mathbf{C}$ then there exist $\mathbf{D}$ in $V$ and monomorphisms $\delta : \mathbf{B} \to \mathbf{D}$, $\varepsilon : \mathbf{C} \to \mathbf{D}$ such that $\delta(x) = \varepsilon(x)$ for all $x \in \mathbf{A}$.

SAP: AP with extra conditions:

$$\delta(x) \leq \varepsilon(y) \iff (\exists z \in \mathbf{A})(x \leq z \text{ and } z \leq y),$$

$$\delta(x) \geq \varepsilon(y) \iff (\exists z \in \mathbf{A})(x \geq z \text{ and } z \geq y).$$

StrAP: AP with an extra condition:

$$\delta(\mathbf{B}) \cap \varepsilon(\mathbf{C}) = \delta(\mathbf{A}).$$

ES*. For any $\mathbf{A}$, $\mathbf{B}$, for any monomorphism $\alpha : \mathbf{A} \to \mathbf{B}$ and for any $x \in \mathbf{B} - \alpha(\mathbf{A})$, such that $\{x\} \cup \alpha(\mathbf{A})$ generates $\mathbf{B}$, there exist $\mathbf{C} \in V$ and monomorphisms $\beta : \mathbf{B} \to \mathbf{C}$ and $\gamma : \mathbf{B} \to \mathbf{C}$ such that $\beta\alpha = \gamma\alpha$ and $\beta(x) \neq \gamma(x)$.

SES. For any $\mathbf{A}, \mathbf{B}$ in $V$, for any monomorphism $\alpha : \mathbf{A} \to \mathbf{B}$ and for any $x \in \mathbf{B} - \alpha(\mathbf{A})$ there exist $\mathbf{C} \in V$ and monomorphisms $\beta : \mathbf{B} \to \mathbf{C}$ and $\gamma : \mathbf{B} \to \mathbf{C}$ such that $\beta\alpha = \gamma\alpha$ and $\beta(x) \neq \gamma(x)$.

In varieties of modal algebras we have:

$$\text{SAP} \Rightarrow \text{StrAP} \Leftrightarrow \text{AP} \,\&\, \text{ES}^* \Rightarrow \text{SES} \Rightarrow \text{ES}^*,$$
$$\text{ES}^* \not\Rightarrow \text{AP and AP} \not\Rightarrow \text{ES}^*.$$

From Proposition 5.9 we get that all the varieties of transitive modal algebras, of Heyting algebras and of relatively pseudocomplemented lattices have ES*. Therefore, in varieties of modal transitive algebras:

$$\text{SAP} \Rightarrow \text{StrAP} \Leftrightarrow \text{AP} \Rightarrow \text{SES and SES} \not\Rightarrow \text{AP}.$$

For varieties of Heyting algebras and of relatively pseudocomplemented lattices we have:

$$\text{SAP} \Leftrightarrow \text{StrAP} \Leftrightarrow \text{AP} \Rightarrow \text{SES and SES} \not\Rightarrow \text{AP}.$$

Using Proposition 6.1, one can rewrite decidability results of Section 5 in terms of varieties of algebras. We say that a property P of varieties is *base-decidable* if there is an algorithm which for any finite set S of identities decides whether a subvariety determined by S has the property P or not.

STATEMENT 6.2. The following properties are base-decidable:

- Tabularity and Pretabularity in varieties of Heyting algebras, relatively pseudocomplemented lattices and of closure algebras.
- Tabularity in varieties of diagonalizable algebras.
- Local tabularity in varieties of closure algebras and of diagonalizable algebras.
- Amalgamation, strong amalgamation and superamalgamation in varieties of Heyting algebras, relatively pseudo-complemented lattices and closure algebras.
- Strong epimorphisms surjectivity in varieties of Heyting algebras, relatively pseudo-complemented lattices, of Grzegorczyk algebras and of epistemic algebras.

By Statement 5.9 and Proposition 6.1, it is evident that ES* is decidable in varieties of transitive algebras, Heyting algebras and relatively pseudocomplemented lattices.

## REFERENCES

[1] E. W. BETH, *On Padoa's method in the theory of definitions*, ***Indagationes Mathematicae***, vol. 15 (1953), no. 4, pp. 330–339.

[2] A. V. CHAGROV, *Nontabularity - pretabularity, antitabularity, coantitabularity*, ***Algebraic and logical constructions***, Kalinin State University, Kalinin, 1989, pp. 105–111.

[3] ———, *Undecidable properties of superintuitionistic logics*, ***Mathematical problems of cybernetics*** (S. V. Jablonsky, editor), vol. 5, Fizmatlit, Moscow, 1994, pp. 67–108.

[4] A. V. CHAGROV and M. ZAKHARYASCHEV, ***Modal logics***, Clarendon Press, Oxford, 1997.

[5] J. CZELAKOWSKI, *Logical matrices and the amalgamation property*, ***Studia Logica***, vol. 41 (1982), no. 4, pp. 329–341.

[6] L. ESAKIA and V. MESKHI, *Five critical systems*, ***Theoria***, vol. 40 (1977), pp. 52–60.

[7] M. J. FISHER and R. E. LADNER, *Propositional dynamic logic of regular programs*, ***Journal of Computer and System Sciences***, vol. 18 (1979), pp. 194–211.

[8] J. Y. Halpern and Y. Moses, *A guide to completeness and complexity for modal logics of knowledge and belief*, ***Artificial Intelligence***, vol. 54 (1992), pp. 319–379.

[9] L. Henkin, D. Monk, and A. Tarski, ***Cylindric algebras, Part II***, North Holland, Amsterdam, 1985.

[10] T. Hosoi, *On intermediate logics I*, ***Journal of the Faculty of Science University of Tokyo, Section Ia***, vol. 14 (1967), pp. 293–312.

[11] D. S. Johnson, *A catalog of complexity classes*, ***Handbook of theoretical computer science, Volume A*** (J. van Leeuwen, editor), Elsevier, 1990, pp. 67–161.

[12] G. Kreisel, *Explicit definability in intuitionistic logic*, ***The Journal of Symbolic Logic***, vol. 25 (1960), pp. 389–390.

[13] A. V. Kuznetsov, *Undecidability of general problems of completeness, decidability and equivalence for propositional calculi*, ***Algebra and Logic***, vol. 2 (1963), pp. 47–66.

[14] ———, *Some properties of the lattice of varieties of pseudo-Boolean algebras*, ***11th Sovjet algebraic colloquium, abstracts (Kishinev, 1971)***, 1971, pp. 255–256.

[15] R. E. Ladner, *The computational complexity of provability in systems of modal propositional logics*, ***SIAM Journal of Computing***, vol. 6 (1977), no. 3, pp. 467–480.

[16] S. Linial and E. L. Post, *Recursive unsolvability of the deducibility, Tarski's completeness and independence of axioms problems in the propositional calculus*, ***Bulletin of American Mathematical Society***, vol. 55 (1949), p. 50.

[17] D. C. Makinson, *On some completeness theorem in modal logic*, ***Zeitschrift fuer Mathematische Logik und Grundlagen der Mathematik***, vol. 12 (1966), pp. 379–384.

[18] L. L. Maksimova, *Pretabular superintuitionistic logics*, ***Algebra and Logic***, vol. 11 (1972), pp. 558–570.

[19] ———, *Modal logics of finite slices*, ***Algebra and Logic***, vol. 14 (1975), pp. 188–197.

[20] ———, *Pretabular extensions of Lewis S4*, ***Algebra and Logic***, vol. 14 (1975), pp. 16–33.

[21] ———, *Craig's interpolation theorem and amalgamable varieties*, ***Doklady AN SSSR***, vol. 237 (1977), pp. 1281–1284.

[22] ———, *Craig's theorem in superintuitionistic logics and amalgamable varieties of Pseudo-Boolean algebras*, ***Algebra i Logika***, vol. 16 (1977), no. 6, pp. 643–681.

[23] ———, *Interpolation theorems in modal logics and amalgamable varieties of Topoboolean algebras*, ***Algebra i Logika***, vol. 18 (1979), no. 5, pp. 556–586.

[24] ———, *On a classification of modal logics*, ***Algebra and Logic***, vol. 18 (1979), no. 3, pp. 328–340.

[25] ———, *Interpolation theorems in modal logics. Sufficient conditions*, ***Algebra i Logika***, vol. 19 (1980), no. 2, pp. 194–213.

[26] ———, *Absence of interpolation in modal companions of Dummett's logic*, ***Algebra i logika***, vol. 21 (1982), pp. 690–694.

[27] ———, *The Beth properties, interpolation, and amalgamability in varieties of modal algebras*, ***Doklady Akademii Nauk SSSR***, vol. 319 (1991), no. 6, pp. 1309–1312.

[28] ———, *An analog of the Beth theorem in normal extensions of the modal logic K4*, ***Siberian Mathematical Journal***, vol. 33 (1992), no. 6, pp. 118–130.

[29] ———, *Explicit and implicit definability in modal and related logics*, ***Bulletin of the Section of Logic***, vol. 27 (1998), no. 1/2, pp. 36–39.

[30] ———, *Intuitionistic logic and implicit definability*, ***Annals of Pure and Applied Logic***, vol. 105 (2000), pp. 83–102.

[31] ———, *Strongly decidable properties of modal and intuitionistic calculi*, ***Logic Journal of the IGPL***, vol. 8 (2000), pp. 797–819.

[32] ———, *Decidability of projective Beth's property in varieties of Heyting algebras*, ***Algebra and Logic***, vol. 40 (2001), pp. 290–301.

[33] ———, *Complexity of interpolation and related problems in positive calculi*, ***The Journal of Symbolic Logic***, vol. 67 (2002), pp. 397–408.

[34] ———, *Implicit definability and positive logics*, ***Algebra and Logic***, vol. 42 (2003), pp. 65–93.

[35] L. L. Maksimova and A. Voronkov, *Complexity of some problems in modal and intuitionistic calculi*, ***Computer science logic (Proc. of 17th Intern. Workshop, CSL 2003, 12th Annual Conference of the EACSL, and 8th Kurt Goedel Colloquium, KGC 2003, Vienna, Austria August 25-30, 2003)*** (M. Baaz and J. M. Makowsky, editors), Lecture Notes in Computer Science, vol. 2803, Springer, 2003, pp. 397–412.

[36] C. H. Papadimitriou, ***Computational complexity***, Addison-Wesley, Reading, Massachusets, 1994.

[37] V. R. Pratt, *Models of program logics*, ***Proceedings 20th IEEE symposium on foundations of computer science***, 1979, pp. 115–122.

[38] W. Rautenberg, *Der Verband der normalen verzweigten Modallogiken*, ***Mathematische Zeitschrift***, vol. 156 (1977), pp. 123–140.

[39] V. Rybakov, *Completeness of modal logics of prefinite width*, ***Matematicheskie Zametki***, vol. 32 (1982), pp. 223–228.

[40] ———, ***Admissibility of logical inference rules***, Elsevier, Amsterdam, New York, 1997.

[41] K. Segerberg, ***An essay in classical modal logics***, Uppsala, 1971.

[42] E. Spaan, ***Complexity of modal logics***, Dissertation, Institute for Logic, Language and Computation, University of Amsterdam, 1992.

[43] R. Statman, *Intuitionistic logic is polynomial-space complete*, ***Theoretical Computer Science***, vol. 9 (1979), pp. 67–72.

[44] M. I. Verhozina, *Intermediate positive logics*, ***Algorithmic problems of algebraic systems***, Irkutsk, 1978, pp. 13–25.

SOBOLEV INSTITUTE OF MATHEMATICS
SIBERIAN BRANCH OF RUSSIAN ACADEMY OF SCIENCES
630090, NOVOSIBIRSK, RUSSIA
*E-mail*: lmaksi@math.nsc.ru

# STABILIZATION—AN ALTERNATIVE TO DOUBLE-NEGATION TRANSLATION FOR CLASSICAL NATURAL DEDUCTION

RALPH MATTHES

**Abstract.** A new proof of strong normalization of Parigot's second-order $\lambda\mu$-calculus is given by a reduction-preserving embedding into system F (second-order polymorphic $\lambda$-calculus). The main idea is to use the least stable supertype for any type. These non-strictly positive inductive types and their associated iteration principle are available in system F and allow to give a translation vaguely related to CPS translations (corresponding to Kolmogorov's double-negation embedding of classical logic into intuitionistic logic). However, they simulate Parigot's $\mu$-reductions whereas CPS translations hide them.

As a major advantage, this embedding does not use the idea of reducing stability ($\neg\neg A \to A$) to that for atomic formulae. Therefore, it even extends to positive fixed-point types.

The article expands on "Parigot's Second-Order $\lambda\mu$-Calculus and Inductive Types" (Conference Proceedings TLCA 2001, Springer LNCS 2044) by the author.

**§1. Introduction.** If $\neg\neg A \to A$ is provable, then $A$ is called stable. Stability of all $A$ is the key characteristic of classical logic. In the framework of intuitionistic logic, classical features can be added in several ways. We concentrate on "reductio ad absurdum" (RAA), i.e., proof by contradiction: If falsity is provable from the negation of $A$, then $A$ is provable.

Natural deduction systems can be formulated — via the Curry-Howard-isomorphism — as lambda calculi. Here, we are interested in second-order logic, hence in second-order (polymorphic) lambda-calculus, also called system F.

System F is a term rewrite system, consequently also its extension by RAA comes with rewrite rules: we may speak about operationalized classical logic. The most widely known such system is Parigot's $\lambda\mu$-calculus, and it is strongly normalizing [19, 20], i.e., admits no infinite reduction sequences starting with a typable term. Its $\mu$-reduction rules describe how RAA for composite types can be reduced to RAA for the constituent types. In the case of implication, this corresponds to the fact that stability of $A \to B$ follows intuitionistically from stability of $B$.

In [14], an extension of system F by a least stable supertype $\sharp A$ — called the stabilization of $A$ — for any type $A$, has been studied for the purpose of reproving strong normalization of Parigot's second-order $\lambda\mu$-calculus.

**Logic Colloquium '03**
Edited by V. Stoltenberg-Hansen and J. Väänänen
Lecture Notes in Logic, 24

In the present article, this approach is refined to also cover $\lambda\mu$-calculus in the formulation of de Groote [5]. Also typing issues are considered, in the sense that typing à la Church is used instead of fully type-decorated terms. An important part of this article is the comparison with other approaches to prove strong normalization of second-order classical natural deduction from that of system F. For direct proofs of strong normalization through logical predicates — even capturing sums/disjunction with permutative/commuting conversions — see the account in [15].

In a nutshell, [14] has been extended

- to Church typing instead of full type annotations (with subtle problems leading to the notion of "refined embedding" — Definition 2 — and associated substitution rules),
- to $\lambda\mu$-calculus in de Groote's formulation (with $\bot$ being a regular type),
- by a comparison of stabilization with double negation,
- by a motivation of $\mu$-reduction rules and a presentation of the method by Joly [11],
- by an explanation why translations fail with Curry-style typing or just with plain classical natural deduction (without Parigot's "special substitution"),
- by a discussion of the recent criticism of CPS-translations by Nakazawa and Tatsuta [17].

The article is structured as follows: In Section 2, system F and second-order classical natural deduction are defined. Section 3 contains the definition of System I$\sharp$ with the least stable supertype $\sharp A$ for every type $A$ and its impredicative encoding in F. It also contains a comparison of $\sharp A$ and $\neg\neg A$. Section 4 is the central part: It is first shown that the more naive system $\mathsf{F}^{\neg}$ of classical natural deduction cannot be embedded into F, then $\lambda\mu$-calculus is defined and successfully embedded in I$\sharp$. Plenty of discussion material is given in Section 5: After a presentation of Joly's embedding, which is based on the idea of reducing stability to that for atomic formulae (= type variables), an extension to fixed points is given where Joly's method cannot be seen to work — unlike our stabilization. Even least fixed points are treated. Finally, the article by Nakazawa and Tatsuta is discussed under the heading "embeddings in continuation-passing style?".

**§2. Second-order natural deduction.** In this article, natural deduction is always represented by the help of lambda terms. For second-order natural deduction, the basic — intuitionistic — system is F [10], which is defined in the first part of this section. Mostly, notation is given that is used throughout the article.

In the second part, F is extended by "reductio ad absurdum" (RAA), which is again just a fixing of notation. In addition, an argument is given why F itself fails to be "classical", and the reduction rules for RAA — which are a variation on those of $\lambda\mu$-calculus [18] — are motivated by the reduction of stability for composite formulas/types to that for atoms/type variables.

**2.1. System F.** This is the polymorphic lambda-calculus whose normalization has first been shown by Girard [10]. In a proof-theoretic reading, F provides proof terms for intuitionistic second-order propositional logic in natural deduction style, and also considers proof transformation rules. Normalization for second-order systems does not yield the subformula property but nevertheless guarantees logical consistency and allows, e.g., to conclude that F is not classical — see below. In the programming language perspective, normalization is the calculation of values for terms inhabiting a concrete datatype. And F is a (functional) programming language in which datatypes can be represented [4].

We define system F and recall some of its basic metatheoretic properties.

*Types.* (Denoted by uppercase letters.) Metavariable $X$ ranges over an infinite set TV of type variables.

$$A, B, C, D ::= X \mid \bot \mid A \to B \mid \forall X \cdot A$$

The occurrences of $X$ in $A$ are bound in $\forall X \cdot A$. We identify types whose de Bruijn representations coincide. Let $\mathsf{FV}(A)$ be the set of free type variables in $A$. The type $\bot$ just serves as a constant in F. It is used to define $\neg A := A \to \bot$. Note that intuitionistic falsum is already present in the pure system: $\forall X \cdot X$ has "ex falsum quodlibet" by trivial universal elimination.

*Terms.* (Denoted by lowercase letters.) The metavariable $x$ ranges over an infinite set of term variables.

$$r, s, t ::= x \mid \lambda x^A \cdot t \mid rs \mid \Lambda X \cdot t \mid rB$$

The occurrences of $x$ in $t$ are bound in $\lambda x^A \cdot t$. Also, the occurrences of type variable $X$ in $t$ are bound in $\Lambda X \cdot t$ We identify $\alpha$-equivalent terms, that is, terms whose bound type and term variables are consistently renamed. Capture-avoiding substitution of a term $s$ for a variable $x$ in term $r$ is denoted by $r[x := s]$. Analogously, we define type substitution $A[X := B]$ and substitution $t[X := B]$ of type $B$ for type variable $X$ in term $t$. We think of terms as parse trees according to the above grammar rules and insert parentheses in case of ambiguities, especially around the application $rs$. But we will omit them as much as possible and assume that application associates to the left. Hence, we write $rs_1 \cdots s_n$ for the parenthesized term $(\cdots(rs_1)\cdots s_n)$ — also for types instead of terms $s_i$. We abbreviate $\mathsf{id}^A := \lambda x^A \cdot x$, the identity on $A$.

*Contexts.* They are finite sets of pairs $(x : A)$ of term variables and types and will be denoted by $\Gamma$. Term variables in a context $\Gamma$ are assumed to be distinct, and the notation $\Gamma, x : A$ shall indicate $\Gamma \cup \{(x : A)\}$ where $x$ does not occur in $\Gamma$. Set $\mathsf{FV}(\Gamma) := \bigcup_{(x:A)\in\Gamma} \mathsf{FV}(A)$.

*Well-typed terms.* $\Gamma \vdash t : A$ is inductively defined by

$$\frac{(x : A) \in \Gamma}{\Gamma \vdash x : A} \qquad \frac{\Gamma, x : A \vdash t : B}{\Gamma \vdash \lambda x^A \cdot t : A \to B} \qquad \frac{\Gamma \vdash r : A \to B \quad \Gamma \vdash s : A}{\Gamma \vdash rs : B}$$

as for simply-typed lambda-calculus, and by the two rules for the universal quantifier whose use is traced in the terms (unlike formulations in a style à la Curry):

$$\frac{\Gamma \vdash t : A}{\Gamma \vdash \Lambda X \cdot t : \forall X \cdot A} \text{ if } X \notin \mathsf{FV}(\Gamma) \qquad \frac{\Gamma \vdash r : \forall X \cdot A}{\Gamma \vdash rB : A[X := B]}$$

A term $t$ is *typable* if there are $\Gamma$ and $A$ such that $\Gamma \vdash t : A$. We write $\vdash t : A$ for $\emptyset \vdash t : A$. If there is a term $t$ with $\vdash t : A$, the type $A$ is *inhabited*. Evidently, if $\Gamma \vdash t : A$ and $\Gamma \subseteq \Gamma'$ for the context $\Gamma'$, then $\Gamma' \vdash t : A$ as well, i.e., *weakening* is an admissible typing rule. In the sequel, this word will be reserved for the rule

$$\frac{\Gamma \vdash t : B}{\Gamma, x : A \vdash t : B} \quad x \text{ not declared in } \Gamma.$$

It is also easy to see that the following *cut* rules are admissible typing rules:

$$\frac{\Gamma, x : A \vdash t : B \quad \Gamma \vdash s : A}{\Gamma \vdash t[x := s] : B} \qquad \frac{\Gamma \vdash t : B}{\Gamma[X := A] \vdash t[X := A] : B[X := A]}$$

Here, $\Gamma[X := A] := x_1 : B_1[X := A], \ldots, x_n : B_n[X := A]$ for the context $\Gamma = x_1 : B_1, \ldots, x_n : B_n$.

*Reduction.* The one-step reduction relation $t \longrightarrow t'$ between terms $t$ and $t'$ is defined as the closure of the following axioms under all term constructors.

$$(\lambda x^A \cdot t)s \longrightarrow_\beta t[x := s]$$
$$(\Lambda X \cdot t)B \longrightarrow_\beta t[X := B]$$

We denote the transitive closure of $\longrightarrow$ by $\longrightarrow^+$ and the reflexive-transitive closure by $\longrightarrow^*$. It is clear that $r \longrightarrow r'$ implies $r[x := s] \longrightarrow r'[x := s]$ and $s[x := r] \longrightarrow^* s[x := r']$.

*Normal terms.* Inductively define the set NF of *normal forms* by:

- If, for all $1 \leq i \leq n$, either $s_i \in \mathsf{NF}$ or $s_i$ is a type, then $xs_1 \cdots s_n \in \mathsf{NF}$.
- If $t \in \mathsf{NF}$ then $\lambda x^A \cdot t \in \mathsf{NF}$ and $\Lambda X \cdot t \in \mathsf{NF}$.

It is easy to see that the terms in NF are exactly those terms $t$ such that no $t'$ exists with $t \longrightarrow t'$, i.e., the normal forms are the *normal terms*.

*Metatheoretic properties.*

LEMMA 1 (Subject Reduction). *If* $\Gamma \vdash t : A$ *and* $t \longrightarrow t'$ *then* $\Gamma \vdash t' : A$.

PROOF. This only requires the cut rules above. Subject reduction would be much more difficult in a formulation with typing à la Curry. ⊣

LEMMA 2 (Local Confluence). F *is locally confluent, i.e., if* $r \longrightarrow r_1, r_2$ *then there is a term* $t$ *such that* $r_1, r_2 \longrightarrow^* t$.

PROOF. There are no critical pairs. ⊣

It is well known that F is even *confluent*, i.e., $\longrightarrow^*$ is locally confluent.

LEMMA 3 (Strong Normalization). *If* $\Gamma \vdash t : A$ *then there is no infinite reduction sequence* $t \longrightarrow t_1 \longrightarrow t_2 \longrightarrow \cdots$. *In other words, every typable term is strongly normalizing.*

This deep result is known since Tait's adaptation [26] of Girard's proof of weak normalization [10].

**2.2. Classical natural deduction.** It is well known that system F is not classical: There is no closed term of type $\neg\neg X \to X$, for $X$ any type variable. (This is equivalent to the non-existence of a closed term of type $\forall X \cdot \neg\neg X \to X$.)

A proof can be obtained as follows. If, to the contrary, there were such a term $r_c$, one would also get a closed term $r_p$ of type $(\neg X \to X) \to X$, which amounts to the Peirce law: Set

$$r_p := \lambda x^{\neg X \to X} \cdot r_c\big(\lambda y^{\neg X} \cdot y(xy)\big).$$

By normalization of F, there would also be a closed *normal* term $r$ of type $(\neg X \to X) \to X$. Several steps of reasoning about normal terms (using their equivalence with the inductively defined set NF of normal *forms*) would yield that $r$ has the form $\lambda x^{\neg X \to X} \cdot x(\lambda z^X \cdot s)$ with $s$ a normal term, and $x : \neg X \to X,\ z : X \vdash s : \bot$. This is plainly impossible, by inspection of normal forms.[1]

Therefore, F has to be extended in order to be classical. Following Prawitz [22, 23], system $\mathsf{F}^\neg$ will now be based on "reductio ad absurdum" (RAA): If one can derive absurdity from the assumption $\neg A$, then $A$ obtains. A term notation for a first-order system in this spirit has been proposed by Rehof and Sørensen [24], and later on even for pure type systems [3]. While those authors use the letter $\Delta$ for the binder, our name $\mu$ is taken from Parigot's $\lambda\mu$-calculus [18].

The term syntax of $\mathsf{F}^\neg$ is that of F, plus

$$r, s, t ::= \cdots \mid \mu x^{\neg A} \cdot r$$

As regards variable binding, $\mu x^{\neg A} \cdot r$ is the same as $\lambda x^{\neg A} \cdot r$.

---

[1] This proof — with the detour through Peirce's law — also works in case that $\bot$ is set to $\forall Y \cdot Y$, i.e., for the system with built-in *ex falsum quodlibet* $\forall X \cdot \bot \to X$.

The new typing rule models RAA:

$$\frac{\Gamma, x : \neg A \vdash r : \bot}{\Gamma \vdash \mu x^{\neg A} . r : A}$$

We also add the axiom schemes of $\mu$-reduction to the reduction system:

$$(\mu x^{\neg(A\to B)} \cdot r)s \longrightarrow_\mu \mu y^{\neg B} \cdot r[x := \lambda z^{A\to B} \cdot y(zs)]$$
$$(\mu x^{\neg\forall X\cdot A} \cdot r)B \longrightarrow_\mu \mu y^{\neg A[X:=B]} \cdot r[x := \lambda z^{\forall X\cdot A} \cdot y(zB)]$$

The new term closure will again be denoted by $\longrightarrow$.

The first rule is present in $\lambda_\Delta$ in [24], the second one is the canonical analogue for the second-order quantification, as has been studied for $\lambda\mu$-calculus in [20].

*Motivation of the $\mu$-rules.* RAA is equivalent to stability, $\forall X \cdot \neg\neg X \to X$, also called "duplex negatio affirmat": If one introduces a constant $\mathsf{stab}^A$ assuming $\neg\neg A \to A$, then $\mu x^{\neg A} \cdot r$ can be defined as $\mathsf{stab}^A(\lambda x^{\neg A} \cdot r)$. In the other direction, one may define the term $\mathsf{stab}^A$ of type $\neg\neg A \to A$ as $\mathsf{stab}^A := \lambda v^{\neg\neg A} \mu x^{\neg A} \cdot vx$.

It is general knowledge that, in minimal logic, stability of $A \to B$ follows from stability of $B$. In a first-order system, this may be used to assume only constants $\mathsf{stab}^X : \neg\neg X \to X$ for type variables $X$, and to define the closed term

$$\mathsf{stab}^{A\to B} := \lambda v^{\neg\neg(A\to B)} \lambda w^A \cdot \mathsf{stab}^B\left(\lambda y^{\neg B} \cdot v(\lambda z^{A\to B} \cdot y(zw))\right)$$

of type $\neg\neg(A \to B) \to A \to B$. The above-mentioned encoding of RAA immediately yields the first $\mu$-rule:

$$\begin{aligned}
&(\mu x^{\neg(A\to B)} \cdot r)s \\
&\quad = \mathsf{stab}^{A\to B}(\lambda x^{\neg(A\to B)} \cdot r)s \\
&\quad \longrightarrow^2 \mathsf{stab}^B\left(\lambda y^{\neg B} \cdot (\lambda x^{\neg(A\to B)} \cdot r)(\lambda z^{A\to B} \cdot y(zs))\right) \\
&\quad = \mu y^{\neg B} \cdot \underbrace{(\lambda x^{\neg(A\to B)} \cdot r)(\lambda z^{A\to B} \cdot y(zs))}_{\longrightarrow r[x:=\lambda z^{A\to B}\cdot y(zs)]}.
\end{aligned}$$

Therefore, the termination of simply-typed lambda-calculus with the first-order $\mu$-reduction rule is very easy.

We might extend our definition to the second order as follows:

$$\mathsf{stab}^{\forall X\cdot A} := \lambda v^{\neg\neg\forall X\cdot A} \Lambda X \cdot \mathsf{stab}^A\left(\lambda y^{\neg A} \cdot v(\lambda z^{\forall X\cdot A} \cdot y(zX))\right),$$

which yields a closed term of type $\neg\neg(\forall X \cdot A) \to \forall X \cdot A$ if $\mathsf{stab}^A$ is closed and of type $\neg\neg A \to A$. Again with our encoding of RAA, we would get

$$
\begin{aligned}
&(\mu x^{\neg\forall X\cdot A}\cdot r)B\\
&\quad = \mathsf{stab}^{\forall X\cdot A}\,(\lambda x^{\neg\forall X\cdot A}\cdot r)B\\
&\quad \longrightarrow^2 \mathsf{stab}^A[X:=B](\lambda y^{\neg A[X:=B]}\\
&\qquad\qquad \cdot(\lambda x^{\neg\forall X\cdot A}\cdot r)(\lambda z^{\forall X\cdot A}\cdot y(zB)))\\
&\quad = \mu y^{\neg A[X:=B]}\cdot \underbrace{(\lambda x^{\neg\forall X\cdot A}\cdot r)(\lambda z^{\forall X\cdot A}\cdot y(zB))}_{\longrightarrow r[x:=\lambda z^{\forall X\cdot A}\cdot y(zB)]}
\end{aligned}
$$

In the passage from the last but one line to the last line, we assumed that $\mathsf{stab}^A[X := B] = \mathsf{stab}^{A[X:=B]}$, which can certainly not be taken for granted! $\mathsf{stab}^X$ in $\mathsf{stab}^X[X := B]$ is a constant while $\mathsf{stab}^B$ is a defined term if $B$ is non-atomic. An easy remedy seems to be a non-standard definition of substitution, namely $\mathsf{stab}^X[X := B] := \mathsf{stab}^B$. But this would give a non-trivial extension of F whose normalization would have to be established, hence no reduction to F would have been obtained.

One might want to repair the situation by assuming stability of $X$ in the context and hence taking variables instead of constants. Unfortunately, the above definition of $\mathsf{stab}^{\forall X\cdot A}$ does not allow $\mathsf{stab}^A$ to be typed with $X$ free in the typing context, hence rules out this idea. Nevertheless, Joly [11] found an embedding of $\mathsf{F}^\neg$ into F that allows to infer strong normalization for $\mathsf{F}^\neg$ from that of F, and that essentially follows our failed motivation of the $\mu$-reduction rules. This will be reported in more detail in Section 5.1.

A less ambitious but correct motivation for the second-order rule can nevertheless be given, just by using the "other direction" in the equivalence of RAA and stability: $\mathsf{stab}^A$ is only an abbreviation for $\lambda v^{\neg\neg A}\mu x^{\neg A}\cdot vx$, and we require that $\mathsf{stab}^{\forall X\cdot A}$ and the term

$$\lambda v^{\neg\neg\forall X\cdot A}\Lambda X\cdot \mathsf{stab}^A\,(\lambda y^{\neg A}\cdot v(\lambda z^{\forall X\cdot A}\cdot y(zX))),$$

that defined $\mathsf{stab}^{\forall X\cdot A}$ above, are equivalent with respect to $=_\beta$, which is the symmetric, transitive and reflexive closure of $\beta$-reduction $\longrightarrow$ of F. We then calculate

$$
\begin{aligned}
&(\mu x^{\neg\forall X\cdot A}\cdot r)B\\
&=_\beta (\mu x^{\neg\forall X\cdot A}\cdot(\lambda x^{\neg\forall X\cdot A}\cdot r)x)B\\
&=_\beta (\lambda v^{\neg\neg\forall X\cdot A}\mu x^{\neg\forall X\cdot A}\cdot vx)(\lambda x^{\neg\forall X\cdot A}\cdot r)B\\
&= \mathsf{stab}^{\forall X\cdot A}\,(\lambda x^{\neg\forall X\cdot A}\cdot r)B\\
&=_\beta (\lambda v^{\neg\neg\forall X\cdot A}\Lambda X\cdot \mathsf{stab}^A\,(\lambda y^{\neg A}\cdot v(\lambda z^{\forall X\cdot A}\cdot y(zX))))(\lambda x^{\neg\forall X\cdot A}\cdot r)B\\
&=_\beta \mathsf{stab}^A[X:=B](\lambda y^{\neg A[X:=B]}\cdot(\lambda x^{\neg\forall X\cdot A}\cdot r)(\lambda z^{\forall X\cdot A}\cdot y(zB)))
\end{aligned}
$$

$$
\begin{aligned}
&= \left(\lambda v^{\neg\neg A[X:=B]} \mu y^{\neg A[X:=B]} . v y\right) \\
&\quad \times \left(\lambda y^{\neg A[X:=B]} . \left(\lambda x^{\neg \forall X \cdot A} . r\right)\left(\lambda z^{\forall X \cdot A} . y(zB)\right)\right) \\
&=_\beta \mu y^{\neg A[X:=B]} . \underbrace{\left(\lambda x^{\neg \forall X \cdot A} . r\right)\left(\lambda z^{\forall X \cdot A} . y(zB)\right)}_{=_\beta r[x:=\lambda z^{\forall X \cdot A} . y(zB)]}
\end{aligned}
$$

Note that this second approach to a motivation does not say anything about normalization. We record that second-order classical natural deduction is not only more expressive than first-order classical natural deduction (this is vacuously true!), but that the reduction to the respective intuitionistic subsystem is essentially more complicated in the second-order case than for first order.

LEMMA 4. *Subject reduction holds for* $\mathsf{F}^{\neg}$.

PROOF. The first attempt at motivating the $\mu$-reduction rules already justifies subject reduction for these rules, the $\beta$-reduction rules can be separately treated for F alone, and the term closure does not raise any difficulty. ⊣

LEMMA 5 (Local Confluence). $\mathsf{F}^{\neg}$ *is locally confluent.*

PROOF. As for F, there are no critical pairs. ⊣

Therefore, strong normalization, as shown below, yields confluence for typable terms by Newman's Lemma. Certainly, the method of complete developments by Takahashi [27] can be extended to show even confluence for all terms.

**§3. Stabilization in iterative style.** In this section, the essential new tool (first published in [14]) for the reduction of second-order classical logic to second-order intuitionistic logic is defined: An extension $\mathsf{I}\sharp$ of F with types of the form $\sharp A$ that represents the *least* stable supertype of $A$. This minimality gives rise to an iteration principle that explains the heading "stabilization in iterative style". Although this is the only way stabilization is treated in the present article, this name is chosen in order to distinguish it from another approach to stabilization capable of capturing sums with permutative conversions, to be reported elsewhere.

The first part is devoted to the definition of $\mathsf{I}\sharp$, and to its (impredicative) justification in terms of F, whose strong normalization is inherited through an embedding. In the second part, we find a retract embedding of $\neg\neg A$ into $\sharp A$, giving a link to the Kolmogorov translation on which CPS-translations are based.

**3.1. Definition of System $\mathsf{I}\sharp$.** We define system $\mathsf{I}\sharp$ as extension of system F by stabilization types with iteration. The type system of $\mathsf{I}\sharp$ is

$$A, B, C, D ::= X \mid \bot \mid A \to B \mid \forall X \cdot A \mid \sharp A$$

The term system and its typing reflects that $\sharp A$ is stable, that $A$ embeds into $\sharp A$ and that $\sharp A$ is the least type with these properties.[2] We extend the grammar of terms with

$$r, s, t ::= \cdots \mid \mathsf{emb} \mid \mathsf{stab} \mid \mathsf{It}^C_\sharp(r, s, t)$$

The definition of $\Gamma \vdash t : A$ for F is extended by three clauses

$$\frac{}{\Gamma \vdash \mathsf{emb} : \forall X \cdot X \to \sharp X} \qquad \frac{}{\Gamma \vdash \mathsf{stab} : \forall X \cdot \neg\neg\sharp X \to \sharp X}$$

$$\frac{\Gamma \vdash r : \sharp A \qquad \Gamma \vdash s_1 : A \to C \qquad \Gamma \vdash s_2 : \neg\neg C \to C}{\Gamma \vdash \mathsf{It}^C_\sharp(r, s_1, s_2) : C}$$

The one-step relation $\longrightarrow$ of $\mathsf{I}\sharp$ is defined with the axiom schemes $\longrightarrow_\beta$ of system F plus the schemes

$$\mathsf{It}^C_\sharp(\mathsf{emb}\, At, s_1, s_2) \longrightarrow_\sharp s_1 t$$

$$\mathsf{It}^C_\sharp(\mathsf{stab}\, At, s_1, s_2) \longrightarrow_\sharp s_2(\lambda y^{\neg C} \cdot t(\lambda z^{\sharp A} \cdot y \mathsf{It}^C_\sharp(z, s_1, s_2)))$$

In the last rule, we assume that $y \neq z$ and that $y, z$ are not free in $s_1, s_2, t$.

These reduction rules also justify to call $s_1$ and $s_2$ the first and second *step term* of $\mathsf{It}^C_\sharp(r, s_1, s_2)$, respectively. The rules will be explained through an embedding in the proof of Lemma 9 below.

LEMMA 6. *Subject reduction holds for* $\mathsf{I}\sharp$.

PROOF. The $\longrightarrow_\sharp$ axioms evidently preserve types, closure under term formation rules works as usual. ⊣

LEMMA 7. $\mathsf{I}\sharp$ *is confluent.*

PROOF. By an immediate extension of the complete developments method by Takahashi [27]. Since there are no critical pairs, local confluence is even trivially true.[3] ⊣

LEMMA 8. $\mathsf{I}\sharp$ *is strongly normalizing.*

The proof will occupy the remainder of this section.

The idea of the proof is as follows: Although the second rule for $\longrightarrow_\sharp$ looks unfamiliar, it is just an instance of iteration over a non-strictly positive inductive type, and hence the reduction behaviour can be simulated within F.

DEFINITION 1. A type-respecting reduction-preserving embedding (embedding for short) of a term rewrite system $S$ with typing into a term rewrite system $S'$ with typing is a function $-'$ (the $-$ sign represents the indefinite

[2] The author is aware that already Paulin-Mohring [21, p. 110] describes a similar idea: An inductively defined predicate may be "made classical" by adding stability as another clause to the definition. This turns the definition into a non-strictly positive one and enforces stability. Note, however, that non-strictly positive inductive definitions lead to inconsistencies in higher-order predicate logic, see the example reported in [21, p. 108]. In the framework of system F, we are in the fortunate situation that arbitrary types may be stabilized without harm to consistency.

[3] If there were also $\eta$ rules, even local confluence would not hold for non-typable terms.

argument of the function $'$) which assigns to every type $A$ of $\mathcal{S}$ a type $A'$ of $\mathcal{S}'$ and to every term $r$ of $\mathcal{S}$ a term $r'$ of $\mathcal{S}'$ such that the following implications hold:

- If $x_1 : A_1, \dots, x_n : A_n \vdash r : A$ in $\mathcal{S}$ then $x_1 : A_1', \dots, x_n : A_n' \vdash r' : A'$ in $\mathcal{S}'$.
- If $r \longrightarrow s$ in $\mathcal{S}$, then $r' \longrightarrow^+ s'$ in $\mathcal{S}'$.

Obviously, if there is an embedding of $\mathcal{S}$ into $\mathcal{S}'$, then strong normalization of $\mathcal{S}'$ is inherited by $\mathcal{S}$. In the sequel, we will write $\Gamma'$ for the context $x_1 : A_1', \dots, x_n : A_n'$ if $\Gamma$ is $x_1 : A_1, \dots, x_n : A_n$.

Lemma 9. *There is an embedding of* $\mathsf{I}\sharp$ *into* $\mathsf{F}$.

Proof. By the very description, $\sharp A$ is nothing but the non-strictly positive inductive type $\mathsf{lfp}\, X \cdot A + \neg\neg X$ with $X \notin \mathsf{FV}(A)$, formulated without the sum type. Its canonical polymorphic encoding would be

$$\forall X \cdot ((A + \neg\neg X) \to X) \to X$$

which can be simplified to

$$\forall X \cdot (A \to X) \to (\neg\neg X \to X) \to X.$$

By iteration on the type $A$ of $\mathsf{I}\sharp$ we now define the type $A'$ of $\mathsf{F}$. This shall be done homomorphically in all cases except for:

$$(\sharp A)' := \forall X \cdot (A' \to X) \to (\neg\neg X \to X) \to X \quad \text{with } X \notin \mathsf{FV}(A).$$

Clearly, $\mathsf{FV}(A') = \mathsf{FV}(A)$ and $(A[X := B])' = A'[X := B']$.

By iteration on the term $r$ of $\mathsf{I}\sharp$ define the term $r'$ of $\mathsf{F}$. (Simultaneously, one has to show that the free variables of $r'$ and $r$ coincide). We only present the non-homomorphic cases.

- $\mathsf{emb}' := \Lambda Y \lambda x^Y \Lambda X \lambda x_1^{Y \to X} \lambda x_2^{\neg\neg X \to X} \cdot x_1 x$
- $\mathsf{stab}' := \Lambda Y \lambda x^{\neg\neg(\sharp Y)'} \Lambda X \lambda x_1^{Y \to X} \lambda x_2^{\neg\neg X \to X} \cdot x_2(\lambda y^{\neg X} \cdot x(\lambda z^{(\sharp Y)'} \cdot y(zXx_1x_2)))$
- $(\mathsf{It}_\sharp^C(r, s_1, s_2))' := r'C's_1's_2'$.

It is fairly easy to see that these settings form an embedding in the sense of Definition 1: The typings go well and, since $(t[x := s])' = t'[x := s']$ and $(t[X := A])' = t'[X := A']$, it is a trivial calculation to prove that

$$((\lambda x^A \cdot t)s)' \longrightarrow^1 (t[x := s])'$$

$$((\Lambda X \cdot t)A)' \longrightarrow^1 (t[X := A])'$$

$$(\mathsf{It}_\sharp^C(\mathsf{emb}\, At, s_1, s_2))' \longrightarrow^5 (s_1 t)'$$

$$(\mathsf{It}_\sharp^C(\mathsf{stab}\, At, s_1, s_2))' \longrightarrow^5 (s_2(\lambda y^{\neg C} \cdot t(\lambda z^{\sharp A} \cdot y \mathsf{It}_\sharp^C(z, s_1, s_2))))'.$$

Here, $\longrightarrow^n$ denotes $n$ steps of reduction $\longrightarrow$ (in $\mathsf{F}$). In fact, the right-hand sides of the $\longrightarrow_\sharp$-rules originally have been found just by reducing the translations of the left-hand sides. ⊣

COROLLARY 1. *System* $\mathsf{I}\sharp$ *is strongly normalizing.*

PROOF. Strong normalization is inherited through our embedding from System F. ⊣

**3.2. Double negation versus stabilization.** The Kolmogorov translation is based on double negation $\neg\neg A$ of $A$ which clearly yields a stable type (since every negated formula is stable already in minimal logic), and $A$ implies $\neg\neg A$. In our embedding of classical natural deduction, we will use stabilization $\sharp A$ of $A$, which is the least stable supertype of $A$ by definition. Hence, one would expect $\sharp A$ to be "smaller" than $\neg\neg A$. Perhaps surprisingly, we will find a retract in the other direction.

LEMMA 10. $\neg\neg A$ *and* $\sharp A$ *are logically equivalent. Moreover and more precisely,* $\neg\neg A$ *can be embedded by a retract into* $\sharp A$*, i.e., there are terms* $p, q$ *of* $\mathsf{I}\sharp$ *with* $\vdash p : \forall X \cdot \sharp X \to \neg\neg X$ *and* $\vdash q : \forall X \cdot \neg\neg X \to \sharp X$ *such that* $p \circ q := \Lambda X \lambda x^{\neg\neg X} \cdot pX(qXx) \longrightarrow^{+}_{\beta\sharp\eta} \Lambda X \lambda x^{\neg\neg X} \cdot x$.

*Here, the index* $\eta$ *indicates that another axiom scheme for the reduction is* $\lambda x^A \cdot rx \longrightarrow_\eta r$ *for* $x$ *not free in* $r$.

PROOF. For $p$, we just have to formalize the above discussion: Set $p := \Lambda X \lambda x^{\sharp X} \cdot \mathsf{It}_\sharp^{\neg\neg X}(x, s_1, s_2)$ with terms $s_1, s_2$ such that $\vdash s_1 : X \to \neg\neg X$ and $\vdash s_2 : \neg\neg\neg\neg X \to \neg\neg X$. This is achieved with:

$$s_1 := \lambda x^X \lambda k^{\neg X} \cdot kx$$
$$s_2 := \lambda y^{\neg\neg\neg\neg X} \lambda z^{\neg X} \cdot y(\lambda u^{\neg\neg X} \cdot uz).$$

Set $q := \Lambda X \lambda x^{\neg\neg X} \cdot \mathsf{stab}\, X t$ with a term $t$ such that $x : \neg\neg X \vdash t : \neg\neg\sharp X$, namely the canonical lifting of $z : X \vdash \mathsf{emb}\, Xz : \sharp X$ over the double negation:

$$t := \lambda y^{\neg\sharp X} \cdot x(\lambda z^X \cdot y(\mathsf{emb}\, Xz)).$$

In the following calculation, we always indicate in the reduction steps which kind of axiom scheme has been used:

$$p \circ q \longrightarrow^4_\beta \Lambda X \lambda x^{\neg\neg X} \cdot \underbrace{\mathsf{It}_\sharp^{\neg\neg X}(\mathsf{stab}\, Xt, s_1, s_2)}_{r :=}$$

$$\begin{aligned} r &\longrightarrow_\sharp s_2(\lambda y^{\neg\neg\neg X} \cdot t(\lambda z_1^{\sharp X} \cdot y \mathsf{It}_\sharp^{\neg\neg X}(z_1, s_1, s_2))) \\ &\longrightarrow_\beta \lambda z^{\neg X} \cdot (\lambda y^{\neg\neg\neg X} \\ &\qquad \cdot t(\lambda z_1^{\sharp X} \cdot y \mathsf{It}_\sharp^{\neg\neg X}(z_1, s_1, s_2)))(\lambda u^{\neg\neg X} \cdot uz) \\ &\longrightarrow_\beta \lambda z^{\neg X} \cdot t(\lambda z_1^{\sharp X} \cdot (\lambda u^{\neg\neg X} \cdot uz) \mathsf{It}_\sharp^{\neg\neg X}(z_1, s_1, s_2)) \\ &\longrightarrow_\beta \lambda z^{\neg X} \cdot t(\lambda z_1^{\sharp X} \cdot \mathsf{It}_\sharp^{\neg\neg X}(z_1, s_1, s_2) z) \end{aligned}$$

$$\begin{aligned}
&\longrightarrow_\beta \lambda z^{\neg X}\cdot x\big(\lambda z_2^X\cdot(\lambda z_1^{\sharp X}\cdot \mathsf{It}_\sharp^{\neg\neg X}(z_1,s_1,s_2)z)(\mathsf{emb}\,Xz_2)\big)\\
&\longrightarrow_\beta \lambda z^{\neg X}\cdot x\big(\lambda z_2^X\cdot \mathsf{It}_\sharp^{\neg\neg X}(\mathsf{emb}\,Xz_2,s_1,s_2)z\big)\\
&\longrightarrow_\sharp \lambda z^{\neg X}\cdot x\big(\lambda z_2^X\cdot s_1z_2z\big)\\
&\longrightarrow_\beta^2 \lambda z^{\neg X}\cdot x\big(\lambda z_2^X\cdot zz_2\big)\\
&\longrightarrow_\eta \lambda z^{\neg X}\cdot xz\longrightarrow_\eta x. \qquad\dashv
\end{aligned}$$

REMARK 1. The reduction behaviour of $q\circ p := \Lambda X\lambda x^{\sharp X}\cdot qX(pXx)$ with $p,q$ taken from the above proof is not perspicuous. There is no indication at all that $q\circ p$ behaves like the identity. E.g.,

$$\begin{aligned}
(q\circ p)X(\mathsf{emb}\,Xx) &\longrightarrow_\beta^4 qX\big(\mathsf{It}_\sharp^{\neg\neg X}(\mathsf{emb}\,Xx,s_1,s_2)\big)\\
&\longrightarrow_\sharp qX(s_1x)\\
&\longrightarrow_\beta qX(\lambda k^{\neg X}\cdot kx)\\
&\longrightarrow_\beta^2 \mathsf{stab}\,Xt[x:=\lambda k^{\neg X}\cdot kx]\\
&\longrightarrow_\beta \mathsf{stab}\,X\big(\lambda y^{\neg\sharp X}\cdot(\lambda z^X\cdot y(\mathsf{emb}\,Xz))x\big)\\
&\longrightarrow_\beta \mathsf{stab}\,X\big(\lambda y^{\neg\sharp X}\cdot y(\mathsf{emb}\,Xx)\big).
\end{aligned}$$

The last term cannot be reduced to emb $Xx$, and the author has no knowledge of a reasonable extension of the reduction system that allows to reduce $q\circ p$ to $\Lambda X\lambda x^{\sharp X}\cdot x$. (Already the rule $\mathsf{stab}\,A(\lambda x^{\neg\sharp A}\cdot xt)\longrightarrow_\eta t$ for $x$ not free in $t$ would break local confluence: Non-canonical stability witnesses that occur as second step terms of $\mathsf{It}_\sharp$ would not enjoy a similar operational rule, hence a coherence problem arises in form of non-confluence.)

We conclude that, in a sense, $\sharp A$ is an operationally larger type than $\neg\neg A$. In Section 4.3, this will be exploited in an embedding of $\lambda\mu$-calculus into $\mathsf{I}\sharp$ that does not erase the $\mu$-reductions.

**§4. A new embedding of classical natural deduction.** We would like to inherit strong normalization of $\mathsf{F}^\neg$ from $\mathsf{F}$, by giving an embedding of $\mathsf{F}^\neg$ into $\mathsf{I}\sharp$. This will fail due to some "administrative" reductions. By switching from $\mathsf{F}^\neg$ to a formulation of $\lambda\mu$-calculus, an embedding can be achieved.

In the first part, we expound the problems with embeddings of $\mathsf{F}^\neg$ into $\mathsf{I}\sharp$: Firstly, typing information is needed in order to *define* the embedding. Secondly, one has to provide a retract embedding of $\bot$ into $\sharp\bot$. Thirdly, the attempted embedding turns out not to *simulate* the $\mu$-reductions.

The second part defines system $\lambda\mu$, our variant of $\lambda\mu$-calculus, the third part exhibits a "refined embedding" of $\lambda\mu$ into $\mathsf{I}\sharp$ and discusses why this would not work for a formulation with typing à la Curry.

**4.1. An unsuccessful attempt at embedding $\mathsf{F}^{\neg}$ into $\mathsf{I}\sharp$.** We aim at an embedding $-'$ of $\mathsf{F}^{\neg}$ into $\mathsf{I}\sharp$. The type translation will at least have the following definition clauses:

$$X' := \sharp X$$
$$\bot' := \sharp\bot$$
$$(A \to B)' := \sharp(A' \to B').$$

In this way, we ensure that every image $A'$ is of the form $\sharp B$, hence the type expressing its stability is inhabited by $\mathsf{stab}\, B$. Since $B$ is unique with $\sharp B = A'$, we define $A^* := B$ for that $B$ and have $A' = \sharp A^*$. For the time being, we do not say anything about universal quantification.

We would like to define a term $r'$ of $\mathsf{I}\sharp$ by iteration on the term $r$ of $\mathsf{F}^{\neg}$ such that $-'$ becomes an embedding of these systems in the sense of Definition 1. Since this will fail, we only discuss the syntax elements $x$ and $\lambda x^A \cdot t$.

We are forced to map variables to variables in order to get a substitution lemma that is needed for the simulation of $\beta$-reduction. The canonical choice is $x' := x$ and will allow us to prove $(t[x := s])' = t'[x := s']$. If we already know that $\Gamma', x : A' \vdash t' : B'$, we want to define $(\lambda x^A \cdot t)'$ such that $\Gamma' \vdash (\lambda x^A \cdot t)' : (A \to B)' = \sharp(A' \to B')$. This would be achieved by setting

$$(\lambda x^A \cdot t)' := \mathsf{emb}(A' \to B')(\lambda x^{A'} \cdot t').$$

Unfortunately, the type $B'$ cannot be inferred from the input term $\lambda x^A \cdot t$, unless one knows the context $\Gamma$. Hence, the notion of embedding in Definition 1 is too narrow.

Definition 2. A context-sensitive type-respecting reduction-preserving embedding (refined embedding for short) of a term rewrite system $\mathcal{S}$ with typing into a term rewrite system $\mathcal{S}'$ with typing is a function $-'$ which assigns to every type $A$ of $\mathcal{S}$ a type $A'$ of $\mathcal{S}'$, and for every context $\Gamma$ of $\mathcal{S}$ a function $-'_\Gamma$ which assigns to every term $r$ of $\mathcal{S}$ typable in context $\Gamma$, a term $r'_\Gamma$ of $\mathcal{S}'$ such that the following implications hold:

- If $\Gamma = x_1 : A_1, \dots, x_n : A_n$ and $\Gamma \vdash r : A$ holds in system $\mathcal{S}$ then $x_1 : A'_1, \dots, x_n : A'_n \vdash r'_\Gamma : A'$ holds in system $\mathcal{S}'$.
- If $r$ is typable in context $\Gamma$ and $r \longrightarrow s$ in $\mathcal{S}$, then $r'_\Gamma \longrightarrow^+ s'_\Gamma$ in $\mathcal{S}'$.

Clearly, the above definition is well-formed in case $\mathcal{S}$ has subject reduction: $s'_\Gamma$ is defined if $r'_\Gamma$ is. We shall use the notation $\Gamma'$ as described for Definition 1.

Even with this refined notion, our attempt at defining an embedding of $\mathsf{F}^{\neg}$ into $\mathsf{I}\sharp$ will fail. Again, we do not treat universal quantification and only discuss the syntax elements $x$, $\lambda x^A \cdot t$, $rs$ and $\mu x^{\neg A} \cdot r$.

We set $x'_\Gamma := x$. If $\lambda x^A \cdot t$ is typable in context $\Gamma$, then $\Gamma, x : A \vdash t : B$ for a unique type $B$. Set $\Delta := \Gamma, x : A$ and

$$(\lambda x^A \cdot t)'_\Gamma := \mathsf{emb}(A' \to B')(\lambda x^{A'} \cdot t'_\Delta).$$

In a proof that this definition respects types (the first condition in the definition), one would have $\Gamma', x : A' \vdash t'_\Delta : B'$ by induction hypothesis, hence $\Gamma' \vdash (\lambda x^A \cdot t)'_\Gamma : (A \to B)'$.

With the (inductive) assumptions $\Gamma' \vdash r'_\Gamma : (A \to B)' = \sharp(A' \to B')$ and $\Gamma' \vdash s'_\Gamma : A'$ in mind, we set

$$(rs)'_\Gamma := \mathrm{It}_\sharp^{B'}(r'_\Gamma, \lambda z^{A' \to B'} \cdot z s'_\Gamma, \mathrm{stab}\, B^*)$$

with $z$ not free in $s$. Since $\Gamma' \vdash \lambda z^{A' \to B'} \cdot z s'_\Gamma : (A' \to B') \to B'$ and $\Gamma' \vdash \mathrm{stab}\, B^* : \neg\neg B' \to B'$, we get $\Gamma' \vdash (rs)'_\Gamma : B'$. The definition is well-formed since $A$ and $B$ are uniquely determined by $\Gamma$ and $r$: The type $A \to B$ is the unique type of $r$ in context $\Gamma$.

The crucial case is as follows: Setting $\Delta := \Gamma, x : \neg A$, we may assume that $\Gamma', x : (\neg A)' \vdash r'_\Delta : \bot'$ and want to define $(\mu x^{\neg A} \cdot r)'_\Gamma$ such that $\Gamma' \vdash (\mu x^{\neg A} \cdot r)'_\Gamma : A'$. Unfolding the definitions, our assumption reads $\Gamma'$, $x : \sharp(A' \to \sharp\bot) \vdash r'_\Delta : \sharp\bot$. First, we need to go back and forth between $\sharp\bot$ and $\bot$:

LEMMA 11. *The type $\bot$ can be embedded by a retract into $\sharp\bot$, i.e., there are terms $p, q$ with $\vdash p : \sharp\bot \to \bot$ and $\vdash q : \bot \to \sharp\bot$ such that $p(qx) \longrightarrow^+ x$. (Note that we do not even need $\eta$ reductions here.)*

PROOF. Set $q := \mathrm{emb}\,\bot$ and $p := \lambda x^{\sharp\bot} \cdot \mathrm{It}_\sharp^\bot(x, \mathrm{id}^\bot, \lambda z^{\neg\neg\bot} \cdot z\, \mathrm{id}^\bot)$. Calculate $p(qx) = p(\mathrm{emb}\,\bot x) \longrightarrow^2 \mathrm{id}^\bot x \longrightarrow x$. ⊣

With these terms $p, q$ we define $q \circ^A y := \lambda z^{A'} \cdot q(yz)$ (one should think of $y : \neg A'$ in the context, hence $q \circ^A y : A' \to \sharp\bot$) and

$$(\mu x^{\neg A} \cdot r)'_\Gamma := \mathrm{stab}\, A^*(\lambda y^{\neg A'} \cdot p(r'_\Delta[x := \mathrm{emb}(A' \to \sharp\bot)(q \circ^A y)])).$$

This is type-correct since $y : \neg A' \vdash \mathrm{emb}(A' \to \sharp\bot)(q \circ^A y) : \sharp(A' \to \sharp\bot)$, hence $\Gamma', y : \neg A' \vdash r'_\Delta[x := \mathrm{emb}(A' \to \sharp\bot)(q \circ^A y)] : \sharp\bot$ by the cut rule, and thus, $\Gamma' \vdash \lambda y^{\neg A'} \cdot p(r'_\Delta[x := \mathrm{emb}(A' \to \sharp\bot)(q \circ^A y)]) : \neg\neg A'$.

For the syntax captured by these definitions, we can prove that weakening and cut also hold for the prospective embedding, in the following sense:

- If $\Gamma \vdash t : B$ and $x$ is not declared in $\Gamma$, then $t'_\Gamma = t'_{\Gamma, x:A}$.
- If $\Gamma, x : A \vdash t : B$ and $\Gamma \vdash s : A$ then $(t[x := s])'_\Gamma = t'_{\Gamma, x:A}[x := s'_\Gamma]$.

Although ordinary $\beta$-reduction would be simulated, we demonstrate that, for $(\mu x^{\neg(A \to B)} \cdot r)s$ typable in $\Gamma$, we do *not* have

$$L := ((\mu x^{\neg(A \to B)} \cdot r)s)'_\Gamma \longrightarrow^+ (\mu y^{\neg B} \cdot r[x := \lambda z^{A \to B} \cdot y(zs)])'_\Gamma =: R.$$

Abbreviate $\mathrm{emb}(A' \to \sharp\bot)$ by $e^A$ and set $\Delta := \Gamma, x : \neg(A \to B)$.

$$
\begin{aligned}
L = {} & \mathsf{It}_\sharp^{B'}\big(\mathsf{stab}(A \to B)^*\big(\lambda y^{\neg(A\to B)'} \cdot p(r'_\Delta[x := e^{A\to B}(q \circ^{A\to B} y)])\big), \\
& \quad \lambda z^{A'\to B'} \cdot zs'_\Gamma, \mathsf{stab}\, B^*\big) \\
& \longrightarrow_\sharp \mathsf{stab}\, B^*\big(\lambda y^{\neg B'} \cdot (\lambda y^{\neg(A\to B)'} \cdot p(r'_\Delta[x := e^{A\to B}(q \circ^{A\to B} y)])) \\
& \qquad \times (\lambda z^{(A\to B)'} \cdot y(zs)'_{\Gamma,z:A\to B})\big) \\
& \longrightarrow_\beta \mathsf{stab}\, B^*\big(\lambda y^{\neg B'} \\
& \qquad \cdot p(r'_\Delta[x := e^{A\to B}(\lambda z^{(A\to B)'} \\
& \qquad\qquad \cdot q((\lambda z^{(A\to B)'} \cdot y(zs)'_{\Gamma,z:A\to B})z))])\big) \\
& \longrightarrow^* \mathsf{stab}\, B^*\big(\lambda y^{\neg B'} \cdot p(r'_\Delta[x := e^{A\to B}(\lambda z^{(A\to B)'} \cdot q(y(zs)'_{\Gamma,z:A\to B}))])\big) \\
& =: L^+
\end{aligned}
$$

$$
\begin{aligned}
R = {} & \mathsf{stab}\, B^*\big(\lambda y^{\neg B'} \cdot p(r'_\Delta[x := \underbrace{(\lambda z^{A\to B} \cdot y(zs))'_{\Gamma,y:\neg B}}_{e^{A\to B}(\lambda z^{(A\to B)'} \cdot (y(zs))'_{\Gamma,y:\neg B,z:A\to B})}][y := e^B(q \circ^B y)])\big) \\
& = \mathsf{stab}\, B^*\big(\lambda y^{\neg B'} \\
& \quad \cdot p(r'_\Delta[x := e^{A\to B}(\lambda z^{(A\to B)'} \cdot (y(zs))'_{\Gamma,y:\neg B,z:A\to B}[y := e^B(q \circ^B y)])])\big) \\
& \longrightarrow^* L^+
\end{aligned}
$$

This is justified by the following reductions for every occurrence of $x$ in $r'_\Delta$ (only in case $x$ not free in $r'_\Delta$, the desired $R = L^+$ holds true):

$$
(y(zs))'_{\Gamma,y:\neg B,z:A\to B} = \mathsf{It}_\sharp^{\sharp\bot}(y, \lambda u^{B'\to\sharp\bot} \cdot u(zs)'_{\Gamma,z:A\to B}, \mathsf{stab}\,\bot),
$$

hence

$$
\begin{aligned}
& (y(zs))'_{\Gamma,y:\neg B,z:A\to B}[y := e^B(q \circ^B y)] \\
& \quad = \mathsf{It}_\sharp^{\sharp\bot}(e^B(q \circ^B y), \lambda u^{B'\to\sharp\bot} \cdot u(zs)'_{\Gamma,z:A\to B}, \mathsf{stab}\,\bot) \\
& \quad \longrightarrow_\sharp (\lambda u^{B'\to\sharp\bot} \cdot u(zs)'_{\Gamma,z:A\to B})(q \circ^B y) \\
& \quad \longrightarrow_\beta (q \circ^B y)(zs)'_{\Gamma,z:A\to B} \\
& \quad \longrightarrow_\beta q(y(zs)'_{\Gamma,z:A\to B}).
\end{aligned}
$$

The author does not see a way to infer strong normalization of $\mathsf{F}^\neg$ from that of $\mathsf{I}\sharp$ due to this failure of simulation. The unwanted "administrative" reductions in $R$ will be avoided if we do not replace $x$ by $\lambda z^{A\to B} \cdot y(zs)$ in the $\mu$-reduction rule, but, loosely speaking, every occurrence of a subterm of the form $xt$ by $y(ts)$. This will only be a reasonable reduction relation if every occurrence of $x$ is in a subterm of that form $xt$. Certainly, this cannot be expected. But we only need it for the variables $x$ that are bound by $\mu$. Hence, we introduce two name spaces for variables, one for the usual variables that may be bound by $\lambda$, hence called $\lambda$-variables, the other one for the variables possibly bound by $\mu$, called $\mu$-variables. And only for those $\mu$-variables $a$, we will ensure that they occur only in a subterm of the form $at$. Essentially, this gives Parigot's $\lambda\mu$-calculus [18]. However, like de Groote [5], we will not be so strict as to require that one can only $\mu$-abstract over terms of that form $at$. We even allow $\bot$ as a type former and hence stay more closely to the more recent presentation by de Groote [7].

REMARK 2. There is a less radical approach for the circumvention of administrative reductions than the introduction of another set of variables, due to Andou [1, 2]. The need above, namely that $\mu$-bound variables $x$ only occur free in subterms of the form $xt$, is turned into the definition of a *regular* term, and only regular terms are studied. (Clearly, if one knows that $x$ is bound by some $\mu x^{\neg A}$, one might replace free occurrences of $x$ by $\lambda y^A \cdot xy$ and arrive at a regular term.) Then, one may define reduction as in $\lambda\mu$-calculus, and regularity is not lost during reduction. However, we would still only obtain a common reduct for the translated terms on both sides of the new $\mu$-reduction rule. For a proper simulation of $\mu$-reduction, it will be essential to translate the type $\neg A$ of such a variable $x$ into $\neg A'$ and not into $(\neg A)'$. Therefore, compositionality considerations require to open up a new namespace of "$\mu$-variables" for the variables that receive this special treatment of their types during translation.

**4.2. Refined classical natural deduction: $\lambda\mu$-calculus.** As mentioned above, Parigot's second-order typed $\lambda\mu$-calculus will be slightly extended in the style of de Groote [7]. However, we do not put the typing context for the $\mu$-variables to the right-hand side of $\vdash$. Moreover, we use raw syntax in Church style.

The system of types of $\lambda\mu$ is just that of $\mathsf{F}$ (which is the same as that for $\mathsf{F}^{\neg}$). The term system of $\mathsf{F}$ is extended as follows: There is an infinite set of $\mu$-variables, whose elements are denoted by $a, b, c$. They are supposed to be disjoint from the variables $x$, now called $\lambda$-variables. The additional formation rules for terms in $\lambda\mu$ are

$$r, s, t ::= \cdots \mid at \mid \mu a^A \cdot r$$

Hence, a $\mu$-variable alone is no term, and every variable that will ever get bound by $\mu$ only occurs as left-hand side $a$ of an application $at$.

The contexts may also contain pairs $(a : A)$ of $\mu$-variables and types, to be interpreted as $a$ having type $\neg A$.

In addition to the typing rules of $\mathsf{F}$, we have the rules involving the $\mu$-variables

$$\frac{\Gamma, a : A \vdash t : A}{\Gamma, a : A \vdash at : \bot} \qquad \frac{\Gamma, a : A \vdash r : \bot}{\Gamma \vdash \mu a^A \cdot r : A}$$

Although a $\mu$-variable is not a term, it can be turned into a term of the appropriate type: We have $a : A \vdash \lambda x^A \cdot ax : \neg A$.

The axiom schemes for $\longrightarrow_\beta$ are taken from F. $\longrightarrow_\mu$ of $\mathsf{F}^\neg$ is changed to a rule that no longer has a lambda abstraction in the right-hand side. We need a special form of substitution which is called structural substitution by de Groote [7].

DEFINITION 3. A *term context* is derived from a term by replacing exactly one occurrence of a $\lambda$-variable in that term with the new symbol $\star$. This occurrence must not be in the scope of any binder (for simplicity). Term contexts will be communicated by the letter $E$, and the result of a textual substitution of a term $r$ for $\star$ in $E$ will be denoted by $E[r]$ (and is a term).

DEFINITION 4 (Special Substitution). For $r$ a term, $\mu$-variables $a, b$ and a term context $E$, define the term $r[a\star := bE]$ as the result of replacing in $r$ recursively every subterm of the form $at$ by $b(E[t])$. The precise definition is by iteration on $r$. Every case is homomorphic with exception of

$$(ct)[a\star := bE] := \begin{cases} b\big(E\big[t[a\star := bE]\big]\big), & \text{if } c = a, \\ c\big(t[a\star := bE]\big), & \text{else.} \end{cases}$$

LEMMA 12. *If $\Gamma, a : A \vdash r : C$ and $\Gamma, z : A \vdash E[z] : B$ then $\Gamma, b : B \vdash r[a\star := bE] : C$.*

Add the axiom schemes of $\mu$-reduction (with a "fresh" $\mu$-variable $b$)

$$(\mu a^{A \to B} \cdot r)s \longrightarrow_\mu \mu b^B \cdot r\big[a\star := b(\star s)\big]$$
$$(\mu a^{\forall X \cdot A} \cdot r)\,B \longrightarrow_\mu \mu b^{A[X:=B]} \cdot r\big[a\star := b(\star B)\big]$$

Evidently, this rule just imposes a bit of a reduction strategy on the reduction relation of $\mathsf{F}^\neg$ since, with the only reasonable interpretation of $r[a := t]$, one has in $\lambda\mu$ that

$$r\big[a := \lambda z^A \cdot b\big(E[z]\big)\big] \longrightarrow_\beta^* r[a\star := bE].$$

Therefore, $\lambda\mu$ embeds into $\mathsf{F}^\neg$ with $-'$ the identity, using the following definition:

DEFINITION 5. An embedding of a $\lambda\mu$-calculus $\mathcal{S}$ into an extension $\mathcal{S}'$ of system F without $\mu$-variables is a function $-'$ which assigns to every type $A$ of $\mathcal{S}$ a type $A'$ of $\mathcal{S}'$ and to every term $r$ of $\mathcal{S}$ a term $r'$ of $\mathcal{S}'$ such that the following implications hold:

- If $x_1 : A_1, \dots, x_n : A_n, a_1 : B_1, \dots, a_k : B_k \vdash r : A$ in $\mathcal{S}$, then one has the typing $x_1 : A_1', \dots, x_n : A_n', a_1 : \neg B_1', \dots, a_k : \neg B_k' \vdash r' : A'$ in $\mathcal{S}'$.
- If $r \longrightarrow s$ in $\mathcal{S}$, then $r' \longrightarrow^+ s'$ in $\mathcal{S}'$.

Here, we generally consider the $\mu$-variables of $\mathcal{S}$ as term variables of $\mathcal{S}'$.

This definition shows the way to overcome the problems reported in Section 4.1: The translated $a$ will be thought of as having type $\neg A'$ and no longer the type $(\neg A)' = \sharp(A' \to \sharp\bot)$.

It is obvious that also an embedding in the sense of this definition allows to derive strong normalization of $S$ from that of $S'$. As for Definition 1, we call the transformed context $\Gamma'$ if the original context was $\Gamma$ in the above implication.

**4.3. Embedding $\lambda\mu$ into $\mathsf{I}\sharp$.** It is known from [20] that $\lambda\mu$ is strongly normalizing. It also follows from strong normalization of $\mathsf{F}^\neg$, shown in [15]. Here, we want to give an elementary proof through an embedding into $\mathsf{I}\sharp$, and hence only have to appeal to the now classical result that $\mathsf{F}$ is strongly normalizing. Another embedding into $\mathsf{F}$ has been found by Joly [11] — see Section 5.1 — but that one does not seem to extend to fixed points, as described in Section 5.2. Translations in continuation-passing style neither reprove strong normalization of $\mathsf{F}$, but are no embedding in the sense of simulation of all reductions — see Section 5.3.

This section presents an extension of [14] from $\lambda\mu$-calculus in a formulation closer to Parigot to the present one with $\bot$ in the type system and in a precise Church-style typing discipline.

The attempted embedding of Section 4.1 is adjusted to system $\lambda\mu$ and works out smoothly. Certainly, again, the naive notion of embedding (this time Definition 5) is too narrow to be useful and has to be relativized to contexts:

DEFINITION 6. A context-sensitive type-respecting reduction-preserving embedding (refined embedding for short) of a $\lambda\mu$-calculus $S$ into an extension $S'$ of system $\mathsf{F}$ without $\mu$-variables is a function $-'$ which assigns to every type $A$ of $S$ a type $A'$ of $S'$, and for every context $\Gamma$ of $S$ a function $-'_\Gamma$ which assigns to every term $r$ of $S$ typable in context $\Gamma$, a term $r'_\Gamma$ of $S'$ such that the following implications hold:

- If $\Gamma = x_1 : A_1, \ldots, x_n : A_n, a_1 : B_1, \ldots, a_k : B_k$ and $\Gamma \vdash r : A$ in $S$ then, for
$$\Gamma' := x_1 : A'_1, \ldots, x_n : A'_n, a_1 : \neg B'_1, \ldots, a_k : \neg B'_k,$$
one has $\Gamma' \vdash r'_\Gamma : A'$ in $S'$.
- If $r$ is typable in context $\Gamma$ and $r \longrightarrow s$ in $S$, then $r'_\Gamma \longrightarrow^+ s'_\Gamma$ in $S'$.

Here, we generally consider the $\mu$-variables of $S$ as term variables of $S'$.

THEOREM 4.1. *There is a refined embedding of $\lambda\mu$ into $\mathsf{I}\sharp$.*

The proof forms the remainder of this section.

DEFINITION 7. By iteration on the type $A$ of $\lambda\mu$ define the type $A^*$ of $\mathsf{I}\sharp$:
$$\begin{aligned} X^* &:= X \\ \bot^* &:= \bot \\ (A \to B)^* &:= \sharp A^* \to \sharp B^* \\ (\forall X \cdot A)^* &:= \forall X \cdot \sharp A^* \end{aligned}$$

Clearly, $\mathsf{FV}(A^*) = \mathsf{FV}(A)$. By induction on $A$, one immediately proves compatibility with substitution: $(A[X := B])^* = A^*[X := B^*]$.

The translation of the types for the embedding is defined by $A' := \sharp A^*$. Trivially, $\mathsf{FV}(A') = \mathsf{FV}(A)$ and

$$X' = \sharp X$$
$$\bot' = \sharp\bot$$
$$(A \to B)' = \sharp(A' \to B')$$
$$(\forall X \cdot A)' = \sharp\forall X \cdot A'$$

Therefore, the first three clauses are as in Section 4.1. Important for the treatment of universal quantification will be the immediate consequence of the above compatibility with substitution:

$$(A[X := B])' = A'[X := B^*].$$

DEFINITION 8. By iteration on the term $r$ of $\lambda\mu$ define the term $r'_\Gamma$ of $\mathsf{I}\sharp$ whenever $r$ is typable in context $\Gamma$ in $\lambda\mu$:

- $x'_\Gamma := x$.
- If $\Gamma \vdash \lambda x^A \cdot t : A \to B$ then $(\lambda x^A \cdot t)'_\Gamma := \mathsf{emb}(A' \to B')(\lambda x^{A'} \cdot t'_{\Gamma,x:A})$.
- If $\Gamma \vdash r : A \to B$ and $\Gamma \vdash s : A$ then
$$(rs)'_\Gamma := \mathsf{It}_\sharp^{B'}(r'_\Gamma, \lambda z^{A' \to B'} \cdot zs'_\Gamma, \mathsf{stab}\, B^*).$$
- If $\Gamma \vdash \Lambda X \cdot t : \forall X \cdot A$ then $(\Lambda X \cdot t)'_\Gamma := \mathsf{emb}(\forall X \cdot A')(\Lambda X \cdot t'_\Gamma)$.
- If $\Gamma \vdash r : \forall X \cdot A$ then
$$(rB)'_\Gamma := \mathsf{It}_\sharp^{(A[X:=B])'}(r'_\Gamma, \lambda z^{\forall X \cdot A'} \cdot zB^*, \mathsf{stab}\,(A[X := B])^*).$$
- If $\Gamma, a : A \vdash t : A$ then $(at)'_{\Gamma,a:A} := q(at'_{\Gamma,a:A})$.
- If $\Gamma \vdash \mu a^A \cdot r : A$ then $(\mu a^A \cdot r)'_\Gamma := \mathsf{stab}\, A^*(\lambda a^{\neg A'} \cdot pr'_{\Gamma,a:A})$.

The first three defining clauses are taken from Section 4.1, the fourth and fifth are the analogues for universal quantification. The last two rules use the terms $q$ and $p$ from Lemma 11 that gave a retract of $\bot$ into $\bot'$. (Note the simplification of the clause for $\mu$ in comparison with that of Section 4.1.)

REMARK 3. If $\Gamma \vdash \mu a^A \cdot bt : A$ then, for some type $B$, this comes from $(b : B) \in \Gamma \cup \{(a : A)\}$ and $\Gamma, a : A \vdash t : B$. Then,

$$\begin{aligned}(\mu a^A \cdot bt)'_\Gamma &= \mathsf{stab}\, A^*(\lambda a^{\neg A'} \cdot p(q(bt'_{\Gamma,a:A}))) \\ &\longrightarrow^* \mathsf{stab}\, A^*(\lambda a^{\neg A'} \cdot bt'_{\Gamma,a:A}),\end{aligned}$$

by Lemma 11. The reduct thus achieved has originally been the definition of the translation of $\mu a^A \cdot bt$ in [14]. It only treated $\lambda\mu$-calculus à la Parigot where $\mu a^A \cdot r$ always has to be of the form $\mu a^A \cdot bt$. It might be interesting to note that the present remark is the only place where the retract property in Lemma 11 is needed. Theorem 4.1 does not depend on it.

LEMMA 13 (Soundness). *If* $\Gamma \vdash r : A$ *then* $\Gamma' \vdash r'_\Gamma : A'$.

PROOF. By induction on $\Gamma \vdash r : A$. For variables, lambda abstraction and application, this has been done in Section 4.1, for type application $rB$, one uses $(A[X := B])' = A'[X := B^*]$. The case of type abstraction $\Lambda X \cdot t$ satisfies the proviso of the respective typing rule since the set of free type variables is not affected by the translation of types. The two rules involving $\mu$-variables are easily treated by help of the induction hypothesis, e.g., $\Gamma', a : \neg A' \vdash (at)'_{\Gamma, a:A} : \sharp\bot$ since $\Gamma', a : \neg A' \vdash a : \neg A'$ and $\Gamma', a : \neg A' \vdash t'_{\Gamma, a:A} : A'$, hence $\Gamma', a : \neg A' \vdash at'_{\Gamma, a:A} : \bot$. ⊣

REMARK 4 (Church versus Curry). The previous lemma would be incorrect for $\lambda\mu$ in a Curry-style typing discipline: There, from $\Gamma \vdash t : A$ and $X$ not free in $\Gamma$, one would derive $\Gamma \vdash t : \forall X \cdot A$— without changing the term $t$. Assume, we had $\Gamma' \vdash t'_\Gamma : A'$ by induction hypothesis. Then, we would be required to show $\Gamma' \vdash t'_\Gamma : (\forall X \cdot A)'$. Certainly, we get $\Gamma' \vdash t'_\Gamma : \forall X \cdot A'$. But $(\forall X \cdot A)' = \sharp(\forall X \cdot A')$, and the passage to the stabilization of a type cannot go unnoticed by the term system.

As an attempt at resolving this problem, one might set $(\forall X \cdot A)' := \forall X \cdot A'$. Among other nuisances, the rule of universal elimination becomes unsound: Assume $\Gamma \vdash t : \forall X \cdot X \to X$. In Curry-style typing, this entails $\Gamma \vdash t : (\forall Y \cdot Y) \to (\forall Y \cdot Y)$ — again, without changing the term $t$. Assume we knew already that $\Gamma' \vdash t'_\Gamma : (\forall X \cdot X \to X)'$. Then, we would need to establish $\Gamma' \vdash t'_\Gamma : ((\forall Y \cdot Y) \to (\forall Y \cdot Y))'$. Obviously, $(\forall X \cdot X \to X)' = \forall X \cdot \sharp(\sharp X \to \sharp X)$ cannot be instantiated to $((\forall Y \cdot Y) \to (\forall Y \cdot Y))' = \sharp((\forall Y \cdot \sharp Y) \to (\forall Y \cdot \sharp Y))$. Hence, also this alternative approach fails. In Section 5.3, it will be mentioned that typing à la Curry also excludes CPS-translations (for second-order systems).

Before we can show that the translation also respects reduction, we need that the admissible typing rules of weakening and cut are respected by the prospective embedding.

LEMMA 14. • *If* $\Gamma \vdash t : B$ *and* $x$ *is not declared in* $\Gamma$, *then* $t'_\Gamma = t'_{\Gamma, x:A}$.
• *If* $\Gamma, x : A \vdash t : B$ *and* $\Gamma \vdash s : A$ *then* $(t[x := s])'_\Gamma = t'_{\Gamma, x:A}[x := s'_\Gamma]$.
• *If* $\Gamma \vdash t : B$ *then* $(t[X := A])'_{\Gamma[X:=A]} = t'_\Gamma[X := A^*]$.

PROOF. Easy (but tedious) induction. ⊣

These rules will guarantee the simulation of the reduction rules of F. For $\mu$-reductions, we need a lemma about the interaction of special substitution with our translation:

LEMMA 15. *If* $\Gamma, a : A \vdash r : C$, *the term context* $E$ *has* $\Gamma, z : A \vdash E[z] : B$ *and* $b$ *is a "fresh"* $\mu$*-variable, then*

$$r'_{\Gamma, a:A}\left[a := \lambda z^{A'} \cdot b\left(E[z]\right)'_{\Gamma, z:A}\right] \longrightarrow^* \left(r[a\star := bE]\right)'_{\Gamma, b:B}.$$

PROOF. By induction on $r$. The only interesting case is with $r = at$, hence with the same $\mu$-variable $a$. Then, $C = \bot$ and $\Gamma, a : A \vdash t : A$. Call the term to the left $L$ and the term to the right $R$. Finally, the term substituted into $L$ will be called $S$. Then,

$$L = q(at'_{\Gamma,a:A})[a := S] = q(St'_{\Gamma,a:A}[a := S]).$$

By induction hypothesis, $t'_{\Gamma,a:A}[a := S] \longrightarrow^* (t[a\star := bE])'_{\Gamma,b:B}$. Therefore,

$$\begin{aligned}
L &\longrightarrow^* q\Big(S\big(t[a\star := bE]\big)'_{\Gamma,b:B}\Big)\\
&= q\Big(\big(\lambda z^{A'} \cdot b(E[z])'_{\Gamma,z:A}\big)\big(t[a\star := bE]\big)'_{\Gamma,b:B}\Big)\\
&\longrightarrow_\beta q\Big(b(E[z])'_{\Gamma,z:A}\Big[z := \big(t[a\star := bE]\big)'_{\Gamma,b:B}\Big]\Big)\\
&= q\Big(b(E[z])'_{\Gamma,b:B,z:A}\Big[z := \big(t[a\star := bE]\big)'_{\Gamma,b:B}\Big]\Big)\\
&= q\Big(b\big((E[z])\,[z := t[a\star := bE]]\big)'_{\Gamma,b:B}\Big)\\
&= q\Big(b\big(E\,[t[a\star := bE]]\big)'_{\Gamma,b:B}\Big).\\
R &= \big((at)[a\star := bE]\big)'_{\Gamma,b:B}\\
&= \big(b\big(E\,[t[a\star := bE]]\big)\big)'_{\Gamma,b:B}\\
&= q\Big(b\big(E\,[t[a\star := bE]]\big)'_{\Gamma,b:B}\Big).
\end{aligned}$$

⊣

REMARK 5. The previous lemma is the reformulation of Lemma 3 in [14] for our slightly more general and more typing-aware situation.

LEMMA 16 (Simulation). *If* $\Gamma \vdash r : A$ *and* $r \longrightarrow s$ *then* $r'_\Gamma \longrightarrow^+ s'_\Gamma$.

PROOF. By induction on $r \longrightarrow s$. Since the defining clauses for $r'_\Gamma$ never erase subterms — unlike several CPS-translations which therefore fail, see Section 5.3 — the closure under the term formation rules is straightforward (also use subject reduction for $\lambda\mu$). Hence, we only discuss the axiom schemes.

$\Gamma \vdash (\lambda x^A \cdot t)s : B$ can only be derived from $\Gamma, x : A \vdash t : B$ and $\Gamma \vdash s : A$, and we have to show that $((\lambda x^A \cdot t)s)'_\Gamma \longrightarrow^+ (t[x := s])'_\Gamma$. The left-hand side is

$$\mathsf{It}_\sharp^{B'}\big(\mathsf{emb}(A' \to B')(\lambda x^{A'} \cdot t'_{\Gamma,x:A}), \lambda z^{A' \to B'} \cdot zs'_\Gamma, \mathsf{stab}\, B^*\big).$$

One step by help of $\longrightarrow_\sharp$ yields $(\lambda z^{A' \to B'} \cdot zs'_\Gamma)(\lambda x^{A'} \cdot t'_{\Gamma,x:A})$, and two $\beta$-reduction steps lead to $t'_{\Gamma,x:A}[x := s'_\Gamma]$ which is the right-hand side by Lemma 14.

If $\Gamma \vdash (\Lambda X \cdot t)B : A[X := B]$ has been derived from $\Gamma \vdash t : A$ and $X \notin \mathsf{FV}(\Gamma)$, then we have to show $((\Lambda X \cdot t)B)'_\Gamma \longrightarrow^+ (t[X := B])'_\Gamma$. The left-hand side is

$$\mathsf{It}_\sharp^{(A[X:=B])'}\big(\mathsf{emb}(\forall X \cdot A')(\Lambda X \cdot t'_\Gamma), \lambda z^{\forall X \cdot A'} \cdot zB^*, \mathsf{stab}(A[X := B])^*\big).$$

It reduces by one step of $\longrightarrow_\sharp$ to $(\lambda z^{\forall X\cdot A'}\cdot zB^*)(\Lambda X\cdot t'_\Gamma)$ and two more $\beta$-reduction steps to $t'_\Gamma[X := B^*]$. Since $X \notin \mathsf{FV}(\Gamma)$, hence $\Gamma[X := B] = \Gamma$, Lemma 14 ensures that this term is the right-hand side.

$\Gamma \vdash (\mu a^{A\to B}\cdot r)s : B$ is necessarily derived from $\Gamma, a : A \to B \vdash r : \bot$ and $\Gamma \vdash s : A$. We have to show $((\mu a^{A\to B}\cdot r)s)'_\Gamma \longrightarrow^+ (\mu b^B \cdot r[a\star := b(\star s)])'_\Gamma$. The left-hand side is

$$\mathsf{It}_\sharp^{B'}\left(\mathsf{stab}(A\to B)^*(\lambda a^{\neg(A\to B)'}\cdot pr'_{\Gamma,a:A\to B}), \lambda z^{A'\to B'}\cdot zs'_\Gamma, \mathsf{stab}\,B^*\right).$$

One $\longrightarrow_\sharp$-reduction step yields

$$\mathsf{stab}\,B^*\left(\lambda b^{\neg B'}\cdot \underbrace{\left(\lambda a^{\neg(A\to B)'}\cdot pr'_{\Gamma,a:A\to B}\right)\left(\lambda z^{(A\to B)'}\cdot b(zs)'_{\Gamma,z:A\to B}\right)}_{\longrightarrow_\beta pr'_{\Gamma,a:A\to B}[a:=\lambda z^{(A\to B)'}\cdot b(zs)'_{\Gamma,z:A\to B}]}\right).$$

Using the previous lemma with $A \to B$ in place of $A$, $C := \bot$ and $E := \star s$, we get

$$r'_{\Gamma,a:A\to B}\left[a := \lambda z^{(A\to B)'}\cdot b(zs)'_{\Gamma,z:A\to B}\right] \longrightarrow^* \left(r\left[a\star := b(\star s)\right]\right)'_{\Gamma,b:B}$$

To conclude,

$$((\mu a^{A\to B}\cdot r)s)'_\Gamma \longrightarrow^+ \mathsf{stab}\,B^*\left(\lambda b^{\neg B'}\cdot p\left(r\left[a\star := b(\star s)\right]\right)'_{\Gamma,b:B}\right).$$

The right-hand side is the one required above, by definition.

$\Gamma \vdash (\mu a^{\forall X\cdot A}\cdot r)B : A[X := B]$ is derived from $\Gamma, a : \forall X\cdot A \vdash r : \bot$, and we have to show that $((\mu a^{\forall X\cdot A}\cdot r)B)'_\Gamma \longrightarrow^+ (\mu b^{A[X:=B]}\cdot r[a\star := b(\star B)])'_\Gamma$. This is done analogously to the previous case, again by help of the previous lemma, this time with $A$ replaced by $\forall X\cdot A$ and $B$ replaced by $A[X := B]$, and $E := \star B$. ⊣

This concludes the proof of Theorem 4.1 and hence the alternative proof of strong normalization of $\lambda\mu$.

**§5. Other approaches and extensions.** In the first part, the embedding of $\mathsf{F}^\neg$ into $\mathsf{F}$ by Joly is presented. It is based on a very different idea than the one of using stabilization through system $\mathsf{I}\sharp$. The second part challenges Joly's method by the addition of positive fixed points. The systems with these fixed points are defined, the probable failure of Joly's method for them is discussed, the success of our method demonstrated, and a further extension by *least* fixed points is dealt with. In the third part, we comment on the recent article by Nakazawa and Tatsuta [17] about failing attempts at proving strong normalization for second-order classical natural deduction by translations in continuation-passing style, and also on their proposed solution.

**5.1. The embedding by Joly.** In this section, an easier embedding — even of $\mathsf{F}^\neg$ instead of $\lambda\mu$ — into $\mathsf{F}$ by Joly [11] is described in some detail, in order to show a further variation on embeddings, and to see its limitations, notably the fact that fixed points cannot be treated (see the next section).

From the discussion in Section 2.2 we know that, from a term witnessing stability of $A$, a term of type $\neg\neg(\forall X \cdot A) \to \forall X \cdot A$ can be defined. If $X$ occurs free in $A$, this term possibly involves the constant $\mathsf{stab}^X$. The idea to resolve this problem is the relativization of universal quantifiers $\forall X \cdot A$ to stable $X$, i.e., $\forall X \cdot A$ will be interpreted by $\forall X \cdot (\neg\neg X \to X) \to A$, and $A$ will be transformed recursively. Then, a term witnessing stability for the interpretation of a type $A$ can be defined from only the stability constants $\mathsf{stab}^X$ with $X$ free in $A$. But there are no stability constants in F! They are even not needed because they can now be replaced by variables of type $\neg\neg X \to X$ for $X \in \mathsf{FV}(A)$. The freedom thus obtained allows Joly to get the substitution property we were lacking in Section 2.2.

**Definition** 9. By iteration on type $A$, define the type $A'$:

$$\begin{aligned} X' &:= X \\ \bot' &:= \bot \\ (A \to B)' &:= A' \to B' \\ (\forall X \cdot A)' &:= \forall X \cdot (\neg\neg X \to X) \to A' \end{aligned}$$

Clearly, $\mathsf{FV}(A') = \mathsf{FV}(A)$ and $(A[X := B])' = A'[X := B']$.

**Definition** 10 (Environment). Let an environment by any injective function $\xi$ from the set TV of type variables into the set of all term variables such that countably infinitely many term variables are not in the range of $\xi$.

Let always denote $\xi, \xi_1, \xi_2$ environments. Environments exist if we assume, e.g., that the type variable set and term variable set are countable.

**Definition** 11 (Modified environment). Let $\xi$ be an environment, $X$ be a type variable and $u$ be a term variable not occurring in the range of $\xi$. Then, $\xi[X \mapsto u]$ is defined to be the environment that is equal to $\xi$ for all variables except $X$, and such that $\xi[X \mapsto u](X) = u$.

Let always denote $\xi[X \mapsto u]$ a modified environment, that is, the injectivity condition is tacitly assumed.

**Definition** 12 (Stability witnesses). By iteration on the type $A$, define the term $\mathsf{stab}^A_\xi$:

$$\begin{aligned} \mathsf{stab}^X_\xi &:= \xi(X) \\ \mathsf{stab}^\bot_\xi &:= \lambda v^{\neg\neg\bot} \cdot v\, \mathsf{id}^\bot \\ \mathsf{stab}^{A\to B}_\xi &:= \lambda v^{\neg\neg(A\to B)'} \lambda w^{A'} \cdot \mathsf{stab}^B_\xi \left(\lambda y^{\neg B'} \cdot v\left(\lambda z^{(A\to B)'} \cdot y(zw)\right)\right) \\ \mathsf{stab}^{\forall X\cdot A}_\xi &:= \lambda v^{\neg\neg(\forall X\cdot A)'} \Lambda X \lambda u^{\neg\neg X \to X} \\ &\qquad \cdot \mathsf{stab}^A_{\xi[X\mapsto u]} \left(\lambda y^{\neg A'} \cdot v\left(\lambda z^{(\forall X\cdot A)'} \cdot y(zXu)\right)\right). \end{aligned}$$

The definition of $\mathsf{stab}_\xi^{A\to B}$ is similar to that in Section 2.2, the one for the universal quantifier is essentially different in its use of the variable $u$ which also modifies the environment.

LEMMA 17 (Coincidence). *Let* $\xi_1(X) = \xi_2(X)$ *for all* $X \in \mathsf{FV}(A)$. *Then* $\mathsf{stab}_{\xi_1}^A = \mathsf{stab}_{\xi_2}^A$.

COROLLARY 2. $\mathsf{stab}_\xi^A = \mathsf{stab}_{\xi[X\mapsto u]}^A$ *for* $X \notin \mathsf{FV}(A)$.

The following lemma states in which sense we defined "stability witnesses".

LEMMA 18 (Correct types). *Let* $\mathsf{FV}(A) \subseteq \{X_1, \ldots, X_n\}$. *Then*

$$\xi(X_1) : \neg\neg X_1 \to X_1, \ldots, \xi(X_n) : \neg\neg X_n \to X_n \vdash \mathsf{stab}_\xi^A : \neg\neg A' \to A'.$$

PROOF. Induction on $A$. ⊣

LEMMA 19 (Substitution). $\mathsf{stab}_\xi^{A[X:=B]} = \mathsf{stab}_{\xi[X\mapsto u]}^A[X := B'][u := \mathsf{stab}_\xi^B]$.

PROOF. Induction on $A$, using injectivity of environments in the case $A = Y \neq X$, and the Coincidence Lemma in the case of universal quantification. ⊣

The definition of the translation of terms of $\mathsf{F}^\neg$ into those of F is always with respect to some environment.

DEFINITION 13. By iteration on the term $r$ of System $\mathsf{F}^\neg$, define the term $r'_\xi$ of System F:

$$\begin{aligned} x'_\xi &:= x \\ (\lambda x^A \cdot r)'_\xi &:= \lambda x^{A'} \cdot r'_\xi \\ (rs)'_\xi &:= r'_\xi s'_\xi \\ (\Lambda X \cdot t)'_\xi &:= \Lambda X \lambda u^{\neg\neg X \to X} \cdot t'_{\xi[X\mapsto u]} \\ (rB)'_\xi &:= r'_\xi B' \mathsf{stab}_\xi^B \\ (\mu x^{\neg A} \cdot r)'_\xi &:= \mathsf{stab}_\xi^A (\lambda x^{\neg A'} \cdot r'_\xi). \end{aligned}$$

DEFINITION 14 (Critical variables). By iteration on the term $r$ of $\mathsf{F}^\neg$, define its (finite) set $\mathsf{CV}(r)$ of critical type variables:

$$\begin{aligned} \mathsf{CV}(x) &:= \emptyset \\ \mathsf{CV}(\lambda x^A \cdot r) &:= \mathsf{CV}(r) \\ \mathsf{CV}(rs) &:= \mathsf{CV}(r) \cup \mathsf{CV}(s) \\ \mathsf{CV}(\Lambda X \cdot t) &:= \mathsf{CV}(r) \setminus \{X\} \\ \mathsf{CV}(rB) &:= \mathsf{CV}(r) \cup \mathsf{FV}(B) \\ \mathsf{CV}(\mu x^{\neg A} \cdot r) &:= \mathsf{FV}(A) \cup \mathsf{CV}(r) \end{aligned}$$

This means, we record all free type variables that occur in type applications and uses of RAA.

LEMMA 20 (Coincidence). *Let* $\xi_1(X) = \xi_2(X)$ *for all* $X \in \mathsf{CV}(r)$. *Then* $r'_{\xi_1} = r'_{\xi_2}$. ⊣

Our translation of $\mathsf{F}^{\neg}$ into $\mathsf{F}$ respects types in the following sense.

LEMMA 21 (Correct types). *Let* $x_1 : A_1, \ldots, x_m : A_m \vdash r : A$ *in System* $\mathsf{F}^{\neg}$, *and assume* $\mathsf{CV}(r) \subseteq \{X_1, \ldots, X_n\}$. *Set*

$$\Delta := \xi(X_1) : \neg\neg X_1 \to X_1, \ldots, \xi(X_n) : \neg\neg X_n \to X_n, x_1 : A'_1, \ldots, x_m : A'_m.$$

*Then,* $\Delta \vdash r'_\xi : A'$ *holds in System* $\mathsf{F}$.

PROOF. Induction on $r$. ⊣

LEMMA 22 (Substitution). *One has* $(r[x := s])'_\xi = r'_\xi[x := s'_\xi]$, *if* $x$ *is not in the range of* $\xi$. *Moreover, for* $u$ *not free in* $r$, *it holds that*

$$(r[X := B])'_\xi = r'_{\xi[X \mapsto u]}[X := B'][u := \mathsf{stab}^B_\xi].$$

PROOF. Induction on $r$. ⊣

LEMMA 23 (Simulation). *If* $r \longrightarrow s$ *in* $\mathsf{F}^{\neg}$ *then* $r'_\xi \longrightarrow^+ s'_\xi$ *in* $\mathsf{F}$.

PROOF. Induction on $\longrightarrow$. Only the axiom schemes of reduction need to be studied since the term closure trivially works by using term closure rules again. The $\beta$-reduction rules of $\mathsf{F}$ are immediately dealt with by the previous lemma. The first $\mu$-reduction rule — for implication — goes just as in the motivation in Section 2.2, decorated with $\xi$ and $-'$. The failed calculation for the $\mu$-reduction rule for the universal quantifier is replaced by the following correct behaviour:

$$\begin{aligned}
((\mu x^{\neg\forall X\cdot A} \cdot r)B)'_\xi &= \mathsf{stab}^{\forall X\cdot A}_\xi\,(\lambda x^{\neg(\forall X\cdot A)'} \cdot r'_\xi)B'\,\mathsf{stab}^B_\xi \\
&\longrightarrow^3 \underbrace{\mathsf{stab}^A_{\xi[X\mapsto u]}[X := B'][u := \mathsf{stab}^B_\xi]}_{\mathsf{stab}^{A[X:=B]}_\xi} \\
&\quad \times (\lambda y^{\neg A'[X:=B']} \cdot (\lambda x^{\neg(\forall X\cdot A)'} \cdot r'_\xi)(\lambda z^{(\forall X\cdot A)'} \cdot y(zB'\,\mathsf{stab}^B_\xi))) \\
&\longrightarrow \mathsf{stab}^{A[X:=B]}_\xi \left(\lambda y^{\neg A[X:=B]'} \cdot r'_\xi[x := \lambda z^{(\forall X\cdot A)'} \cdot y\,\underbrace{(zB'\,\mathsf{stab}^B_\xi)}_{(zB)'_\xi}]\right) \\
&= (\mu y^{\neg A[X:=B]} \cdot r[x := \lambda z^{\forall X\cdot A} \cdot y(zB)])'_\xi,
\end{aligned}$$

by the previous lemma. ⊣

By Lemma 21 and the previous lemma, $\mathsf{F}^{\neg}$ immediately inherits strong normalization from system $\mathsf{F}$.

Note that these results do not constitute an embedding in the sense of any of Definitions 1, 2 or 5/6. The passage from the first to the second definition is driven by the need to refer to the typing when defining the translation of a term. In our present Definition 13, this is clearly not called for. The definitions 5 and 6 are tailor-made for $\lambda\mu$-calculus. For the present embedding, we would have to allow a change in the typing context — even according to the critical variables. It does not look worthwhile introducing a general definition for so specific a case.

**5.2. Extension by fixed points.** We extend $\lambda\mu$ by positive fixed points, and get System $\lambda\mu^{\mathsf{fix}}$: The inductive definition of the set of types is extended by the clause $\mathsf{fix}\, X \cdot A$, with the proviso that $X$ occurs only positively in $A$. More precisely, we define the set $\mathsf{TP}$ of types and for every type $A \in \mathsf{TP}$ the sets $\mathsf{TV}_+(A)$ and $\mathsf{TV}_-(A)$ of type variables that occur only positively in $A$ or only negatively in $A$, respectively. Let always range $p$ (polarity) over $\{-,+\}$ and set $-- := +$ and $-+ := -$.

- $X \in \mathsf{TP}$, $\mathsf{TV}_-(X) := \mathsf{TV} \setminus \{X\}$, $\mathsf{TV}_+(X) := \mathsf{TV}$.
- $\bot \in \mathsf{TP}$, $\mathsf{TV}_p(\bot) := \mathsf{TV}$.
- If $A, B \in \mathsf{TP}$, then $A \to B \in \mathsf{TP}$ and $\mathsf{TV}_p(A \to B) := \mathsf{TV}_{-p}(A) \cap \mathsf{TV}_p(B)$.
- If $A \in \mathsf{TP}$, then $\forall X \cdot A \in \mathsf{TP}$ and $\mathsf{TV}_p(\forall X \cdot A) := \mathsf{TV}_p(A) \cup \{X\}$.
- If $A \in \mathsf{TP}$ and $X \in \mathsf{TV}_+(A)$ then $\mathsf{fix}\, X \cdot A \in \mathsf{TP}$ (only here is a positivity condition) and $\mathsf{TV}_p(\mathsf{fix}\, X \cdot A) := \mathsf{TV}_p(A) \cup \{X\}$.

An important example is the impredicative definition of disjoint sums:

$$A + B := \forall X \cdot (A \to X) \to (B \to X) \to X$$

for $X$ not free in $A$ or $B$. Trivially, if $A, B \in \mathsf{TP}$ then $A + B \in \mathsf{TP}$. Moreover, $\mathsf{TV}_p(A + B) = \mathsf{TV}_p(A) \cap \mathsf{TV}_p(B)$.

Our definition allows interleaving of fixed points, e.g., $\mathsf{fix}\, X \cdot \mathsf{fix}\, Y \cdot X + Y \in \mathsf{TP}$ since $X \in \mathsf{TV}_+(\mathsf{fix}\, Y \cdot X + Y)$. Note that this is more than just nesting of fixed points: The outer fix binds the parameter $X$ of the inner fixed point. Here, we already use non-strict positivity, since $Y \in \mathsf{TV}_+(X + Y)$ is only derivable by help of $\mathsf{TV}_-$. Had we not encoded sums but taken them as primitives, we would have nevertheless obtained genuine examples of non-strict positivity, e.g., $\mathsf{fix}\, X \cdot A + \neg\neg X$ for $X \notin \mathsf{FV}(A)$ since $X \in \mathsf{TV}_+(\neg\neg X)$ which comes from $X \in \mathsf{TV}_-(\neg X)$ (which in turn rests on $X \in \mathsf{TV}_+(X)$). Note that $X \notin \mathsf{FV}(A)$ implies $X \in \mathsf{TV}_-(A) \cap \mathsf{TV}_+(A)$. In general, and loosely speaking, a non-strictly positive occurrence is to the left of an even number of $\to$. A strictly-positive occurrence is never to the left of $\to$.

The term system of $\lambda\mu$ is extended by fixed-point folding and unfolding, which yields the following term grammar for $\lambda\mu^{\mathsf{fix}}$:

$$r, s, t ::= \cdots \mid \mathsf{in}_{X \cdot A} t \mid \mathsf{out}_{X \cdot A} r.$$

Variable binding for the indices $X \cdot A$ is assumed as if they were $\lambda X \cdot A$. The $\lambda$ is just left out for stylistic reasons.

The definition of $\Gamma \vdash t : A$ is extended by the two clauses

$$\frac{\Gamma \vdash t : A[X := \mathsf{fix}\, X \cdot A]}{\Gamma \vdash \mathsf{in}_{X \cdot A} t : \mathsf{fix}\, X \cdot A} \qquad \frac{\Gamma \vdash r : \mathsf{fix}\, X \cdot A}{\Gamma \vdash \mathsf{out}_{X \cdot A} r : A[X := \mathsf{fix}\, X \cdot A]}$$

and there is a new axiom scheme of reduction

$$\mathsf{out}_{X\cdot A}\big(\mathsf{in}_{X\cdot A}t\big) \longrightarrow_{\mathsf{fix}} t.$$

Moreover, there is a new $\mu$-rule:

$$\mathsf{out}_{X\cdot A}\big(\mu a^{\mathsf{fix}\,X\cdot A}\cdot r\big) \longrightarrow_{\mu} \mu b^{A[X:=\mathsf{fix}\,X\cdot A]}\cdot r\big[a\star := b\big(\mathsf{out}_{X\cdot A}\star\big)\big].$$

It follows the pattern of the other two $\mu$-reduction rules and clearly also enjoys subject reduction.

The new $\mu$-reduction rule corresponds to a "reduction" of stability of $\mathsf{fix}\,X\cdot A$ to that of $A[X := \mathsf{fix}\,X\cdot A]$. The latter type can be more complex than the former which rules out the method by Joly, discussed in the previous section: The only candidate for $(\mathsf{fix}\,X\cdot A)'$ is $\mathsf{fix}\,X\cdot A'$. (A relativization to stable $X$ as for the universal quantifier would not yield a positive type.) With Lemma 18 on the proper types for $\mathsf{stab}_\xi^A$ in mind, we would like to define

$$\begin{aligned}&\mathsf{stab}_\xi^{\mathsf{fix}\,X\cdot A}\\ &:= \lambda v^{\neg\neg(\mathsf{fix}\,X\cdot A)'}\\ &\quad\cdot\, \mathsf{in}_{X\cdot A'}\big(\mathsf{stab}_\xi^{A[X:=\mathsf{fix}\,X\cdot A]}\big(\lambda y^{\neg A'[X:=\mathsf{fix}\,X\cdot A']}\cdot v\big(\lambda z^{(\mathsf{fix}\,X\cdot A)'}\cdot y(\mathsf{out}_{X\cdot A'}z)\big)\big)\big).\end{aligned}$$

Clearly, this cannot be a definition: For $A = X$, this would mean that $\mathsf{stab}_\xi^{\mathsf{fix}\,X\cdot X}$ is defined by an expression that involves $\mathsf{stab}_\xi^{\mathsf{fix}\,X\cdot X}$ again. There is nothing like the solution to such fixed-point equations for *terms* in our systems. For more complex $A$, the situation would even be worse in that $A[X := \mathsf{fix}\,X\cdot A]$ would be more complex than $\mathsf{fix}\,X\cdot A$. The author does not envision a solution to this problem, which therefore limits the applicability of Joly's method.

System F with *non-interleaving* positive fixed points essentially has been studied by Geuvers [9] under the name $F_{ret}$, and strong normalization has been shown by him through an embedding into Mendler's system [16]. A direct proof of strong normalization by saturated sets has been given by the author [13] under the name NPF. No embedding into system F exists [25]. One expects that this negative result also holds for any reasonable extension of the two systems by some $\eta$-rules.

Here, we show that strong normalization also holds with the reductions for classical logic and arbitrary positive fixed points. Using a model construction involving saturated sets, this has been performed already in the respective extension of the Curry-style formulation of $\mathsf{F}^\neg$ by positive fixed points (even in the presence of sums with permutative conversions) in [15].

The point here is to give a straightforward extension of the embedding of $\lambda\mu$ into F in Section 4.3 to an embedding of $\lambda\mu^{\mathsf{fix}}$ into $\mathsf{F}^{\mathsf{fix}}$. As expected, $\mathsf{F}^{\mathsf{fix}}$ shall denote the extension of F to the set TP of positive types, and with the additional axiom for $\longrightarrow_{\mathsf{fix}}$. System $\mathsf{F}^{\mathsf{fix}}$ is strongly normalizing, to be proven by an easy adaptation of the proof in [13] or by omitting everything on RAA and sums in the proof in [15].

To begin with, we also need the extension $\mathsf{I}\sharp^{\mathsf{fix}}$ of system $\mathsf{I}\sharp$ by positive fixed points: For this, we have to stipulate that if $A$ is a positive type, then so is $\sharp A$, and set $\mathsf{TV}_p(\sharp A) := \mathsf{TV}_p(A)$. The reduction axioms are those of $\mathsf{I}\sharp$, plus the above axiom for $\longrightarrow_{\mathsf{fix}}$.

It is obvious that the embedding in Section 3.1 of $\mathsf{I}\sharp$ into $\mathsf{F}$ immediately extends to an embedding of $\mathsf{I}\sharp^{\mathsf{fix}}$ into $\mathsf{F}^{\mathsf{fix}}$: the fixed-point rules are translated homomorphically. (For the type translation, this is legitimate since positive and negative occurrences are maintained by it, especially by the definition of $(\sharp A)'$.) Hence, we are left with the task to extend Theorem 4.1 to $\lambda\mu^{\mathsf{fix}}$ and $\mathsf{I}\sharp^{\mathsf{fix}}$.

Definition 7 is extended by

$$(\mathsf{fix}\, X \cdot A)^* := \mathsf{fix}\, X \cdot \sharp A^*$$

In order to show that $A^*$ is also a positive type, one simultaneously has to show the obvious fact that $\mathsf{TV}_p(A^*) = \mathsf{TV}_p(A)$. By induction on $A$, one gets $\mathsf{FV}(A^*) = \mathsf{FV}(A)$ and $(A[X := B])^* = A^*[X := B^*]$.

As before, the translation of types is defined by $A' := \sharp A^*$, hence

$$(\mathsf{fix}\, X \cdot A)' = \sharp\, \mathsf{fix}\, X \cdot A'$$

We also get the substitution property

$$(A[X := B])' = A'[X := B^*],$$

which instantiates to the crucial equation

$$(A[X := \mathsf{fix}\, X \cdot A])' = A'[X := \mathsf{fix}\, X \cdot A'].$$

Definition 8 is extended by

- If $\Gamma \vdash t : A[X := \mathsf{fix}\, X \cdot A]$ then $(\mathsf{in}_{X \cdot A} t)'_\Gamma := \mathsf{emb}(\mathsf{fix}\, X \cdot A')(\mathsf{in}_{X \cdot A'} t'_\Gamma)$.
- If $\Gamma \vdash r : \mathsf{fix}\, X \cdot A$ then $(\mathsf{out}_{X \cdot A} r)_\Gamma$ is defined as

$$\mathsf{It}_\sharp^{(A[X:=\mathsf{fix}\, X \cdot A])'}\left(r'_\Gamma, \lambda z^{\mathsf{fix}\, X \cdot A'} \cdot \mathsf{out}_{X \cdot A'} z, \mathsf{stab}\left(A[X := \mathsf{fix}\, X \cdot A]\right)^*\right).$$

In a straightforward manner, the proofs of Lemma 13, Lemma 14 and Lemma 15 can be extended to the present systems. For simulation (Lemma 16), we again remark that the term translation never erases subterms. Consequently, only the simulation of the two new reduction rules within $\mathsf{I}\sharp^{\mathsf{fix}}$ has to be verified. For the rule for $\longrightarrow_{\mathsf{fix}}$, this is a trivial sequence of three reduction steps; the $\mu$-reduction rule for the fixed points has to be treated analogously to that for implication in the proof of Lemma 16.

Composing the two embeddings, we get the following:

THEOREM 5.1. *There is a refined embedding of $\lambda\mu^{\mathsf{fix}}$ into $\mathsf{F}^{\mathsf{fix}}$. Hence, $\lambda\mu^{\mathsf{fix}}$ is strongly normalizing.*

There is an objection to the usefulness of $\lambda\mu^{\mathsf{fix}}$, though: The fixed point $\mathsf{fix}\, X \cdot A$ is just meant to be an arbitrary solution to the informal fixed-point equation $F \simeq A[X := F]$. Recursion principles would need more information, namely, that we had the least or the greatest solution of that equation. As an example, we consider primitive recursion à la Mendler [16], which we extend

- to arbitrary least fixed points — not just positive ones — following the observation made in [28] (journal version: [29]), and
- by the appropriate $\mu$-reduction rule.

For our extension $\lambda\mu^{\mathsf{lfp}}$, we extend $\lambda\mu$ by types of the form $\mathsf{lfp}\, X \cdot A$ for arbitrary $X$ and $A$. That they model least fixed points, comes from the new term and typing rules: Extend the term grammar by

$$r, s, t ::= \cdots \mid \mathsf{in}_{X\cdot A} t \mid \mathsf{MRec}^B_{X\cdot A}(r, s),$$

and the typing rules by

$$\frac{\Gamma \vdash t : A[X := \mathsf{lfp}\, X \cdot A]}{\Gamma \vdash \mathsf{in}_{X\cdot A} t : \mathsf{lfp}\, X \cdot A}$$

$$\frac{\Gamma \vdash r : \mathsf{lfp}\, X \cdot A \quad s : \forall X \cdot (X \to \mathsf{lfp}\, X \cdot A) \to (X \to B) \to A \to B}{\Gamma \vdash \mathsf{MRec}^B_{X\cdot A}(r, s) : B}$$

Clearly, the introduction rule is just taken from $\lambda\mu^{\mathsf{fix}}$, but the rule for the Mendler recursor MRec is new (to our discourse). The reduction axiom for primitive recursion is

$$\mathsf{MRec}^B_{X\cdot A}(\mathsf{in}_{X\cdot A} t, s) \longrightarrow_{\mathsf{lfp}} s(\mathsf{lfp}\, X \cdot A)\, \mathsf{id}^{\mathsf{lfp}\, X\cdot A}\, (\lambda x^{\mathsf{lfp}\, X\cdot A} \cdot \mathsf{MRec}^B_{X\cdot A}(x, s))\, t$$

For explanations, consult [29, p. 326] or [12, chapter 6]. (Very simple instances of the schema are the primitive recursive functionals of Gödel's system T.)

The new $\mu$-reduction rule is as expected:

$$\mathsf{MRec}^B_{X\cdot A}(\mu a^{\mathsf{lfp}\, X\cdot A} \cdot r, s) \longrightarrow_{\mu} \mu b^B \cdot r[a\star := b(\mathsf{MRec}^B_{X\cdot A}(\star, s))].$$

Note, however, that there is no relation whatsoever between the type $B$ and $\mathsf{lfp}\, X \cdot A$, hence no "reduction" of stability in any sense. Nevertheless, we get the following theorem:

**Theorem 5.2.** *There is a refined embedding of $\lambda\mu^{\mathsf{lfp}}$ into $\mathsf{F}^{\mathsf{fix}}$. Hence, $\lambda\mu^{\mathsf{lfp}}$ is strongly normalizing.*

Proof. By the previous theorem, it suffices to find an embedding of $\lambda\mu^{\mathsf{lfp}}$ into $\lambda\mu^{\mathsf{fix}}$. All type formation rules are treated homomorphically, with exception of $\mathsf{lfp}\, X \cdot A$. If already $A$ has been translated to $A'$, then $\mathsf{lfp}\, X \cdot A$ is translated to $(\mathsf{lfp}\, X \cdot A)' := \mathsf{fix}\, Y \cdot \hat{A}$, with

$$\hat{A} := \forall Z \cdot (\forall X \cdot (X \to Y) \to (X \to Z) \to A' \to Z) \to Z.$$

This definition can be obtained by applying the propositions 1, 5 and 9 in [29] and some simple isomorphisms in order to get rid of existential quantification and products.

The interesting clauses for the term translation are:

- $(\mathsf{MRec}^B_{X\cdot A}(r,s))' := \mathsf{out}_{Y\cdot \hat{A}}r'B's'$
- $(\mathsf{in}_{X\cdot A}t)' := \mathsf{in}_{Y\cdot \hat{A}}(\Lambda Z\lambda z^{\forall X\cdot(X\to(\mathsf{lfp}\,X\cdot A)')\to(X\to Z)\to A'\to Z}$
  $\cdot z(\mathsf{lfp}\,X\cdot A)'\,\mathsf{id}^{(\mathsf{lfp}\,X\cdot A)'}(\lambda x^{(\mathsf{lfp}\,X\cdot A)'}\cdot(\mathsf{MRec}^Z_{X\cdot A}(x,z))')t')$

It is fairly easy to check the statement on types in Definition 1 for this setting. Also, the compatibility of the term translation with the notions of term substitution, type substitution and special substitution is easily established. Given these compatibilities, simulation for the two reduction rules concerning MRec is just a matter of calculation. In both cases, three reduction steps lead from the translation of the left-hand side to that of the right-hand side; for the $\mu$-reduction rule, it is just three $\mu$-reductions, pertaining to the application of $\mathsf{out}_{Y\cdot\hat{A}}$, of $B'$ and of $s'$. ⊣

Note that it seems impossible to extend the translation by stabilization from Section 4.3 by these least fixed points (also in the target system $\mathsf{I}\sharp$). So, we made essential use of impredicativity, which cannot be too inconvenient since Mendler's style anyway rests on impredicativity. For formulations of primitive recursion in the style of universal algebra (see, e.g., [9]), see the treatment and discussion in [14, section 7].

**5.3. Embeddings in continuation-passing style?** Nakazawa and Tatsuta [17] show that a number of published proofs of strong normalization of classical natural deduction by translations in continuation-passing style (CPS) fail. They explicity mention [20, 6, 8].

It should be recalled that CPS translations do not model the $\mu$-reductions but only allow to deduce strong normalization from that of F and separately of the $\mu$-reductions (which, according to [20], strongly normalize even in the untyped case).

In the cited articles, the term closure does not preserve simulation of reduction steps. This is due to "erasing-continuations" [17]: Vacuous $\mu$-abstractions may devour an argument term, hence ordinary $\beta$-reductions in that argument are not simulated but erased.

In more recent work [7], de Groote distinguishes the vacuous abstraction from the other ones and gives different translations, involving $\beta$-expansions in the vacuous case. Unfortunately, the case distinction is not invariant under $\beta$-reduction: $\mu$-variables may get lost. Simulation fails, e.g., for

$$\mu a^A\cdot(\lambda z^\perp\cdot x)(ay)\longrightarrow \mu a^A\cdot x$$

(with $x, y, z$ different) in that the translated terms only have a common reduct.

One would have hoped to see the promised proof by "optimized CPS-translation" of strong normalization of domain-free classical pure type systems [3, section 6.1], but there does not seem to be a full version of that article.

Nakazawa and Tatsuta give a new CPS-translation that simulates $\beta$-reductions and erases $\mu$-reductions, as expected from CPS-translations. They use the notion of augmentation, which gives rise to a quite complicated form of embedding. Unfortunately, the whole article treats a formulation of second-order $\lambda\mu$-calculus with typing à la Curry. As for the embedding of this article (see remark 4) and for the CPS-translation in [5, Proposition 5.2] (see the remarks between the statement and the proof), this does not work. In [17], Proposition 4.6 fails. A counterexample (in their notation) is provided by the typing $\lambda x \cdot x : \emptyset \vdash \forall X \cdot X \to X, \emptyset$. The lemma claims that $(\lambda x \cdot x)^* : (f : \neg(\forall X \cdot X \to X)^*) \vdash \bot$. In our notation, this amounts to the typing $f : \neg\forall X \cdot \neg\neg(\neg\neg X \to \neg\neg X) \vdash f(\lambda x \lambda g \cdot xg) : \bot$, which is certainly impossible in Curry-style system F.

It seems highly plausible that the Church-style formulation can be treated by their method (a draft has been provided by Nakazawa and Tatsuta after the present author informed them of this counterexample). But their term translation is more demanding than ours: Simulation is only proven for a clever choice among an infinite set of "augmentations" for every term. In effect, a hypothetical reduction sequence in $\lambda\mu$ which contains infinitely many $\beta$-reduction steps is translated into one in F, where the choices depend on the earlier reduction steps. This works well for the sake of inheriting strong normalization, but does not seem to be very explicative.

The author conjectures that the method by Nakazawa and Tatsuta will smoothly extend to fixed points, as introduced in the previous section.

**§6. Conclusion.** A new translation of classical second-order logic into intuitionistic second-order logic has been described: Stabilization. It has been put forward as an alternative to double-negation translations (translations in continuation-passing style). It is also faithful to the operational behaviour of those systems, and therefore allows to infer strong normalization of the classical systems from the intuitionistic systems.

For the pure system, also the embedding by Joly is available, but does not seem to extend to fixed points, hence fixed points are a true application of our stabilization method.

Published proofs of strong normalization of classical second-order logic by double-negation translations seem to have failed altogether, but there is the recent work by Nakazawa and Tatsuta which does perhaps not fully qualify as an "embedding".

**Acknowledgments.** I am thankful for fruitful discussions with Andreas Abel and Klaus Aehlig. I also thank the anonymous referee for inducing a further thought process.

REFERENCES

[1] Yuuki Andou, *A normalization-procedure for the first order classical natural deduction with full logical symbols*, ***Tsukuba Journal of Mathematics***, vol. 19 (1995), no. 1, pp. 153–162.

[2] ———, *Church-Rosser property of a simple reduction for full first-order classical natural deduction*, ***Annals of Pure and Applied Logic***, vol. 119 (2003), pp. 225–237.

[3] Gilles Barthe, John Hatcliff, and Morten Heine Sørensen, *A notion of classical pure type system (preliminary version)*, ***Proceedings of the thirteenth conference on the mathematical foundations of programming semantics*** (Stephen Brookes and Michael Mislove, editors), Electronic Notes in Theoretical Computer Science, vol. 6, Elsevier, 1997, 56 pp.

[4] Corrado Böhm and Alessandro Berarducci, *Automatic synthesis of typed λ-programs on term algebras*, ***Theoretical Computer Science***, vol. 39 (1985), pp. 135–154.

[5] Philippe de Groote, *A CPS-translation of the λμ-calculus*, ***Trees in algebra and programming - CAAP'94, 19th international colloquium*** (Edinburgh) (Sophie Tison, editor), Lecture Notes in Computer Science, vol. 787, Springer Verlag, 1994, pp. 85–99.

[6] ———, *A simple calculus of exception handling*, ***Proceedings of the second international conference on typed lambda calculi and applications (TLCA '95), Edinburgh, United Kingdom, April 1995*** (Mariangiola Dezani-Ciancaglini and Gordon Plotkin, editors), Lecture Notes in Computer Science, vol. 902, Springer Verlag, 1995, pp. 201–215.

[7] ———, *Strong normalization of classical natural deduction with disjunction*, ***Proceedings of TLCA 2001*** (Samson Abramsky, editor), Lecture Notes in Computer Science, vol. 2044, Springer Verlag, 2001, pp. 182–196.

[8] Ken-Etsu Fujita, *Domain-free λμ-calculus*, ***RAIRO - Theoretical Informatics and Applications***, vol. 34 (2000), pp. 433–466.

[9] Herman Geuvers, *Inductive and coinductive types with iteration and recursion*, ***Proceedings of the workshop on types for proofs and programs, Båstad, Sweden*** (Bengt Nordström, Kent Pettersson, and Gordon Plotkin, editors), 1992, only published via `ftp://ftp.cs.chalmers.se/pub/cs-reports/baastad.92/proc.dvi.Z`, pp. 193–217.

[10] Jean-Yves Girard, ***Interprétation fonctionnelle et élimination des coupures dans l'arithmétique d'ordre supérieur***, Thèse de Doctorat d'État, Université de Paris VII, 1972.

[11] Thierry Joly, *An embedding of 2nd order classical logic into functional arithmetic FA2*, ***Comptes Rendus de l'Académie des Sciences. Série I. Matheématique***, vol. 325 (1997), pp. 1–4.

[12] Ralph Matthes, ***Extensions of system F by iteration and primitive recursion on monotone inductive types***, Doktorarbeit (PhD thesis), University of Munich, 1998, available via the homepage `http://www.tcs.informatik.uni-muenchen.de/~matthes/`.

[13] ———, *Monotone fixed-point types and strong normalization*, ***Computer science logic, 12th international workshop, Brno, Czech republic, August 24–28, 1998, proceedings*** (Georg Gottlob, Etienne Grandjean, and Katrin Seyr, editors), Lecture Notes in Computer Science, vol. 1584, Springer Verlag, 1999, pp. 298–312.

[14] ———, *Parigot's second order λμ-calculus and inductive types*, ***Proceedings of TLCA 2001*** (Samson Abramsky, editor), Lecture Notes in Computer Science, vol. 2044, Springer Verlag, 2001, pp. 329–343.

[15] ———, *Non-strictly positive fixed-points for classical natural deduction*, ***Annals of Pure and Applied Logic***, vol. 133 (2005), pp. 205–230.

[16] Nax P. Mendler, *Recursive types and type constraints in second-order lambda calculus*, ***Proceedings of the second annual IEEE symposium on logic in computer science, Ithaca, N.Y.***, IEEE Computer Society Press, 1987, pp. 30–36.

[17] Koji Nakazawa and Makoto Tatsuta, *Strong normalization proof with CPS-translation for second order classical natural deduction*, ***The Journal of Symbolic Logic***, vol. 68 (2003), no. 3, pp. 851–859, Corrigendum: vol. 68 (2003), no. 4, pp. 1415–1416.

[18] Michel Parigot, *λμ-calculus: an algorithmic interpretation of classical natural deduction*, ***Logic programming and automated reasoning, international conference LPAR'92, St. Petersburg,***

***Russia*** (Andrei Voronkov, editor), Lecture Notes in Computer Science, vol. 624, Springer Verlag, 1992, pp. 190–201.

[19] ———, *Strong normalization for second order classical natural deduction*, ***Proceedings, eighth annual IEEE symposium on logic in computer science*** (Montreal, Canada), IEEE Computer Society Press, 1993, pp. 39–46.

[20] ———, *Proofs of strong normalisation for second order classical natural deduction*, ***The Journal of Symbolic Logic***, vol. 62 (1997), no. 4, pp. 1461–1479.

[21] CHRISTINE PAULIN-MOHRING, ***Définitions inductives en théorie des types d'ordre supérieur***, Habilitation à diriger les recherches, Université Claude Bernard Lyon I, 1996.

[22] DAG PRAWITZ, ***Natural deduction. a proof-theoretical study.***, Almquist and Wiksell, 1965.

[23] ———, *Ideas and results in proof theory*, ***Proceedings of the second Scandianvian logic symposium*** (Jens E. Fenstad, editor), North-Holland, Amsterdam, 1971, pp. 235–307.

[24] JAKOB REHOF and MORTEN HEINE SØRENSEN, *The $\lambda_\Delta$-calculus*, ***Theoretical aspects of computer software, international conference TACS '94, Sendai, Japan, proceedings*** (Masami Hagiya and John C. Mitchell, editors), Lecture Notes in Computer Science, vol. 789, Springer Verlag, 1994, pp. 516–542.

[25] ZDZISŁAW SPŁAWSKI and PAWEŁ URZYCZYN, *Type Fixpoints: Iteration vs. Recursion*, ***SIGPLAN Notices***, vol. 34 (1999), no. 9, pp. 102–113, Proceedings of the 1999 International Conference on Functional Programming (ICFP), Paris, France.

[26] WILLIAM W. TAIT, *A realizability interpretation of the theory of species*, ***Logic colloquium Boston 1971/72*** (Rohit Parikh, editor), Lecture Notes in Mathematics, vol. 453, Springer Verlag, 1975, pp. 240–251.

[27] MASAKO TAKAHASHI, *Parallel reduction in $\lambda$-calculus*, ***Information and Computation***, vol. 118 (1995), no. 1, pp. 120–127.

[28] TARMO UUSTALU and VARMO VENE, *A cube of proof systems for the intuitionistic predicate $\mu$-, $\nu$-logic*, ***Selected papers of the 8th nordic workshop on programming theory (NWPT '96), Oslo, Norway, December 1996*** (Magne Haveraaen and Olaf Owe, editors), Research Reports, Department of Informatics, University of Oslo, vol. 248, May 1997, pp. 237–246.

[29] ———, *Least and greatest fixed points in intuitionistic natural deduction*, ***Theoretical Computer Science***, vol. 272 (2002), pp. 315–339.

LEHR- UND FORSCHUNGSEINHEIT FÜR THEORETISCHE INFORMATIK
INSTITUT FÜR INFORMATIK DER UNIVERSITÄT MÜNCHEN
OETTINGENSTRAßE 67, D-80538 MÜNCHEN, GERMANY

*Current address*: IRIT - Université Paul Sabatier, Équipe ACADIE, 118 route de Narbonne, F-31062 Toulouse Cedex 9, France

*E-mail*: matthes@irit.fr

# DEFINABILITY AND REDUCIBILITY IN HIGHER TYPES OVER THE REALS

DAG NORMANN

**Abstract.** We define the concept of *reducible* sets of total, continuous functionals of finite type over the reals. We show that all Polish spaces are homeomorphic to reducible subsets of $\mathbb{R} \to \mathbb{R}$. We further show that the class of topological spaces homeomorphic to reducible sets is closed under the formation of function spaces and, to some extent, the formation of definable subspaces.

The Approximation Lemma will be an important tool that also is of independent interest. This lemma states that continuous functions defined on a subspace may, under certain conditions, be approximated by continuous functions defined on the full space.

**§1. Introduction.** In this paper we will be concerned with hierarchies of functionals of finite types over the reals. In general we are interested in such hierarchies because they play a part in understanding computability issues for the real numbers. Thus, even if we focus on hereditarily total functionals, we want, in some sense, to restrict ourselves to "continuous objects" where some concepts of computability make sense.

If we consider the natural numbers as the base type, a corresponding hierarchy is well understood, with numerous characterisations. The functionals of this hierarchy are known as the *Kleene-Kreisel continuous functionals*. If we consider $\mathbb{R}$ as the base type, the situation is not that clear. There is no canonical way to represent the reals as a datatype, and it is not clear that different representations do not lead to different hierarchies of hereditarily total objects. This is discussed in Bauer, Escardó and Simpson [2] and in Normann [13].

In Section 2, we will define the hierarchy of hereditarily total functionals over the reals, based on a standard algebraic domain representation of the real line, and on a standard domain-theoretical way of forming function spaces. This hierarchy also have numerous characterisations. It is equivalent to the one studied by Weihrauch and his group in the project of the Type Two Theory of Effectivity, see Weihrauch [18]. The equivalence is proved combining the characterisations due to Schröder [15] of the Weihrauch hierarchy and to Normann [12] of the hierarchy in this paper as the hierarchy obtained in the category of Kuratowski limit spaces. In his thesis [7], De Jaeger gave a characterisation in the category of filter spaces.

Logic Colloquium '03
Edited by V. Stoltenberg-Hansen and J. Väänänen
Lecture Notes in Logic, 24

In Section 3 we prove the Approximation Lemma, If $A$ is a subspace of a topological space $X$ and $f : A \to \mathbb{R}$ is continuous, we may not in general extend $f$ to a continuous function $g : X \to \mathbb{R}$. We show that in cases relevant to us we may find a sequence of continuous $g_i : X \to \mathbb{R}$ that converges to $f$ on $A$. What this means will be made precise in the text. This lemma will be used in cases where an extention lemma would be useful but is not available.

In Section 4 we will define the concept of a *reducible set*, which technically will be a subset $A$ of a special kind of what we will call $Ct_{\mathbb{R}}(\sigma)$ for some finite type $\sigma$. We use the terminology "reducible" because, by definition, elementhood in $A$ can be reduced to totality for certain partial objects of some fixed type $\tau$ depending on $A$. The aim of the paper is to demonstrate that the class of spaces homeomorphic to reducible sets is rich and with strong closure properties, while computability issues for each such space may be inherited from computability issues for the typed hierarchies over the reals.

Following the terminology introduced in Normann [11] we call a concept of computability *external* if it is imposed upon us from the outside, e.g. via an effective enumeration of the finitary elements of an algebraic domain and *internal* if it is defined via some [programming-] language interpreted over the structure at hand, and thus independent of the actual representation. This distinction is much like the distinction between concrete and abstract concepts of computability discussed by Tucker and Zucker in [17].

A *topological algebra* will be a topological space with some continuous functions. Technically, there will be a signature $\Sigma$ of function symbols $f$ of given arity, and then the algebra $\mathcal{A}$ will consist of a topological space $A$ together with continuous interpretations $f^{\mathcal{A}} : A^n \to A$ of each function symbol $f$ with arity $n$.

Examples of topological algebras are the natural numbers with zero and successor and discrete topology, $\mathbb{R}$ with plus, times, exponentials, trigonometric functions etc. and Eucledian topology, various Banach spaces etc.

Datatypes may be modeled as topological algebras. When this is the case, it is natural to enrich the algebra to one that also contains elements representing partial information. One way to do this is to use an algebraic or continuous domain, and represent the original space as a *quotient space*, as a set of *total elements* or both.

If a topological algebra $\mathcal{A}$ has a domain that is a reducible set, we automatically have access to internal notions of computability like Escardó's Real PCF [4]. This will of course be external from the point of view of $\mathcal{A}$, and it remains to be seen to which extent this approach can be used to give a semantics for typed computability over some topological algebras.

Blanck [3] showed how complete metric spaces in general can be represented as domains, and how effective metric spaces can be represented as effective domains. Polish spaces are characterised as the $G_\delta$-subspaces of $[0, 1]^{\mathbb{N}}$. In Section 4 we will use this to show that Polish spaces are homeomorphic to reducible subsets of $\mathbb{R} \to \mathbb{R}$. Moreover we will show that certain definable subsets of reducible sets will be reducible, and that the class of reducible sets, up to homeomorphic equivalence, is closed under the formation of function spaces. Thus our conclusion will be that the hierarchy of hereditarily total functionals over the reals is rich, and that many questions about higher type computability in analysis in general may be reduced to questions about this hierarchy.

## §2. Preliminaries.

**2.1. Types, domains and functionals.** We will consider functionals of finite types over one base type 0.

DEFINITION 1. 0 is a type term.
If $\sigma$ and $\tau$ are type terms, then $(\sigma \to \tau)$ is a type term.

We will use the standard observation that any type term will be of the form

$$(\sigma_1 \to (\sigma_2 \to \cdots (\sigma_k \to 0) \cdots))$$

where $k \geq 0$, and we will write $\sigma_1, \sigma_2, \ldots, \sigma_k \to 0$ for this.

We will work in the category of separable algebraic domains, and give a brief introduction:

Let $\mathcal{D}$ be the class of partial orderings $(X, \sqsubseteq)$ such that each directed set is bounded and each bounded set $Y$ has a least upper bound $\bigsqcup Y$.

Let $(X, \sqsubseteq) \in \mathcal{D}$. An element $x_0 \in X$ is *finitary* or *compact* if for each bounded set $Y \subseteq X$, if $x_0 \sqsubseteq \bigsqcup Y$ then there is a finite subset $Y_0 \subseteq Y$ such that $x_0 \sqsubseteq \bigsqcup Y_0$.

An algebraic domain is an element $(X, \sqsubseteq)$ of $\mathcal{D}$ such that each $x \in X$ is the least upper bound of the finitary objects below it.

If $x_0 \in X$ is finitary, then $B^X(x_0) = \{y \in X \mid x_0 \sqsubseteq y\}$ is a *basic open set* in the *Scott Topology*. If $(X, \sqsubseteq_X)$ and $(Y, \sqsubseteq_Y)$ are two algebraic domains, the set of continuous functions from $X$ to $Y$ ordered pointwise form a new algebraic domain. If we let the morphisms be the continuous functions, the algebraic domains then form a Cartesian closed category.

We will assume some familiarity with the theory for algebraic domains, see e.g. Stoltenberg-Hansen et al. [16], Abramsky and Jung [1] or Gierz et al. [5].

We will consider a typed hierarchy over the reals:

DEFINITION 2. a) We define the algebraic domain $R(\sigma)$ for each type term $\sigma$ as follows:

- $R_0(0) = \{\mathbb{R}\} \cup \{[p,q] \mid p \in \mathbb{Q} \wedge q \in \mathbb{Q} \wedge p \leq q\}$ ordered by reverse inclusion.
- $R(0)$ is the set of ideals in $R_0(0)$ ordered by inclusion.
- $R(\sigma \to \tau) = R(\sigma) \to R(\tau)$ in the category of algebraic domains.

b) We define the set $\bar{R}(\sigma)$ of hereditarily total objects of type $\sigma$ by recursion on $\sigma$ as follows:

- $\bar{R}(0)$ is the set of ideals $\alpha \subseteq R_0(0)$ such that $\sqcap\alpha$ consists of one real number.
- $\bar{R}(\sigma \to \tau) = \{x \in R(\sigma \to \tau) \mid \forall y \in \bar{R}(\sigma)(x(y) \in \bar{R}(\tau))\}$.

The following was proved by Longo and Moggi [10] for the natural numbers. The proof works for this hierarchy as well, see e.g. Normann [12].

PROPOSITION 1. *When $x_1$ and $x_2$ are elements of $\bar{R}(\sigma)$, let $x_1 \approx x_2$ when $x_1 \sqcap x_2 \in \bar{R}(\sigma)$.*
*Then $\approx$ is an equivalence relation and, when $\sigma = \sigma_1 \to \tau$, we have*

$$x_1 \approx x_2 \Leftrightarrow \forall y_1 \in \bar{R}(\sigma_1)\forall y_2 \in \bar{R}(\sigma_1)(y_1 \approx y_2 \to x_1(y_1) \approx x_2(y_2)).$$

The proof is fairly simple by induction on $\sigma$.
As a consequence, we may view the hierarchy of quotients as a typed hierarchy of extensional functionals:

DEFINITION 3. For each type term $\sigma$ we define the map $\rho_\sigma$ defined on $\bar{R}(\sigma)$ and the set $Ct_{\mathbb{R}}(\sigma)$ of values of $\rho_\sigma$ by

- If $\alpha$ is an ideal in $\bar{R}(0)$, then $\rho_0(\alpha)$ is the unique element in $\sqcap\alpha$.

$$Ct_{\mathbb{R}}(0) = \mathbb{R}.$$

- If $x \in \bar{R}(\sigma \to \tau)$, we let

$$\rho_{\sigma\to\tau}(x) : Ct_{\mathbb{R}}(\sigma) \to Ct_{\mathbb{R}}(\tau)$$

  be defined by

$$\rho_{\sigma\to\tau}(x)(\rho_\sigma(y)) = \rho_\tau(x(y)).$$

PROPOSITION 2. *Each $\rho_\sigma$ is well defined, and for $x_1 \in \bar{R}(\sigma)$ and $x_2 \in \bar{R}(\sigma)$ we have that*

$$x_1 \approx x_2 \Leftrightarrow \rho_\sigma(x_1) = \rho_\sigma(x_2).$$

All domains $R(\sigma)$ are effective domains. This means that the set of finitary objects is countable, and that there is an enumeration of the finitary objects such that $\sqsubseteq$ and the boundedness-relation are effective, and that the least upper bound operator for finite bounded sets of finitary objects is effective.

PROPOSITION 3. *Let $\sigma$ be a type term.*
*Uniformly in $\sigma$ and a finitary object $x_0 \in R(\sigma)$ there is an extension $x \in \bar{R}(\sigma)$ of $x_0$.*

This was first proved in Normann [12]. De Jaeger's characterisation in [7] of this hierarchy in the category of filter spaces is an alternative source for the proof of this proposition, which we call the *Density Theorem*.

**2.2. Topology.** The topology on $Ct_{\mathbb{R}}(\sigma)$ is inherited from the quotient topology of $(\bar{R}(\sigma), \approx)$ via $\rho_\sigma$. This will be a sequential topology, i.e. it is the finest topology where all convergent sequences converge. The limit structure on $Ct_{\mathbb{R}}(\sigma \to \tau)$ can alternatively be defined in the category of limit spaces by

$$f = \lim_{n\to\infty} f_n \Leftrightarrow \forall a, \{a_n\}_{n\in\mathbb{N}} \in Ct_{\mathbb{R}}(\sigma)(a = \lim_{n\to\infty} a_n \Rightarrow f(a) = \lim_{n\to\infty} f_n(a_n)).$$

If $A$ and $B$ are metric spaces, they are topological limit spaces, and then $f = \lim_{n\to\infty} f_n$ in $A \to B$ exactly when $f$ is the pointwise limit of the equicontinuous sequence $\{f_n\}_{n\in\mathbb{N}}$.

Let $A \subseteq Ct_{\mathbb{R}}(\sigma)$. There are two natural ways to define a topology on $A$. Let $T_1$ be the topology inherited from $Ct_{\mathbb{R}}(\sigma)$.

Let $\bar{A} = \{x \in \bar{R}(\sigma) \mid \rho_\sigma(x) \in A\}$.

Let $T_2$ be the quotient topology inherited from the topology on $\bar{A}$. In general, $T_2$ will be a finer topology than $T_1$.

LEMMA 1. *If $A$ is closed or if $A$ is open, then $T_1 = T_2$.*

The proof is easy and is left for the reader.

DEFINITION 4. We let $T_2$ be the *induced topology* on $A \subseteq Ct_{\mathbb{R}}(k)$.

We then have

LEMMA 2. *The topology on $A \subseteq Ct_{\mathbb{R}}(\sigma)$ is generated by its convergent sequences.*

PROPOSITION 4. *Let $A \subseteq Ct_{\mathbb{R}}(\sigma)$, and let $f : A \to Ct_{\mathbb{R}}(\tau)$ be continuous. Then there is an $\hat{f} \in R(\sigma \to \tau)$ such that*

$$\forall x \in \bar{R}(\sigma)(x \in \bar{A} \Rightarrow \hat{f}(x) \in \bar{R}(\tau) \wedge \rho_\tau(\hat{f}(x)) = f(\rho_\sigma(x)))$$

We omit the proof. This proposition is a special case of Lemma 12 in Normann [12], where $A$ may be replaced by any $\bar{X} \subseteq X$, $X$ is a separable domain and $\bar{X}$ is closed upwards. We use the lemma for $X = R(\sigma)$ and $\bar{X} = \bar{A}$.

If $A \subseteq Ct_{\mathbb{R}}(\sigma)$ and $B \subseteq Ct_{\mathbb{R}}(\tau)$, we let $\bar{R}(A \to B)$ be the set of $\hat{f} \in R(\sigma \to \tau)$ such that

$$\forall x \in \bar{R}(\sigma)(\rho_\sigma(x) \in A \Rightarrow \hat{f}(x) \in \bar{R}(\tau) \wedge \rho_\tau(\hat{f}(x)) \in B).$$

Consistent elements of $\bar{R}(\tau)$ will be equivalent, so when $\hat{f} \in \bar{R}(A \to B)$, $\hat{f}$ determines a continuous function $f : A \to B$. We let $\rho_{A\to B}(\hat{f}) = f$ in this case. Both $A$ and $B$ will be topological limit spaces, so there is a limit space structure on $A \to B$.

LEMMA 3. *The limit space structure on $A \to B$ defines the identification topology induced by $\rho_{A\to B}$.*

PROOF. We have to prove that the following are equivalent:

1. $f = \lim_{n\to\infty} f_n$ in $A \to B$ in the identification topology.
2. There is a convergent sequence $\hat{f} = \lim_{n\to\infty} \hat{f}_n$ in $\bar{R}(A \to B)$ such that $f = \rho_{A\to B}(\hat{f})$ and $f_n = \rho_{A\to B}(\hat{f}_n)$ for each $n$.
3. Whenever $a = \lim_{n\to\infty} a_n$ in $A$, then $f(a) = \lim_{n\to\infty} f_n(a_n)$ in $B$,

and use the fact that the identification topology will be generated from its set of convergent sequences.

2. ⇒ 1. is trivial.

1. ⇒ 3. follows from the fact that application is continuous in the identification topologies on $(A \to B) \times A$ and $B$.

3. ⇒ 2. is a consequence of Lemma 12 from Normann [12] as referred to after Proposition 4. ⊣

**2.3. Coding.** In this section we will show that in many cases we may restrict our attention to the pure types:

DEFINITION 5. As type terms, let 0 denote 0 as before, and let $k + 1$ denote $(k \to 0)$. These types are called the *pure types.*

It is well known that there exist embeddings and projections between the pure types. Thus the next definition is mainly meant to settle the notation:

DEFINITION 6. Let $i, k \in \mathbb{N}$. We will define the maps $\Phi_{i,k} \in \bar{R}(i \to k)$ as follows:

- $\Phi_{k,k}$ is the identity on $R(k)$.
- $\Phi_{0,1}(x) = \lambda y.x$.
- $\Phi_{1,0}(f) = f(0)$.
- $\Phi_{k,k+1}(x) = \lambda y.x(\Phi_{k,k-1}(y))$ when $k > 0$.
- $\Phi_{k+1,k}(F) = \lambda z.F(\Phi_{k-1,k}(z))$ when $k > 0$.
- If $|i - k| > 1$, define $\Phi_{i,k}$ as the shortest possible composition of those above.

The following is folklore:

PROPOSITION 5. *If $i < k$, then $\Phi_{k,i}(\Phi_{i,k}(x)) = x$ for each $x \in R(i)$. Moreover, the maps $\Phi_{i,k}$ are total for all $i$ and $k$.*

It is well known that $\mathbb{N}$ and $\mathbb{N}^k$ can be put in a 1-1 correspondence via a computable bijection. This can be used to justify that we only restrict attention to pure higher types, since mixed types of total functionals over $\mathbb{N}$ can be reduced to pure types.

A *retraction* on a topological space $X$ is a continuous map $f : X \to X$ such that $f = f^2$. A *retract* of $X$ will then be the image of a retraction on $X$. A retract of a Hausdorff-space will be closed. Thus the induced topology on any retract of $Ct_{\mathbb{R}}(\sigma)$ will be the subspace topology.

It is well known that $\mathbb{R}$ and $\mathbb{R}^2$ are not homeomorphic, and that $\mathbb{R}^2$ is not homeomorphic to any retract of $\mathbb{R}$. This shows that we do not have the same nice coding mechanisms at bottom level for the reals as for the natural numbers. We will see that much of the machinery needed for coding can be obtained at type 1, and that it then extends to higher types.

There will be alternative ways to define topologies on a cartesian product $\prod_{i=1}^{n} Ct_{\mathbb{R}}(k_i)$, like for subsets of $Ct_{\mathbb{R}}(k)$. The alternatives are the product topology and the one induced from $\prod_{i=1}^{n} \bar{R}(k_i)$. We will always use the latter. E.g. application will not be continuous otherwise.

Lemma 4. *$(\mathbb{R} \to \mathbb{R}) \times \mathbb{R}$ is homeomorphic to a retract of $(\mathbb{R} \to \mathbb{R})$.*

Proof. Let $f : \mathbb{R} \to \mathbb{R}$ and $a \in \mathbb{R}$ be given.

We let $g = \langle f, a\rangle$ be defined by

$$g(x) = f(x) \quad \text{when } x \leq 0.$$

$$g(x) = f(x-1) + a \quad \text{when } x \geq 1.$$

$$g(x) = x \cdot a \quad \text{if } 0 \leq x \leq 1.$$

It is easy to see that the image of this map is a retract of $\mathbb{R} \to \mathbb{R}$. ⊣

By the same method we see that $(\mathbb{R} \to \mathbb{R}) \times (\mathbb{N} \to \mathbb{R})$ and $(\mathbb{R} \to \mathbb{R})^2$ can be realized as retracts of $\mathbb{R} \to \mathbb{R}$, just cut the graphs into pieces, and glue the pieces together into one function.

This method actually extends to higher types:

Let $k > 1$. Let $z_0$ be the constant zero element of type $k-2$ ($z_0 = 0$ if $k = 2$).

Let $V_k = \{F \in Ct_{\mathbb{R}}(k-1) \mid F(z_0) = 0\}$, $\mathbf{1}_{k-1}$ the constant 1 functional of type $k-1$. Then every $F \in Ct_{\mathbb{R}}(k-1)$ can be uniquely described as $F = F_0 + a \cdot \mathbf{1}_{k-1}$, where $a \in \mathbb{R}$ and $F_0 \in V_k$.

There is a canonical bijection $\Theta$ between $Ct_{\mathbb{R}}(k)$ and $V_k \to (\mathbb{R} \to \mathbb{R})$ by

$$\Theta(\phi)(F_0)(a) = \phi(F_0 + a \cdot \mathbf{1}_{k-1}).$$

As an example, let us extend

$$\langle\, ,\, \rangle : (\mathbb{R} \to \mathbb{R}) \times \mathbb{R} \to (\mathbb{R} \to \mathbb{R})$$

constructed above to a function

$$\langle\, ,\, \rangle_k : Ct_{\mathbb{R}}(k) \times \mathbb{R} \to Ct_{\mathbb{R}}(k)$$

for $k \geq 2$.

Let $\phi \in Ct_{\mathbb{R}}(k)$, $a \in \mathbb{R}$. Let

$$\langle \phi, a\rangle_k(F_0 + x \cdot \mathbf{1}_{k-1}) = \langle \lambda b \in \mathbb{R}.\Theta(\phi)(F_0, b), a\rangle(x)$$

where $F_0 \in V_k$ and $x \in \mathbb{R}$. Simple calculation shows that this works.

This was just an example. Use of the same method shows:

Theorem 1. *Let $1 \leq n \leq k$.*
*Then $Ct_{\mathbb{R}}(k) \times Ct_{\mathbb{R}}(n)$ is homeomorphic to a retract of $Ct_{\mathbb{R}}(k)$.*

**§3. The approximation lemma.** The Approximation Lemma may be used instead of an extension theorem in many situations. Our main application will be in Section 4.2. Further applications will be found in the forthcoming Normann [14].

If $A \subseteq Ct_{\mathbb{R}}(k)$ and $f : A \to \mathbb{R}$ is continuous, we will use the density theorem for $Ct_{\mathbb{R}}(k+1)$ to show that $f$ can be approximated on $A$ by a sequence $\{f_n\}_{n\in\mathbb{N}}$ of functions in $Ct_{\mathbb{R}}(k+1)$. In general, $f$ cannot be extended to a total, continuous function on $Ct_{\mathbb{R}}(k)$.

We need a more accurate description of the finitary elements of $R(k+1)$ in the proof of the approximation lemma. If $\sigma$ is a finitary element in $R(k)$ and $[p,q] \in R_0(0)$, then the pair $(\sigma, [p,q])$ defines the function $f_{(\sigma,[p,q])}$ defined by

- $f_{(\sigma,[p,q])}(x) = [p,q]$ if $\sigma \sqsubseteq x$
- $f_{(\sigma,[p,q])}(x) = \bot (= \mathbb{R})$ otherwise.

The Approximation Lemma will be

THEOREM 2. *Let $A \subseteq Ct_{\mathbb{R}}(k)$. Then continuously in $f : A \to \mathbb{R}$ there is a sequence $\{f_n\}_{n\in\mathbb{N}}$ of functions $f_n : Ct_{\mathbb{R}}(k) \to \mathbb{R}$ such that whenever $x \in A$, each $x_n \in Ct_{\mathbb{R}}(k)$ and $x = \lim_{n\to\infty} x_n$, then $f(x) = \lim_{n\to\infty} f_n(x_n)$.*

PROOF. Let $\bar{A} = \{x \in \bar{R}(k) \mid \rho_k(x) \in A\}$.

As discussed in Section 2.2, if $f : A \to \mathbb{R}$ is continuous, there will be an $\hat{f} \in R(k+1)$ such that $\hat{f}$ maps $\bar{A}$ to $\bar{R}(0)$ and such that $f(\rho_k(x)) = \rho_0(\hat{f}(x))$ whenever $x \in \hat{A}$.

The naïve idea behind the construction is as follows:

Let $\{x_i\}_{x\in\mathbb{N}}$ be an enumeration of a dense subset of $\bar{A}$.

Depending on $\hat{f}(x_0), \dots, \hat{f}(x_n)$ we will let certain finitary elements in $R(k+1)$ be the $n$th approximation to $f$ with some probability. Taking the weighted sum of the total extensions of these finitary elements will give us the approximation $f_n$ to $f$.

First we will develop some general machinery independent of $f$:

Let $\Sigma_A$ be the set of finitary elements $\sigma$ in $R(k)$ such that $\sigma$ has an extension in $\bar{A}$.

The *basic pairs* will be the set of pairs $(\sigma, [p,q])$ where $\sigma \in \Sigma_A$ and $p < q$ are rational numbers. Let $\{(\sigma_i, [p_i, q_i])\}_{i\in\mathbb{N}}$ be an enumeration of the basic pairs.

Let $X \subseteq \{(\sigma_i, [p_i, q_i])\}_{i\le n}$. $X$ need not be consistent, i.e. there may be $i, j \le n$ such that $\sigma_i$ and $\sigma_j$ have a joint extension in $R(k)$, but $[p_i, q_i] \cap [p_j, q_j] = \emptyset$. We will employ the construction behind the lifting theorem in Normann [12] to modify $X$ such that it becomes consistent. So let $X$ be as above.

Let $\mathrm{Mod}(n, X)$ be the set of $(\sigma_j, [p_j, q_j])$ such that for some $m \le n$

- $j \le n$
- $(\sigma_m, [p_m, q_m]) \in X$
- $\sigma_m \sqsubseteq \sigma_j$

- whenever $i < m$ and $[p_i, q_i] \cap [p_j, q_j] = \emptyset$ then $\sigma_i$ and $\sigma_j$ are inconsistent (as elements in $R(k)$).

CLAIM. $\mathrm{Mod}(n, X)$ is consistent.

PROOF. Let $(\sigma_{j_i}, [p_{j_i}, q_{j_i}]) \in \mathrm{Mod}(n, X)$ for $i = 1, 2$ and let $m_1 \leq n$ and $m_2 \leq n$ witness that these basic pairs are in $\mathrm{Mod}(n, X)$. We may assume that $m_1 \leq m_2$.

If $m_1 = m_2 = m$ we have that $[p_m, q_m] \subseteq [p_{j_1}, q_{j_1}] \cap [p_{j_2}, q_{j_2}]$, so there is nothing to worry about.

If $m_1 < m_2$ and $[p_{j_1}, q_{j_1}] \cap [p_{j_2}, q_{j_2}] = \emptyset$, then in particular $[p_{m_1}, q_{m_1}] \cap [p_{j_2}, q_{j_2}] = \emptyset$, and then $\sigma_{j_2}$ is inconsistent with $\sigma_{m_1}$. It follows that $\sigma_{j_2}$ is inconsistent with $\sigma_{j_1}$.

This ends the proof of the claim. ⊣

Using Proposition 3 we let $\alpha_{n,X}$ be an extension of $\mathrm{Mod}(n, X)$ to an element in $\bar{R}(k+1)$.

Let $\Pi$ be the set of pairs $(i, j)$ such that $i < j$ and such that $\sigma_i$ and $\sigma_j$ have a joint extension in $\bar{A}$ while $[p_i, q_i] \cap [p_j, q_j] = \emptyset$. Let $\Pi_n$ be the set of pairs $(i, j) \in \Pi$ with $j \leq n$.

Let $(i, j) \in \Pi$. Given $f$, it is impossible that $f$ extends both $(\sigma_i, [p_i, q_i])$ and $(\sigma_j, [p_j, q_j])$, but we cannot, in a continuous way, give an absolute preference to one of them. We will let $f$ induce a probability measure on the set $\Delta_n$ of preference maps defined on $\Pi_n$. From each $\delta \in \Delta_n$, we will define the set $X_\delta$ being the approximation to some function suggested by $\delta$. Finally we will let $f_n$ be the weighted sum of the corresponding total objects $\alpha_{n,X_\delta}$.

It is about time to be more precise.

Let $\Delta_n$ be the set of maps $\delta$ defined on $\Pi_n$ such that $\delta(i, j) \in \{i, j\}$.

For each $\delta \in \Delta_n$, we let $X_\delta$ be the set of basic pairs $(\sigma_i, [p_i, q_i])$ such that $i \leq n$ and such that $\delta(i, j) = i$ whenever $(i, j) \in \Pi_n$ and such that $\delta(j, i) = i$ whenever $(j, i) \in \Pi_n$. Thus $X_\delta$ is the set of $(\sigma_i, [p_i, q_i])$ such that $i$ is always preferred by $\delta$ when this is an option.

Now, if $(i, j) \in \Pi_n$, let $x_{i,j} \in \bar{A}$ be a joint extension of $\sigma_i$ and $\sigma_j$, and let $y_{i,j} = \rho_k(x_{i,j})$.

Let $g_{i,j} : \mathbb{R} \to [0, 1]$ be continuous such that $g_{i,j}^{-1}(0) = [p_i, q_i]$ and $g_{i,j}^{-1}(1) = [p_j, q_j]$.

Given $f$, let the probability of $\delta(i, j) = j$ be $g(f(y_{i,j}))$ and let the probability of $\delta(i, j) = i$ be $1 - g(f(y_{i,j}))$.

We may view $\Delta_n$ as a product

$$\Delta_n = \prod_{(i,j) \in \Pi_n} \{i, j\}$$

so $f$ induces the product probability measure $\mu_{f,n}$ on $\Delta_n$.

Finally, we let

$$f_n = \sum_{\delta \in \Delta_n} \mu_{f,n} \cdot \alpha_{n,X_\delta}.$$

Note that the only element of the construction that depends on $f$ is the probability measure $\mu_{f,n}$. By construction, $\mu_{f,n}$ depends in a continuous way on $f$, and thus $f_n$ depends in a continuous way on $f$. Actually, the construction can be seen as carried out for $f \in R(k) \to R(0)$ such that

$$\forall x \in \bar{R}(k)(\rho_k(x) \in A \Rightarrow \rho_0(f(x)) \in \mathbb{R})$$

and then we construct $f_n \in \bar{R}(k+1)$ continuously in $f$.

It remains to show the claim on sequential continuity.

Any convergent sequence in $Ct_{\mathbb{R}}(k)$ will be the $\rho_k$-image of a convergent sequence in $\bar{R}(k)$, so let $x = \lim_{n\to\infty} x_n$ be a convergent sequence from $\bar{R}(k)$ with limit in $\bar{A}$.

Let $\varepsilon > 0$ be given.

Let $(\sigma, [p,q]) \sqsubseteq f$ be such that $q - p < \varepsilon$ and $\sigma \sqsubseteq x$.

For some $i$, $(\sigma, [p,q]) = (\sigma_i, [p_i, q_i])$. Let this $i$ be fixed for the rest of the proof.

CLAIM. Let $n \geq i$.

$$\mu_{f,n}(\{\delta \in \Delta_n \mid (\sigma, [p,q]) \in X_\delta\}) = 1.$$

PROOF. If $(\sigma, [p,q]) \notin X_\delta$, there must be some $j \leq n$ such that $(i,j) \in \Pi$ (or $(j,i) \in \Pi$, but we will only consider the first case) and $\delta(i,j) = j$.

But then $\mu_{f,n}(\{\delta\})$ is a product where one factor is 0, since $f(x_{i,j}) \in [p_i, q_i]$. This proves the claim. ⊣

The modification of $X_\delta$ to $\text{mod}(n, X_\delta)$ is like the construction used to prove that a continuous function from $A$ to $\mathbb{R}$ can be realised as a partial continuous function from $R(k)$ to $\mathbb{R}$. The rest of this proof is an adjustment of that argument:

Let $j < i$ be such that $[p_i, q_i] \cap [p_j, q_j] = \emptyset$.

Then $x$ must be inconsistent with $\sigma_j$, since otherwise $x \sqcup \sigma_j$ is a common extension $y$ of $\sigma_i$ and $\sigma_j$ with $\rho_k(x) = \rho_k(y) \in A$.

Then there is an approximation $\tau$ to $x$ inconsistent with $\sigma_j$. We may chose $\tau$ such that $\tau$ is inconsistent with all relevant $\sigma_j$ for $j < i$.

For some $m_0$, $(\tau, [p,q]) = (\sigma_{m_0}, [p_{m_0}, q_{m_0}])$, and for some $m_1$, $n \geq m_1 \Rightarrow \tau \sqsubseteq x_n$.

Choose $n \geq \max\{i, m_0, m_1\}$ and $\delta \in \Delta_n$ such that $\mu_{f,n}(\{\delta\}) > 0$.

Then we see that $(\tau, [p,q]) \in \text{mod}(n, X_\delta)$.

$f_n$ will be the weighted sum of functions $\alpha_{n,X_\delta}$ each of them sending $x_n$ into $[p,q]$, so $f_n(x_n) \in [p,q]$.

This shows that $f(x) = \lim_{n\to\infty} f_n(x_n)$, and the proof is complete. ⊣

Extending a well known terminology from metric spaces, we say that $f$ is the *equicontinuous and pointwise limit* of $\{f_n\}_{n\in\mathbb{N}}$ on $A$ if whenever $x = \lim_{n\to\infty} x_n$ on $A$ then $f(x) = \lim_{n\to\infty} f_n(x_n)$.

Then we have

COROLLARY 1. *Let $A \subseteq Ct_{\mathbb{R}}(k)$ and let $f : A \to Ct_{\mathbb{R}}(m)$ be continuous. Then there are functions $f_n : Ct_{\mathbb{R}}(k) \to Ct_{\mathbb{R}}(m)$ uniformly continuous in $f$, such that $f = \lim_{n\to\infty} f_n$ pointwise and equicontinuously on $A$.*

PROOF. Apply the approximation lemma to $A \times Ct_{\mathbb{R}}(m-1)$ in case $m > 0$. ⊣

## §4. Reducibility.

**4.1. Reductions.** Let $N(0) = \mathbb{N}_\perp$ and let $N(\sigma \to \tau) = N(\sigma) \to N(\tau)$ in the category of algebraic domains. We may construct the Kleene-Kreisel continuous functionals $Ct_{\mathbb{N}}(\sigma)$ of type $\sigma$ over the natural numbers as the extensional collapse of the hereditarily total functionals in this hierarchy, in analogy with the construction of $Ct_{\mathbb{R}}(\sigma)$. The corresponding sets $\bar{N}(\sigma)$ will be complete $\Pi^1_k$ when $k > 0$ and the type rank of $\sigma$ is $k+1$. This fact, and the constructions behind it, turned out, together with the density theorem, to be powerful tools while investigating the Kleene-Kreisel continuous functionals. In this section we will develop the analogue machinery for the $Ct_{\mathbb{R}}(\sigma)$-hierarchy.

Our definition of reduction has its root in realisability semantics, in particular in Kreisel's "constructive interpretation" of statements of analysis, i.e. of second order number theory, see Kreisel [9], and the related representation theorem for analytical (second order definable) subsets of $\mathbb{N}^{\mathbb{N}}$.

If $A \subseteq \mathbb{N}^{\mathbb{N}}$ is analytical, we may find a pure type $k$ and a representation

$$x \in A \Leftrightarrow \forall F \in Ct_{\mathbb{N}}(k) \exists n R(x, F, n)$$

where $R$ is primitive recursive. Then $x \in A$ if and only if

$$G_x = \lambda F \mu n R(x, F, n)$$

is total.

$G_x$ is a reduction of $A$ to totality of objects of type $k+1$ in such a way that lack of totality actually means that the value $\infty$ is obtained for some arguments. It is this kind of reductions we will generalise to the $Ct_{\mathbb{R}}$-hierarchy.

DEFINITION 7. Let $\mathbb{R}^+ = [0, \infty)$, i.e. the set of non-negative reals. Let $R_0^+$ consist of all closed non-empty intervals with non-negative rational endpoints, including the unbounded ones, and let $R_0^+$ be ordered by reverse inclusion.

The ideals in $R_0^+$ form an algebraic domain $R^+$.

We let $R^+(0) = R^+$ and we let $R^+(k+1) = R(k) \to R^+$.

Some ideals in $R^+$ will contain a bounded interval, and then there will be a corresponding ideal in $R(0)$. By abuse of notation, we will consider these ideals to be equal, and we let $\bar{R}^+ = R^+ \cap \bar{R}(0)$.

Moreover, there will be an ideal generated by $\{[n, \infty) \mid n \in \mathbb{N}\}$. We will denote this ideal by $\infty$ and let $\bar{R}^+_\infty = \bar{R}^+ \cup \{\infty\}$.

We let $x \in R^+(k+1)$ be *total* if $x(y) \in \bar{R}^+$ whenever $y \in \bar{R}(k)$ and we let $x \in R^+(k+1)$ be *weakly total* if $x(y) \in \bar{R}^+_\infty$ whenever $y \in \bar{R}(k)$.

DEFINITION 8. Let $A \subseteq Ct_{\mathbb{R}}(k)$ and let $\phi : R(k) \to R^+(k'+1)$ be continuous. We call $\phi$ a *reduction* if

- $\phi(x)$ is weakly total for each $x \in \bar{R}(k)$.
- For each $x \in \bar{R}(k)$ we have that $\rho_k(x) \in A \Leftrightarrow \phi(x)$ is total.

We then say that $A$ is *reducible* to $Ct_{\mathbb{R}}(k'+1)$.

LEMMA 5. *If $A$ is reducible to $Ct_{\mathbb{R}}(k'+1)$, then $A$ is reducible to $Ct_{\mathbb{R}}(k''+1)$ whenever $k' \le k''$.*

PROOF. Recall the projections $\Phi_{k'',k'}$ from Section 2.3. If $\phi$ is a reduction of $A$ to $Ct_{\mathbb{R}}(k'+1)$, then

$$\psi(x) = \lambda y \in R(k'').\phi(x)(\Phi_{k'',k'}(y))$$

is a reduction of $A$ to $Ct_{\mathbb{R}}(k''+1)$. ⊣

LEMMA 6. *If $A \subseteq Ct_{\mathbb{R}}(k)$ and $B \subseteq Ct_{\mathbb{R}}(k)$ are reducible to $Ct_{\mathbb{R}}(k'+1)$, then $A \cap B$ is reducible to $Ct_{\mathbb{R}}(k'+1)$.*

PROOF. Let $\phi_A$ and $\phi_B$ be the reductions.

Then $\psi(x) = \phi_A(x) + \phi_B(x)$ is a reduction of $A \cap B$.

For the rest of this paper, let $\{\xi^k_n\}_{n \in \mathbb{N}}$ be an effective enumeration of the dense subset of $\bar{R}(k)$ obtained from the proof of the density theorem. ⊣

DEFINITION 9. Let $f \in Ct_{\mathbb{R}}(k+1)$. The *trace* of $f$ is the function $h_f \in \mathbb{R}^{\mathbb{N}}$ defined by

$$h_f(n) = f(\xi^k_n).$$

$\mathbb{R}^{\mathbb{N}}$ is a metric space. We will use a bounded metric $d$ on $\mathbb{R}^{\mathbb{N}}$, e.g. the product metric induced by $\min\{1, |x-y|\}$ on $\mathbb{R}$.

Via the trace, this metric induce a metric for a weak topology on $Ct_{\mathbb{R}}(k+1)$.

The trace map $x \mapsto h_x$ is clearly 1-1 on $Ct_{\mathbb{R}}(k+1)$, since a continuous function is determined by its values on a dense set. Of course, $h_x$ is also well defined for weakly total $x \in R^+(k+1)$, then $h_x \in [0, \infty]^{\mathbb{N}}$. For technical reasons, because we want to perform standard algebraic operations on $h_x$ also in this situation, we will consider weakly total $x$ where $h_x \in \mathbb{R}^{\mathbb{N}}$:

DEFINITION 10. $x \in R^+(k+1)$ is *adequate* if $x$ is weakly total and $x(\xi^k_n) \in \bar{R}(0)$ for each $n \in \mathbb{N}$.

LEMMA 7. *There is a continuous map $\Xi_k : R^+(k+1) \to R^+(k+1)$ such that*

- *If $x$ is weakly total, then $\Xi_k(x)$ is adequate.*
- *$x$ is total if and only if $\Xi_k(x)$ is total.*

PROOF. We will give separate proofs for $k = 0$ and $k > 0$.

$k = 0$: $\xi_n^0 \in \mathbb{Q}$ for each $n \in \mathbb{N}$. Let $a \in [0, 1]$ be irrational. Let $x \in R(0) \to R^+(0)$ and $y \in R(0)$.

We let $\Xi_0(x)(y)$ be defined as follows: Let $y_0$ be the maximal integer with $y_0 \leq y$, and let $z = y - y_0$. (If the data is not sufficiently accurate to identify $y_0$, the outcome will be $\bot$.)

If $z \leq a$, let

$$\Xi_0(x)(y) = \min\left\{\frac{a}{a - z}, \sup\{x(v) \mid y_0 \leq v \leq y\}\right\}.$$

If $a \leq z$, let

$$\Xi_0(x)(y) = \min\left\{\frac{1 - a}{z - a}, \sup\{x(v) \mid y \leq v \leq y_0 + 1\}\right\}.$$

It is easy to verify that $\Xi_0$ is continuous and satisfies the required properties.

We have formulated the construction as if each input is total, assuming that it is clear what to do when a subconstruction based on partial input gives a partial output.

Now let $k > 0$. We will use that $Ct_{\mathbb{R}}(k) \times Ct_{\mathbb{R}}(k)$ is homeomorphic to a retract of $Ct_{\mathbb{R}}(k)$ via $\phi : R(k) \times R(k) \to R(k)$ and $(\psi_0, \psi_1) : R(k) \to R(k) \times R(k)$. Let $y_0 \in \bar{R}(k)$ be such that $y_0$ is inconsistent with $\psi_0(\xi_n^k)$ for each $n$. Let

$$\Xi_k(x)(y) = \min\left\{\frac{1}{d(h_{y_0}, h_{\psi_0(y)})}, x(\psi_1(y))\right\}.$$

The general comment about the construction for $k = 0$ is still valid. ⊣

DEFINITION 11. A reduction $\phi \in R(k + 1 \to k' + 1)$ is *adequate* if $\phi(x)$ is adequate for each $x \in \bar{R}(k + 1)$.

By the previous lemma, each reduction of a set $A$ may be transformed into an adequate reduction.

LEMMA 8. *If $A \subseteq Ct_{\mathbb{R}}(k)$ can be reduced to $Ct_{\mathbb{R}}(k' + 1)$, then $Ct_{\mathbb{R}}(k) \setminus A$ can be reduced to $Ct_{\mathbb{R}}(k' + 2)$.*

PROOF. Let $\phi : R(k) \to R(k' + 1)$ be an adequate reduction of $A$. Let

$$\psi(x) = \lambda y \in R(k' + 1)\frac{1}{d(h_y, h_{\phi(x)})}.$$

Let $x \in \bar{R}(k)$. If $\rho_k(x) \in A$, then $\phi(x)$ is total. Let $y = \phi(x)$. Then $\psi(x)(y) = \infty$.

If $\rho_k(x) \notin A$, then $h_{\phi(x)}$ is total since $\phi(x)$ is adequate. Since $\phi(x)$ is not total, it follows that $h_{\phi(x)} \neq h_y$ for all total $y$ (since otherwise $y(z) = \infty$ whenever $\phi(x)(z) = \infty$ for total $z$). It follows that $\psi(x)$ is total. ⊣

REMARK 1. $\frac{1}{d(h_y,h_{\phi(x)})}$ is a substitute for $\mu n(h_y(n) \neq h_{\phi(x)}(n))$ which is used in a similar context to represent negation in Kreisel's representation theorem. So this construction represents functional interpretation of negation.

LEMMA 9. *Inequality on $Ct_{\mathbb{R}}(k)$ is reducible to $Ct_{\mathbb{R}}(0)$.*

PROOF.

$$x = y \Leftrightarrow h_x = h_y \Leftrightarrow \frac{1}{d(h_x, h_y)} = \infty. \qquad \dashv$$

The concept of reduction may, as above, be extended to several variables. So $A \subseteq Ct_{\mathbb{R}}(k_1) \times \cdots \times Ct_{\mathbb{R}}(k_n)$ is reducible to $Ct_{\mathbb{R}}(k)$ if there is a continuous map

$$\phi : R(k_1) \times \cdots \times R(k_n) \to R^+(k)$$

such that whenever $(x_1, \ldots, x_n) \in \bar{R}(k_1) \times \cdots \times \bar{R}(k_n)$ we have that

- $\phi(x_1, \ldots, x_n)$ is weakly total
- $\phi(x_1, \ldots, x_n)$ is total $\Leftrightarrow (\rho_{k_1}(x_1), \ldots, \rho_{k_n}(x_n)) \in A$.

We then observe

LEMMA 10. *If $A \subseteq Ct_{\mathbb{R}}(k_1) \times \cdots \times Ct_{\mathbb{R}}(k_n) \times Ct_{\mathbb{R}}(k_{n+1})$ is reducible, and*

$$(x_1, \ldots, x_n) \in B \Leftrightarrow \forall y \in Ct_{\mathbb{R}}(k_{n+1})(x_1, \ldots, x_n, y) \in A,$$

*then $B$ is reducible.*

EXAMPLE 1. The set of linear operators in $Ct_{\mathbb{R}}(2)$ is reducible.

**4.2. Function spaces.** We will now see that the set of reducible sets is, up to homeomorphism, closed under the formation of function spaces.

THEOREM 3. *Let $A \subseteq Ct_{\mathbb{R}}(k)$ be reducible via $\phi$ to $Ct_{\mathbb{R}}(k'+1)$. Let $B \subseteq Ct_{\mathbb{R}}(k'')$. Then $A \to B$ with the sequential topology is homeomorphic to a set*

$$D \subseteq Ct_{\mathbb{R}}(\max\{k+1, k'+1, k''\}).$$

*Moreover, if also $B$ is reducible, then $D$ is reducible.*

PROOF. We write $\xi_n$ instead of $\xi_n^{k'}$. Assume that $\phi$ is adequate. Then $\phi(x)(\xi_n) \in \bar{R}(0)$ for all $n$ and all $x \in \bar{R}(k)$.

Let $\hat{f} \in \bar{R}(A \to B)$ (defined in Section 2.2).

By the Approximation Lemma there is, continuously in $\hat{f}$, a sequence $\{\hat{f}_n\}_{n \in \mathbb{N}}$ such that $f = \lim_{n\to\infty} f_n$ pointwise and equicontinuously on $A$, where $f = \rho_{A\to B}(\hat{f})$ and $f_n = \rho_{k\to k''}(\hat{f}_n)$. Now, let $x \in \bar{R}(k)$ and $y \in \bar{R}(k'+1)$ be given. We define the measure $\mu_{x,y}$ on $\mathbb{N}$ as follows:

If $\sum_{i \leq n} |\phi(x)(\xi_i) - y(\xi_i)| < 1$, let

$$\mu_{x,y}(n) = |\phi(x)(\xi_n) - y(\xi_n)|.$$

If $\sum_{i \leq n} |\phi(x)(\xi_n) - y(\xi_i)| \geq 1$, let

$$\mu_{x,y}(n+1) = 0.$$

If $\sum_{i<n} |\phi(x)(\xi_i) - y(\xi_i)| < 1$, but $\sum_{i\leq n} |\phi(x)(\xi_i) - y(\xi_i)| \geq 1$, let

$$\mu_{x,y}(n) = 1 - \sum_{i<n} |\phi(x)(\xi_i) - y(\xi_i)|.$$

Let $\psi(\hat{f})(x,y) = \hat{f}(x)$ if $(\forall i \in \mathbb{N})(\phi(x)(\xi_i) = y(\xi_i))$, and let $\psi(\hat{f})(x,y) = \sum_{i=0}^{\infty} \mu_{x,y}(i) \cdot \hat{f}_i(x)$ otherwise.

Claim 1. $\psi$ is total and continuous in $(\hat{f}, x, y) \in \bar{R}(A \to B) \times \bar{R}(k) \times \bar{R}(k'+1)$.

Proof. Let $z \in \bar{R}(k''-1)$. (We may ignore $z$ in this argument if $k'' = 0$.) First assume that $x$ and $y$ are such that for some $n$, $\phi(x)(\xi_n) \neq y(\xi_n)$. Then, for infinitely many $i$,

$$|\phi(x)(\xi_i) - y(\xi_i)| > \frac{|\phi(x)(\xi_n) - y(\xi_n)|}{2}$$

so

$$\sum_{n=0}^{\infty} |\phi(x)(\xi_n) - y(\xi_n)| = \infty.$$

Let $\varepsilon > 0$ be given and assume that $\varepsilon < 1$. Let $n_0$ be such that

$$\sum_{i=0}^{n_0} |\phi(x)(\xi_i) - y(\xi_i)| > 1.$$

Then

$$\psi(\hat{f})(x,y) = \sum_{i=0}^{n_0} \mu_{x,y}(i) \cdot \hat{f}_i(x) \in \mathbb{R}.$$

Thus $\psi(\hat{f})(x,y)$ is total in this case.

There will be approximations $\sigma_0$ and $\tau_0$ to $x$ and $y$ such that the *partial real*

$$\sum_{i=0}^{n_0} |\phi(\sigma_0)(\xi_i) - \tau_0(\xi_i)| > 1.$$

From now on, in this proof, we let $\hat{f}'$ range over $\bar{R}(A \to B)$, $x'$ over $\bar{R}(k)$, $y'$ over $\bar{R}(k'+1)$ and $z'$ over $\bar{R}(k''-1)$.

There is an approximation $\delta$ to $\hat{f}$, $\sigma_1$ to $x$ and $\pi_1$ to $z$ such that for all $i \leq n_0$, all $\hat{f}'$ extending $\delta$, all $x'$ extending $\sigma_1$ and all $z'$ extending $\pi_1$ we have that

$$|\hat{f}_i(x)(z) - \hat{f}'_i(x')(z')| < \frac{\varepsilon}{2}.$$

Let

$$M = 1 + \max\left(\{\hat{f}_i(x)(z) \mid i \leq n_0\} \cup \{\hat{f}(x)(z)\}\right).$$

Given two measures $\mu_1$ and $\mu_2$ on $\mathbb{N}$, we let the *symmetric difference* $\mu_1 \bigtriangleup \mu_2$

be defined by

$$(\mu_1 \triangle \mu_2)(i) = |\mu_1(i) - \mu_2(i)|.$$

Then $0 \leq (\mu_1 \triangle \mu_2)(\mathbb{N}) \leq 2$.

By continuity of the construction there is an approximation $\sigma_3$ to $x$ and an approximation $\tau_3$ to $y$ such that whenever $x'$ extends $\sigma_3$ and $y'$ extends $\tau_3$, then

$$(\mu_{x,y} \triangle \mu_{x',y'})(\mathbb{N}) < \frac{\varepsilon}{2M}.$$

It follows that if $\hat{f}'$ extends $\delta$, $x'$ extends $\sigma_1 \sqcup \sigma_2 \sqcup \sigma_3$, $y'$ extends $\tau_1 \sqcup \tau_3$ and $z'$ extends $\pi_1$, then

$$\begin{aligned}
&|\psi(\hat{f})(x,y)(z) - \psi(\hat{f}')(x',y')(z')| \\
&\quad = \left|\sum_{i=0}^{\infty} \mu_{x,y}(i) \cdot \hat{f}_i(x)(z) - \sum_{i=0}^{\infty} \mu_{x',y'}(i) \cdot \hat{f}'_i(x')(z')\right| \\
&\quad \leq \sum_{i=0}^{\infty} |\mu_{x,y}(i) \cdot \hat{f}_i(x)(z) - \mu_{x',y'}(i) \cdot \hat{f}'_i(x')(z')| \\
&\quad = \sum_{i=0}^{n_0} |\mu_{x,y}(i) \cdot \hat{f}_i(x)(z) - \mu_{x',y'}(i) \cdot \hat{f}'_i(x')(z')| \\
&\quad \leq (\mu_{x,y} \triangle \mu_{x',y'})(\mathbb{N}) \cdot \max\{|\hat{f}_i(x)(z) - \hat{f}'_i(x')(z')| \mid i \leq n_0\} \\
&\qquad + \max\{|\hat{f}_i(x)(z) - \hat{f}'_i(x')(z')| \mid i \leq n_0\} \\
&\quad \leq \frac{\varepsilon}{2} + \frac{\varepsilon}{2}.
\end{aligned}$$

This shows continuity in this case. Totality in this case is trivial.

Now assume that $\phi(x)(\xi_n) = y(\xi_n)$ for all $n$. Then $\rho_k(x) \in A$. Let $z \in \bar{R}(k'' - 1)$ (which we still may ignore if $k'' = 0$).

Let $\varepsilon > 0$ be given.

Let $n_1 \in \mathbb{N}$, $\delta_1 \sqsubseteq \hat{f}$, $\sigma_1 \sqsubseteq x$ and $\pi_1 \sqsubseteq z$ be such that if $n \geq n_1$, $\sigma_1 \sqsubseteq x'$, $\pi_1 \sqsubseteq z'$ and $\delta_1 \sqsubseteq f'$ then

$$|\hat{f}'_n(x')(z') - \hat{f}(x)(z)| < \frac{\varepsilon}{2}.$$

Let $M = \max\{\hat{f}_n(x), \hat{f}(x) \mid n \leq n_1\} + 1$.

Let $\sigma_2 \sqsubseteq x$ and $\tau_2 \sqsubseteq y$ be such that if $\sigma_2 \sqsubseteq x'$ and $\tau_2 \sqsubseteq y'$ then

$$\sum_{n \leq n_1} |\phi(x')(\xi_n) - y(\xi_n)| < \frac{\varepsilon}{4M}.$$

It follows that if $\delta_1 \sqsubseteq \hat{f}'$, $\sigma_1 \sqcup \sigma_2 \sqsubseteq x'$, $\tau_2 \sqsubseteq y'$ and $\pi_1 \sqsubseteq z'$ then

$$\begin{aligned}
&|\psi(\hat{f})(x,y)(z) - \psi(\hat{f}')(x',y')(z')| \\
&= \left| \hat{f}(x)(z) - \sum_{i=0}^{\infty} \mu_{x',y'}(i) \cdot \hat{f}_i'(x')(z') \right| \\
&\leq \sum_{i=0}^{\infty} \mu_{x',y'}(i) \cdot |\hat{f}(x)(z) - \hat{f}_i'(x')(z')| \\
&\leq \sum_{i=0}^{\infty} \mu_{x',y'}(i) \cdot |\hat{f}(x)(z) - \hat{f}_i'(x')(z')| \\
&= \sum_{i=0}^{n_1} \mu_{x',y'}(i) \cdot |\hat{f}(x)(z) - \hat{f}_i'(x')(z')| \\
&\quad + \sum_{i=n_1+1}^{\infty} \mu_{x',y'}(i) \cdot |\hat{f}(x)(z) - \hat{f}_i'(x')(z')| \\
&\leq \frac{\varepsilon}{4M} \cdot 2M + \frac{\varepsilon}{2} = \varepsilon.
\end{aligned}$$

This proves continuity in this case, and Claim 1 is proved. ⊣

From now on, let $\psi \in R((k \to k'') \to (k, k'+1 \to k''))$ be continuous and as constructed on $\bar{R}(A \to B) \times \bar{R}(k) \times \bar{R}(k'+1)$. Let

$$\tilde{C} = \{\lambda x \in R(k).\lambda y \in R(k'+1).\psi(\hat{f})(x,y) \mid f \in \bar{R}(A \to B)\}.$$

and let

$$C = \{\rho_{k,k'+1\to k''}(g) \mid g \in \tilde{C}\}.$$

Let $\bar{C}$ be the set of elements in $\bar{R}(k, k'+1 \to k'')$ equivalent to an element in $\tilde{C}$.

For $u \in \bar{C}$, let $\hat{\imath}_u(x) = u(x, \phi(x))$. Then $\hat{\imath}_u \in \bar{R}(A \to B)$. The function $u \mapsto \hat{\imath}_u$ is continuous, and whenever $\hat{f} \in \bar{R}(A \to B)$ we have that $\hat{\imath}_{\psi(\hat{f})}$ and $\hat{f}$ are consistent, so

$$\rho_{A\to B}(\hat{\imath}_{\psi(\hat{f})}) = \rho_{A\to B}(\hat{f}).$$

But then $C$ and $A \to B$ will be homeomorphic via the continuous maps represented by $\psi$ and $u \mapsto \hat{\imath}_u$.

$C$ is a set in $(Ct_{\mathbb{R}}(k) \times Ct_{\mathbb{R}}(k'+1)) \to Ct_{\mathbb{R}}(k'')$.

This set is homeomorphic to a retract of $Ct_{\mathbb{R}}(\max\{k+1, k'+2, k''\})$, and we let $D$ be the image of $C$ under this homeomorphism.

Now assume that $B$ can be reduced to $Ct_{\mathbb{R}}(k''')$ via $\phi'$. In order to prove that $D$ also is reducible, we show that $D$ is definable in the appropriate way. We will introduce some notation and some functions:

Let $n = \max\{k+1, k'+2, k''\}$.

Let

$$\Xi : ((R(k) \times R(k'+1)) \to R(k'')) \to R(n)$$

and

$$\Xi^d : R(n) \to ((R(k) \times R(k'+1)) \to R(k''))$$

be total such that $\Xi^d(\Xi(g)) \approx g$ for all total $g$ in $R(k, k'+1 \to k'')$.

Then

$$D = \{\rho_n(\Xi(g)) \mid \rho_{k,k'+1\to k''}(g) \in C\}.$$

For $h \in \bar{R}(n)$ we then have

(1) $$\rho_n(h) \in D \Leftrightarrow \rho_{k,k'+1\to k''}(\Xi^d(h)) \in C \wedge h \approx \Xi(\Xi^d(h)).$$

The second part of (1) is reducible. The first part of (1) is equivalent to

(2) $$(\exists f \in R(k) \to R(k''))(\rho_{A\to B}(f) \in A \to B \wedge (\Xi^d(h) \approx \psi(f))).$$

Let $g_h = \Xi^d(h)$. If there is a $f$ satisfying (2), then it can be recovered up to equivalence from $h$ as $f_h(x) = g_h(x, \phi(x))$, so (2) is equivalent to

(3) $$f_h \in \bar{R}(A \to B) \wedge g_h \approx \psi(f_h).$$

The second part of (3) is reducible. The first part of (3) is equivalent to

(4) $$\forall x \in \bar{R}(k)(\rho_k(x) \notin A \vee \phi'(f_h(x)) \in \bar{R}(k''')),$$

which is equivalent to

(5)

$$(\forall x \in \bar{R}(k''))(\forall y \in \bar{R}(k''))\exists m(\phi(x)(\xi^{k'}_m) \neq y(\xi^{k'}_m) \vee \phi'(f_h(x)) \in \bar{R}(k''')).$$

Using the approximation lemma, let the sequence $\{f_{h,n}\}_{n\in\mathbb{N}}$ be obtained by applying the construction in the proof of the lemma to $f_h$. Let $\mu^h_{x,y}$ be defined from $f_h$ as above. We let

$$\psi'(h)(x,y) = \sum_{i=0}^{\infty} \mu^h_{x,y}(i) \cdot \phi'(f_{h,i}(x))$$

if $\exists m(\phi(x)(\xi^{k'}_m) \neq y(\xi^{k'}_m))$, and we let

$$\psi'(h)(x,y) = \phi'(f_h(x))$$

otherwise.

We can show in analogy with previous arguments that $\psi'$ is continuous and is a reduction of

$$\{\rho_n(h) \mid h \in \bar{R}(n) \wedge f_h \in \bar{R}(A \to B)\}.$$

This ends the proof of the theorem. Though the constructions of the reductions are explicit, we have made no effort in bringing the type level down. ⊣

As mentioned before, there is a connection between our reductions and constructions of realizers, but we have made no deep exploration of this connection.

§5. **Polish spaces.** A Polish space is a separable topological space that admits a complete metric. We recommend Hoffmann-Jørgensen [6] or Kechris [8] as standard references on Polish spaces.

A useful characterisation is that the Polish spaces are exactly the topological spaces homeomorphic to $G_\delta$ subsets of $[0,1]^{\mathbb{N}}$. Moreover, a $G_\delta$ subspace of a Polish space will be Polish.

$[0,1]^{\mathbb{N}}$ is homeomorphic to a $G_\delta$ subspace of $\mathbb{R} \to \mathbb{R}$, which itself is Polish, so the Polish spaces may as well be characterised via the $G_\delta$ subspaces of $\mathbb{R} \to \mathbb{R}$.

LEMMA 11. *Let $A \subseteq Ct_{\mathbb{R}}(1)$.*
*Then $A$ is reducible to $Ct_{\mathbb{R}}(1)$ if and only if $A$ is $G_\delta$.*

PROOF. Let $A$ be $G_\delta$, i.e.

$$A = \bigcap_{n \in \mathbb{N}} A_n$$

where each $A_n$ is open. We may assume that $A_{n+1} \subseteq A_n$.

Let $B_n = (\mathbb{R} \to \mathbb{R}) \setminus A_n$. Consider the sequence $\{\frac{1}{d(f,B_n)}\}_{n \in \mathbb{N}}$ for $f \in \mathbb{R} \to \mathbb{R}$. If $f \in A$, this defines a sequence of reals, while if $f \notin A$ this sequence is $\infty$ from one $n$ on.

We code this sequence into a function $\phi(f)$ by

- $\phi(f)(x) = 0$ for $x \leq 0$.
- $\phi(f)(k \cdot \frac{\pi}{2}) = \frac{1}{d(f,B_n)}$.
- Between $k \cdot \frac{\pi}{2}$ and $(k+1) \cdot \frac{\pi}{2}$ the function $\phi(f)$ increases like the tangent function until it reaches the value $\frac{1}{d(f,B_{n+1})}$.

Then $A$ is reducible via $\phi$.

In order to prove the converse, let $A$ be reducible to $Ct_{\mathbb{R}}(1)$ via $\phi$. Then

$$f \in A \Leftrightarrow \forall x(\phi(f)(x) < \infty) \Leftrightarrow \forall n \exists m \forall x(|x| \leq n \Rightarrow |\phi(f)(x)| < m).$$

It is well known that the topology on $\mathbb{R} \to \mathbb{R}$ is the compact-open topology, so

$$Q_{n,m} = \{f \mid |x| \leq n \Rightarrow |\phi(f)(x)| < m\}$$

is open. The lemma follows. ⊣

We then have all the ingredients needed to prove

THEOREM 4. *Let $P_1, \ldots, P_n$ be Polish spaces.*

*Let $X$ be a higher type space obtained in the category of limit spaces by closing $\{P_1, \ldots, P_n\}$ under the formation of function spaces. Then $X$ is homeomorphic to a reducible subset of some $Ct_{\mathbb{R}}(k)$.*

REMARK 2. If we start with Polish algebras instead, we may also close the hierarchy under definable subspaces.

**Acknowledgment.** I am grateful to the anonymous referee for valuable comments on the exposition and for pointing out several typos in the original manuscript. In particular I am grateful for the suggestion of making the Approximation Lemma more visible.

REFERENCES

[1] S. ABRAMSKY and A. JUNG, *Domain theory*, ***Handbook of logic in computer science*** (S. Abramsky, D. M. Gabbay, and T. S. E. Maibaum, editors), vol. 3, Clarendon Press, 1994.

[2] A. BAUER, M. ESCARDÓ, and A. SIMPSON, *Comparing functional paradigms for exact real-number computation*, ***Proceedings ICALP 2002***, LNCS, vol. 2380, Springer, 2002, pp. 488–500.

[3] J. BLANCK, *Domain representability of metric spaces*, ***Annals of Pure and Applied Logic***, vol. 83 (1997), pp. 225–247.

[4] M. H. ESCARDÓ, *PCF extended with real numbers*, ***Theoretical Computer Science***, vol. 162 (1996), no. (1), pp. 79–115.

[5] GIERZ, HOFMANN, KEIMEL, LAWSON, MISLOVE, and SCOTT, ***Continuous lattices and domains***, Cambridge University Press, 2003.

[6] J. HOFFMANN-JØRGENSEN, ***The theory of analytic spaces***, Various Publication Series No. 10, Aarhus University, 1970.

[7] F. DE JAEGER, ***Calculabilité sur les réels***, Thesis, Paris VII, 2003.

[8] A. S. KECHRIS, ***Classical descriptive set theory***, Springer, New York, 1995.

[9] G. KREISEL, *Interpretation of analysis by means of functionals of finite type*, ***Constructivity in mathematics*** (A. Heyting, editor), North-Holland, (1959), pp. 101–128.

[10] G. LONGO and E. MOGGI, *The hereditary partial effective functionals and recursion theory in higher types*, ***The Journal of Symbolic Logic***, vol. 49 (1984), pp. 1319–1332.

[11] D. NORMANN, *External and internal algorithms on the continuous functionals*, ***Patras logic symposium*** (G. Metakides, editor), North-Holland, 1982, pp. 137–144.

[12] ———, *The continuous functionals of finite types over the reals*, ***Domains and processes*** (K. Keimel, G. Q. Zhang, Y. Liu, and Y. Chen, editors), Kluwer Academic Publishers, 2001, pp. 103–124.

[13] ———, *Hierarchies of total functionals over the reals*, ***Theoretical Computer Science***, vol. 316 (2004), pp. 137–151.

[14] ———, *Comparing hierarchies of total functionals*, in preparation.

[15] M. SCHRÖDER, *Admissible representations of limit spaces*, ***Computability and complexity in analysis*** (J. Blanck, V. Brattka, P. Hertling, and K. Weihrauch, editors), vol. 237, Informatik Berichte, 2000, pp. 369–388.

[16] V. STOLTENBERG-HANSEN, I. LINDSTRÖM, and E. R. GRIFFOR, ***Mathematical theory of domains***, Cambridge Tracts in Theoretical Computer Science, vol. 22, Cambridge University Press, 1994.

[17] J. V. Tucker and J. I. Zucker, *Abstract versus concrete models of computation on partial metric algebras*, *ACM Transactions on Computational Logic*, vol. 5 (2004), pp. 611–668.

[18] K. Weihrauch, ***Computable analysis***, Texts in Theoretical Computer Science, Springer Verlag, Berlin, 2000.

DEPARTMENT OF MATHEMATICS
THE UNIVERSITY OF OSLO
P.O. BOX 1053 BLINDERN
N-0316 OSLO, NORWAY
*E-mail*: dnormann@math.uio.no

# PREDICATIVITY PROBLEMS IN POINT-FREE TOPOLOGY

ERIK PALMGREN†

**Abstract.** When developing topology in a constructive setting, it has turned out to be advantageous to start out with some representation of the frame of open sets and then derive points and point-functions from this. This is known as the point-free approach to topology. The most prominent of such approaches, locale theory, uses impredicative constructions based on complete lattices which are not available in generalised predicative systems such as Martin-Löf type theories or constructive set theory. The formal topology of Martin-Löf and Sambin was designed to solve the basic predicativity problems in locale theory. Further predicativity problems arise from point related notions and fundamental topological constructions. In this article some such problems are considered, and some solutions are given.

**§1. Introduction.** Predicativity has been an important issue in the foundations of mathematics and proof theory since the beginning of the last century. The notion received a particular technical and philosophically well-motivated meaning in the proof theory of ramified second order arithmetic, as analysed by Feferman and Schütte in the 1960s. On the other hand, in the foundations for constructive mathematics, which we shall be concerned with here, the notion has come to encompass a wide class of theories whose basic objects and logic have a clear inductive structure. Such theories have been called *generalised predicative* by Aczel. Modern type-theoretic frameworks make it possible to attempt a precise criterion for a predicative system, namely that the introduction rule of each type should be consistent with a recursion or elimination principle. We refer to Palmgren [1998], Dybjer [2000] and Dybjer and Setzer [2003] for further discussions.

Russell's ramified theory of types is the first example of a predicative theory intended for the formalisation of mathematics. It suffers from a well-known defect when it comes to the real numbers. If we consider a real number as a symmetric Dedekind cut, it is given by a pair of predicates $(L, U)$, of a certain degree of the ramification, on the rational numbers. This has the consequence that the real numbers fall into a hierarchy

$$\mathbb{R}_1 \subseteq \mathbb{R}_2 \subseteq \mathbb{R}_3 \subseteq \cdots \subseteq \mathbb{R}_n \subseteq \cdots$$

†The author is supported by a grant from the Swedish Research Council (VR).

**Logic Colloquium '03**
Edited by V. Stoltenberg-Hansen and J. Väänänen
Lecture Notes in Logic, 24

of number systems, which fail to be order-complete. Instead of giving up the idea of predicativity, one may settle for weaker completeness properties. Weyl [1918] investigated the real numbers of the first level. Recent investigations of similar restrictions on set existence have been carried out in reverse mathematics (Simpson [1999]). These pertain to the classical setting. The situation is quite different in the constructive setting.

Constructive set theory CZF (Aczel [1986]) and type theory (Martin-Löf [1984]) and Feferman's Explicit Mathematics are three main formal systems built on predicative principles. Of these the first two have been used extensively for the formalisation of constructive mathematics. Using the logical principle of dependent choice, which is a theorem of the type theory and is true in the standard model of CZF, one may construct real numbers as Cauchy sequences of rational numbers. These real numbers are then Cauchy-complete, but order-complete only in a restricted sense (cf. Bishop and Bridges [1985]). Constructively, not every subset has a characteristic function, so even with dependent choice these completeness properties are not equivalent, unlike the classical setting. Topology in constructive mathematics following Bishop [1967] has essentially been restricted to metric spaces. More general topologies can be treated using non-classical or non-recursive axioms of Brouwer, but the results may then contradict classical or recursive mathematics, which is unsatisfactory for many reasons. The point-free topologies in the form of locale theory (Johnstone [1982]) and formal topology (Sambin [1987]) give a satisfactory treatment of general topology, which is also consistent with classical and effective interpretations. However, there are again many problems of predicativity. It is the purpose of this paper to survey some solutions and remaining problems in the point-free approach. We also present some new main results, Theorem 4.5 and 5.2. Some basic references in formal topology are Sambin [1987], [2003], Aczel [2002], Coquand [1995], Coquand, Sambin, Smith, and Valentini [2003], Negri and Soravia [1999] and Sigstam and Stoltenberg-Hansen [1997].

**§2. Sets in type theory.** The established way of understanding a set in Bishop's constructive mathematics is as a type, or "preset", together with an equivalence relation $A = (\underline{A}, =_A)$. A function $f$ from the set $A$ to the set $B$ is an operation from $\underline{A}$ to $\underline{B}$ respecting the equivalence relations. The resulting category of sets and functions, if built on a suitable type theory, can be seen to have many properties in common with toposes, such as existence of products, disjoint sums and quotients. We shall here assume that our type theory is Martin-Löf type theory with a cumulative sequence of universes (Martin-Löf 1998/1972)

$$\mathbb{U}_1 \subseteq \mathbb{U}_2 \subseteq \mathbb{U}_3 \subseteq \cdots \subseteq \mathbb{U}_n \subseteq \cdots$$

so that each type is contained in some $\mathbb{U}_n$, where $n$ is an index external to the theory. Moreover, for each $n$: $\mathbb{U}_n \in \mathbb{U}_{n+1}$ and $\mathbb{U}_n$ is closed under the type formers for dependent products ($\Pi$) and dependent sums ($\Sigma$), binary sums $(+)$, identity types (Id) and well-founded trees ($W$). However, we do not assume the universe to be closed under full power sets. Instead, because of the hierarchical structure of universes, and hence the truth-values, there will be a ramification structure of subsets, which we now describe. A set $A$ is an *$(m, n)$-set* if $\underline{A} \in \mathbb{U}_m$ and $=_A$ is a propositional function $\underline{A} \times \underline{A} \longrightarrow \mathbb{U}_n$. If $m = n$, then $A$ is called an *$n$-set*, and if $m = n + 1$, $A$ is an *$n$-class*. A set $S$ together with an injection $i : S \longrightarrow A$ is a *subset* of $A$. The subset $(S, i)$ is *included* in another subset $(T, j)$, symbolically $(S, i) \subseteq (T, j)$, if there is a function $f : S \longrightarrow T$ with $j \circ f = i$. If $S$ is an $(m, n)$-set, then $(S, i)$ is an $(m, n)$-subset of $A$. The derived notions *$m$-subset* and *$m$-subclass* are defined in the obvious way. Note that if $A$ is an $m$-class, $X$ is an $m$-set, and $f : X \longrightarrow A$ is a function, then the image of $f$ in $A$, given by $\mathrm{Im}(f) = (A, =')$, where $x =' y$ iff $f(x) = f(y)$, and where the injection is $f$, is an $m$-subset. This is the crucial replacement property of classes relative to sets. The type of $k$-sets $S = (\underline{S}, =_S)$ belongs to $\mathbb{U}_{k+1}$ and is denoted $\mathrm{Sets}_k$. Let $A$ be an $(m, n)$-set. The set $\mathcal{P}_k(A)$ of $k$-subsets of $A$ is given by the type

$$(\Sigma S \in \mathrm{Sets}_k)(\Sigma i : \underline{S} \longrightarrow \underline{A})[i : S \longrightarrow A \text{ is an injection}]$$

which is in $\mathbb{U}_{\max(k+1,m,n)}$, and the equivalence relation $(S, i, p) = (S', i', p')$ is given by the conjunction $((S, i) \subseteq (S', i')) \wedge ((S', i') \subseteq (S, i))$ which is a proposition in $\mathbb{U}_{\max(k,n)}$. The set $\mathcal{P}_k(A)$ is thus a $(\max(k+1, m, n), \max(k, n))$-set. In particular, if $A$ is a $k$-set, or a $k$-class, $\mathcal{P}_k(A)$ is a $k$-class.

Following Aczel [2002] and Gambino [2002] we define a *partially ordered $k$-class* $(X, \leq)$ to consist of a $k$-class $X$ and a $k$-subclass $\leq$ of $X \times X$, where $\leq$ is a partial order on $X$. Define $(X, \leq)$ to be *complete* if every $k$-subset of $X$ has a supremum.

EXAMPLE 2.1. If $A$ is a $k$-set, or a $k$-class, then $(\mathcal{P}_k(A), \subseteq)$ is a complete partially ordered $k$-class.

A category $\mathbb{C}$ is *locally $n$-small*, if the morphisms between any two fixed objects form an $n$-set.

We remark that there is a predicative version of topos (Moerdijk and Palmgren [2002]) which is an abstraction of the notion of sets in the mentioned type theory. In actual development of mathematics within type theory it seems rarely to be necessary to go beyond the first few levels of sets.

§3. **Formal topologies.** The fundamentals of locale theory have been developed predicatively in CZF (Aczel [2002], Gambino [2002]). We could use the notions of $k$-set and $k$-class introduced in the previous section to translate this development into type theory. However, we shall here stay closer to the usual presentations of formal topologies following Martin-Löf and Sambin, in that frames are not used explicitly. We use the term *formal space* almost interchangeably with *formal topology*, but prefer the former term when points have a significant role.

DEFINITION 3.1. Let $(X, \leq)$ be a pre-ordered $k$-set, whose elements will be called *formal neighbourhoods*. Then for $k$-subsets $U$ and $V$ of $X$ define their *formal intersection*, to be the $k$-subset

$$U \wedge V = \{x \in X : (\exists u \in U)(\exists v \in V)\, x \leq u \,\&\, x \leq v\}.$$

A *$k$-formal topology* $X = (X, \leq, \mathcal{A})$ arises when $X$ is equipped with a *covering operator* $\mathcal{A} : \mathcal{P}_k(X) \longrightarrow \mathcal{P}_k(X)$ which is monotone, inflationary, idempotent and which satisfies

$$\mathcal{A}(U) \cap \mathcal{A}(V) = \mathcal{A}(U \wedge V).$$

We often write $a \lhd_{\mathcal{A}} U$ for $a \in \mathcal{A}(U)$, to be read *$a$ is covered by $U$*. A subset $U$ of $X$ is *$\leq$-filtering*, if for any $x, y \in U$ there is some $z \in U$ with $z \leq x$ and $z \leq y$. A *$k$-point* $\alpha$ of the topology $X$ is an inhabited $k$-subset of $X$ which is $\leq$-filtering and such that $U \cap \alpha$ is inhabited whenever $U \in \mathcal{P}_k(X)$ is such that $\mathcal{A}(U) \cap \alpha$ is inhabited. For $k \geq m$, a $k$-formal topology $X$ is said to be *$m$-set presented,* if there is a family $I(a)$ $(a \in X)$ of $m$-sets, and a family of $m$-subsets $C(a, i)$, $(a \in X, i \in I(a))$ such that

$$\mathcal{A}(U) = \{x \in X : (\exists i \in I(x))\, C(x, i) \subseteq U\}.$$

Let $X$ be a $k$-formal topology. The $k$-class of *saturated* $k$-subsets is

$$\mathrm{Sat}_k(X) =_{\mathrm{def}} \{V \in \mathcal{P}_k(X) : \mathcal{A}(V) = V\}.$$

Noting that the supremum of a family of saturated subsets is

$$\mathcal{A}(\cup_{i \in I} V_i),$$

it is straightforward to prove

THEOREM 3.2. *For each $k$-formal topology,* $(\mathrm{Sat}_k(X), \subseteq)$ *is a complete $k$-frame, i.e. complete lattice with $k$-set indexed infinite distributive law.*

Let $X = (X, \leq, \mathcal{A})$ be a $k$-formal topology. Define a $(k+1)$-formal topology $X^+ = (X, \leq, \mathcal{A}^+)$ by extending $\mathcal{A}$ to an operator $\mathcal{A}^+$ on $\mathcal{P}_{k+1}(X)$ as follows

$$\mathcal{A}^+(U) = \{x \in X : (\exists V \in \mathcal{P}_k(X))\, V \subseteq U \,\&\, x \in \mathcal{A}(V)\}.$$

The notation $\mathcal{A}^{+m}$ means that $\mathcal{A}$ has been extended $m$ steps in this way. The collection of $k$-points of $X$ is a $k$-class, which is denoted $\mathrm{Pt}_k(X)$. Then we have the following relations between a level and the next.

PROPOSITION 3.3. *(i)* $\mathcal{A}^+$ *is a covering operator for* $(X, \leq)$ *which extends* $\mathcal{A}$, *in the sense that* $\mathcal{A}^+(U) = \mathcal{A}(U)$, *for all* $U \in \mathcal{P}_k(X)$. *Consequently,* $X^+$ *is a* $(k+1)$*-formal topology.*

*(ii)* $\mathrm{Pt}_k(X) = \mathrm{Pt}_{k+1}(X^+) \cap \mathcal{P}_k(X)$.

Since $\mathcal{A}^+$ extends $\mathcal{A}$ it follows that $X^+$ is compact iff $X$ is compact. (Recall that compactness means that if each neighbourhood is covered by some fixed set $V$ of neighbourhoods, then there is a finite list of neighbourhoods in $V$ that covers any neighbourhood.) We also have $x \wedge a \in \mathcal{A}^+(\emptyset)$ iff $x \wedge a \in \mathcal{A}(\emptyset)$, so $a \ll_{\mathcal{A}^+} b$ iff $a \ll_{\mathcal{A}} b$. From this it follows that $X^+$ is regular iff $X$ is regular. Recall also that a formal topology $(X, \leq, \mathcal{A})$ is regular if $z \lhd_{\mathcal{A}} \{x \in X : x \ll_{\mathcal{A}} z\}$ for each $z \in X$, and where $x \ll_{\mathcal{A}} z$ ($x$ is *well covered by* $z$) means that any neighbourhood is covered by $\{z\} \cup \{a \in A : x \wedge a \lhd_{\mathcal{A}} \emptyset\}$. The latter is the neighbourhoods "apart from $x$", denoted $x^*$.

A predicativity problem similar to Russell's is now: for a given $k$-formal topology $X$, does the notion of point stabilize at some level? We call a $k$-formal topology $T_1$ if each of its points $\alpha$ is (absolutely) maximal, in the sense that $\beta = \alpha$ whenever $\beta \supseteq \alpha$ is a $(k+m)$-point in $X^{+m}$. By Proposition 3.3(ii), the collection of points stabilise at level $k$ for such a topology. Another question is whether the points of a $k$-formal topology form a $k$-set, and not only a $k$-class. This fails for one of the simplest examples: when $X$ is the Sierpinski topology; see e.g. (Palmgren [2002a]). The phenomenon is explained by the fact that there exist partial points (i.e. positively non-maximal) with respect to inclusion. They may then be used to represent the class of proposition as points. On the other hand, in case $X$ is a 1-set-presented $T_1$-topology, it is shown in the same paper that its points $X$ are isomorphic to a set in $\mathbb{U}_1$. It had already been established that the points of a locally compact regular topology form a set (Curi [2002]), and later that the same is true for regular topologies (Aczel [2002]). The conditions on these topologies pertain to the covers only, and are thus more in the spirit of point-free topology than the $T_1$-condition.

A crucial method for constructing new formal topologies is the inductive generation of covers or operators; see Johnstone [1982] and Coquand, Sambin, Smith, and Valentini [2003]. Here is a method similar to (Aczel [2002]).

For any pre-ordered $k$-set $(X, \leq)$ and a pair of functions $g : I \longrightarrow X$ and $G : I \longrightarrow \mathcal{P}_k(X)$, where $I$ is a $k$-set (the pair $(g, G)$ is called the *generator*) define an operator $\mathcal{A} : \mathcal{P}_k(X) \longrightarrow \mathcal{P}_k(X)$ as follows. For each $k$-subset $U \subseteq X$, let $\mathcal{A}(U)$ be the smallest subset $V \subseteq X$ including $U$ and such that

- $b \in V, a \leq b$ implies $a \in V$,
- $a_1, \ldots, a_n \in X, n \geq 0, G_i \wedge \{a_1\} \wedge \cdots \wedge \{a_n\} \subseteq V$ implies $\{g_i\} \wedge \{a_1\} \wedge \cdots \wedge \{a_n\} \subseteq V$.

By assuming that $\mathbb{U}_k$ contains well-founded tree-types ($W$-types) one may prove that this set $V$ is a $k$-subset of $X$. Then we have the following theorem which entails an induction principle for covers.

THEOREM 3.4. *The structure $(X, \leq, \mathcal{A})$ defined as above is a k-formal topology whose covering operator satisfies*

$$g_i \in \mathcal{A}(G_i) \quad (i \in I) \tag{1}$$

*and if $\mathcal{B}$ is any other covering operator on $\mathcal{P}_n(X)$, $n \geq k$, satisfying this condition, then $\mathcal{A}^{+(n-k)}(U) \subseteq \mathcal{B}(U)$ for all $U \in \mathcal{P}_n(X)$.*

**§4. Morphisms of formal topology.** A continuous function $f$ between topological spaces induces a relation $U \subseteq f^{-1}[V]$ among the open sets of the respective spaces. The notion of morphism between formal topologies is an abstraction of this relation (Sambin [1987]). Since morphisms are generally represented by relations, rather than functions, we now encounter further predicativity problems in the category of formal topologies. For a relation $R \subseteq X \times Y$ the *inverse image of $V \subseteq Y$ under the relation $R$* is, as usual, $R^{-1}[V] =_{\text{def}} \{a \in X : (\exists b \in V)\, a\, R\, b\}$. We write $R^{-1}b$ for $R^{-1}[\{b\}]$.

DEFINITION 4.1. Let $X = (X, \leq_X, \mathcal{A}_X)$ and $Y = (Y, \leq_Y, \mathcal{A}_Y)$ be $k$-formal topologies. A relation $R \in \mathcal{P}_k(X \times Y)$ is an *approximable mapping*, or *formal morphism* (of level $k$) from $X$ to $Y$ if

(A1) $R^{-1}[\mathcal{A}_Y(U)] \subseteq \mathcal{A}_X(R^{-1}[U])$,
(A2) $\mathcal{A}_X(R^{-1}b) \subseteq R^{-1}b$,
(A3) $X = \mathcal{A}_X(R^{-1}[Y])$,
(A4) $R^{-1}b \cap R^{-1}c \subseteq \mathcal{A}_X(R^{-1}[b \wedge c])$.

The collection of $k$-level morphisms $X \longrightarrow Y$ is a $k$-class, which we denote $\mathrm{Hom}_k(X, Y)$. A formal morphism $R : X \longrightarrow Y$ gives rise to a $k$-frame morphism $\mathrm{Sat}_k(Y) \longrightarrow \mathrm{Sat}_k(X)$

$$f_R(U) = \mathcal{A}(R^{-1}(U)).$$

The preservation of suprema is shown by applying (A1) and noting that $R^{-1}$ commutes with unions. The corresponding point-function $\mathrm{Pt}_k(X) \longrightarrow \mathrm{Pt}_k(Y)$ is given by

$$\mathrm{Pt}_k(R)(\alpha) = \{y \in Y : (\exists x \in \alpha)\, x\, R\, y\}.$$

Let $\mathbf{FTop}_k$ be the category of $k$-formal topologies and formal morphisms of level $k$. For a formal topology $X$ we define the identity $I : X \longrightarrow X$ by

$$aIb \iff a \in \mathcal{A}_X\{b\}.$$

For morphisms $R_1 : X_1 \longrightarrow X_2$ and $R_2 : X_2 \longrightarrow X_3$ between formal topologies, define the composition

$$a(R_2 \circ R_1)b \iff a \in \mathcal{A}_{X_1}(R_1^{-1}[R_2^{-1}b]).$$

This is a morphism $(R_2 \circ R_1) : X_1 \longrightarrow X_3$.

EXAMPLE 4.2. For any $k$-formal topology $X$ the morphisms in $\mathrm{Hom}_k(1, X)$ stand in bijective correspondence with $\mathrm{Pt}_k(X)$, via

$$R \mapsto \{a \in X : \star R\, a\}.$$

Since these points need not form a $k$-set (for e.g. the $X$ = Sierpinski topology) the same is true for the morphisms of level $k$. The object 1 above is the terminal $k$-formal topology given by a singleton set $\{\star\}$ and the identity operator.

This example shows that $\mathbf{FTop}_k$ is not locally $k$-small. The analogy of points and morphisms leads to the consideration of the inclusion order between morphisms. The morphism $R \in \mathrm{Hom}_k(X, Y)$ is *maximal*, if for any $R' \in \mathrm{Hom}_{k+m}(X^{+m}, Y^{+m})$ with $R' \supseteq R$ (as graphs) we have $R' = R$. We have as in Proposition 3.3

$$\mathrm{Hom}_k(X, Y) = \mathrm{Hom}_{k+1}\left(X^+, Y^+\right) \cap \mathcal{P}_k(X \times Y).$$

The Hom-sets stabilise at $k$ when all $k$-morphisms are maximal.

The following is a generalisation of the result by Sambin that the points of a regular topology are maximal.

THEOREM 4.3. *Let $X$ be a $k$-formal topology and $Y$ be a regular $k$-formal topology. Then all the morphisms in $\mathrm{Hom}_k(X, Y)$ are maximal.*

PROOF. Suppose $R \in \mathrm{Hom}_k(X, Y)$ and $R' \in \mathrm{Hom}_{k+m}(X^{+m}, Y^{+m})$, $m \geq 0$, with $R' \supseteq R$. We show that $R' \subseteq R$ follows from the implication

$$a\, R'\, c,\ c \ll b \Longrightarrow a\, R\, b. \tag{2}$$

Since $Y$ is regular, we have $b \lhd_Y \{c \in Y : c \ll b\}$, and hence using (2)

$$\begin{aligned} \mathcal{A}_{X^{+m}}(R'^{-1}b) &\subseteq \mathcal{A}_{X^{+m}}(R'^{-1}\{c \in Y : c \ll b\}) \\ &\subseteq \mathcal{A}_{X^{+m}}(\{a' \in X : a'\, R\, b\}) = \mathcal{A}_{X^{+m}}(R^{-1}b). \end{aligned}$$

Now if $a\, R'\, b$, then $a \in \mathcal{A}_{X^{+m}}(R'^{-1}b)$, and hence by the inclusions above, $a\, R\, b$.

Proof of (2): From $c \ll b$, i.e. $Y \lhd \{b\} \cup c^*$, and (A3), it follows that for any $a$

$$a \lhd_X R^{-1}[\{b\} \cup c^*] \wedge a.$$

The latter is covered by $R^{-1}b$: since if $x\, R\, d$, $d \in c^*$ and $x \leq a$, we have also $x\, R'\, d$ by the assumption that $R \subseteq R'$. If $a\, R'\, c$, then by localisation $x\, R'\, d \wedge c$. Hence $x$ is covered by $\emptyset$, and therefore trivially covered by $R^{-1}b$. Thus $a\, R\, b$. ⊣

COROLLARY 4.4. *For the formal topology $\mathcal{R}$ of real numbers, the 1-class $\mathrm{Hom}_1(\mathcal{R}, \mathcal{R})$ is actually a 1-set.*

PROOF. Each $F \in \mathrm{Hom}_1(\mathcal{R}, \mathcal{R})$ gives a continuous function $f = \mathrm{Pt}_1(F) : \mathrm{Pt}_1(\mathcal{R}) \longrightarrow \mathrm{Pt}_1(\mathcal{R})$ in the sense of Bishop (Palmgren [2003]). It is straightforward to prove that $A_f \supseteq F$, where $A_f$ is the morphism given by

$$(a,b)\, A_f\, (c,d) \Longleftrightarrow f(a,b) \subseteq (c,d).$$

Since morphisms are all maximal, by the theorem above, we get $A_f = F$. Each morphism is thereby represented by a function. ⊣

Assuming that universes $\mathbb{U}_k$ are closed under formation of the next regular universe — a constructive version of the Hartogs function (Palmgren [2002b]) — we can show a stronger result.

THEOREM 4.5. *Suppose that $\mathbb{U}_1$ is closed under the formation of regular universes. For 1-set presented 1-formal topologies $X$ and $Y$, where $Y$ is a regular topology, the 1-class* $\mathrm{Hom}_1(X, Y)$ *is a 1-set.*

PROOF. Suppose that $X$ and $Y$ are 1-set presented by $C(x,i)$ $(i \in I(x))$ and $D(y, j)$ $(j \in J(y))$ respectively. Then $R \in \mathrm{Hom}_1(X, Y)$ iff for all $x \in X$, $y \in Y$

(A1') $x\,R\,y \Rightarrow (\forall j \in I(y))(\exists i \in I(x))\,(\forall u \in C(x,i))\,(\exists v \in D(y,j))\,u\,R\,v$,
(A2') $(\forall i \in I(x))[(\forall u \in C(x,i))\,u\,R\,y \Rightarrow x\,R\,y]$,
(A3') $(\exists i \in I(x))(\forall u \in C(x,i))(\exists v \in Y)\,u\,R\,v$,
(A4') $(\forall b,c \in Y)[x\,R\,b, x\,R\,c \Rightarrow (\exists i \in I(x))(\forall u \in C(x,i))(\exists y \leq b,c)\,u\,R\,y]$.

Since $R$ is maximal, according to Theorem 4.3, we may try to reconstruct $R$ inductively from below. All the morphisms $R$ are contained in $\mathcal{P}_1(X \times Y)$. We shall show that they are actually contained in a more restricted regular power set $\mathcal{R}(X \times Y)$, which is a 1-set. This is built from a certain regular universe $\mathcal{U}$. Using the $W$-type construction we can find $W \in \mathcal{U}$ which has for generating rules

$$0 \in W \qquad \frac{\alpha \in W}{S(\alpha) \in W} \qquad \frac{\alpha \in W \quad \beta \in W}{(\alpha,\beta) \in W}$$

and

$$\text{(3)} \qquad \frac{x \in X \quad i \in I(x) \quad f : C(x,i) \longrightarrow W}{\sup_{x,i}(f) \in W}.$$

Then construct a family $R_\alpha \in \mathcal{R}(X \times Y)$ $(\alpha \in W)$ so that $R_\alpha \subseteq R$, which satisfies the conditions obtained by modifying (A1')–(A4') as follows. Replace the successive $R$'s by $R_\alpha$ and $R_{S(\alpha)}$ in (A1'), by $R_{f(u)}$ and $R_{\sup_{x,i}(f)}$ in (A2'), where $f : C(x,i) \longrightarrow W$, moreover by $R_0$ in (A3') and by $R_\alpha$, $R_\beta$ and $R_{(\alpha,\beta)}$ in (A4'). These may be constructed by recursion on $W$, and using the type-theoretic choice principles. ($R$ is thus used as "template" for the sequence.) Let

$$R' = \bigcup_{\alpha \in W} R_\alpha \in \mathcal{R}(X \times Y).$$

Then $R' \subseteq R$. By the construction of the sequence $R_\alpha$ is straightforward to see that $R'$ satisfies the original (A1')–(A4'). For (A2') use the type-theoretic choice principle and (3). Thus $R'$ is a morphism $X \longrightarrow Y$ and by maximality $R' = R$. Since $R$ was arbitrary, the result is proved. ⊣

This theorem shows that the subcategory of set-presented regular topologies is locally small. The theorem is an improvement of the result of Curi [2003] which shows that $\mathrm{Hom}_1(X, Y)$ is a set when $X$ is locally compact and $Y$ is regular. See also Sigstam [1995] for a recursive version of Hyland's construction of exponential locales.

§5. **Some categorical constructions.** Important basic constructions of classical topological spaces (cf. Armstrong [1983]) may be expressed in category-theoretic language: products, disjoint unions (coproducts), quotient spaces and identification spaces (coequalisers and pushouts). We may then ask for the same categorical constructions in the category of formal topologies. This leads to predicativity problems again. For classical constructions in locales, see Borceux [1994].

The only known way of constructing the product of two formal topologies is to assume that they are inductively defined (Coquand, Sambin, Smith, and Valentini [2003]). Equalizers can be constructed using a similar method, which we sketch below. Coproducts do not need this assumption. However quotients or coequalizers seem to present serious problems. The standard locale-theoretic solution is to consider the equalizer in frames. Simply translating this into formal topologies, means that we must consider the class of saturated sets as new basic neighbourhoods. This uses up one universe, and seems to make it impossible to prove the universal property predicatively. Some subtler construction, or restriction on topologies or morphisms, need to be applied[1]. An equivalent problem is to construct pushouts. Here we have indeed a nontrivial special case which may easily be converted into a predicative proof, using e.g. Palmgren [2003]. For a locale $L$ and $a \in L$ consider the locale $L_a = \{x \in L : x \geq a\}$, and the locale morphism $L_a \longrightarrow L$ given by the frame morphism $x \mapsto x \vee a$. This is the embedding of the closed subspace $L_a$ determined by the complement of $a$ into $L$.

PROPOSITION 5.1 (gluing for closed subspaces). *Let $L$ be a locale and $a, b \in L$ with $a \wedge b = 0$. Then the following is a pushout*

$$\begin{array}{ccc} L_{(a \vee b)} & \longrightarrow & L_b \\ \big\downarrow & & \big\downarrow \\ L_a & \longrightarrow & L \end{array}$$

[1] Added in proof: a predicative construction of general coequalisers has been made (December 2004); see E. Palmgren, Coequalisers in formal topology. Uppsala University, Department of Mathematics, Report 2005:17.

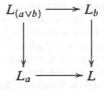

This pushout construction may, for instance, be used to join, continuously, two functions defined on adjacent intervals. Using equalizers we may construct zero sets or graphs of functions as subspaces.

THEOREM 5.2. *In $FTop_1$ the equalizer of $F_1, F_2 : X \longrightarrow Y$ exists when $X$ is 1-set presented.*

PROOF. Let $C(x,i)(i \in I(x))$ be a 1-set presentation of $X$. Then $\lhd_X$ is the minimal cover relation so that $x \lhd_X C(x,i)$ for all $x \in X$ and $i \in I(x)$. Let $E = (X, \leq_X, \mathcal{E})$ where is $\mathcal{E}$ is the operator generated by the pairs $(g, G)$ given by

(G1) $(x, C(x,i))$ for $x \in X$, $i \in I(x)$,
(G2) $(a, F_2^{-1}b)$ for $b \in B$, $a \in F_1^{-1}b$,
(G3) $(a, F_1^{-1}b)$ for $b \in B$, $a \in F_2^{-1}b$.

Then $\mathcal{E}(F_1^{-1}b) = \mathcal{E}(F_2^{-1}b)$. The equalizing morphism $F : E \longrightarrow X$ is given by

$$a\,F\,b \Longleftrightarrow a \lhd_E \{b\}.$$

To verify condition (A1) one proves that $b\,K\,V \Leftrightarrow \forall a\,(a\,F\,b \longrightarrow a \lhd_E F^{-1}V)$. defines cover relation on $X$. Then by minimality: $b \lhd_X V$ implies $b\,K\,V$ which verifies (A1). The conditions (A2)–(A4) are straightforward to check, as well proving that $F_1 \circ F = F_2 \circ F$. If $G : C \longrightarrow X$ is any other morphism making $F_1 \circ G = F_2 \circ G$, then $H : C \longrightarrow E$ defined by $H = G$ is the unique morphism so that $F \circ H = G$. To verify (A1) for $H$ one need to show that

$$b\,L\,V \Leftrightarrow_{\text{def}} \forall a\,\left(a\,H\,b \longrightarrow a \lhd_C H^{-1}V\right)$$

defines a cover relation. ⊣

## REFERENCES

P. ACZEL [1986], *The type-theoretic interpretation of constructive set theory: inductive definitions, **Logic, methodology and philosophy of science VII*** (R.B. Marcus et al., editors), North-Holland, Amsterdam.

P. ACZEL [2002], *Aspects of general topology in constructive set theory*, to appear in Annals of Pure and Applied Logic.

M.A. ARMSTRONG [1983], ***Basic topology***, Springer.

E. BISHOP [1967], ***Foundations of constructive analysis***, McGraw-Hill, New York.

E. BISHOP AND D. BRIDGES [1985], ***Constructive analysis***, Springer, Berlin.

F. BORCEUX [1994], ***Handbook of categorical algebra***, vol. 3, Cambridge University Press.

T. COQUAND [1995], *A constructive topological proof of van der Waerden's theorem*, ***Journal of Pure and Applied Algebra***, vol. 105, no. 3, pp. 251–259.

T. COQUAND, G. SAMBIN, J. SMITH, AND S. VALENTINI [2003], *Inductively generated formal topologies*, ***Annals of Pure and Applied Logic***, vol. 124, pp. 71–106.

G. CURI [2002], *Compact Hausdorff spaces are data types*, Preliminary version, July 30, 2002. To appear in Annals of Pure and Applied Logic.

G. CURI [2003], ***Exact approximations to Stone-Cech compactification***, Manuscript.

P. Dybjer [2000], *A general formulation of simultaneous inductive-recursive definitions in type theory*, ***The Journal of Symbolic Logic***, vol. 65, pp. 525–549.

P. Dybjer and A. Setzer [2003], *Induction-recursion and initial algebras*, ***Annals of Pure and Applied Logic***, vol. 124, pp. 1–47.

N. Gambino [2002], ***Sheaf interpretations for generalised predicative intuitionistic systems***, Ph.D. thesis, Manchester University, Manchester.

P.T. Johnstone [1982], ***Stone spaces***, Cambridge University Press.

P. Martin-Löf [1984], ***Intuitionistic type theory***, Notes by G. Sambin of a series of lectures given in Padua, June 1980, Studies in Proof Theory 1, Bibliopolis, Naples.

P. Martin-Löf [1998], *An intuitionistic theory of types*, ***Twenty-five years of constructive type theory*** (G. Sambin and J.M. Smith, editors), Oxford, (First published as a preprint from the Department of Mathematics, University of Stockholm, 1972.).

I. Moerdijk and E. Palmgren [2002], *Type theories, toposes and constructive set theory: predicative aspects of AST*, ***Annals of Pure and Applied Logic***, vol. 114, pp. 155–201.

S. Negri and D. Soravia [1999], *The continuum as a formal space*, ***Archive for Mathematical Logic***, vol. 38, no. 7, pp. 423–447.

E. Palmgren [1998], *On universes in type theory*, ***Twenty-five years of constructive type theory*** (G. Sambin and J.M. Smith, editors), Oxford University Press.

E. Palmgren [2002a], *Maximal and partial points in formal spaces*, Report 2002:23, Uppsala University, Department of Mathematics, URL: `www.math.uu.se`. To appear in Annals of Pure and Applied Logic.

E. Palmgren [2002b], *Regular universes and formal spaces*, Report 2002:42, Uppsala University, Department of Mathematics, URL: `www.math.uu.se`. To appear in Annals of Pure and Applied Logic.

E. Palmgren [2003], *Continuity on the real line and in formal spaces*, Report 2003:32, Uppsala University, Department of Mathematics, URL: `www.math.uu.se`.

G. Sambin [1987], *Intuitionistic formal spaces — a first communication*, ***Mathematical logic and its applications*** (D. Skordev, editor), Plenum Press, pp. 187–204.

G. Sambin [2003], *Some points in formal topology*, ***Theoretical Computer Science***, vol. 305, pp. 347–408.

I. Sigstam [1995], *Formal spaces and their effective presentations*, ***Archive for Mathematical Logic***, vol. 34, pp. 211–246.

I. Sigstam and V. Stoltenberg-Hansen [1997], *Representability of locally compact spaces by domains and formal spaces*, ***Theoretical Computer Science***, vol. 179, pp. 319–331.

S.G. Simpson [1999], ***Subsystems of second order arithmetic***, Springer.

H. Weyl [1918], ***Das Kontinuum: Kritische Untersuchungen über die Grundlagen der Analysis***, Veit, Leipzig.

DEPARTMENT OF MATHEMATICS
UPPSALA UNIVERSITY
PO BOX 480, SE-751 06 UPPSALA, SWEDEN
*E-mail*: palmgren@math.uu.se

# RANK INEQUALITIES IN THE THEORY OF DIFFERENTIALLY CLOSED FIELDS

WAI YAN PONG

Let $m \geq 1$ be an integer. The theory of differentially closed fields with $m$ commuting derivations ($m$-$\mathrm{DCF}_0$) has been actively studied recently [11, 14, 2, 1] and [12]. Readers who are interested in the model theory of differential fields and its applications can consult [8, 9, 13] and [21]. In this article, I limit myself to give a "road-map" of the proof of the following result:

THEOREM. *Let $p$ be a complete $n$-type in $m$-$\mathrm{DCF}_0$ over a differential field $K$. Suppose $d$ and $e$ are the typical differential dimension and the differential type of $p$ respectively and that $e \geq 1$. Then $1 \leq d \leq n$ and*

$$(*) \qquad \omega^e d \leq \mathrm{RU}(p) \leq \mathrm{RM}(p) \leq \mathrm{RH}(p) \leq \mathrm{RD}(p) < \omega^e(d+1).$$

For 1-types over an ordinary differential field $K$, the inequalities are due to Poizat [15]. Their generalization to $n$-types are fairly easy and were obtained by the author in [17] with RH replaced by $\Delta$-dimension and $\mathrm{RD}(p)$ replaced by $\omega d + b$ where $dT + b$ is the Kolchin polynomial of the type $p$. Since both $d$ and $b$ in that case are natural numbers so the map $dT + b \mapsto \omega d + b$ induces a well-order on the set of Kolchin polynomials. At that time, however, I did not realize that this assignment can be regarded as a generalization of RD. Later Benoist generalized the definitions of both RH and RD to $n$-types over ordinary differential fields and proved the inequalities for these ranks in [3]. His definition of the rank RD is different from ours, we will address this issue in Section 2. Let us also note that each of the inequalities appeared can be strict (even in the ordinary case). Examples that illustrate these phenomena are nontrivial except for the first inequality (see § 3). In particular, the question of whether RU = RM in $\mathrm{DCF}_0$ was open for quite a while until it was settled by Hrushovski and Scanlon in [5].

The proofs of these inequalities become considerably harder when the differential field is equipped with more than one derivation. McGrail in her thesis (see [11]) succeeded in proving the theorem without RD and with RH replaced by the $\Delta$-dimension. So our primary goal here is to clarify the relationship between the rank RD, RH and $\Delta$-dimension. The use of Kolchin polynomials in McGrail works is prominent. In fact, with a perfect hindsight

**Logic Colloquium '03**
Edited by V. Stoltenberg-Hansen and J. Väänänen
Lecture Notes in Logic, 24

one sees that she had already developed the results of RD that are necessary in proving these inequalities.

One of the difficulties in generalizing the definition of RD to the several derivations case is that the coefficients of the Kolchin polynomials in this case are merely rational numbers. Therefore, it is no longer obvious that the simple idea of assigning the Kolchin polynomials to their values at $\omega$ will yield a well ordering on them. It was not until a meeting in 2000 that I first learned that the Kolchin polynomials are well ordered by dominance. Again with a perfect hindsight, this result of Sit [20] can be seen as an easily consequence of the results in [11] (Lemma 4.2.13 and Proposition 4.2.15). At the same meeting, Scanlon suggested that one should investigate the "meanings" of the coefficients of the Kolchin polynomials. Around the same time, an application of McGrail's version of the theorem also prompted me to take a closer look of it again: it follows immediately from these inequalities that RU = RM for types with RM of the form $\omega^e d$ with $e \geq 1$. This fact is a crucial ingredient in showing that RU = RM for generic types of definable groups in $m$-$\mathrm{DCF}_0$ [14]. On the other hand, Benoist in [3] showed that RM and RH of a definable group in $\mathrm{DCF}_0$ can be different. Moreover, he showed that the notion of RM-generic and RH-generic are different.

In the first version of [2], I defined the rank RD using sequence of minimizing coefficients (see § 2) and proved a result about the RD of the fibers in definable families. The proofs there used some algebra of monomial ideals. However, it is Aschenbrenner who first recognized the full potential of relating the study of the Kolchin polynomials to that of the monomial ideals. Later through our study of Noetherian orderings we are able to give a conceptual proof of Sit's Theorem and a definition of RD that works in the several derivations case [1, 2].

**§1. Preliminaries.** We start with a brief review of the facts about differential algebra and the model theory of differential fields that will be used later. Readers may find it helpful to have [6, 7] and [11] at hand.

By a *differential ring* we mean a commutative ring with 1 equipped with $m$ commuting derivations. We use $\Delta = \{\delta_1, \ldots, \delta_m\}$ to denote the set of derivations. A *Ritt ring* is a differential ring containing the field of rational numbers. A *differential field* is a Ritt ring which is also a field. In particular, it has characteristic 0. We use the adjective "ordinary" to emphasize the case $m = 1$. Let $R$ be a differential ring, an ideal $I$ of $R$ is a *differential ideal* if it is closed under the derivations, i.e. $\delta_i I \subseteq I$ for all $1 \leq i \leq m$. An ideal $I$ is *perfect* if it equals to its own radical. In a Ritt ring, the radical of a differential ideal is still a differential ideal [7, Theorem 3.2.12]. Moreover, any maximal differential ideal of a Ritt ring is prime and the nilradical of a Ritt ring is the intersection of all prime differential ideals [7, Corollary 3.2.20]. In particular, every Ritt ring possesses a differential prime ideal. We use Spec $R$

and $\Delta$-Spec $R$ to denote the set of prime ideals and the set of differential prime ideals of $R$ respectively.

Let $\Theta$ be the free commutative monoid generated by the derivations. A typical element of $\Theta$ is of the form $\delta_1^{e_1}\cdots\delta_m^{e_m}$ where the $e_i$'s are non-negative integers. The sum of the $e_i$'s is called the *order* of $\delta_1^{e_1}\cdots\delta_m^{e_m}$. Let $a$ be an element of a differential ring $R$. A *derivative* of $a$ is an element of $R$ of the form $\theta a$ for some $\theta\in\Theta$. The *order* of a derivative $b$ of $a$ is defined to be the smallest integer $s$ such that $b=\theta a$ for some $\theta\in\Theta$ of order $s$. Let $L$ be a differential field extension of a differential field $K$. A subset $A$ of $L$ is *differentially independent* over $K$ if the set of derivatives of elements of $A$ is algebraically independent over $K$. In the case where $A$ is a singleton $\{a\}$, we say that the element $a$ is *differentially transcendental* over $K$. An element is *differentially algebraic* over $K$ if is not differentially transcendental over $K$.

Let $R$ be a differential ring. The *differential polynomial ring over R with variables* $y_1,\ldots,y_n$ is the polynomial ring

$$R\{y_1,\ldots,y_n\}:=R[\theta y_j:\theta\in\Theta,1\leq j\leq n].$$

The derivations $\delta_1,\ldots,\delta_m$ on $R$ extend naturally to $R\{\bar{y}\}=R\{y_1,\ldots,y_n\}$ making it into a differential ring extension of $R$. In general differential polynomial rings, unlike their algebraic counterparts, are not Noetherian. However, it follows from the Ritt-Raudenbush basis theorem (or the differential basis theorem) [7, Theorem 3.2.23, 5.3.17] that the set of perfect differential ideals of a differential polynomial ring over a differential field satisfies the ascending chain condition.

Let $\mathfrak{L}_m$ be the language of fields $\{+,-,\cdot,^{-1},0,1\}$ together with the set of unary function symbols $\Delta=\{\delta_1,\ldots,\delta_m\}$. From the model theoretic point of view, a differential field can be regarded as an $\mathfrak{L}_m$ structure with the symbols interpreted as the usual field operations and the $\delta_i$'s as derivations. A differential field is *differentially closed* if it is existentially closed in the sense of model theory. It turns out that the class of differentially closed fields is elementary. We use $m$-DCF$_0$ to denote the common first order theory of differentially closed fields in $\mathfrak{L}_m$. A relatively simple set of axioms for $m$-DCF$_0$ can be found in [11, §3]. Recently, a "coordinate free" approach to the axiomatization of the class of differentially closed fields is given by Pierce [12].

The theory $m$-DCF$_0$ admits quantifiers elimination [11, Theorem 3.1.7]. A consequence of this fact is the "type-ideal correspondence". Let $p\in S_n(K)$ be a complete $n$-type over a differential field $K$. Denote by $I_p$ the set

$$\{f\in K\{y_1,\ldots,y_n\}:\text{“}f(y_1,\ldots,y_n)=0\text{”}\in p\}.$$

One checks directly that $I_p$ is a prime differential ideal. It follows from the quantifier elimination of $m$-DCF$_0$ that the association $p\leftrightarrow I_p$ is a 1-1 correspondence between $S_n(K)$ and $\Delta$-Spec $K\{y_1,\ldots,y_n\}$. Moreover, one can verify readily that a tuple is a realization of a type $p\in S(K)$ if and only if its vanishing ideal over $K$ is $I_p$. Using the type-ideal correspondence and the differential basis theorem, one can show that $m$-DCF$_0$ is $\omega$-stable by a type-counting argument [11, Theorem 3.2.1].

It is convenient to fix a universal domain $U$ of $m$-DCF$_0$ and consider all differential fields other than $U$ as its small subfields. For each $n \in \mathbb{N}$, we equip $U^n$ with the *Kolchin topology*. The closed sets in this topology are of the form

$$V(S) = \{a \in U^n : f(a) = 0, \text{ for all } f \in S\}$$

where $S$ is a subset of $U\{y_1, \ldots, y_n\}$. It is easy to check that $V(S) = V(\{S\})$ where $\{S\}$ is the perfect differential ideal in $U\{y_1, \ldots, y_n\}$ generated by $S$. On the other hand, given $X$ a Kolchin closed subset of $U^n$, the set

$$I(X) = \{f \in U\{y_1, \ldots, y_n\} : f(a) = 0, \text{ for all } a \in X\}$$

is a perfect differential ideal of $U\{y_1, \ldots, y_n\}$. A Kolchin closed set is *irreducible* if it not empty and is not a union of two proper nonempty closed subsets. Let $X$ be a Kolchin closed set, a maximal irreducible Kolchin closed subset of $X$ is called an *irreducible component* of $X$. Let $K$ be a differential field, a Kolchin closed subset of $U^n$ is $K$*-closed* if it is of the form $V(S)$ for some $S \subseteq K\{y_1, \ldots, y_n\}$. It is not hard to see that the $K$-closed sets form a topology on $U^n$. We understand the terms $K$*-irreducible* and $K$*-irreducible component* in the obvious way. Let $L$ be a differential field extension of $K$, then the $K$-closed subsets of $U^n$ induces a topology on $L^n$. We write $V_L(S)$ for the set $L^n \cap V(S)$ and $I_L(V)$ for the perfect differential ideal $I(V) \cap L\{y_1, \ldots, y_n\}$ of $L\{y_1, \ldots, y_n\}$. In the case when $L$ is differentially closed, the differential nullstellensatz [6, Corollary 1, Theorem 2, p. 148] states that $X \mapsto I_L(X)$ is an inclusion reserving 1-1 correspondence between the $L$-closed subsets of $L^n$ and the perfect differential ideals of $L\{y_1, \ldots, y_n\}$.

**§2. RH, RD and Δ-dimension.** All results in this section, besides the definition of RD and Proposition 2.10, are either folklore or appear in some form in [4, 11].

In Section 1, we have remarked that the differential spectrum of a differential polynomial ring over a differential field satisfies the ascending chain condition. In general, we can make the following definition:

DEFINITION 2.1. Let $R$ be a differential ring. Suppose $\Delta$-Spec $R$ satisfies the ascending chain condition then for $P \in \Delta$-Spec $R$, we define inductively the *Δ-dimension of* $P$, denoted by $\Delta$-dim $P$, to be the ordinal

$$\sup\{\Delta\text{-dim } Q + 1 : Q \in \Delta\text{-Spec } R, Q \supsetneq P\}.$$

In particular, maximal elements of $\Delta$-Spec $R$ have $\Delta$-dimension 0. For $p \in S_n(K)$, we define its $\Delta$-dimension, $\Delta$-dim $p$, to be

$$\sup\{\Delta\text{-dim } P : L \supseteq K, P \in \Delta\text{-Spec } L\{\bar{y}\}, P \cap K\{\bar{y}\} = I_p\}.$$

Poizat introduced the differential height (RH) for 1-types over ordinary differential fields in [15]. Benoist generalized this definition to $n$-types over ordinary differential fields in [4] (see also [3]). His definition works equally well in the several derivations case without any modification.

DEFINITION 2.2. Let $p \in S(K)$, the ordinal $\mathrm{RH}(p)$ is defined inductively as follows:

- For $\alpha$ a limit ordinal, $\mathrm{RH}(p) \geq \alpha$ if $\mathrm{RH}(p) \geq \beta$ for all $\beta < \alpha$.
- $\mathrm{RH}(p) \geq \alpha + 1$ if there exist $L$ a differential field extension of $K$ and $q, r \in S(L)$ such that $q$ is an extension of $p$ to $L$, $I_q \subsetneq I_r$ and $\mathrm{RH}(r) \geq \alpha$.
- $\mathrm{RH}(p) = \alpha$ if $\mathrm{RH}(p) \geq \alpha$ and $\mathrm{RH}(p) \not\geq \alpha + 1$.

The following proposition records some basic properties of RH and $\Delta$-dimension. The first two of them are satisfied by every notion of rank [16, Chapter 17].

PROPOSITION 2.3. *Let $L \supseteq K$ be differential fields.*

1. *Both RH and $\Delta$-dimension are invariant under automorphisms of differential fields.*
2. *Let $p \in S(K)$ and $q$ be an extension of $p$ over $L$, then* $\mathrm{RH}(p) \geq \mathrm{RH}(q)$.
3. *Suppose $p_1, p_2 \in S(K)$ and $I_{p_1} \subsetneq I_{p_2}$ then* $\mathrm{RH}(p_1) > \mathrm{RH}(p_2)$.
4. $\Delta\text{-dim}\, I_p \leq \mathrm{RH}(p)$.

PROOF. Both (1) and (2) follow immediately from the definitions. For (3), one simply takes $L = K$, $q = p_1$ and $r = p_2$ in the definition of RH. To show (4), suppose $\Delta\text{-dim}\, I_p \geq \alpha$; then for each $\beta < \alpha$ there exists $I_q \supsetneq I_p$ such that $\Delta\text{-dim}\, I_q \geq \beta$. By induction hypothesis, $\mathrm{RH}(q) \geq \beta$. Hence by (3), $\mathrm{RH}(p) \geq \beta + 1$. Since $\beta < \alpha$ is arbitrary, we conclude that $\mathrm{RH}(p) \geq \alpha$. ⊣

The differential order (RD) was also introduced by Poizat in for 1-types over ordinary differential fields in [15] (see also [16, Chapter 6]). In that case, $\mathrm{RD}(a/K)$ is defined to be the order of the minimal polynomial of the vanishing ideal of $a$ over $K$. When $a$ is differentially algebraic over $K$, $\mathrm{RD}(a/K)$ coincides with the transcendence degree over $K$ of the differential field generated by $a$ and $K$. Recently, a generalization of RD to $n$-types in $m$-$\mathrm{DCF}_0$ has been worked out by Aschenbrenner and the author in [2]. Here we will only give a quick introduction to this rank and refer our readers to [1] and [2] for a thorough treatment.

A polynomial $f(T) \in \mathbb{Q}[T]$ is *numerical* if $f(s)$ is an integer for all sufficiently large integer $s$. For $f(T), g(T) \in \mathbb{Q}[T]$, we say that $f$ *dominates* $g$, denoted by $f \geq g$, if $f(s) \geq g(s)$ for all sufficient large integer $s$. It is straight forward to check that dominance is a total order on $\mathbb{Q}[T]$.

Let $a$ be a tuple in some differential extension of $K$. Denote by $K\langle a\rangle_s$ the field generated over $K$ by the derivatives of $a$ of order at most $s$. For sufficiently large $s$, the transcendence degree of $K\langle a\rangle_s$ over $K$ is given by a numerical polynomial [6, Theorem 6, p. 115]. We call this polynomial the *Kolchin polynomial of $a$ over $K$* and denote it by $\chi_{a/K}$. It is easy to see that $\chi_{a/K}$ is completely determined by the type of $a$ over $K$; so we can define, $\chi_p$, the *Kolchin polynomial of a type $p \in S(K)$* to be $\chi_{a/K}$ where $a$ is any realization of $p$. Using the type-ideal correspondence (see § 1), we define the *Kolchin polynomial of a prime differential ideal* to be the Kolchin polynomial of its corresponding type. Let $p$ be an $n$-type over $K$. Since $\chi_p$ is a numerical polynomial, it can be written as $\sum_{i=0}^{e} a_i \binom{T+i}{i}$ where $a_i \in \mathbb{Z}$ (e.g. see [19, Lemma 1, Ch.II B]). We call $e$, the degree of the polynomial $\chi_p$, the *differential type of $p$* and $a_e$ the *typical differential dimension of $p$*.[1] By Theorem 6 of [6, p. 115], $e \leq m$ the number of derivations and $0 \leq a_e \leq n$. Moreover, $a_e = 0$ if and only if $p$ is an algebraic type. By a theorem of Sit [20, Proposition 5] (also see [1]), the set of Kolchin polynomials is well-ordered by dominance. The order type of the set of Kolchin polynomials is $\omega^{m+1}$ [1, §3].

DEFINITION 2.4. The differential order of a tuple $a$ over $K$, denoted by $\mathrm{RD}(a/K)$ is defined to be the ordinal corresponding to $\chi_{a/K}$ under the dominance order. Similarly, we define the *differential order of a type* (*a differential prime ideal*) to be the ordinal corresponds to its Kolchin polynomial under the dominance order.

For each Kolchin polynomial $\chi$, there is a tuple $(b_e, \ldots, b_0) \in \mathbb{N}^e$ where $e$ is the degree of $\chi$ such that $\mathrm{RD}(\chi) = \sum_{i=0}^{e} \omega^i b_i$ [1, § 3]. The tuple $(b_e, \ldots, b_0)$ is called the *sequence of minimizing coefficients* of $\chi$ (see [7, Definition 2.4.9 and Proposition 2.4.10]). As an example, let us compute the differential order of a linear Kolchin polynomial $\chi(T) = dT + b$. Write $\chi(T)$ as $d\binom{T+1}{1} + (b-d)$. Then according to [7, Definition 2.4.9], the sequence of minimizing coefficients of $\chi$ is $(d, v)$ where

$$v = \chi(T+d) - \binom{T+d+2}{2} + \binom{T+2}{2} = \binom{d}{2} + (b-d).$$

So $\mathrm{RD}(\chi) = \omega d + \binom{d}{2} + (b-d)$. In particular, if $a$ is differentially transcendental over an ordinary differential field $K$, then $\chi_{a/K} = T+1$ and hence $\mathrm{RD}(a/K) = \omega$.

Let us explain the relationship between our definition of RD and the one given by Benoist in [3]. As we have seen, the RD of $dT+b$ is $\omega d + \binom{b}{2} + (b-d)$ while according to Benoist's definition it will be $\omega d + (b-d)$. The discrepancy here is due to the fact that the Kolchin polynomials considered in [3] only come from types over ordinary differential fields. They form a proper subset of the set of all linear Kolchin polynomials, for example in the ordinary case the

[1] The differential type and the typical differential dimension of a type over $K$ are called the $K$-type and $K$-degree in [11]. However, the latter terminology may cause confusion when used in conjunction with various model-theoretic notions.

constant term of a Kolchin polynomial is always non-negative, in fact it is never smaller than the leading coefficient.

Let us also point out that Kolchin polynomial is not a differential bi-rational invariant, i.e. two tuples may generate the same differential field over $K$ yet their Kolchin polynomials over $K$ are different. This can be seen from the following "silly" example: Let $a$ be differentially transcendental over an ordinary differential field $K$. Clearly $a$ and the pair $(a, a')$ are differential bi-rational with each other. But their Kolchin polynomials over $K$ are $T + 1$ and $T + 2$ respectively. This phenomenon can also arise from the interplay between the derivations. An example can be found in [7, Example 2.4.5].

Our next proposition gathers some basic properties of RD.

PROPOSITION 2.5. *Let $p$ and $q$ be complete types over $K$ and $L$ respectively.*

1. *$q$ extends $p$ if and only if $I_q$ lies over $I_p$.*
2. *If $q$ extends $p$ then $\chi_p \geq \chi_q$ hence* $\mathrm{RD}(p) \geq \mathrm{RD}(q)$.
3. *Let $P \in \Delta\text{-Spec}\, K\{\bar{y}\}$. There are finitely many minimal prime differential ideals in $L\{\bar{y}\}$ containing the perfect ideal generated by $P$ in $L\{\bar{y}\}$. If $Q$ is one of them then $Q$ lies over $P$ and $\chi_P = \chi_Q$.*

PROOF. Statement (1) follows immediately from quantifier elimination. Statement (3) is the characteristic 0 case of [6, Proposition 3(b), p. 131]. Finally, if $q$ is an extension of $p$, by (1) $I_q$ lies over $I_p$. It follows from (3) and the type-ideal correspondence that there exists a type $p'$ over $L$ such that $I_{p'} \subseteq I_q$ is lying over $I_p$. Moreover $\chi_{p'} = \chi_p$. So by [6, Proposition 2, p. 130], $\chi_q \leq \chi_{p'} = \chi_p$. ⊣

A prime differential ideal $Q$ satisfying (3) in the above proposition is called *a prime differential component* of $P$ over $L$ or simply *an $L$-component* of $P$.

LEMMA 2.6. *Let $L$ be a $|K|^+$-saturated differentially closed field containing $K$. Suppose $V$ is a $K$-irreducible closed set. Then the $L$-irreducible components of $V$ are conjugate with each other under* $\mathrm{Aut}_K(L)$, *the differential automorphisms group of $L$ over $K$.*

PROOF. Since $V$ is defined over $K$, $\mathrm{Aut}_K(L)$ acts on the $L$-irreducible components of $V$. There are only finitely many $L$-irreducible components of $V$, therefore the union of the elements of each orbit is an $L$-closed set. These unions are stabilized by $\mathrm{Aut}_K(L)$ and since $L$ is $|K|^+$-saturated, we conclude that they are all $K$-closed. The union of all these $K$-closed sets is $V$ itself and since $V$ is $K$-irreducible therefore $\mathrm{Aut}_K(L)$ must act transitively on the $L$-irreducible components of $V$. ⊣

Using the differential nullstellensatz, we obtain the following algebraic version of Lemma 2.6: Let $L/K$ be the same as stated in the lemma. Let $P$ be a prime differential ideal of $K\{y_1, \dots, y_n\}$, then $\mathrm{Aut}_K(L)$ acts transitively on the $L$-components of $P$. One can also prove this directly by using the several derivations version of Corollary 3.6 in [8].

We say that the *Going-Down theorem for (differential) prime ideals* holds for a (differential) ring extension $A \subseteq B$ if for $P_2 \subsetneq P_1$ (differential) prime ideals of $A$ and $Q_1$ a (differential) prime ideal of $B$ lying over $P_1$, there exists $Q_2$ a (differential) prime ideal of $B$ contained in $Q_1$ lying over $P_2$.

PROPOSITION 2.7. *Let $L/K$ be a differential field extension. The Going-Down theorem for differential prime ideals holds for the differential ring extension $K\{y_1, \ldots, y_n\} \subseteq L\{y_1, \ldots, y_n\}$.*

PROOF. Since the natural inclusion $K \hookrightarrow L$ is a flat map and flatness is stable under base change, therefore $B := L\{y_1, \ldots, y_n\} = L \otimes_K K\{y_1, \ldots, y_n\}$ is flat over $A := K\{y_1, \ldots, y_n\}$. So the Going-Down Theorem for prime ideal holds for this extension [10, Theorem 9.5]. Therefore, if $P_2 \subsetneq P_1$ are two prime differential ideals in $A$ and $Q_1$ is a prime differential ideal in $B$ lying over $P_1$, then the differential ring $B_{Q_1} \otimes_{A_{P_1}} \kappa(P_2 A_{P_1})$ has a prime ideal and hence it is not the zero ring. Thus the canonical map from $Q$ into this ring is injective so it is a Ritt ring and therefore possesses a prime differential ideal (see § 1). The preimage of this prime differential ideal in $B$ is contained in $Q_1$ and lying over $P_2$. ⊣

Our study of RH and RD leads to the following characterization of forking in $m$-DCF$_0$ (see also [11, Theorem 4.3.10]).

PROPOSITION 2.8 (Characterization of Forking). *Let $K \subseteq L$ be an extension of differential fields. Let $p \in S(K)$ and $q \in S(L)$ be an extension of $p$. Then the following are equivalent:*

1. *$q$ is a non-forking extension of $p$.*
2. $\chi_q = \chi_p$.
3. $\mathrm{RD}(p) = \mathrm{RD}(q)$.
4. *$I_q$ is an $L$-component of $I_p$.*
5. $\mathrm{RH}(p) = \mathrm{RH}(q)$.

PROOF. (1) $\iff$ (2) follows from (2) of Proposition 2.5 and the equivalence of (1) and (3) in [11, Theorem 4.3.10].

(2) $\Rightarrow$ (3). Follows from definition of RD.

(3) $\Rightarrow$ (4). Suppose (4) is not true, then $I_p \subseteq I_r \subsetneq I_q$ for some $r \in S(L)$. By Proposition 2.5 (1), we have $I_p = I_q \cap K\{\bar{y}\} = I_r \cap K\{\bar{y}\}$ and $r$ is an extension of $p$; but then by [6, Proposition 2, p. 130] $\chi_q < \chi_r \leq \chi_p$ hence $\mathrm{RD}(q) < \mathrm{RD}(p)$.

(4) $\Rightarrow$ (5). Since RH can only go down under extension, we can assume $L$ is a $|K|^+$-saturated differentially closed field. We prove by induction that

$$\mathrm{RH}(p) \geq \alpha \Rightarrow \mathrm{RH}(q) \geq \alpha.$$

The limit case is clear. Suppose $\mathrm{RH}(p) \geq \alpha + 1$. Then there exist $K' \supseteq K$ and $r, s \in S(K')$ such that $r$ is an extension of $p$ with $I_r \subsetneq I_s$ and $\mathrm{RH}(s) \geq \alpha$. Let $L'$ be a differentially closed field containing both $K'$ and $L$. Let $I_{s'}$ be an

$L'$-component of $I_s$. By the induction hypothesis, $\mathrm{RH}(s') = \mathrm{RH}(s) \geq \alpha$. Let $I_{q'}$ be an $L'$-component of $I_p$ contained in $I_{s'}$. The containment must be strict, otherwise $I_{q'}$ lies over $I_s \supsetneq I_r$ and therefore by Proposition 2.7 (differential polynomial ring version of the Going-Down theorem), $I_{q'}$ cannot be an $L'$-component of $I_p$. Thus $\mathrm{RH}(q') \geq \alpha + 1$. Let $q'' = q'|_L$ then it follows from Proposition 2.7 again then $I_{q''} = I_{q'} \cap L\{\bar{y}\}$ must be an $L$-component of $I_p$. By Lemma 2.6, $I_{q''}$ and $I_q$ are conjugates hence so are $q''$ and $q$. Therefore,

$$\mathrm{RH}(q) = \mathrm{RH}(q'') \geq \mathrm{RH}(q') \geq \alpha + 1.$$

(5) $\Rightarrow$ (2). Let $I_r$ be an $L$-component of $I_p$ containing in $I_q$. In particular, $r$ extends $p$. If $I_r \subsetneq I_q$, then $\mathrm{RH}(q) < \mathrm{RH}(r) \leq \mathrm{RH}(p)$ contradicting (5). This shows that $I_q$ is an $L$-component of $I_p$ thus $\chi_q = \chi_p$ by (3) of Proposition 2.5. ⊣

If $L$ is differentially closed, then by the differential nullstellensatz, one can replace condition (4) by "$V_L(I_q)$ is an $L$-irreducible component of $V_L(I_p)$". The main part of the following proof is taken from the proof of [4, Proposition 0.1.3], again the argument given there works equally well in the several derivations case.

**Proposition 2.9.** *Let $p \in S(K)$, if $K$ is $\omega$-saturated, then* $\Delta\text{-}\dim I_p = \mathrm{RH}(p)$.

**Proof.** By (4) of Proposition 2.3, it suffices to show that $\Delta\text{-}\dim I_p \geq \mathrm{RH}(p)$. The limit case is immediate. Suppose $\mathrm{RH}(p) \geq \alpha + 1$ then there exist $q, r$ over, $L$, some extension of $K$ such that $q$ extends $p$ and $I_q \subsetneq I_r$ with $\mathrm{RH}(r) \geq \alpha$. By taking nonforking extensions if necessary, we can assume $L$ is $\omega$-saturated. Suppose $b$ is a tuple from $L$ consisting of the coefficients of a basis of $I_r$, say $\phi(y, b)$ is a conjunction of differential polynomial equations defining $V(I_r)$. Here we simply say that $b$ is a tuple of parameters of $I_r$. Let $a$ be a tuple of parameters of $I_p$ from $K$. Then it is easy to see that

1. "$V(I_p) \supsetneq V(I_r)$" is expressible by a formula in $\mathrm{tp}(b/a)$ and,
2. "$V(I_r)$ is irreducible" is expressible by an infinite collection of formulas in $\mathrm{tp}(b)$.

By $\omega$-saturation, let $b'$ be a tuple in $K$ realizing the type $\mathrm{tp}(b/a)$. Hence by (2) the realizations of $\phi(y, b')$ in $K$ and $L$ are irreducible closed sets. By differential nullstellensatz, they determine prime differential ideals and hence types $s'$ and $s$ over $L$ and $K$ respectively with $s'$ extending $s$. By condition (1) and the differential nullstellensatz, we have $I_s \supsetneq I_p$. Since $b$ and $b'$ have the same type over $a$ and $L$ is $\omega$-saturated, therefore $b$ and $b'$ hence $r$ and $s'$ are conjugates under an automorphism of $L$. Therefore, $\mathrm{RH}(s) \geq \mathrm{RH}(s') = \mathrm{RH}(r) \geq \alpha$. Hence $\Delta\text{-}\dim I_p > \Delta\text{-}\dim I_s \geq \alpha$ by induction hypothesis. ⊣

Finally we show that in $m$-$DCF_0$ $\Delta$-dimension and RH are the same for complete types.

PROPOSITION 2.10. $\Delta\text{-dim}\, p = \mathrm{RH}(p)$

PROOF. By definition, for every $\beta < \Delta\text{-dim}\, p$, there exists some prime differential ideal $P$ lying over $I_p$ such that $\Delta\text{-dim}\, P \geq \beta + 1$. Since $P = I_{p'}$ for some $p'$ extending $p$, therefore by (2) and (4) of Proposition 2.3,

$$\beta + 1 \leq \Delta\text{-dim}\, I_{p'} \leq \mathrm{RH}(p') \leq \mathrm{RH}(p).$$

We now show the reverse inequality by induction. Again the limit case is easy. Suppose $\mathrm{RH}(p) \geq \alpha + 1$. Then there are types $q$, $r$ over some $L \supseteq K$ such that $q$ extends $p$, $I_q \subsetneq I_r$ and $\mathrm{RH}(r) \geq \alpha$. Let $L'$ be an $\omega$-saturated extension of $L$ and $r'$ be a nonforking extension of $r$ over $L'$. Let $q'$ be a type over $L'$ such that $V(I_{q'})$ is an $L'$-irreducible component of $V(I_q)$ containing $V(I_{r'})$. Hence $I_{q'} \subsetneq I_{r'}$, by Proposition 2.9 and Proposition 2.8 we have

$$\Delta\text{-dim}\, p \geq \Delta\text{-dim}\, I_{q'} > \Delta\text{-dim}\, I_{r'} = \mathrm{RH}(r') = \mathrm{RH}(r) \geq \alpha.$$

This concludes the proof. ⊣

§3. **Rank inequalities.** Let us recall the theorem that we want to prove.

THEOREM 3.1. *Let $p$ be a complete $n$-type in $m$-$DCF_0$ over a differential field $K$. Suppose $d$ and $e$ are the typical differential dimension and the differential type of $p$ respectively and that $e \geq 1$. Then $1 \leq d \leq n$ and*

$$(*) \qquad \omega^e d \leq \mathrm{RU}(p) \leq \mathrm{RM}(p) \leq \mathrm{RH}(p) \leq \mathrm{RD}(p) < \omega^e(d+1).$$

PROOF. Most of the work has been done in [11]. We have proved in Proposition 2.10 that $\Delta$-dim $=$ RH. The inequality $\mathrm{RD}(p) < \omega^e(d+1)$ is clear from the sequence of minimizing coefficient approach (§ 2). Therefore, it remains to show that $\mathrm{RH}(p) \leq \mathrm{RD}(p)$. We prove this by induction. The limit case is clear. Suppose $\mathrm{RH}(p) \geq \alpha + 1$. Then there are types $q$ and $r$ with $q$ an extension of $p$ and $I_q \subsetneq I_r$ and $\mathrm{RH}(r) \geq \alpha$. By the induction hypothesis, $\mathrm{RD}(r) \geq \alpha$. Since $I_q \subsetneq I_r$, $\chi_q > \chi_r$ [6, Proposition 2, p. 130] and hence $\mathrm{RD}(q) > \mathrm{RD}(r)$. By (2) of Proposition 2.5, $\mathrm{RD}(r) < \mathrm{RD}(q)$ so we have

$$\mathrm{RD}(p) \geq \mathrm{RD}(q) > \mathrm{RD}(r) \geq \alpha.$$

This completes the proof. ⊣

EXAMPLES 3.2. We conclude this article with some known examples in $DCF_0$ showing that each of the inequalities in Theorem 3.1 can be strict. These can certainly be viewed as evidence in supporting the claim made by Sack in the introduction of [18]: the least misleading example of a totally transcendental theory is the theory of differentially closed fields of characteristic 0.

1. It is easy to see that the first inequality can be strict. For example, let $a$ be differentially transcendental over $\mathbb{Q}$ and $b$ be a generic constant over $\mathbb{Q}(a)$. Then $\mathrm{tp}(a, b)$ has U-rank $\omega + 1$. However, both its differential type and typical differential dimension are 1.
2. Let $a$ be differentially transcendental over $\mathbb{Q}$ and $E_a$ be the elliptic curve with $j$-invariant $a$. Let $b$ be a realization of the generic type (over $a$) of the Manin kernel of $E_a$. Then $\mathrm{tp}(a, b)$ has U-rank $\omega$ but Morley rank $\omega + 1$. This example is due to Hrushovski, see [17, addendum] for details. In [5], Hrushovski and Scanlon showed that in $\mathrm{DCF}_0$ the Morley rank can be strictly greater than the U-rank even for types of finite Morley rank.
3. The type of a generic solution to the equation $y\delta^2 y - \delta y = 0$ has Morley rank 1 and differential height 2. This example is due to Poizat, for details see [8, p. 64].
4. The type of a generic solution of the Painlevé equation $\delta^2 y = 6y^2 + a$ where $a$ is an element such that $\delta a = 1$ has differential height 1 and differential order 2. The analysis of this Painlevé equation is due to Kolchin, see [8, p. 66] for details.

**Acknowledgment.** I thank the referee for many useful suggestions and comments.

REFERENCES

[1] M. Aschenbrenner and W.Y. Pong, *Orderings of monomial ideals*, ***Fundamenta Mathematicae***, vol. 181 (2004), pp. 27–74.

[2] ———, *The differential order*, preprint.

[3] F. Benoist, *Rangs et types de rang maximum dans les corps diff rentiellement clos*, ***The Journal of Symbolic Logic***, vol. 67 (2002), no. 3, pp. 1178–1188.

[4] ———, *Théorie des modles des corps différentiellement clos, Mémoire de Magistre de l'ENS*, Paris, 2000.

[5] E. Hrushovski and T. Scanlon, *Lascar and Morley ranks differ in differentially closed fields*, ***The Journal of Symbolic Logic***, vol. 64 (1999), no. 3, pp. 1280–1284.

[6] E. Kolchin, ***Differential algebra and algebraic groups***, Pure and Applied Mathematics, vol. 54, Academic Press, 1973.

[7] M.V. Kondratieva, A.B. Levin, A.V. Mikhalev, and E.V. Pankratiev, ***Differential and difference dimension polynomials***, Mathematics and Its Applications, vol. 461, Kluwer Academic Publishers, 1999.

[8] D. Marker, *Model theory of differential fields*, ***Model theory of fields***, Lecture Notes in Logic, vol. 5, Springer-Verlag, Berlin, 1996.

[9] ———, *Model theory of differential fields*, ***Model theory, algebra, and geometry***, Math. Sci. Res. Inst. Publ., vol. 39, Cambridge University Press, Cambridge, 2000, pp. 53–63.

[10] H. Matsumura, ***Commutative ring theory***, 2nd ed., Cambridge Studies in Advance Mathematics, vol. 8, Cambridge University Press, Cambridge, 1989.

[11] T. McGrail, *The model theory of differential fields with finitely many commuting derivations*, ***The Journal of Symbolic Logic***, vol. 65 (2000), no. 2, pp. 885–913.

[12] D. Pierce, *Differential forms in the model theory of differetial fields*, ***The Journal of Symbolic Logic***, vol. 68 (2003), no. 3, pp. 923–945.

[13] A. Pillay, *Differential fields*, ***Lectures on algebraic model theory***, Fields Institute Monographs, vol. 15, American Mathematical Society, Providence, RI, 2002, pp. 1–45.

[14] A. Pillay and W.Y. Pong, *On Lascar rank and Morley rank of definable groups in differentially closed fields*, ***The Journal of Symbolic Logic***, vol. 67 (2002), no. 3, pp. 1189–1196.

[15] B. Poizat, *Rangs des types dans les corps diffirentiels*, Groupe d'Itude de Thiories Stables, 1re annie (Univ. Pierre et Marie Curie, Paris, 1977/78), Exp. No. 6.

[16] ———, ***A course in model theory. An introduction to contemporary mathematical logic***, Universitext, Springer-Verlag, New York, 2000, Translated from the French by Moses Klein and revised by the author.

[17] W.Y. Pong, *Some applications of ordinal dimensions to the theory of differentially closed fields*, ***The Journal of Symbolic Logic***, vol. 65 (2000), no. 1, pp. 347–356.

[18] G. Sacks, ***Saturated model theory***, Mathematics Lecture Note Series, W. A. Benjamin, Inc., Reading, Mass., 1972.

[19] J.P. Serre, ***Local algebra***, Springer Monographs in Mathematics, Springer-Verlag, Berlin, 2000, Translated from the French by CheeWhye Chin and revised by the author.

[20] W. Sit, *Well-ordering of certain numerical polynomials*, ***Transactions of the American Mathematical Society***, vol. 212 (1975), pp. 37–45.

[21] C. Wood, *Differentially closed fields*, ***Model theory and algebraic geometry***, Lecture Notes in Mathematics, vol. 1696, Springer, Berlin, 1998, pp. 129–141.

DEPARTMENT OF MATHEMATICS
CALIFORNIA STATE UNIVERSITY
DOMINGUEZ HILLS
1000 E. VICTORIA STREET
CARSON, CA 90747, USA
*E-mail*: wpong@csudh.edu

# CONSISTENCY AND GAMES — IN SEARCH OF NEW COMBINATORIAL PRINCIPLES

PAVEL PUDLÁK†

**Abstract.** We show that a semantical interpretation of Herbrand's disjunctions can be used to obtain $\Pi_2$ independent sentences whose nature is more combinatorial than the nature of the usual consistency statements. Then we apply this method to Bounded Arithmetic and present $\forall\Sigma_1^b$ combinatorial sentences that characterize all $\forall\Sigma_1^b$ sentences provable in $S_2$. We use the concept of a two player game to describe these sentences.

**§1. Introduction.** One of the main problems in proof complexity is to prove that certain theories are not equal. The theories that we would like to separate correspond to complexity classes that are conjectured to be distinct. The only tool for such results that we currently have is diagonalization, the same as in computational complexity. In first order logic diagonalization is essentially Gödel's incompleteness theorem. Most researchers believe that conjectures such as $P \neq NP$ cannot be proved by diagonalization, and the same opinion is shared about separations of subtheories of Bounded Arithmetic. Yet it is worthwhile to study the sentences used in Gödel's theorem. Recall that the Gödel sentence for a theory $T$ is equivalent to the consistency of $T$, provided that $T$ is sufficiently strong (this is the content of the second incompleteness theorem). For weak theories, such as Bounded Arithmetic, or Robinson's Arithmetic, these sentences must also be relatively weak, hence their combinatorial meaning should not be very complicated. If we could find some combinatorial interpretations of them, we would learn a lot about provability in Bounded Arithmetic.

The usual consistency statements are not very transparent because the concept of a proof in the usual Hilbert style proof system is fairly complicated. A much more combinatorial concept is the Herbrand proof (the proof based on the characterization of the first order provability given by Herbrand's Theorem). Therefore we have concentrated on studying Herbrand proofs. In

†Partially supported by grants EAA1019401 of the AV ČR, and grants 201/01/1195 401/03/H047 of the GA ČR.

Logic Colloquium '03
Edited by V. Stoltenberg-Hansen and J. Väänänen
Lecture Notes in Logic, 24

this article we shall present a sort of semantical interpretation of such proofs. Using the incompleteness theorem and this interpretation we obtain an independent sentence which is more "mathematical" than the usual consistency statements. However, we have to pay for it by increasing the complexity to $\Pi_2$. Independent combinatorial sentences in Peano Arithmetic and stronger theories have been studied since 1977 when Paris and Harrington published the seminal paper [24]. All the known such sentences are $\Pi_2$. In section 5, we shall describe a sentence independent of Peano Arithmetic. It is by far not so nice as those of Paris and Harrington and others, but the advantage of our method is that it is universally applicable. What is, perhaps, more important is that the reason why we have to increase the complexity to $\Pi_2$ is more apparent in our proof.

The main result of this article is an application of this framework to Bounded Arithmetic. Since it is unlikely that the separation problems could be solved using diagonalization, we consider a related problem of characterizing low complexity theorems of subtheories of Bounded Arithmetic. We shall present $\forall\Sigma_1^b$ sentences $A_k$ such that $A_k$ is provable in $S_2^k$ and all $\forall\Sigma_1^b$ theorems of $S_2^k$ are provable in $S_2^1 + A_{k+4}$. The sentences are still fairly complicated, so it will be necessary to simplify them before we will be able to use them to conditional separations of theories. The problem of finding simpler versions of these sentences is not a purely technical point. It requires a better understanding of the dependencies of terms on variables in Herbrand disjunctions. In this article we have restricted ourselves to Herbrand proofs, but the same can be achieved using the Mid Sequent Theorem. Therefore a better understanding of the combinatorial principles $A_k$ is also closely connected with better understanding of the structure of proofs in the sequent calculus.

In the presentations of the combinatorial principles we shall use the concept of a two player game. This is a natural way to interpret a sequence of alternating quantifiers. When the concept of a game is applied to Herbrand disjunctions new phenomena arise. The game corresponding to the original formula splits into several copies. What is the most interesting fact is that the games are not played sequentially. If we insisted that they should be played sequentially we would get only very weak principles; thus the power is in the parallelism.

As the principle $A_k$ is too complicated to be used to prove separations (based on oracles or on assumptions from complexity theory), we suggest a more uniform version $B_k$ in Section 7. It has some nice properties, but, so far, we are neither able to prove that these sentences characterize the $\forall\Sigma_1^b$ consequences of $S_2$, nor to prove the independence of the oracle versions of them. In the last section we shall suggest some potential applications of our results.

Related problems were studied by several authors. Buss, Chiari, F. Ferreira, Hanika, Impagliazzo and Krajíček characterized low complexity theorems of

$S_2^i$ for some small $i$ and proved some conditional separations [8, 10, 12, 13, 16, 18, 19]. Adamowicz and Kolodziejczyk [3] used Herbrand's theorem in a similar way as in this paper, but they did not interpret the sentences in terms of games.

**§2. Games and matches.** A game of finite length $k$ is

$$G \subseteq M_1 \times M_2 \times \cdots \times M_k.$$

$M_i$ are sets of possible moves in the $i$th step of the game. If all sets $M_i$ are finite, we say that the game is finite. The elements of $M_i$ are legal *moves* in the $i$th step. The game is played by two players who alternate in choosing elements form sets $M_i$. $G$ is the set of winning positions of the first player. We shall also consider games of infinite lengths in which $k$ is $\omega$. More generally, a game of length $k$ is a pair $(M, G)$ where $M$ is an arbitrary nonempty set of strings of length $k$ and $G \subseteq M$. A legal position $(x_1, \ldots, x_i)$ is an initial segment of a string from $M$, and a legal move from this position is an $x_{i+1}$ such that $(x_1, \ldots, x_i, x_{i+1})$ is a legal position.

Every game of finite length is determined, which means that there exists a winning strategy either for the first player or for the second. However, the strategy may be very difficult to describe even if the whole game is finite and it is described very explicitly. From the point of view of computational complexity, the natural setting is a game $G$ defined by a boolean circuit $C$. Here we assume that the elements of each $M_i$ are strings of 0s and 1s of length $m_i$ and the size of the circuit $C$ is polynomial in the input size $\sum_i m_i$. In such a situation the number of plays (i.e., legal final positions) is exponential in the size of the description of the game. It is well-known that then it is **PSPACE**-complete to determine which player has a wining strategy. Consequently, to describe the moves of a winning strategy is **PSPACE**-hard.

Thus if we restrict the computational power of the players to polynomial size boolean circuits, for some games there may be no winning strategies in this restricted sense. It seems that some of the difficult games that people play (chess, go, checkers) are of this nature. To show that this is a very interesting field of research let us mention a simple example based on randomized strategies:

> *It is conceivable that there exists a game G described by a polynomial size boolean circuit and a strategy σ for the first player such that σ is defined by a probabilistic polynomial size circuit and with probability tending to 1 strategy σ beats all strategies for the second player which are defined by probabilistic polynomial size circuits. In the absolute sense, however, it is the second player who has a winning strategy.*

Such an example can be constructed assuming that one-way functions exist, which is the basic conjecture of cryptography. Let $F : \{0,1\}^n \to \{0,1\}^n$ be a one way function. Define $G = \{(y,z,x);\ F(x) = y \wedge F(z) \neq y\}$. The strategy $\sigma$ for the first player is to choose randomly uniformly an $x$, and play $F(x)$ in the first move and $x$ in the third move. The assumption about $F$ guarantees that the second player has a very small chance to find a $z$ such that $F(z) = y$.

In the rest of the paper we shall consider only deterministic strategies.

We shall also consider systems of games, which we shall call *matches*. As in the real world, playing a match means that we play several games and then the winner is determined by the results of the games. Unlike in the real world, in the matches that we shall consider the players may start next game without finishing the previous one (however, they have to finish all games eventually). Formally a match is determined by:

1. a set of boards with a finite length game assigned to every board; the games assigned to different boards may be the same;
2. a schedule that determines a linear order in which the moves of the games will be played; the only restriction is that the moves in every game are played in the order given by this game;
3. a rule to determine the winner from the information about who won the games.

A match, like a game, is played by two players which we shall call *Alice* and *Bob*. Given a game $G$ in a match, it is necessary to define who will start the game (the person who starts will also be called *the first player*). Therefore we shall assume that the rules of every game include also a rule that Alice starts or that Bob starts. Then for every game $G$ we have the *complementary game*, denoted by $G^{\perp}$, in which the roles of Alice and Bob are switched.

We can treat a match also as a finite length game. In particular, we can talk about winning strategies for matches. However it turns out that the most interesting is the structure of a match, which often determines who has a winning strategy without our knowing who has the winning strategies for the games.

**§3. Herbrand's proofs as winning strategies.** We shall use first order logic without equality. As usual, it means that our theories may contain equality, but it is treated as an ordinary binary relation. Consider a sentence of the form

$$\exists x_1 \forall y_1 \exists x_2 \forall y_2 \cdots \gamma(x_1, y_1, x_2, y_2, \dots), \tag{1}$$

where $\gamma$ is open. We shall use the convention that $a_1, a_2, \dots$ represents a *finite* sequence, unless stated otherwise. (This is especially useful when denoting quantifier prefixes such as above, because the last quantifier may be any of the two.) According to Herbrand's theorem this sentence is provable in the predicate calculus if and only if a disjunction of the form

$$(2)\qquad \bigvee_{i=1}^{\ell} \gamma(t_{i1}, f_1(t_{i1}), t_{i2}, f_2(t_{i1}, t_{i2}), \dots),$$

is provable in the propositional calculus, where $t_{i1}, t_{i2}, \dots$ are terms in the language of $\gamma$ augmented with new function symbols $f_1, f_2, \dots$.

Let such a disjunction be given. For every term $f_j(t_{i1}, \dots, t_{ij})$ we introduce a new variable $z_{ij}$. Notice that some pairs of terms $f_j(t_{i1}, \dots, t_{ij})$, $f_{j'}(t_{i'1}, \dots, t_{i'j'})$ with $(i, j) \neq (i', j')$ may be equal; in such a case we will use the same variable. Then starting with the longest term $f_j(t_{i1}, \dots, t_{ij})$, we replace gradually all these terms by the corresponding variables in all terms of the disjunction. Substitute $z_{11}$ for all variables other than $z_{ij}$ and for all remaining subterms of the form $f_i(t)$. Thus we eliminate all occurrences of the new function symbols. The resulting disjunction has the form

$$(3)\qquad \bigvee_{i=1}^{\ell} \gamma(\tilde{t}_{i1}, z_{i1}, \tilde{t}_{i2}, z_{i2}, \dots),$$

where $\tilde{t}_{ij}$ are terms resulting from $t_{ij}$ when we apply the substitution described above. Let us stress that $z_{ij}$ with distinct indices do not necessarily represent distinct variables. We have to substitute the same variable for all occurrences of the same term, otherwise the logical validity of the formulas will not be always preserved. This disjunction has two properties:

- (1) logically follows from the universal closure of the formula (3);
- if (2) is a propositional tautology, then so is (3).

The first property is a consequence of the following syntactical property of (3): we can order the terms and variables $\tilde{t}_{ij}, z_{pk}$ in such a way that

1. the terms and variables of the $i$th disjunct occur in the following order $\tilde{t}_{i1} \prec z_{i1} \prec \tilde{t}_{i2} \prec z_{i2} \prec \cdots$;
2. $\tilde{t}_{ij}$ may contain only variables which in the ordering $\prec$ occur before it.

The second property follows from the fact that for every term $t$ in (2) we have substituted the same term for all occurrences of $t$ in (2). The latter condition can be expressed by saying that (3) satisfies the following syntactical condition:

3. If $\tilde{t}_{i1} = \tilde{t}_{i'1}, \dots, \tilde{t}_{ij} = \tilde{t}_{i'j}$, then $z_{ij+1} = z_{i'j+1}$.

We shall not prove it as all this is well-known. We only remark that it can also be proved using the Mid Sequent Theorem (see [30]). We shall call (3) a *Herbrand disjunction* for (1). If the Herbrand disjunction is a propositional tautology we shall call it a *Herbrand proof* of the sentence (1).

Our aim is to interpret this syntactical object in combinatorial way. To this end we shall use games. We shall interpret $\gamma$ as a definition of the game $G$ in which the first player wins in a finial position $a_1, b_1, a_2, b_2, \dots$ iff $\gamma(a_1, b_1, a_2, b_2, \dots)$ is true. Let us first consider one term of the disjunction

$\gamma(\tilde{t}_{i1}, z_{i1}, \tilde{t}_{i2}, z_{i2}, \dots)$. We shall think of terms $\tilde{t}_{ij}$ as defining a strategy for Alice to play the game $G$ as the first player. The variables $z_{ij}$ are moves of Bob who plays as the second.

Now having $\ell$ formulas of the form $\gamma(\tilde{t}_{i1}, z_{i1}, \tilde{t}_{i2}, z_{i2}, \dots)$, we shall think of it as a match in which Alice and Bob play the same game $\gamma$ simultaneously on $\ell$ boards. At each moment one player makes one move on one board. The order of moves is given by the ordering $\prec$. Condition 1. above says that they play in the correct order on each board, Condition 2. says that the terms $\tilde{t}_{ij}$ define a strategy for Alice.

A strange thing in this interpretation is that Alice's moves on one board may depend on what Bob played on other boards. However, recall that we shall consider situations in which the players do not have access to the winning strategies. In the situation where they cannot play the best strategy, any information about the game may be useful. A player may learn a good move from his opponent and use it on another board. Let us view it from the viewpoint of Alice. If Bob plays well, then in every move he reveals nontrivial information. On the other hand, if he plays poorly, then it should not be difficult to beat him. Thus Alice has a good chance to win on at least one board.

Let us now interpret the logical validity of disjunction (3). It means that the strategy defined by the terms ensures Alice to win on at least one board. Hence if we define that this is the goal of the match, the terms define a winning strategy for the match.

In the same way we can analyze other forms of quantifier prefixes, namely, prefixes in which the first quantifier is universal and prefixes in which there are consecutive occurrences of the same quantifier. Let us consider one special case, the quantifier prefix $\forall\exists\forall$. (We shall always associate Alice with $\exists$ and Bob with $\forall$, thus the order in which they play is Bob-Alice-Bob.) Then the disjunction (3) has the form:

$$\bigvee_{i=1}^{\ell} \gamma(z_0, \tilde{t}_i, z_i). \tag{4}$$

(Notice that Bob starts with the same move on all boards. This is a special case of Rule $B$ that will be defined in Section 4.) To get an ordering satisfying the conditions above, we shall first consider the partial ordering of terms $t_i$ by their size, then we extend it to a total ordering and order the terms $\tilde{t}_i$ in this way. W.l.o.g. assume that we get $\tilde{t}_1 \prec \tilde{t}_2 \prec \cdots \prec \tilde{t}_\ell$. Then we extend it to

$$z_0 \prec \tilde{t}_1 \prec z_1 \prec \tilde{t}_2 \prec z_2 \prec \cdots \prec \tilde{t}_\ell \prec z_\ell. \tag{5}$$

In general, $z_i$ may not occur in $\tilde{t}_{i+1}$, if so, then it is possible to move $z_i$ behind $\tilde{t}_{i+1}$, but such an ordering may not be always possible. The ordering in (5) is the most general one. In terms of games it means that we allow Alice to use as much information as possible. The Herbrand disjunction has the form

$$\bigvee_{i=1}^{\ell} \gamma(z_0, \tilde{t}_i(z_0, z_1, \ldots, z_{i-1}), z_i), \tag{6}$$

where $\tilde{t}_i(z_0, z_1, \ldots, z_{i-1})$ means that only variables $z_0, z_1, \ldots, z_{i-1}$ may occur in $\tilde{t}_i$.

For $\forall\exists\forall\exists$ we get a very similar formula:

$$\bigvee_{i=1}^{\ell} \bigvee_{j=1}^{k_i} \gamma(z_0, \tilde{t}_i(z_0, z_1, \ldots, z_{i-1}), z_i, \tilde{s}_{ij}(z_0, z_1, \ldots, z_\ell)).$$

An explanation in terms of games is as follows. The most general ordering is an ordering in which Alice has as much advantage as possible. If she is to play the last move in the game $\gamma$, then for her it is the best if she can postpone her last moves on all boards after all moves of Bob.

The fact that there are the most general orderings for the prefixes contained in $\forall\exists\forall\exists$ case is well-known and it is often used. For $\exists\forall\exists\forall$ and prefixes with more quantifier alternations there does not seem to exist the most general ordering. This is the traditional stumbling block to the attempts to generalize results to longer quantifier prefixes. However, if we do not insist on the same number of terms in the disjunction, then we can find a trivial normal form: simply take the disjunction of all Herbrand disjunctions of a given length.

**§4. $\Pi_1$ and $\Pi_2$ unprovable sentences.** Our main goal is to understand the statements expressing consistency in such a way that we will be able to find a combinatorial interpretation of them. Herbrand's theorem alone transforms the consistency statement, which speaks about proofs in first order logic, into a more combinatorial one, which only refers to propositional provability. In the previous section we have suggested an interpretation of Herbrand disjunctions as statements about games. Now we shall explain it in greater detail.

Suppose we have a consistent theory $T$ axiomatized by a sentence $\Phi$ whose prenex form

$$\forall x_1 \exists y_1 \forall x_2 \exists y_2 \cdots \phi(x_1, y_1, x_2, y_2, \ldots), \tag{7}$$

has $k$ existential quantifiers. The inconsistency of $T$ means that the negation of this sentence, which is

$$\exists x_1 \forall y_1 \exists x_2 \forall y_2 \cdots \neg\phi(x_1, y_1, x_2, y_2, \ldots), \tag{8}$$

is provable in the predicate calculus. Provability in the predicate calculus is equivalent to the Herbrand provability, thus if (8) were provable, there would be a Herbrand disjunction based on this formula which would be a tautology. Let

$$\Delta(z_1, \ldots, z_m)$$

be such a Herbrand disjunction, where all free variables are displayed. As $T$ is consistent, it is not a propositional tautology. We want to find a semantical interpretation of this disjunction. Thus we interpret it in terms of games and we obtain a match. In this match Alice and Bob play $\neg\phi(x_1, y_1, x_2, y_2, \ldots)$ on several boards and Alice wins the match if she wins on at least one board. The terms of the disjunction define a strategy for Alice. Now instead of saying that $\Delta(z_1, \ldots, z_m)$ is not a propositional tautology we would like to say that Bob can beat Alice's strategy. What is the difference between these two statements? The first one is:

The formula $\Delta(z_1, \ldots, z_m)$ is not a propositional tautology.

The second one is

$$\exists z_1 \cdots \exists z_m \neg\Delta(z_1, \ldots, z_m).$$

Clearly, the second one is stronger, because it says that not only $\Delta(z_1, \ldots, z_m)$ is not a propositional tautology, but one can find a falsifying assignment to the atomic formulas *by a suitable substitution of values for* $z_1, \ldots, z_m$. In fact, it is not obvious that the latter should be true, it depends on the model in which we interpret it.

Let us see now what happens if we modify the statement that $T$ is consistent in this way. Let $HD(\Xi, \Delta)$ denote the formalization of '$\Delta$ *is a Herbrand disjunction for* $\Xi$'. We get the following:

$$\forall\Delta(HD(\neg\Phi, \Delta) \to \exists z_1 \cdots \exists z_m \neg\Delta(z_1, \ldots, z_m)).$$

This is not a first order formula; in order to write this formula in the first order logic, we need a formula that expresses the truth of existential formulas. Let us assume that we can formalize it and denote such a formula by $Tr_\exists$. Then a more precise statement reads:

$$\forall\Delta(HD(\neg\Phi, \Delta) \to Tr_\exists(\exists z_1 \cdots \exists z_m \neg\Delta(z_1, \ldots, z_m))). \tag{9}$$

Now we would like to give a game-theoretical interpretation of this sentence. The Herbrand disjunction defines a match in which Alice and Bob play several copies of the game defined by $\neg\phi(x_1, y_1, x_2, y_2, \ldots)$. The order of moves is determined (in general, not uniquely) by the dependences of terms on variables. Furthermore, we have to satisfy the syntactical condition 3. from Section 3. (Recall that the condition expressed the fact that the same variable $z_{ij}$ was substituted for all occurrences of the same term.) As our aim is to have a semantical interpretation, we replace it by the following condition, which, in general, is stronger than the corresponding syntactical one.

**Rule $\mathcal{B}$.** *If the same position occurs on several boards, then Bob must play the same move on all these boards.*

A special instance of this rule is the following. If Bob is the first player in the game on which the match is based, then he must start with the same move on all boards. By imposing this rule we make Bob's task harder, hence the statement stronger. But if $\Phi$ is true, then he has a winning strategy, so using the same winning strategy on all boards, he can obey Rule $\mathcal{B}$ and beat Alice. Let us stress that we will not consider such a restriction for Alice, it would make the principle too weak. (Remember: $\mathcal{B}$ stands for Bob.)

Thus we get the following principle.

**The Match Principle for $\Phi$.** *For every match based on the game defined by $\neg\phi(x_1, y_1, x_2, y_2, \dots)$ and for every strategy of Alice computable by terms of the language of $T$, one can find moves of Bob which satisfy Rule $\mathcal{B}$ and which beat Alice (i.e., Bob wins on all boards).*

By interpreting the syntactical object, the Herbrand proof, semantically, we have gained a statement that looks more natural, more "mathematical". But we have also lost something: we have transformed the $\Pi_1$ sentence expressing the consistency of $T$ into a $\Pi_2$ sentence. The fact that, for a sufficiently strong theory, the complexity of this statement cannot be reduced to $\Pi_1$ is an immediate corollary of the following proposition.

PROPOSITION 4.1. *Suppose that $T$ is axiomatized by the sentence $\Phi$ and contains at least the subtheory $I\Sigma_1$ of PA. Then, provably in $T$, the match principle for $\Phi$ is equivalent to the $\Sigma_1$ reflection principle for $T$.*

PROOF. Recall that the $\Sigma_1$ reflection principle for $T$ is the formalization of the sentence: *'For all $\Sigma_1$ sentences $\psi$, if $T$ proves $\psi$, then $\psi$ is true.'* A little bit more formally the sentence is

$$\forall x \in \Sigma_1(Pr_T(x) \rightarrow Tr_{\Sigma_1}(x)).$$

Since every bounded formula is equivalent to an existential formula, provably in $I\Sigma_1$ (as Matiyasevich's Theorem is provable in $I\Sigma_1$), it suffices to consider only existential sentences.

1. First we prove that the match principle for $\Phi$ implies the $\Sigma_1$-reflection principle for $T$. Since $T$ contains $I\Sigma_1$, the provability of a sentence $\Psi$ in $T$ is equivalent to Herbrand provability of $\Phi \rightarrow \Psi$.

We shall argue in $T$. Suppose that $\Psi$ is an arbitrary sentence of the form $\exists x_1 \exists x_2 \cdots \exists x_q\ \psi(x_1, x_2, \dots, x_q)$ with $\psi$ open. Consider a Herbrand proof of $\Phi \rightarrow \Psi$. We can present it in the form

$$\bigvee_j \psi(t_{1j}, t_{2,j}, \dots, t_{q,j}) \vee \Delta,$$

where $\Delta$ is a Herbrand disjunction for $\neg\Phi$. Assuming the match principle for $\Phi$ we can find elements such that after substituting them for the free variables of $\Delta$, formula $\Delta$ becomes false. Hence

$$\bigvee_j \psi(\tilde{t}_{1j}, \tilde{t}_{2,j}, \ldots, \tilde{t}_{q,j}),$$

must be true, where $\tilde{t}_{ij}$ denote $t_{ij}$ after this substitution. Thus $\exists x_1 \exists x_2 \cdots \exists x_q$ $\psi(x_1, x_2, \ldots, x_q)$ is true.

2. Let us now prove the opposite. The match principle for $\Psi$ has the form $\forall n \exists m\ \Gamma_1(n, m)$, where $n$ is a number that codes a match and a strategy of Alice, and $n$ codes moves of Bob. Again, we can replace this $\Sigma_1$ sentence by an equivalent existential sentence

$$\forall n \exists m_1 \cdots \exists m_k \Gamma(n, m_1, \ldots, m_k). \tag{10}$$

LEMMA 4.2. (1) *For every* $n$, $\exists m_1 \cdots \exists m_k\ \Gamma(\boldsymbol{n}, m_1, \ldots, m_k)$ *is provable in* $T$ ($\boldsymbol{n}$ *denotes the* $n$*th numeral*).

(2) *This proof can be formalized in* $I\Sigma_1$.

Assuming the $\Sigma_1$ reflection principle for $T$, we get immediately (10), so it only remains to prove the lemma. The idea of the proof of (1) is based on the observation that $\Phi$ expresses the fact that Bob has a winning strategy for the game defined by $\phi$. Since $n$ is an actual number, we have a concrete match and we can use this strategy to define moves of Bob one by one on all boards so that Bob wins on all boards. This gives a construction of a proof of $\exists m_1 \cdots \exists m_k \Gamma(\boldsymbol{n}, m_1, \ldots, m_k)$ in $T$ from the match coded by the number $n$.

More formally, let $n$ be given. The number $n$ determines a match and a strategy, which is defined by terms. Using this match and the terms we construct a proof of $\exists m_1 \cdots \exists m_k\ \Gamma(\boldsymbol{n}, m_1, \ldots, m_k)$ as follows.

First suppose that the match has a single board. Since the game is defined by $\phi$ and the axiom of theory $T$ is $\Phi$, $T$ proves that Bob can always win. Thus, if we plug in the moves of Alice we get gradually

$$\exists y_1 \ldots \exists y_i \forall x_{i+1} \exists y_{i+1} \forall x_{i+2} \exists y_{i+2} \ldots \phi(a_1, y_1, a_2, y_2, \ldots, a_i, y_i, x_{i+1}, y_{i+1}, x_{i+2}, y_{i+2}, \ldots), \tag{11}$$

where $a_1, \ldots, a_i$ are moves of Alice determined by the given strategy. Eventually we get

$$\exists y_1 \exists y_2 \cdots \phi(a_1, y_1, a_2, y_2, \ldots),$$

which is (equivalent to) the instance of the principle

$$\exists m_1 \cdots \exists m_k \Gamma(\boldsymbol{n}, m_1, \ldots, m_k).$$

When the match has more boards we shall keep one formula of the form (11) for each board and gradually replace the universal quantifiers by the moves of Alice determined by her strategy. The order in which we change the formulas follows the order of the moves of the match. (Recall that the match is given by a concrete number, so also the number of boards is a concrete number.)

Notice that what we are essentially doing is showing the well-known fact that the every prenex formula implies the existential closure of any of its Herbrand disjunctions.

To prove (2) just observe that the above proof is described very explicitly. To prove its existence we only refer to syntactical operations with proofs and Matiyasevich's Theorem (the latter can be avoided with some more work). ⊣

This proof actually shows more: we can add arbitrary universal sentences to $T$ without changing the independent sentence. In particular this applies also to the equality axioms, thus we can ignore them completely.

The following proposition is an application of the above semantical interpretation of Herbrand disjunctions which shows that certain orderings are not universal for Herbrand's theorem.

PROPOSITION 4.3. *There exists a provable* $\exists\forall\exists$ *sentence which does not have a Herbrand proof of the form*

$$(12)\quad \phi(t_1, z_1, s_1(z_1)) \vee \phi(t_2(z_1), z_2, s_2(z_1, z_2)) \vee \cdots \vee \phi(t_\ell(z_1, \ldots, z_{\ell-1}), z_\ell, s_\ell(z_1, \ldots, z_{\ell-1}, z_\ell)).$$

PROOF. Suppose the contrary, namely, that every $\exists\forall\exists$ provable sentence has a Herbrand proof of this form. Let $\Psi$ be a $\Pi_2$ sentence expressing this fact. Consider the theory $T = I\Sigma_1 + \Psi$. This theory is finitely axiomatizable and has a $\Pi_3$ axiomatization. If we expand the language of arithmetic with suitable function symbols, we get a $\forall\exists\forall$ axiomatization.[1] So let $\Gamma$ be a $\forall\exists\forall$ sentence axiomatizing $T$. In $T$ we know that if $\neg\Gamma$ is provable, then it also has a Herbrand proof of the form (12). But in $T$ we can also prove that Bob can beat any strategy of Alice in every match of the form (12). The point is that in $T$ we have $\Gamma$ which is $\exists x \forall z \exists y \neg\phi(x, z, y)$. It says that Bob has a winning strategy for the game used in the match. Then, using $\Sigma_1$ induction, we can prove that in fact Bob can win all games in such a sequence of games. In more detail: by induction on $k = 1, \ldots, \ell$, we prove

$$\exists z_1 \cdots \exists z_k \neg\phi(t_1, z_1, s_1(z_1)) \wedge \cdots \wedge \neg\phi(t_k(z_1, \ldots, z_{k-1}), z_k, s_k(z_1, \ldots, z_{k-1}, z_k)).$$

Thus $T$ would prove its own consistency, which means that $T$ would be inconsistent. Hence $\Psi$ must be false. ⊣

In terms of games this is an example that shows that Alice can really get an advantage if the games in a match are not played sequentially, one after the other, even if they are only of length 3. As we have already noted, for Alice it is better to postpone the last moves on all boards to the very end of the match.

[1] We cannot reduce the complexity of $T$ further, because if we skolemize the induction axiom we only get induction for $\Sigma_1$ arithmetical formulas, but we do not get the induction for $\Sigma_1$ formulas in the language expanded by the Skolem function.

The most intriguing question is whether we can obtain a $\Pi_1$ sentence in a similar way. When trying to preserve the property of being $\Pi_1$ we have to use essentially the structure of the formulas in the Herbrand disjunction, we cannot simply say that they are some arithmetical formulas. But then we can easily get into troubles, because a single equation of the form $p(\bar{x}) = 0$, with $p$ a multivariate polynomial with integer coefficients, can express very complex facts. For instance, the existence of a solution may be equivalent to the inconsistency of the theory. Thus our feeling is that in order to get an independent $\Pi_1$ sentence we need an essentially new idea.

**§5. A sentence independent of Peano arithmetic.** We shall apply the ideas above to the specific case of Peano Arithmetic. We shall assume that it is axiomatized by open axioms plus induction axioms for all formulas. We assume that the relation $<$ is present in the language. Thus on top of the usual operations we shall need the operation of the predecessor and subtraction truncated to nonnegative numbers. We already know that the particular choice of the language and open axioms is not important, thus we shall not specify it more.

We shall use the induction axiom for a formula $\Phi(x)$ in the form

$$\forall x(\Phi(\mathbf{0}) \wedge \forall y(\Phi(y) \rightarrow \Phi(S(y))) \rightarrow \Phi(x)).$$

It is well known that one does not need parameters in $\Phi$, so the only free variable is $x$. The negation of this sentence can be written as

$$\exists x \forall y(\Phi(\mathbf{0}) \wedge (\Phi(y) \rightarrow \Phi(S(y))) \wedge \neg\Phi(x)).$$

We shall further simplify this formula by assuming that $\Phi(\mathbf{0})$ and $\neg\Phi(x)$ is ensured by a syntactical condition on $\Phi$. More precisely, we shall use formulas with two free variables $\Phi(x, y)$ that have the form

$$(\Phi'(y) \vee y = \mathbf{0}) \wedge y \neq x.$$

Hence the negation of the induction axiom attains the following simple form

$$\exists x > \mathbf{0}\, \forall y(\Phi(x, y) \rightarrow \Phi(x, S(y))). \tag{13}$$

Suppose $\Phi(x, y)$ is $Q\bar{z}\phi(x, y, \bar{z})$, where $Q$ denotes a quantifier prefix with variables $\bar{z}$. Thus a prenex form of (7) is

$$\exists x > \mathbf{0}\, \forall y Q^{\perp}\bar{z} Q\bar{u}(\phi(x, y, \bar{z}) \rightarrow \phi(x, S(y), \bar{u})), \tag{14}$$

where $Q^{\perp}$ denotes the quantifier prefix dual to $Q$. Let us interpret it in terms of games. We shall assume that the quantifier prefix starts with $\forall$. Let us call the games determined by $\phi(x, y, \bar{z})$ *the small games*. Thus the game defined by formula (14) has two small games. It can be described in words as follows.

**Games of type $G_{IND}$.** *Small games indexed by two parameters $x$ and $y$ are given such that if $0 = y \neq x$ then the first player always wins, and if $y = x$ then the second player always wins. First Alice chooses an $x > 0$ and Bob* *chooses a $y$. Then there are two rounds. In the first round they play the small game with index $(x, y)$ and with Bob as the first player. In the second round they play the small game with index $(x, y + 1)$ and Alice is the first player. Bob wins if he beats Alice in both rounds, otherwise Alice wins.*

Now we can apply the framework developed in the previous section to formula (14) and we get a sentence independent from PA. The sentence is:

**The match principle for induction axioms.** *For every match based on a game of type $G_{IND}$ and every strategy of Alice, Bob can beat Alice's strategy on at least one board.*

This principle is independent from PA. It is not much more than a restatement of Herbrand's theorem for PA, but we can make it nicer. So far we have tacitly assumed that the small games are defined by a quantifier free arithmetical formula and the strategies of Alice are defined by arithmetical terms. It is clear that the principle is valid without such restrictions, but it seems that in order to formalize it in PA we have to use definable strategies and games. Yet there is a way out. The idea is to let Alice *pretend* to know the small games and to know a strategy to play the match, while letting Bob know nothing. Now Bob can win in two ways: either by beating Alice, or by catching Alice lying, which means that he will find her answers inconsistent with her claim that she uses particular small games and plays a fixed strategy.

Let us state it more precisely. Let a schedule $R$ to play a match based on a game of type $G_{IND}$ be given. Alice and Bob play the match without knowing the small games. For all boards, as soon as they finish a round, Alice decides who won in the small game. We shall call it *the virtual match* for $R$.

There is an easy strategy for Alice to win the virtual match, but the point is that they will play the match many times and Alice must pretend that the small games are the same and that she always uses the same strategy. More precisely this means that if Bob plays an initial part of the match in the same way as before, she must respond in the same way as before, and whenever the final position in a small game is the same as before, she must decide about the winner in the same way as before. Recall that Bob has to obey Rule $\mathcal{B}$. This rule is sort of the opposite to the first restriction on Alice: Bob must play in the same way in the same positions in each match, but in different matches he can use different moves.

Now we can state the principle.

**The strong match principle for induction axioms.** *For every schedule $R$ to play a match based on a game of type $G_{IND}$ and for every virtual match for $R$, there exist an $m$ and a strategy for Bob to win at least once if they play the virtual match $m$ times and Alice plays consistently.*

PROPOSITION 5.1. *The strong match principle for induction axioms is a true principle, unprovable in Peano Arithmetic.*

PROOF. 1. The proof of the principle is a simple application of König's lemma. Suppose that the strong game principle fails. Bob's strategy will be to try systematically all possible combinations of moves in the match. We fix an enumeration of all infinitely many possible moves of Bob. Given an $m$ he will play the first $m$ moves of this enumeration. We assume that for every $m$ Alice has a winning strategy for playing the match $m$ times. By König's lemma, Alice has a strategy to consistently play all infinitely many games that Bob wants to try so that she never loses. Let us fix also this strategy and let us play these two strategies against each other. Let $S$ be the sequence of plays thus obtained. $S$ contains, in particular, the information about who wins in small games. Let an index $(x, y)$ be given. We define a small game for this index by taking all strings that occurred as the final positions with index $(x, y)$ in $S$ as the legal end positions $M_{x,y}$, and we take as $G_{x,y}$ those that Alice proclaimed to be the winning ones. The game $(M_{x,y}, G_{x,y})$ is determined, thus Bob can use a winning strategy either when playing as the first player or when playing as the second player. Hence, if Alice plays $x$ on board $j$, he can take the largest $y < x$ such that the first player has a winning strategy and win on this board. Thus he can win on all boards. We need only to check that if Bob plays in such a way, all the resulting end positions in small games will occur in $S$. Indeed, since $S$ was obtained by Bob's systematic trying all possible combinations of moves, also the play based on the optimal strategy must occur in $S$. Thus it is not possible that Alice has a winning strategy for every $m$.

2. It is clear that the strong match principle is stronger than principle with concrete games and strategies. Since the latter is unprovable the former is unprovable too. ⊣

**§6. Games in bounded arithmetic.** Bounded Arithmetic started with the seminal papers of Parikh [23] and Cook [9]. The currently most studied systems are $I\Delta_0 + \Omega_1$ of Paris and Wilkie [31], and $S_2$ and $T_2$ of Buss [6]. $S_2$ and $T_2$ are equivalent theories and they are conservative extensions of $I\Delta_0 + \Omega_1$. The importance of these theories stems from the fact that they are proof complexity counterparts of the Polynomial Hierarchy. These theories are too weak to prove the equivalence of the provability in the standard calculi (such as the sequent calculus) and Herbrand's provability.[2] However, for suitable formalizations, Gödel's second incompleteness theorem can be proved for Herbrand's proofs too. This was shown by Adamowicz [1]; further results of this type were proved by Adamowicz and Zbierski [4], Adamowicz [2], Willard [32], and Salehi [29].

In Bounded Arithmetic the semantical interpretation of Herbrand's disjunctions does not produce a $\Pi_1$ independent sentence, but for reasons different

[2]There exist sequences of first order tautologies which have Hilbert style proofs of polynomial length, but the sizes of the shortest Herbrand proofs grow superexponentially.

from those in PA. The formulas in the induction schema are bounded, thus the games are finite. The problem is, however, that terms define functions that we cannot interpret as functions. Provably total functions of $S_2$ increase the lengths of numbers at most polynomially, but the values of terms constructed from # grow faster. Therefore we cannot define the values of terms, i.e., expressed in the language of games, we cannot define Alice's moves. One may propose to restrict the concept of Herbrand's provability so that only terms whose values grow polynomially be allowed in Herbrand disjunctions, but then we are not be able to prove the incompleteness theorem. This route was studied by Krajíček and Takeuti [22] who defined a concept of a restricted consistency. One of the main restrictions was a bound on the terms occurring in proofs. They were able use these statements to prove certain separation results, but failed to prove the Second Incompleteness Theorem for such proofs. Thus it seems essential for the diagonalization used in the proof of the Second Incompleteness Theorem that we can speak about objects that we cannot interpret. In the previous sections we have avoided this problem by using $\Pi_2$ sentences. These sentences say that some fast growing function is total. Such sentences are not interesting in Bounded Arithmetic. Therefore, in this section we shall consider a related, but different problem: the problem of characterizing low complexity consequences of subtheories of $S_2$. Having nice characterizations may eventually help us to prove nonconservativity results for subtheories of $S_2$, which is an important problem in Bounded Arithmetic (we will say more about this problem in Section 8). We shall present a combinatorial principle based on games that axiomatizes all $\forall\exists$ theorems of $S_2$. The principle has a parameter $k$ which roughly corresponds to the fragments $S_2^k$.

Before we start we have to recall some basic definitions and results. $S_2$ contains the standard arithmetical operations plus several others. In particular, it contains the length function $|x|$ and the smash operation $x\#y$ whose interpretations in the standard model are $\lceil \log_2(x+1) \rceil$ and $2^{|x|\cdot|y|}$. The smash function is included to enable definitions of operations that increase the lengths of numbers polynomially. The classes of bounded formulas $\Sigma_i^b$ are defined in analogy with the arithmetical hierarchy except that occurrences of sharply bounded quantifiers are ignored. Sharply bounded quantifiers are bounded quantifiers whose bounds are of the form $x \leq |t|$. The intuition is that we can search all elements less than or equal to $|t|$ using a polynomial time algorithm.

Several types of induction schemas are used in Bounded Arithmetic. One of these is called *LIND*:

$$\forall x, p(\Phi(0, p) \wedge \forall y(\Phi(y, p) \rightarrow \Phi(S(y), p)) \rightarrow \Phi(|x|, p)).$$

$T_2$ is axiomatized by a finite set of basic open axioms called *BASIC* together with the usual induction schema (denoted *IND*) for bounded formulas. $S_2$ is the same except that *LIND* is used in place of *IND*. The subtheories $T_2^i$ and

$S_2^i$, $i = 1, 2, \ldots$, are defined by restricting the *IND* and *LIND* schemas to $\Sigma_i^b$ formulas. Having a parameter $p$ in the schemas is important for the definition of subtheories. The relation between the sets of provable sentences in these subtheories is:

$$S_2^i \subseteq T_2^i \subseteq S_2^{i+1}.$$

In particular $S_2 \equiv T_2$.

$PV$, introduced by Cook [9], is an equational theory whose function symbols define (in the standard model) all polynomial time computable functions and no other functions. (Of course, different function symbols may define the same function.) It is often convenient to have all these function symbols; therefore, we enlarge all the subtheories by $PV$. In such theories quantifier free formulas will define all relations computable in polynomial time and only those. In $S_2^1$ one can define all polynomial time computable functions using $\Sigma_1^b$ formulas. $S_2^1$ is then shown to be a $\forall\Sigma_1^b$ conservative extension of $PV$ (more precisely, of the first order theory axiomatized by the equalities of $PV$). Therefore, by adding $PV$ to the fragments above we obtain conservative extensions of the original systems. The enlarged systems are denoted by $T_2^i(PV)$ and $S_2^i(PV)$ (cf. [21, 17]). It is well-known that in these theories $\Sigma_i^b$ and $\Pi_i^b$ formulas are equivalent to *strict* $\Sigma_i^b$, respectively $\Pi_i^b$, formulas, which means that no sharply bounded quantifier is allowed to precede a quantifier that is not sharply bounded. In the presence of all $PV$ terms we can eliminate also the remaining sharply bounded quantifiers. Blocks of quantifiers of the same type can be replaced by a single one (using the pairing function, as usual) already in the weakest subtheory. Moreover, it is possible to axiomatize the subtheories $T_2^i$ and $S_2^i$ by the corresponding schemas restricted to strict $\Sigma_i^b$, or strict $\Pi_i^b$ formulas.

We shall use the basic result known as *Parikh's Theorem*, which holds for all the theories mentioned above. It says that if $T$ proves a sentence $\forall x \exists y\, \varphi(x, y)$, where $\varphi$ is a bounded formula, then $T$ proves $\forall x \exists y \leq t(x)\, \varphi(x, y)$ for some term $t(x)$.

Furthermore we need the following result of Buss [7]. It says that the *Strong Replacement Schema* for $\Sigma_i^b$ formulas

$$\exists z \forall x \leq |\tau|((\exists y \leq \sigma\, \alpha(x, y)) \equiv ((z)_x \leq \sigma \wedge \alpha(x, (z)_x))),$$

is provable in $S_2^i$. Here $\tau$ and $\sigma$ denote terms and $(z)_x$ denotes the $x$th element of the sequence coded by $z$.

Recall that $S_2^{i+1}$ is $\forall\Sigma_{i+1}^b$ conservative over $T_2^i$. For these and other theorems that we shall use see also [6, 17, 11].

Now we turn our attention to games. The games in the following principles will be defined by boolean circuits, as described in Section 2. Thus by saying *'a game is given'* we mean that a boolean circuit for the game is given. For example, a $PV$ quantifier-free formula of the form $\varphi(x, y_1, z_1, y_2, z_2, \dots)$ defines a game for every fixed $x$. Because $\varphi(x, y_1, z_1, y_2, z_2, \dots)$ is a polynomial time relation, there exists a circuit whose size is polynomial in $n = |x|$ for every number $x$. The circuit has $2k$ inputs for $n$-bit strings which correspond to the moves in a game of length $2k$. The transformation of a polynomial time relation into a series of circuits whose size grows polynomially is essentially Cook's Theorem. This theorem is provable in $S^1_2$; the proof is implicitly contained in the proofs of theorems about translations of arithmetical formulas into propositional formulas, see [20, 17]. Similarly, by saying that *'a strategy for Alice is given'* we shall mean that boolean circuits computing the moves of Alice are given.

In the principle we will have several games represented by circuits. We do not insist that different circuits define different sets of winning positions. In general the number of boards will be larger than the number of games, so one game will be assigned to several boards. For every game, Rule $\mathcal{B}$ will apply to the boards on which the game is played. The complementary game $G^\perp$ of a game $G$ will be determined by the same circuit, except that the roles of who starts will be switched.

Formally such a match is given by numbers $k, \ell$ and $n$, a string $w$ of numbers of length $k\ell$ in which every number $1 \le j \le \ell$ occurs exactly $k$ times, circuits $C_1, \dots, C_n$ defining games of length $k$, and an assignment $h : \{1, \dots, \ell\} \to \{1, \dots, n\} \times \{A, B\}$. $k$ is the length of the games, $\ell$ is the number of boards, $w$ is the schedule (the order of moves), $h$ is the assignment of games to boards, where $h(j) = (i, A)$ (respectively $h(j) = (i, B)$) means that game $i$ is assigned to board $j$ and Alice (respectively Bob) starts.

**Principle $A_k$.** *Let a match be given in which games of length k are played and in which Alice wins if she wins two complementary games. Let a strategy for Alice to play this match be given. Then it is possible to find moves of Bob which beat this strategy.*

The principle says, roughly speaking, that in every such match Bob is the person who has a winning strategy, but it is important how this statement is formalized. We do not say explicitly that Bob has a winning strategy, because we do not know how to express it in Bounded Arithmetic. Instead, we say that every given strategy of Alice can be beaten. We shall illustrate this concept on a couple of examples.

EXAMPLE 1. Consider the match based on a single game $G$ of length 3 and its dual $G^\perp$ in which the order of moves is as follows:

| | | | | | | | | | | | | |
|---|---|---|---|---|---|---|---|---|---|---|---|---|
| $G$: | $a_1$ | | | | $b_3$ | | | | $a_5$ | | | |
| $G^\perp$: | | $b_1$ | | | | | $a_4$ | | | $b_5$ | | |
| $G$: | | | | $a_2$ | | | | $b_4$ | | | $a_6$ | |
| $G^\perp$: | | | $b_2$ | | | $a_3$ | | | | | | $b_6$ |

Let $\gamma$ be a formula defining the game $G$, let $s_1, \dots, s_6$ be functions defining Alice's strategy for the match. Then the special case of Principle $A_3$ determined by this match is the following sentence:

$$\begin{aligned}\exists b_1 \cdots \exists b_6 ((\neg\gamma(s_1, b_3, s_5(b_1, \dots, b_4)) \wedge \neg\gamma(s_2(b_1, b_2), b_4, s_6(b_1, \dots, b_5))) \\ \vee (\gamma(b_1, s_4(b_1, b_2, b_3), b_5) \wedge \gamma(b_2, s_3(b_1, b_2, b_3), b_6)) \\ \wedge b_1 = b_2 \wedge (s_1 = s_2(b_1, b_2) \rightarrow b_3 = b_4) \wedge (s_4(b_1, b_2, b_3) \\ = s_3(b_1, b_2, b_3) \rightarrow b_5 = b_6)).\end{aligned}$$

The first two lines express that there exist moves of Bob such that he either wins on the two boards with $G$ or on the two boards with $G^\perp$. The last line expresses Rule $\mathcal{B}$ for this match (in order to make the rule quite explicit we use two variable $b_1$ and $b_2$ instead of just one).

EXAMPLE 2. Let us now consider a similar match with two games $G_1$ and $G_2$ of length 3.

| | | | | | | | | | | | | |
|---|---|---|---|---|---|---|---|---|---|---|---|---|
| $G_1$ : | $a_1$ | | | | $b_3$ | | | | $a_5$ | | | |
| $G_1^\perp$ : | | $b_1$ | | | | $a_4$ | | | | $b_5$ | | |
| $G_2$ : | | | | $a_2$ | | | | $b_4$ | | | $a_6$ | |
| $G_2^\perp$ : | | | $b_2$ | | | | $a_3$ | | | | | $b_6$ |

The corresponding special case of Principle $A_3$ is:

$$\begin{aligned}\exists b_1 \cdots \exists b_6 ((\neg\gamma(s_1, b_3, s_5(b_1, \dots, b_4) \vee \gamma(b_1, s_4(b_1, b_2, b_3), b_5))) \\ \wedge (\neg\gamma(s_2(b_1, b_2), b_4, s_6(b_1, \dots, b_5)) \vee \gamma(b_2, s_3(b_1, b_2, b_3), b_6))).\end{aligned}$$

Notice that since every game occurs only once, Rule $\mathcal{B}$ is empty in this case. It is important to realize that the distinction of games $G_1$ and $G_2$ is purely formal. If the sets of winning positions of $G_1$ and $G_2$ are the same we still get a sentence different from the one in the Example 1.

THEOREM 6.1.

(1) *For all $k \geq 1$, Principle $A_k$ is a $\forall\Sigma_1^b$ formula and it is provable in $S_2^k(PV)$;*

(2) *For all $k \geq 2$, $S_2^1(PV)$ plus $A_{k+4}$ proves all $\forall\exists$ theorems of $S_2^k(PV)$.*

PROOF. (i) The proof is similar to the proof of Lemma 4.2. As our means are more restricted now, we have to describe the proof more explicitly. First, let us recall the basic idea. We want to define Bob's moves as follows. If there exists a winning strategy from the position on a board that he should play, then we let him play a move that preserves this property. Furthermore, when he encounters the same position in the same game as it was on another board before, he will play in the same way, which will ensure Rule $\mathcal{B}$. Otherwise he plays arbitrarily.

To define the moves formally, we need the following generalization of the Strong Replacement, the *Schema of Dependent Choices*. We shall use the following notation: $z_u$ denotes the $u+1$-st element of the sequence coded by

$z$; $(z_0 \cdots z_u)$ denotes the number that codes the sequence coded by $z$ truncated to the first $u+1$ elements.

$$\begin{aligned} &\forall x\{\forall u \le |x| \exists v \le x\, \alpha(x,u,v) \\ (15) \qquad &\to \exists z\,[\forall u \le |x|\, \alpha(x,u,z_u) \wedge \forall u < |x|(\exists v \le x(\alpha(x,u+1,v) \\ &\wedge \beta(x,(z_0 \cdots z_u),v)) \to \beta(x,(z_0 \cdots z_u),z_{u+1}))]\}. \end{aligned}$$

This is provable for every $\Sigma_k^b$ formulas $\alpha$ and $\beta$ in $T_2^k$. It is essentially the definition of $\Sigma_{k+1}^b$ computations, thus the proof of this schema is implicitly contained in the proof of definability of these computations.

First we shall use the Schema of Dependent Choices to define moves Alice and Bob. The match and Alice's strategy are coded by the number $x$. The formula $\alpha(x,u,z_u)$ expresses that $z_u$ is a legal move in the match $x$ in the step $u+1$.

The formula $\beta(x,(z_0 \cdots z_u),v)$ expresses that after playing $z_0,\dots,z_u$,

1. if it is Alice's turn, then $v$ is a move played according to her strategy;
2. if it is Bob's turn, then $v$ is a "winning move" in the step $u+1$, or that it is the same move as he played on some board with the same game and same position before, in case Rule $\mathcal{B}$ applies to this move.

The condition that $v$ is a winning move can be expressed as the following formula:

$$\forall b \le |x| \bigvee_{i=1}^{k} \forall x_{i+2} \exists x_{i+3} \forall x_{i+4} \exists x_{i+5} \cdots \gamma(x,b,(z_0 \cdots z_u),v,x_{i+2},x_{i+3},\dots),$$

where $\gamma(x,b,(z_0 \cdots z_u),v,x_{i+2},x_{i+3},\dots)$ expresses that in the match coded by $x$ after playing $z_0 \cdots z_u$ it is Bob's turn and he should play on the board $b$, and if $z_{\ell_1},\dots,z_{\ell_{i-1}}$ are the previous moves on board $b$, then

$$(z_{\ell_1},\dots,z_{\ell_{i-1}},v,x_{i+2},x_{i+3},\dots,x_k)$$

is a winning position for Bob on board $b$.

The formula $\alpha$ is $\Sigma_1^b$, the formula $\beta$ is $\Sigma_k^b$ ($\Pi_{k-1}^b$ if $k > 2$). Hence this instance of the schema is provable in $S_2^k$.

Now, having defined the moves $z_0,\dots,z_m$, we prove the following. Let $b$ be a fixed board on which they play a game $G$ such that 'Bob has a winning strategy for $G$'. The last condition means that the sentence

$$\exists x_1 \forall x_2 \exists x_3 \forall x_4 \cdots \gamma(x_1,\dots,x_k),$$

in case Bob starts, and

$$\forall x_1 \exists x_2 \forall x_3 \exists x_4 \cdots \neg\gamma(x_1,\dots,x_k),$$

in case Alice starts, is satisfied (where $\gamma$ defines $G$). Then at every stage of the play Bob has a winning strategy in the current position on $b$. Since $k$ is a fixed number and they play only $k$ moves on the board $b$, we do not need any induction to this end. Thus we get that Bob wins on all such boards. Since for every pair of complementary games $G$ and $G^{\perp}$, Bob has a winning strategy for $G$ or for $G^{\perp}$ (in the above sense), the sequence of moves $z_0, \ldots, z_m$ proves that Bob can beat Alice's strategy.

(ii) We need a parameter-free version of the schema *LIND*. If we simply omitted parameters, we would get a very weak sentence, because all quantifiers in $\Phi(|x|)$ would be sharply bounded. Thus instead we introduce the following schema, which we shall call *LI*.

$$(16) \quad \forall x\{\Phi(1) \wedge \forall y \leq x(|y| < |x| \wedge \Phi(2^{|y|}) \rightarrow \Phi(2^{S(|y|)})) \rightarrow \Phi(2^{|x|})\}.$$

In the sequel we shall omit the bound $y \leq x$ which is in the above formula only in order to formally satisfy the definition of a bounded formula. We shall use this schema without parameters which means that $x$ is the only free variable of $\Phi(x)$. The function $2^{|x|}$ is definable, for instance, by $x\#1$. We also need the above schema with logarithmic number of possible parameters. This will be formalized by:

$$\forall x, p\{\Phi(1, 2^{|p|}) \wedge \forall y(|y| < |x| \wedge \Phi(2^{|y|}, 2^{|p|}) \rightarrow \Phi(2^{S(|y|)}, 2^{|p|})) \rightarrow \Phi(2^{|x|}, 2^{|p|})\}.$$

In the following three lemmas we shall assume $PV$ to be the base theory in which the implications are provable, and $i \geq 1$.

LEMMA 6.2. *Parameter-free schema LI for strict $\Pi^b_{i+2}$ formulas implies LI for strict $\Pi^b_i$ formulas with logarithmic number of parameters. Moreover, for every finite set of strict $\Pi^b_i$ formulas there exists a single strict $\Pi^b_{i+2}$ formula with only one free variable such that the LI axioms with logarithmic number of parameters for these strict $\Pi^b_i$ formulas are provable from LI for the $\Pi^b_{i+2}$ formula.*

PROOF. Let a $\Pi^b_i$ formula $\Psi(x, p)$ be given. Define a $\Pi^b_{i+2}$ formula $\Phi(z)$ by:

$$\forall w \forall v\{\langle v, w\rangle \leq |z| \wedge \Psi(1, 2^v) \wedge \forall y(|y| < |w| \wedge \Psi(2^y, 2^v) \rightarrow \Psi(2^{S(y)}, 2^v)) \rightarrow \Psi(2^w, 2^v)\},$$

where $\langle v, w\rangle$ denotes the usual pairing function. Now apply *LI* to $\Phi(z)$. $\Phi(1)$ is trivially true because $\langle 0, 0\rangle = 0$. The implication $\Phi(2^{|z|}) \rightarrow \Phi(2^{S(|z|)})$ follows from the properties of the pairing function. Indeed, if $\langle v, w\rangle = z$ and we increase $z$ to $z + 1$, then we have either $z + 1 = \langle v + 1, 0\rangle$, in which case we use $\Psi(1, v + 1)$ in the antecedent to derive it in the consequent, or $z + 1 = \langle v - 1, w + 1\rangle$ in which case we use the induction assumption and $\Phi(2^w, 2^{v-1}) \rightarrow \Phi(2^{S(w)}, 2^{v-1})$ to derive $\Psi(2^{w+1}, 2^{v-1})$. Thus we get $\forall z \Phi(2^{|z|})$.

To get

$$\forall x, p\{\Psi(1, 2^{|p|}) \wedge \forall y(|y| < |x| \wedge \Psi(2^{|y|}, 2^{|p|}) \\ \rightarrow \Psi(2^{S(|y|)}, 2^{|p|})) \rightarrow \Psi(2^{|x|}, 2^{|p|})\},$$

take $z = 2^{|\langle p, x\rangle|}$.

To prove the stronger version of the lemma just observe that we can encode $\Pi_i^b$ formulas $\Psi_1(2^{|x|}, 2^{|p|}), \dots, \Psi_\ell(2^{|x|}, 2^{|p|})$ by a single $\Pi_i^b$ formula $\Psi(2^{|x|}, 2^{|p|})$ of the same complexity such that for $i = 1, \dots, \ell$ and $S_2^1$ proves $\Psi(2^{|x|}, 2^{\langle |p|, |i| \rangle}) \equiv \Psi_i(2^{|x|}, 2^{|p|})$, and $\exists q(|q| = \langle |p|, |i| \rangle)$. ⊣

LEMMA 6.3. *The schema LI for strict $\Pi_{i+1}^b$ formulas with logarithmic number of parameters implies parameter-free induction IND for strict $\Pi_i^b$ formulas.*

PROOF. This is only a modification of a well-known proof that $\Pi_{i+1}^b LIND$ implies $\Pi_i^b IND$, but we must check that logarithmic number of parameters suffices for the parameter-free version of the induction.

Let a $\Pi_i^b$ formula $\Psi(x)$, with a single free variable $x$, be given. Suppose

$$\Psi(0) \wedge \forall y(\Psi(y) \rightarrow \Psi(S(y))) \tag{17}$$

is true. Let $\Psi'(x) := \forall y \le x \Psi(y)$. Then (17) is true also for $\Psi'$. Suppose that $\Psi(a)$ is false for some $a$. Then we also have $\neg\Psi'(2^{|a|})$. Let

$$\Phi(y, z) := \forall x \le z(\Psi'(x) \rightarrow \Psi'(x + y)).$$

We have $\Phi(1, 2^{|a|})$ by (17). Suppose that $\Phi(y, 2^{|a|})$ is true. Then, for every $x$ such that $\Psi'(x)$, we have $\Psi'(x + y)$, hence also $x + y < 2^{|a|}$. Thus applying $\Phi(y, 2^{|a|})$ we get $\Psi'(x + y + y)$. This proves $\forall y(|y| < |a| \wedge \Phi(2^{|y|}, 2^{|a|}) \rightarrow \Phi(2^{S(|y|)}, 2^{|a|}))$. Hence, using $LI$ for $\Phi$ we get $\Phi(2^{|a|}, 2^{|a|})$. Since we assume $\Psi'(0)$ and $\neg\Psi'(2^{|a|})$, this is a contradiction. So we have shown $\forall x \Psi(x)$. ⊣

LEMMA 6.4. *Parameter-free induction IND for strict $\Pi_{i+2}^b$ formulas implies induction IND for strict $\Pi_i^b$ formulas (with parameters).*

PROOF. Let a $\Pi_i^b$ formula $\Psi(x, p)$, with $x$ and $p$ the only free variables, be given. Define a $\Pi_{i+2}^b$ formula $\Phi(z)$ by:

$$\forall w \le z \forall v \le z(\langle v, w\rangle \le z \wedge \Psi(0, v) \\ \wedge \forall y < w(\Psi(y, v) \rightarrow \Psi(S(y), v)) \rightarrow \Psi(w, v)),$$

where $\langle v, w\rangle$ denotes the usual pairing function. Now apply induction to $\Phi(z)$. $\Phi(0)$ is trivially true. The implication $\Phi(z) \rightarrow \Phi(S(z))$ follows from the properties of the pairing function in the same way as in the proof of Lemma 6.2. Thus we get $\forall z \Phi(z)$. To get

$$\Psi(0, p) \wedge \forall y(\Psi(y, p) \rightarrow \Psi(S(y), p)) \rightarrow \Psi(x, p),$$

take $z = \langle p, x\rangle$. ⊣

Now we can start proving (ii) of the theorem. Let $k \geq 2$. Suppose that $S_2^k(PV)$ proves a sentence $\Psi := \forall x \exists y\ \psi(x, y)$, where $\psi$ is open. Then $\Psi$ is provable:

1. in $T_2^{k-1}(PV)$, because by Parikh's Theorem we can add a bound to the existential quantifier and $S_2^k(PV)$ is $\forall\Sigma_k^b$ conservative over $T_2^{k-1}(PV)$,
2. in parameter-free $IND$ for strict $\Pi_{k+1}^b$ formulas, by Lemma 6.4 and the well-known fact that $T_2^{k-1}(PV)$ can be axiomatized by $IND$ for strict $\Pi_{k-1}^b$ formulas,
3. from $LI$ for $\Pi_{k+2}^b$ formulas with logarithmic number of parameters, by Lemma 6.3,
4. from a single instance of parameter-free $LI$ for a strict $\Pi_{k+4}$ formula $\Phi$, by Lemma 6.2.

Thus we have in $PV$

$$\Lambda_\Phi \rightarrow \forall x \exists y\ \psi(x, y),$$

where $\Lambda_\Phi$ is (16). By renaming the variable $x$ to $u$, $\Lambda_\Phi$ gets the form $\forall u\ \lambda_\Phi(u)$ with $\lambda_\Phi(u)$ a bounded formula. By Parikh's Theorem, we have in $PV$

$$\forall x \exists u \leq s(x) \exists y (\lambda_\Phi(u) \rightarrow \psi(x, y)),$$

for some term $s(x)$. Hence in the predicate calculus we have

$$\Theta \rightarrow \forall x \exists u \leq s(x) \exists y (\lambda_\Phi(u) \rightarrow \psi(x, y)),$$

where $\Theta$ is a conjunction of some axioms of PV and some equality axioms. We move $\Theta$ to $\psi$; thus we get

$$\forall x \exists u \leq s(x) \exists y \exists y_1' \cdots \exists y_h' (\lambda_\Phi(u) \rightarrow (\theta(y_1', \ldots, y_h') \rightarrow \psi(x, y))),$$

where $\theta(y_1' \cdots y_h')$ is the open part of $\Theta$. To prove $\exists y\ \psi(x, y)$ for a given $x$, it suffices to prove $\exists y \exists y_1' \cdots \exists y_h' (\theta(y_1', \ldots, y_h') \rightarrow \psi(x, y))$, because $\theta(y_1' \cdots y_h')$ is a conjunction of some equality axioms and axioms of $PV$. Since it is a formula of the same type as $\exists y\ \psi(x, y)$, we shall omit $\Theta$ in the rest of the proof.

Furthermore, we can assume that $\Phi$ is in the prenex form with $k+4$ bounded quantifiers and with no other quantifiers. Let $\Phi(z)$ have the form $Q\bar{w}\ \varphi(z, \bar{w})$, with $\varphi$ quantifier-free. Thus we assume that the following formula is provable in the predicate calculus:

$$\begin{gathered}\forall x \exists u \forall z \exists y Q\bar{w} Q^\perp \bar{w}' Q\bar{w}'' Q^\perp \bar{w}''' \\ \{u > s(x) \vee [\varphi(1, \bar{w}) \wedge (|z| < |u| \wedge \varphi(2^{|z|}, \bar{w}') \rightarrow \varphi(2^{S(|z|)}, \bar{w}'')) \\ \wedge \neg\varphi(2^{|u|}, \bar{w}''')] \vee \psi(x, y)\}.\end{gathered}$$

Now we apply Herbrand's theorem and group the formulas that come from the induction axiom together and those that come from the formula $\forall x \exists y\ \psi(x, y)$

together. Thus the Herbrand disjunction gets the form:

$$\Delta(x, |z_1|, \dots, |z_\ell|, \bar{q}) \vee \bigvee_i \psi(x, t_i), \tag{18}$$

where $z_1, \dots, z_\ell$ are variables that come from $z$, $\bar{q}$ are variables that come from the universally quantified variables in $\bar{w}, \bar{w}', \bar{w}'', \bar{w}'''$, and $t_i$ are terms which may depend on all variables.

Now we shall describe a proof which can be formalized in $S^1_2(PV)$. First we shall present the proof and then explain how it can be formalized.

Let $n$ be given, let $m = |s(n)|$. First we substitute $\boldsymbol{n}$ for $x$ in $\Delta(x, z_1, \dots, z_\ell, \bar{q})$. Then for $j = 0, \dots, m-1$ we consider substitutions that substitute $\boldsymbol{j}$ simultaneously for all $|z_1|, \dots, |z_\ell|$. Recall that $m$ is logarithmic in $n$, thus we can list all these formulas. If we find an assignment $\bar{b}$ to the free variables $\bar{q}$ such that formula $\Delta(\boldsymbol{n}, \boldsymbol{j}, \dots, \boldsymbol{j}, \bar{\boldsymbol{b}})$ is falsified for some $j$, then $\bigvee_i \psi(\boldsymbol{n}, t_i[\boldsymbol{n}, \boldsymbol{j}, \bar{\boldsymbol{b}}])$ will be true, where $t_i[\boldsymbol{n}, \boldsymbol{j}, \bar{\boldsymbol{b}}]$ denotes the result of substituting $n$ for $x$, $j$ for $z_1, \dots, z_\ell$ and $\bar{b}$ for $\bar{q}$ in $t_i$. Thus we get $\exists y\, \psi(\boldsymbol{n}, y)$. So it remains only to prove that we can find such an assignment using Principle $A_{k+4}$.

The disjunction $\Delta(\boldsymbol{n}, \boldsymbol{j}, \dots, \boldsymbol{j}, \bar{b})$ has the following form:

$$\bigvee_{i=1}^{\ell} \{\varphi(\mathbf{1}, W_{i1}) \wedge (\boldsymbol{j} \geq r_i(\boldsymbol{n}, \boldsymbol{j}, \bar{q}) \vee \neg\varphi(\mathbf{2}^{\boldsymbol{j}}, W_{i2}) \vee \varphi(\mathbf{2}^{S(\boldsymbol{j})}, W_{i3})) \\ \wedge r_i(\boldsymbol{n}, \boldsymbol{j}, \bar{q}) \leq s(\boldsymbol{n}) \wedge \neg\varphi(\mathbf{2}^{|r_i(\boldsymbol{n},\boldsymbol{j},\bar{q})|}, W_{i4})\}. \tag{19}$$

In this formula, $W_{i1}, \dots, W_{i4}$ denote strings of terms and variables corresponding to the existentially, respectively, universally quantified variables of the quantifier prefixes $\boldsymbol{Q}\bar{w}$, $\boldsymbol{Q}^{\perp}\bar{w}'$, $\boldsymbol{Q}\bar{w}''$, $\boldsymbol{Q}^{\perp}\bar{w}'''$ of the occurrences of $\Phi$. The terms $r_i(\boldsymbol{n}, \boldsymbol{j}, \bar{q})$ come from $u$. We put the whole string of variables $\bar{q}$ there, but in fact $r_i$ does not depend on the variables of $W_{i1}, W_{i2}, W_{i2}, W_{i4}$.

Let us interpret the disjunction (19) in terms of games. The game $G_e$ will be the game defined by $\varphi(2^{|e|}, \bar{w})$. Thus the terms in (19) define a strategy for Alice to play games $G_e$ and $G_e^{\perp}$ on $4\ell$ boards. There is a problem, however, because the games corresponding to the formulas $\neg\varphi(\mathbf{2}^{|r_i(\boldsymbol{n},\boldsymbol{j},\bar{q})|}, W_{i4})$ are not fixed, they are determined by Alice's moves $|r_i(\boldsymbol{n}, \boldsymbol{j}, \bar{q})|$. Recall that we have $|r_i(\boldsymbol{n}, \boldsymbol{j}, \bar{q})| \leq |s(\boldsymbol{n})| = \boldsymbol{m}$. So we resolve this problem by introducing $m+1$ boards with games $G_0^{\perp}, \dots, G_m^{\perp}$ for every such formula. Then we interpret the term $|r_i(\boldsymbol{n}, \boldsymbol{j}, \bar{q})|$ as Alice's decision on which board of these $m+1$ boards she will concentrate, while ignoring the others. Formally, she will play according to the terms in $W_{i4}$ on the chosen board and she will play 0 in every move on all $m$ other boards.

A satisfying assignment for $\Delta(\boldsymbol{n}, \boldsymbol{j}, \dots, \boldsymbol{j}, \bar{q})$ is an assignment $\bar{b}$ that makes one of the formula in the disjunction true. If we interpret the assignment $\bar{b}$ as Bob's moves, it means that Alice wins the following games on some boards:

1. $G_0$, and
2. $G^{\perp}_{r(j)}$, for some $r(j) \leq m$, and
3. if $j < r(j)$, then she also wins $G^{\perp}_j$ or $G_{j+1}$.

The match and the strategy of Alice to which we apply Principle $A_{k+4}$ is determined by the conjunction

$$(20) \qquad \bigwedge_{j=0}^{m-1} \Delta(n,j,\ldots,j,\bar{q}).$$

This match is the union of the matches described above for all $j = 0,\ldots,m-1$. (The same is true about the strategies, they use only Bob's moves on the components. This fact is not relevant for what we want to prove, but it might be used to obtain some strengthening of the theorem. It would also suffice to require Rule $\mathcal{B}$ to be satisfied only locally on each block.)

Suppose that Alice wins this match. Let $r = \min_j r(j)$. Then there are boards on which she wins

1. $G_0$, and
2. $G^{\perp}_j$ or $G_{j+1}$ for all $j = 0,\ldots,r-1$, and
3. $G^{\perp}_r$.

Thus Alice always wins a pair of complementary games. Hence Principle $A_{k+4}$ implies that there is a falsifying assignment to (20), and we get $\Psi$.

It remains to show that the above argument can be done in $S^1_2(PV)$.

First we should realize that the Herbrand proof (18) is of constant size. For formulas of constant size (with terms of constant size) we can define satisfiability and prove the Tarski conditions without using any induction. In the proof we used the notation

$$\alpha(n_1,\ldots,n_h)$$

which means that numerals for $n_1,\ldots,n_h$ are substituted in $\alpha$. This is convenient if we need to write complicated formulas, but for formalizing the proof it is better to use a relation of satisfiability. So the above formula should be

$$\mathrm{Sat}(\alpha; n_1,\ldots,n_h),$$

where Sat is a suitable formula defining satisfiability for a class of open formulas. Since we have to introduce $m+1$ boards for the last term in (19), and then in (20) we have $m$ terms, we need to extend the satisfaction relation to formulas that are unbounded disjunctions and unbounded conjunctions of open formulas of a fixed size. Since $m$ is of logarithmic size, this presents no problems: the Tarski conditions are proved using $\Sigma^b_1 LIND$. The construction of the match uses only elementary operations with strings and graphs and all objects are of polynomial size in the $\log n$. It is well-known that all such elementary operations can be formalized in $S^1_2(PV)$ in such a way that the basic properties of these constructions are provable. The transition from $\varphi(2^{|e|},\bar{w})$

to a boolean circuit for the game $G_e$ is done by the formalization of the proof of Cook's Theorem. It is well-known that $S_2^1(PV)$ proves the existence of the minimum of a set of logarithmic size (the set is assumed to be coded by a number). Hence we can prove the existence of $r = \min_j r(j)$. To find a pair of complementary games in which Alice wins we only need to find the largest $j$, $j < r$, such that she wins only the second game from the pair $G_j^{\perp}, G_{j+1}$. This follows from the same principle as the previous thing.

Thus we have proved (ii). ⊣

We believe that the gap of size 4 between the upper and the lower bounds on the Principle $A_k$ should be possible to reduce, maybe, even to zero. There are several possible avenues to achieve it, but the solution may also be their combination. The first one is to extend the techniques of [14, 5] to get a more suitable conservativity result between the parameter free and general induction schemas. The results of Bloch [5] about the conservativity of general theories over the parameter-free versions cannot be applied directly, because they use general $\Sigma_k^b$ formulas, whereas we need strict $\Sigma_k^b$ formulas. One can also try to extend the simulation of proofs by the principle to an induction schema with (logarithmic number of) parameters. Another approach is to use the quantified propositional sequent calculus. In [20] we showed that for $2 \le j \le i$, the $\forall\Sigma_j^b$ theorems of $T_2^i$ (which are the same as those of $S_2^{i+1}$) can be characterized by certain reflection principles for subsystems of this calculus. The deadline for submitting the paper does not provide us enough time to optimize our result.

We can make this principle more uniform by taking universal schedules. We shall not do it in general, instead we shall consider only the two simplest instances. For such small values of $k$ one can state this principle in a fairly explicit way by taking the order of moves that is the most favorable for Alice. The principle says that Alice must always loose, thus an ordering that is the most favorable for Alice produces the strongest principle. Due to the gap between the theories considered in (i) and (ii) of Theorem 6.1, we do not know if $A_1$ and $A_2$ imply anything interesting; they just may be all provable in $S_2^1$. However, the explicit presentation of them below may help us to clarify their status. In particular, we may compare them with other characterizations of $\forall\exists$ theorems of subtheories of $S_2$.

The general idea used here is to schedule the moves of Bob as early as possible. Thus first to go are all Bob's moves on the boards on which he starts. As soon as Alice plays a move on a board $B$, then immediately Bob answers on $B$. In both instances we will assume that we have $n$ pairs of complementary games $G_i, G_i^{\perp}$; in all $G_i$ Bob is the player who starts. Every $G_i$ will be played on $n$ boards and every $G_i^{\perp}$ will be played on $n$ boards too. However, we shall state the principles without talking about games and strategies. Let $\mathrm{Circ}(n, m)$ denote the set of all circuits computing boolean function with $n$ input bits and $m$ output bits.

**Principle $A_1$.** *Suppose that circuits $C_1, \dots, C_n \in \mathrm{Circ}(n, 1)$, and $\alpha_{ij} \in \mathrm{Circ}(n^3, n)$, for $i, j = 1, \dots, n$, are given. Then there exists an assignment $\bar{b} = (b_1, \dots, b_n)$, $b_i \in \{0, 1\}^n$, such that for all $i, j \in \{1, \dots, n\}$*

$$C_i(b_i) \geq C_i(\alpha_{ij}(\bar{b})). \tag{21}$$

Let us note that the condition on the assignment $\bar{b}$ can also be stated as follows. For all $i \in \{1, \dots, n\}$,

1. either $C_i(b_{ij}) = 1$,
2. or $C_i(\alpha_{ij}(\bar{b})) = 0$ for all $j \in \{1, \dots, n\}$.

We already know that $A_1$ is provable in $S^1_2$ from the general Theorem 6.1, but it is instructive to prove this special case. The proof is very simple: the $\Sigma^b_1$ Strong Replacement gives us immediately an assignment $\bar{b}$ such that

$$\forall i, j((\exists x(C_i(x) = 1)) \equiv C_i(b_i) = 1). \tag{22}$$

This $\bar{b}$, clearly, satisfies (21).

Recall that Buss's Witnessing Theorem for $S^1_2$ implies that there is a polynomial time algorithm to construct an assignment $\bar{b}$ satisfying (21) from given circuits $C_1, \dots, C_n$ and $\alpha_{ij}$. To find this algorithm is fairly simple and we leave it to the reader. But it is interesting to observe that the proof above gives no hint at all how to do it—an assignment satisfying (22) cannot be constructed in polynomial time, unless $\boldsymbol{P} = \boldsymbol{NP}$!

**Principle $A_2$.** *Suppose that circuits $C_1, \dots, C_n \in \mathrm{Circ}(2n, 1)$, $\alpha_{i,j} \in \mathrm{Circ}(n^3 + (j-1)n^2, n)$, and $\alpha'_{ij} \in \mathrm{Circ}(2n^3, n)$, for $i, j = 1, \dots, n$ are given. Then there exists an assignment $(\bar{b}, \bar{b'})$, $\bar{b} = (b_i)^n_{i=1}$, $\bar{b'} = (b'_{ij})^n_{i,j=1}$, $b_i, b'_{ij} \in \{0, 1\}^n$, such that for all $i, j, k \in \{1, \dots, n\}$,*

$$C_i(b_i, \alpha'_{ij}(\bar{b}, \bar{b'})) \geq C_i(\alpha_{ik}(\bar{b}, \bar{b'}_1, \dots, \bar{b'}_{k-1}), b'_{ik}),$$

*and*

$$\alpha_{ij}(\bar{b}, \bar{b'}_1, \dots, \bar{b'}_{j-1}) = \alpha_{ik}(\bar{b}, \bar{b'}_1, \dots, \bar{b'}_{k-1}) \Rightarrow b'_{ij} = b'_{ik},$$

*where $\bar{b'}_j$ denotes $(b'_{1j}, b'_{2j}, \dots, b'_{nj})$.*

Here is an explanation in terms of games. First Bob plays all first moves on the boards with $G_1, \dots, G_n$, i.e., on the boards where he starts. Then Alice must play. She can postpone her moves in $G_i$ to the very end, because there are no other moves in these games than hers. Thus she has to play some $G_i^{\perp}$. As soon as she plays Bob responds. If we denote by $B_{ij}$ (respectively $B'_{ij}$) boards with $G_{ij}$ (respectively $G_{ij}^{\perp}$), then Alice will play in this order:

$$B'_{11}, B'_{21}, \dots, B'_{n1}, B'_{12}, B'_{22}, \dots, B'_{n2}, \dots\dots, B'_{1n}, B'_{2n}, \dots, B'_{nn}.$$

(This order is not forced, but it is universal in the sense that every order on a smaller set can be embedded into it. Here we use the simple principle that given an alphabet $c_1, \ldots, c_m$, every word of length $\ell$ can be obtained from the word $(c_1c_2\cdots c_m)^\ell$ by erasing certain letters.) Finally Alice plays her second moves on all boards $B_{ij}$.

F. Ferreira found a characterization of $\forall\exists$ theorems of $T^2_2$ [10]. As it is also based on Herbrand's theorem, the best way to describe it is to use games. Below we shall assume that 0-1 strings are encoded by numbers in the natural way; all zeros string is the number $0$, $00\cdots01$ is $1$, $00\cdots10$ is $2$ etc.

Let a circuit $C \in \mathrm{Circ}(2n, 1)$ be given such that $C(0, y) = 1$ for all $y \in \{0,1\}^n$. Alice and Bob play the following match. Bob chooses $b \in \{0,1\}^n$. Then in several rounds Alice alternatively tries to find

1. an $a_i$ such that $C(b, a_i) = 0$,
2. or an $a'_i$ such that $a'_i > b$ (where we compare the numbers that represent the strings) and Bob cannot respond with a $b'_i$ such that $C(a'_i, b'_i) = 0$.

So the order of moves is $b, a_1, a'_1, b'_1, a_2, a'_2, b'_2, \ldots$. If she succeeds to find such an $a_i$ or $a'_i$ in at most $n$ rounds, she wins. Otherwise Bob wins.

**Principle $M_2$.** *For every $n$, every circuit $C \in \mathrm{Circ}(2n, 1)$, and every strategy of Alice to play the above match, it is possible to find moves of Bob that beat Alice's strategy.*

THEOREM 6.5 ([10]). *Principle $M_2$ is a $\forall\Sigma^b_1$ sentence provable in $T^2_2$ and every $\forall\Sigma^b_1$ theorem of $T^2_2$ is provable in $S^1_2 + M_2$.*

Due to the $\forall\Sigma^b_1$-conservativity, the theorem is also true with $T^2_2$ replaced by $S^3_2$.

**§7. Matches with synchronized games.** The principle introduced in the previous section is still too complicated. We hope that eventually we will be able to prove that some nice special case of it is as strong as the general case. In this section we propose such a special case, but we are not able to prove that it is as strong as Principles $A_k$ or that it implies all $\forall\Sigma^b_1$ theorems of $S^i_2$ for some $i > 1$.

We know that principles based on playing games sequentially, i.e., one after the other, are too weak. This follows from the proof of Proposition 4.3. The other extreme is to play games in the *synchronized way,* i.e., to play the first move on all boards at once, then to play the second move on all boards at once, and so on. Another question is whether we need to have different games in the match or it is enough to have only one game and its complement. We shall adopt the latter plus another restriction in the following principle.

**Principle $B_{\ell,k}$.** *Let a game $G$ of length $k$ be given in which Bob starts. Let two strategies $\sigma$ and $\sigma'$ for Alice be given such that $\sigma$ is a strategy to play $G$ on $\ell$ boards in the synchronized way, and $\sigma'$ is a strategy to play $G^\perp$ on $\ell$ boards in the synchronized way. Then either there exist Bob's moves to beat $\sigma$ on all boards or there exist Bob's moves to beat $\sigma'$ on all boards.*

**Principle $B_k$** *is the sentence* $\forall x\ B_{|x|,k}$.[3]

To get a match of the type used in $A_k$, use $2\ell$ boards, play $G$ on $\ell$ boards and $G^{\perp}$ on the other $\ell$ boards and think of $\sigma, \sigma'$ as a strategy for this match. Thus the additional restriction mentioned above is the splitting of the match in which both $G$ and $G^{\perp}$ are played into two: one for $G$ and one for $G^{\perp}$. Hence $B_k$ is a special case of $A_k$.

Since we want to get rid of Rule $\mathcal{B}$, we shall restrict the class of strategies that Alice can use. We shall require that when playing the first moves on the boards on which she starts, she must play different moves on all boards, and, similarly, when she plays the second moves on the boards on which Bob starts, she must play different moves on all boards. Then what remains from Rule $\mathcal{B}$ is only the requirement that Bob must play the same move on all boards on which he starts. *Thus Principle* $B_k$ *means the above principle with this additional restriction.*

There are indications that $B_k$ may be weaker $A_k$. In particular, we only know that $A_k$ is provable in $S_2^k$. However, it follows from an unpublished result of J. Håstad that in $S_2^k$ one can prove $B_{k+3}$. The only hardness result that we have so far is rather weak. We studied an infinite version of this principle in [28]. The infinite principle is exactly the same except that the length of the games is $\omega$, thus we can call it $B_{\ell,\omega}$. It is well-known that there are infinite games which are not determined, i.e., none of the two players has a winning strategy. If the games are not determined, such principles may fail. Indeed, we showed that already $B_{2,\omega}$ is false [28]. That said, we do not know anything about provability of $\forall y\ B_{2,|y|}$ in $S_2$ or its subtheories.

Though we do not know precisely the power of this principle it is interesting to study it because it has interesting properties. The most interesting of these is the fact that, for a given $\ell$ and $k$, one can fix the strategies $\sigma$ and $\sigma'$. The idea is that the most general strategies are the *copycat* strategies which simply copy the moves of the opponent. We shall describe it below and show that the principle restricted to copycat strategies is equivalent to the general one.

We need to introduce suitable notation. Once we have it, the proofs will be easy. Let $\ell$ and $k$ be fixed. We assume that $k$ is a constant, and $\ell$ can be arbitrary. In $B_{\ell,k}$ we need a game and two strategies. We shall represent it by the following structure:

$$(M_1, M_2, \ldots, M_k; G, \sigma_1, \ldots, \sigma_k).$$

The sets $M_i$ are possible moves in the step $i$ of the game, $G$ is the set of winning positions, and

$$\sigma_j : M_1^{\ell} \times \cdots \times M_{j-1}^{\ell} \to M_j^{\ell}.$$

[3]We can only code games with logarithmic number of boards, therefore we use $|x|$.

We assume that if $j$ is odd, then $\sigma_j$ depends only on even moves, and if $j$ is even, then $\sigma_j$ depends only on odd moves. The strategies are $\sigma = (\sigma_2, \sigma_4, \dots)$ and $\sigma' = (\sigma_1, \sigma_3, \dots)$.

We shall imagine playing $\ell$ games of length $k$ as constructing an $\ell \times k$ matrix, where the rows correspond to boards, and the columns correspond to moves. The mapping $\sigma_j$ tells us how to construct the $j$th column from the first $j-1$ columns.

Our first observation is that we can assume w.l.o.g. that $M_1 = \{1, \dots, \ell\}$. Indeed, by renaming moves we can assume that if Alice starts then she plays $i$ on the $i$th board (formally, $\sigma_1 = (1, \dots, \ell)$; recall that Alice has to play different moves on different boards when she starts). Then we redefine $G$ by prohibiting to start with an element not in $\{1, \dots, \ell\}$. Since this restriction concerns only Bob, the restricted principle is at least as strong as the unrestricted one.

The sets of moves for the copycat strategy are defined as follows.

1. $N_1 = N_2 = \{1, \dots, \ell\}$;
2. for $j > 2$, $N_j = N_{j-1}^{\ell-1}$.

The strategies are defined by $\rho_1, \dots, \rho_k$, such that

1. $\rho_1$ and $\rho_2$ simply print $i$ in the $i$th board; thus $\rho_2$ does not depend on the first moves;
2. for $j > 2$, on the $i$th board $\rho_j$ prints the $j-1$-st moves from all other boards; in terms of matrices, $\rho_j$ produces the $j$th column such that in every row $i$ the entry is the string of all elements from the $j-1$-st column with the $i$th element omitted.

These functions determine strategies $\rho = (\rho_2, \rho_4, \dots)$ and $\rho' = (\rho_1, \rho_3, \dots)$.

The idea of the copycat strategy is that it records information about all boards on each board, without any redundancy. Thus when Bob starts with the same moves on all $\ell$ boards, Alice does not have to copy the moves from other boards to the given board. What she does instead is encoding the number of the board into the position on the board. Similarly, the reason for omitting the $i$th element when copying the previous moves onto the $i$th board is that this information is already present on the $i$th board.

Given a matrix $U$, we shall denote by $\text{row}_i(U)$ (respectively $\text{col}_j(U)$) the $i$th row (respectively $j$th column) of $U$. Thus $\text{row}_i(U)$ represents the position on the $i$th board and $\text{col}_j(U)$ represents the moves in the $j$th step. We shall consider matrices of the dimensions $\ell \times j$, $1 \le j \le k$, that will represent positions of playing $\ell$ copies of a game for $j$ steps. We shall say that $U$ *was played using* $\rho$ (respectively using $\rho'$) if for every even (respectively odd) $j' \le j$,

$$\text{col}_{j'}(U) = \rho_{j'}(\text{col}_1(U) \cdots \text{col}_{j'-1}(U))$$
$$(\text{respectively}, \text{col}_{j'}(U) = \rho'_{j'}(\text{col}_1(U) \cdots \text{col}_{j'-1}(U))).$$

If $U$ was played using $\rho$, we also assume that the entries in the first column are all the same. Recall that $\rho$ and $\rho'$ are strategies of Alice; Bob can play in arbitrary way (except that he has to obey Rule $\mathcal{B}$).

LEMMA 7.1. *For every $1 \leq j \leq k$ and $w \in N_1 \times \cdots \times N_j$ there exists a unique $i$ and a unique $\ell \times j$ matrix $U$ such that*

1. $\mathrm{row}_i(U) = w$, *and*
2. *if $j$ is even then $U$ was played using $\rho$, if $j$ is odd then $U$ was played using $\rho'$.*

PROOF. For $j = 1, 2$ it follows immediately from definitions. For $j > 2$ we shall use induction. Let $w = w_1 \cdots w_j$. By the induction assumption we have a unique $i$ and a unique $U'$ for $w_1 \cdots w_{j-2}$. Then $w_{j-1}$ and $w_j$ determine uniquely the $j-1$-st column of $U$. To get the $j$th column of $U$ apply $\rho_j$. ⊣

THEOREM 7.2. *Let $k \geq 1$ be fixed. Then it is provable in $S^1_2$ that $B_{\ell,k}$ follows from its special case where the strategies are fixed to be the copycat strategies.*

PROOF. Let $(M_1, M_2, \ldots, M_k; G, \sigma_1, \ldots, \sigma_k)$ be a general structure. We shall construct a homomorphism $\mu_k$ from $(N_1, N_2, \ldots, M_k; \rho_1, \ldots, \rho_k)$ into $(M_1, M_2, \ldots, M_k; \sigma_1, \ldots, \sigma_k)$. Then we take the inverse image of $G$ and obtain an $H \subseteq N_1 \times \cdots \times N_k$. Thus we get a game for the match with the copycat strategies. If we find moves for Bob that beat $\rho$ or $\rho'$, then their homomorphic image will be moves for Bob that beat $\sigma$ or $\sigma'$.

The homomorphism $\mu_k$ will be defined by defining its initial segments $\mu_j : N_1 \times \cdots \times N_j \to M_1 \times \cdots \times M_j$ by recursion. $\mu_1$ is the identity mapping. Let $j > 1$, let $w \in N_1 \times \cdots \times N_j$. Take the unique $i$ and the unique $\ell \times j$ matrix $U$ with the properties guaranteed by the lemma. Let $U'$ be $U$ less the last column. Then we put

$$\mu_j(w) = \mu_j(\mathrm{row}_i(U)) = \mu_{j-1}(\mathrm{row}_i(U'))\,\sigma_{ij}\begin{pmatrix} \mu_{j-1}(\mathrm{row}_1(U')) \\ \vdots \\ \mu_{j-1}(\mathrm{row}_\ell(U')), \end{pmatrix} \tag{23}$$

where $\sigma_{ij}$ denotes the entry in the $i$th row of the column determined by $\sigma_j$. We define

$$H = \{w \in N_1 \times \cdots \times N_k;\ \mu_k(w) \in G\}.$$

Now we claim that if $U$ is a $\ell \times k$ matrix played using $\rho$ (respectively $\rho'$), then its image

$$W = \begin{pmatrix} \mu_k(\mathrm{row}_1(U)) \\ \vdots \\ \mu_k(\mathrm{row}_\ell(U)), \end{pmatrix}$$

is a matrix played using $\sigma$ (respectively $\sigma'$). This easily follows by induction from (23). (Notice that, e.g., if $U$ was played using $\rho$, then we shall use initial parts of $U$ when defining *even* columns of $W$ using the formula (23). However, for defining odd columns we shall use completely different matrices.)

Thus we have reduced the task of finding Bob's moves that beat Alice in the general match to the same task in the game with the copycat strategies. It remains to argue that the reduction can be done in $S^1_2$. Since all the arguments and transformations were completely elementary, the problems boils down to showing that in order to define $\mu_k(w)$ we only need a polynomial number of steps. In (23) we need to know $\mu_{j-1}$ only for $\ell$ values when defining one value of $\mu_j$. Hence to define $\mu_k(w)$ we need only $O(\ell^k)$ steps. Since $k$ is constant this is polynomial. ⊣

COROLLARY 7.3. *For every* $k \geq 1$, $S^1_2$ *proves* $\forall x\, \boldsymbol{B}_{\lceil \sqrt[k]{\log\log x}\rceil,k}$.

PROOF. If $\ell = \lceil \sqrt[k]{\log\log x}\rceil$, then the size of $N_1 \times \cdots \times N_k$ is polynomial in the length of $x$. Hence a winning strategy for one of the players can be coded. Then we can use this strategy to defines Bob's moves in the match. ⊣

This theorem enables us to present Principle $\boldsymbol{B}_{\ell,k}$ in terms of a problem about coloring a concrete hypergraph. Given $\ell$ and $k$ we define a hypergraph $\Gamma_{\ell,k}$ as follows. The hypergraph has edges of size $\ell$ of two types: blue and red. The set of vertices is $V = N_1 \times \cdots \times N_k$ as defined before. A blue edge is an $\ell$ element subset of $V$ whose elements are rows of a matrix played using $\rho$; a red edge is an $\ell$ element subset of $V$ whose elements are rows of a matrix played using $\rho'$. An $H \subseteq V$ determines a coloring in which $v$ is blue if $v \in H$, otherwise $v$ is red. Now we can state Principle $\boldsymbol{B}_{\ell,k}$ in the following equivalent form.

**An equivalent form of Principle $\boldsymbol{B}_{\ell,k}$.** *There exists no coloring of* $\Gamma_{\ell,k}$ *such that every blue hyperedge contains a blue vertex and every red hyperedge contains a red vertex.*

Krajíček observed (see [17], Theorem 14.2.3) that given any ***NP***-complete problem $X$ and any $\forall\Sigma^b_0$ sentence $\Phi$, we can represent $\Phi$ as the statement $\forall n\ s(n) \notin X$ for some concrete sequence $s(n)$, $n = 1, 2, \ldots$. (This can be shown by a more or less straightforward application of Cook's theorem.) Thus we can take the consistency statement for the theory in question and we get a "combinatorial" unprovable sentence. However, the sequence $s(n)$ is some strange sequence with a very complicated definition, thus it is difficult to find any "mathematical" interpretation of such sentences. Our sequence of hypergraphs is fairly explicitly defined. In fact, there is a simple definition of these hypergraphs by recursion on $k$. But our statement has two drawbacks. Firstly, it is $\forall\Sigma^b_1$, because the hypergraphs are not represented by vertices and hyperedges, they are defined by circuits, and so are the colorings. Secondly, which is much worse, we are not able to prove that our principle is unprovable in any reasonably strong theory.

We conclude this section by noting that the theorem that the infinite principle $\boldsymbol{B}_{2,\omega}$ is false was proved using the infinite graph $\Gamma_{2,\omega}$. This implies that there is no way to use this result to derive any hardness of the finite principle $\forall x\, \boldsymbol{B}_{2,|x|}$, because we can show that the latter principle restricted to the copycat strategies is easy (this result will be published elsewhere).

**§8. The separation problem for subtheories of bounded arithmetic.** One of the main problems in proof complexity is to prove that the hierarchy of subtheories of $S_2$ is proper. For example, if we proved that $S_2^{i+1} \neq S_2^i$, we would know that $S_2^i$ does not prove $\Sigma_{i+1}^b = \Sigma_i^b$. The problem of proving $S_2^i \nvdash \Sigma_{i+1}^b = \Sigma_i^b$ is formally weaker than the problem of proving $\Sigma_{i+1}^b \neq \Sigma_i^b$. Yet our experience suggests that it may be as hard as the latter problem. Problems in proof complexity seem to be so closely tight with the corresponding problems in computational complexity that any such separation of theories would be as big a breakthrough as the separation of the corresponding complexity classes.

In this situation we can only hope to prove *conditional results*. These are results based on some complexity theoretical assumptions. The first such result was the result of Krajíček, Pudlák and Takeuti [21] showing that the hierarchy of subtheories $S_2^i$, $i = 1, 2, \ldots$ is strictly increasing provided that the Polynomial Hierarchy $\Sigma_i^b$, $i = 1, 2, \ldots$ is. More precisely, we proved that $S_2^{i+1}$ is stronger than $T_2^i$ if $\Sigma_{i+2}^p \neq \Pi_{i+2}^p$. A similar result was proved by Krajíček [16]. He proved that $T_2^i$ is stronger than $S_2^i$ if polynomial time computations with polynomial number of queries to a $\Sigma_i^p$ oracle can recognize more languages than polynomial time computations with logarithmic number of queries to a $\Sigma_i^p$ oracle. This confirms the conjecture that the hierarchy $S_2^1 \subseteq T_2^1 \subseteq S_2^2 \subseteq T_2^2 \subseteq \cdots$ is strict everywhere; still we want to know more. Assuming that the Polynomial Hierarchy is strictly increasing, we know that there exists a $\forall\Sigma_{i+2}^b$ sentence provable in $S_2^{i+1}$, but not in $T_2^i$. This is the best possible result, because $S_2^{i+1}$ is $\forall\Sigma_{i+1}^b$ conservative over $T_2^i$. For pairs $T_2^i, S_2^i$ the complexity of the known separating sentences also increases with $i$, but no conservativity result is known. In fact, we conjecture that they can be separated already by $\forall\Sigma_0$ sentences (if we consider theories augmented with $PV$, then they should be separable by purely universal sentences). At present we do not have any results that would support this conjecture.

Separation of subtheories of $S_2$ by low complexity sentences is an important open problem in proof complexity. Our feeling is that this is more closely related to the separation of the corresponding classes in complexity theory than merely proving that the theories are different. The main reason we think so is that the problem is closely connected with the problem of showing that propositional proof systems associated with these theories form an increasing hierarchy. (It is outside the scope of this article to explain what the proof system associated with a theory is; we refer the reader to [17, 27].)

The first thing that we need in order to solve this problem is to have sentences that could separate the theories. As far as $\forall\Sigma_0$ sentences are concerned, we do have sentences that very likely separate $S_2^{i+1}$ from $S_2^i$, but we do not like them. The sentences express the consistency of subsystems of the quantified propositional sequent calculus. As usual, we are not able to interpret them in a clear combinatorial way. Apart from them we do not have other $\forall\Sigma_0$ sentences that would be good candidates for separating sentences. The next highest level is $\forall\Sigma_1$ sentences. We have shown here that a semantical interpretation of the Herbrand disjunctions produces such sentences.

For the two lowest levels very explicit sentences were found. We have already mentioned $\forall\Sigma_1^b$ sentences of Ferreira. For $T_2^1$, $\forall\Sigma_1^b$ sentences which axiomatize all $\forall\Sigma_1^b$ theorems of this theory were found by Buss and Krajíček [8]. Their sentences express the fact that every polynomial local search problem has a solution. We shall state their principle, again, using circuits.

**Principle *PLS*.** *For every circuits $D, F$ and $C$, $D \in$ Circ$(n,1)$, $F, C \in$ Circ$(n,n)$ such that $D(0) = 1$, there exists $b \in \{0,1\}^n$ such that either $C(F(b)) \leq C(b)$ (where we compare the numbers that represent the strings) or $D(F(b)) = 0$.*

THEOREM 8.1 ([8]). *Principle **PLS** is a $\forall\Sigma_1^b$ sentence provable in $T_2^1$ and every $\forall\Sigma_1^b$ theorem of $T_2^1$ is provable in $S_2^1 + \boldsymbol{PLS}$.*

Having such characterizations of $\forall\Sigma_1^b$ theorems we can transform the logical problem of showing (conditional) separations into a purely computational problem. Given a sentence $\forall\Sigma_1^b$ of the form $\forall x \exists y \leq t\, \phi(x,y)$, with $\phi$ sharply bounded (or simply open, if we include $PV$ into our theories), we can consider the associated *search problem,* which is the problem, from a given $x$, to construct a $y$ such that $y \leq t \wedge \phi(x,y)$. The reduction is based on the fact that the complexity of the search problem is closely related to the provability of the sentence.

Let us start with the simplest problem: to prove that a $\forall\Sigma_1^b$ sentence is not provable in $S_2^1$. Recall that Buss's Witnessing Theorem [6] says that if such a sentence is provable in $S_2^1$, then the associated search problem can be solved in polynomial time. Thus to prove the unprovability in $S_2^1$, it suffices to show the opposite. Of course, we have to assume something, because a proof that such an algorithm does not exists gives $\boldsymbol{P} \neq \boldsymbol{NP}$.

Now consider a general situation. Let $T$ be an arbitrary subtheory of $S_2$. Suppose we have a $\forall\Sigma_1^b$ sentence $\Phi := \forall x \exists y \leq t\, \phi(x,y)$, with $\phi$ sharply bounded, such that all $\forall\Sigma_1^b$ theorems of a theory $T$ are provable in $PV + \Phi$. Suppose that we want to prove that another $\forall\Sigma_1^b$ sentence $\Psi := \forall x \exists y \leq s\, \psi(x,y)$, with $\psi$ sharply bounded, is not provable in $T$. Then we can use the following reduction.

Let $\Psi$ be provable in $T$. Then we have

$$PV \vdash \forall x \exists y \leq t\, \phi(x,y) \rightarrow \forall x \exists y \leq s\, \psi(x,y). \tag{24}$$

Let us assume that the bounds $y \leq t$ and $y \leq s$ are already in the formulas $\phi$

and $\psi$. Applying Herbrand's theorem for $\forall\exists\forall$ prenex sentences (see formula (6), or [21]) we get some $PV$ terms $t_1, \ldots, t_\ell$ such that

$$\mathbb{N} \models \bigvee_{i=1}^{\ell} (\phi(t_i, z_i) \rightarrow \psi(z_0, s_i)),$$

where terms $t_i$ and $s_i$ may depend only on variables $z_0, \ldots, z_{i-1}$. We can write it in the following equivalent form:

$$\text{(25)} \qquad \mathbb{N} \models \bigwedge_i \phi(t_i, z_i) \rightarrow \bigvee_i \psi(z_0, s_i).$$

Let us write the antecedent with the dependences on variables explicitly displayed.

$$\phi(t_1(z_0), z_1) \wedge \phi(t_2(z_0, z_1), z_2) \wedge \cdots \wedge \phi(t_\ell(z_0, z_1, \ldots, z_{\ell-1}), z_\ell).$$

We shall interpret the terms $t_1, \ldots, t_\ell$ as a polynomial time oracle algorithm that uses an input $z_0$ and answers $z_1, \ldots, z_\ell$ of an oracle for the search problem associated with $\Phi$. This algorithm makes the antecedent of (25) true. Then the terms $s_1, \ldots, s_\ell$ produce a solution to the search problem associated with $\Psi$. Thus we can state the reduction to a computational problem as the following lemma.

LEMMA 8.2. *Suppose that $PV \vdash \Phi \rightarrow \Psi$. Then there exists a polynomial time oracle algorithm and a number $\ell$ such that the algorithm finds a solution of the search problem associated with $\Psi$ using $\ell$ adaptive queries to the oracle for the search problem associated with $\Phi$.*

(*'Adaptive'* means that queries may depend on the answers to the previously asked queries.)

To assume that the base theory is only $PV$ may be too restrictive. If we assume instead that all $\forall\Sigma_1^b$ theorems of a theory $T$ are provable in $S_2^1 + \Phi$, then the same is true except that the number of queries is not restricted (hence in every computation it is implicitly bounded by a polynomial). This follows from [26, Theorem 3]. For more results in this area see [12].

As an example consider the conjecture that $S_2$ is not $\forall\Sigma_1^b$ conservative over any $S_2^i$. To prove this conjecture, it suffices to prove that for every $k$ there exists $k'$ such that the search problem associated with $A_{k'}$ cannot be solved in polynomial time using an oracle for the search problem associated with $A_k$.

Another context in which we can study the problem about the conservativity of subtheories of $S_2$ is the realm of *relativized* theories which are theories with an *oracle*. We adopt the terminology of computational complexity, though oracles in proof complexity are different. Given a subtheory $T$ of $PA$ in which the induction axiom schema is restricted to a class of formulas $\Theta$, we define $T[R]$, the relativization of $T$, to be the theory whose language is augmented by a unary predicate $R$ and whose induction schema is extended to $\Theta[R]$. If

$T$ uses a special form of induction, such as $LIND$, then so does $T[R]$. (This is not a precise definition applicable to every theory, but it suffices for all that we need here.) It is also possible to view $T[R]$ as a very weak extension of $T$ to a second order theory in which $R$ acts as a free second order variable.

Let us stress that we do not accept any other axioms about $R$. That is why such an oracle is different from oracles in computational complexity—when considering a Turing machine with an oracle, the oracle is always a concrete set. Nothing prevents us from adding specific axioms about $R$; the reason why we do not do it is that we want to show independence results. When we prove that $\Phi[R]$ is not provable in $T[R]$, we know that there exists no *uniform way* of proving all sentences $\Phi[R/\psi]$ for $\psi$ such that $\Phi[R/\psi] \in \Theta$.

As in computational complexity, also in proof complexity we can prove unconditional separation results for relativized theories. When we have a conditional separation of theories and a relativization that makes the condition true is known, we can often use it to obtain an unconditional separation of the relativizations of the theories. Thus an oracle that makes the Polynomial Hierarchy strict enabled us to prove $T_2^i[R] \neq S_2^{i+1}[R]$ for all $i \geq 1$, cf. [21]. In a similar vein, Krajíček proved $S_2^i[R] \neq T_2^i[R]$, cf. [16].

The first result concerning relativized $\forall\Sigma_1^b$ nonconservativity appeared already in the seminal book of Buss [6]. It is the result that $T_2^1[R]$ is not $\forall\Sigma_1^b[R]$ conservative over $S_2^1[R]$. This also follows from Theorem 8.1 and a construction of an oracle such that relativized $PLS$ problems cannot be solved in the relativized polynomial time. Another result is due to Krajíček [15] and Pudlák [26] (see also [18]) who proved that $T_2^2[R]$ is not $\forall\Sigma_1^b[R]$ conservative over $S_2^2[R]$. Impagliazzo and Krajíček [13] proved that $T_1^i[R]$ is not $\forall\Sigma_1^b[R]$ conservative over $S_1^i[R]$ for all $i \geq 1$. The theories $S_1^i$ and $T_1^i$ are $S_2^i$ and $T_2^i$ without the # function. For further results see [19, 12].

It is well-known that these separation problems can be reduced to lower bounds on the size of propositional proofs in bounded depth Frege systems [25]. For example, to prove that $T_2^j[R]$ is not $\forall\Sigma_1^b[R]$ conservative over $T_2^i[R]$, it would suffice to prove more than quasipolynomial lower bound on the lengths of proofs of the propositional translations of $\boldsymbol{A}_j$ in the depth $i$ Frege system. As $\boldsymbol{A}_j$ is rather complicated, one may also try $\boldsymbol{B}_j$. We defined $\boldsymbol{B}_j$ hoping that the propositional translations of these principles would be the counterparts of the Sipser functions which separate the levels of the $AC^0$ hierarchy.

We conclude this section by observing that the provability of a strong version of $\boldsymbol{B}_k$ in $S_2^1$ (strong in the same sense as in Section 5) is equivalent to the provability of the relativized principle $\boldsymbol{B}_k[R]$ in $S_2^1[R]$. Having a possibility to translate relativized independence to unrelativized would be great, provided that the translations would be also $\Pi_1$ statements. But our strong version is $\Pi_2$; moreover, it concerns only $S_2^1$.

We shall consider $B_{\ell,k}$ only with the copycat strategies. Thus given $k$ and $\ell$ everything is fixed, except the game $G$. As in Section 5, in the strong principle we do not specify $G$ and instead we let Alice to decide who won after they play $k$ moves on a board. Again, Alice and Bob play several matches one after the other, and Alice ought to play consistently: once she declares a winner in a final position, then if the position occurs again, she must declare the same person the winner.

**Strong Principle $B_k$.** *For every virtual match with games of length k there exists an m and a strategy for Bob to play the match m times such that the strategy guarantees Bob's winning at least one of these m matches.*

As before, we assume that Bob's strategy is given by a boolean circuit. The principle is trivially true, because the games are finite and thus there are circuits computing the winning strategies. Let us stress that we do not claim that the circuits are of polynomial size, most likely they are not. In the relativized principle $B_k[R]$ everything is the same except for the circuits. The circuits get answers from the oracle $R$ for unit cost.

PROPOSITION 8.3. *$S^1_2$ proves Strong Principle $B_k$ if and only if $S^1_2[R]$ proves Principle $B_k[R]$.*

PROOF. ($\Rightarrow$) Suppose $S^1_2$ proves Strong Principle $B_k$. Let $\mathcal{M}^R$ be a model of $S^1_2[R]$ and let $\mathcal{M}$ be $\mathcal{M}^R$ with the interpretation of $R$ omitted. Let a match $M(s, C)$ based on a schedule $s$ and a circuit $C$ be in $\mathcal{M}^R$. Let a strategy $\rho$ for Alice to play $M(s, C)$ be in $\mathcal{M}^R$. Suppose that the game in $M(s, C)$ defined by $C$ is of length $k$. Let $M(s)$ be the corresponding virtual match. Then $M$ is in $\mathcal{M}$. Thus in $\mathcal{M}$ we have a strategy $\sigma$ for Bob to win at least one copy of $M$ in a series of $m$ virtual matches, for some $m$. We can use this strategy also in $\mathcal{M}^R$, and we can play it against $\rho$. If $\sigma$ beats $\rho$ we are done, so suppose not. Thus $\rho$ beats $\sigma$. Take the sequence of moves produced in this way. This is a sequence that is coded by a number, so it is also in $\mathcal{M}$. But this is impossible, because we assumed that $\sigma$ is a winning strategy for the series of virtual matches.

($\Leftarrow$) Suppose $S^1_2[R]$ proves Principle $B_k[R]$. We shall consider only matches $M(s, G)$ with a schedule $s$ and the game $G$ of length $k$ defined as follows: $w$ is a winning position in $G$ if $\langle s, w\rangle \in R$. By the relativized Buss's Witnessing Theorem, there exists an oracle polynomial time algorithm $A$ which queries $R$ and finds Bob's moves that beat Alice in the match $M(s, G)$. Now we want to use this algorithm as Bob's strategy to win a series of virtual matches. The only problem is that, according to the rules, Bob is not allowed to directly ask Alice who is the winner in a given position $w$. But due to Lemma 7.1, Bob can play on $\ell$ boards so that $w$ appears as the final position on one of them. Hence Bob's strategy $\sigma$ will be to get the answers to the queries in this way and then play the moves that beat Alice. We have to show that $\sigma$ beats Alice provably in $S^1_2$. Suppose not. Take a model $\mathcal{M}$ and $\ell$ in $\mathcal{M}$ such that $\sigma$ fails.

The moves of Alice that beat $\sigma$ form a set coded by a number. We shall use the moves to define the interpretation of $R$. We put exactly those pairs $\langle s, w\rangle$ in $R$ for which Alice said that $w$ was a winning position in the game. We have $LIND$ for $\Sigma_1^b[R]$ formulas in this model, because the interpretation of $R$ is a set coded in $\mathcal{M}$. But if we run $A$ in the expanded model, $A$ will not do what it should, which is a contradiction. Hence it is provable in $S_2^1$ that $\sigma$ beats Alice. ⊣

**Acknowledgment.** I would like to thank to Jan Krajíček for suggesting some improvements, and to Johan Håstad and Alexander Razborov for useful discussions. I am also grateful to an anonymous referee for pointing out several errors and suggesting improvements of the presentation of our results.

## REFERENCES

[1] Z. Adamowicz, *On tableaux consistency in weak theories*, Preprint 618 (2001) of the Mathematical Institute of the Polish Academy of Sciences.

[2] ———, *Herbrand consistency and bounded arithmetic*, ***Fundamenta Mathematicae***, vol. 171 (2002), pp. 279–292.

[3] Z. Adamowicz and L. Kolodziejczyk, *Well-behaved principle alternative to bounded induction*, ***Theoretical Computer Science***, vol. 322 (2004), pp. 5–16.

[4] Z. Adamowicz and P. Zbierski, *On Herbrand consistency in weak arithmetic*, ***Archive for Mathematical Logic***, vol. 40 (2001), pp. 399–413.

[5] S. A. Bloch, ***Divide and conquer in parallel complexity and proof theory***, Ph.D. thesis, University of California, San Diego, 1992.

[6] S. R. Buss, ***Bounded arithmetic***, Bibliopolis, 1986.

[7] ———, *Axiomatizations and conservation results for fragments of bounded arithmetic*, ***Logic and computations***, Contemporary Mathematics, vol. 106, American Mathematical Society, 1990, pp. 57–84.

[8] S. R. Buss and J. Krajíček, *An application of boolean complexity to separation problems in bounded arithmetic*, ***Proceedings of the London Mathematical Society***, vol. 69 (1994), no. 3, pp. 1–21.

[9] S. A. Cook, *Feasibly constructive proofs and the propositional calculus*, ***Proceedings of the 7th annual symposium on theory of computing***, ACM Press, 1975, pp. 83–97.

[10] F. Ferreira, *What are the $\forall\Sigma_1^b$-consequences of $T_2^1$ and $T_2^2$?*, ***Annals of Pure and Applied Logic***, vol. 75 (1995), pp. 79–88.

[11] P. Hájek and P. Pudlák, ***Metamathematics of first-order arithmetic***, Perspectives in Mathematical Logic, Springer-Verlag, 1993.

[12] J. Hanika, ***Search problems in bounded arithmetic***, Ph.D. thesis, Charles University, Prague, 2004.

[13] R. Impagliazzo and J. Krajíček, *A note on conservativity relations among bounded arithmetic theories*, ***Mathematical Logic Quarterly***, vol. 48 (2002), no. 3, pp. 375–377.

[14] R. Kaye, J. Paris, and C. Dimitracopoulos, *On parameter schemas*, ***The Journal of Symbolic Logic***, vol. 53 (1988), no. 4, pp. 1082–1097.

[15] J. Krajíček, *No counter-example interpretation and interactive computations*, ***Logic from computer science*** (Y. N. Moschovakis, editor), Springer-Verlag, 1992, pp. 287–293.

[16] ———, *Fragments of bounded arithmetic and bounded query classes*, ***Transactions of the AMS***, vol. 338 (1993), no. 2, pp. 587–98.

[17] ———, ***Bounded arithmetic, propositional logic, and complexity theory***, Encyclopedia of Mathematics and its Applications, vol. 60, Cambridge Univ. Press, 1995.

[18] J. Krajíček and M. Chiari, *Witnessing functions in bounded arithmetic and search problems*, ***The Journal of Symbolic Logic***, vol. 63 (1998), no. 3, pp. 1095–1115.

[19] ———, *Lifting independence results in bounded arithmetic*, ***Archive for Mathematical Logic***, vol. 38 (1999), pp. 123–138.

[20] J. Krajíček and P. Pudlák, *Quantified propositional calculi and fragments of bounded arithmetic*, ***Zeitschrift für Mathematische Logik und Grundlagen der Mathematik***, vol. 36 (1990), pp. 29–46.

[21] J. Krajíček, P. Pudlák, and G. Takeuti, *Bounded arithmetic and the polynomial hierarchy*, ***Annals of Pure and Applied Logic***, vol. 52 (1991), pp. 143–153.

[22] J. Krajíček and G. Takeuti, *On induction-free provability*, ***Annals of Mathematics and Artificial Intelligence***, vol. 6 (1992), pp. 107–126.

[23] R. Parikh, *Existence and feasibility in arithmetic*, ***The Journal of Symbolic Logic***, vol. 36 (1971), pp. 494–508.

[24] J. B. Paris and L. Harrington, *A mathematical incompleteness in Peano arithmetic*, ***Handbook of mathematical logic***, North-Holland, 1977, pp. 1133–1142.

[25] J. B. Paris and A. J. Wilkie, *Counting problems in bounded arithmetic*, ***Methods in mathematical logic***, LNM, vol. 1130, Springer-Verlag, 1985, pp. 317–340.

[26] P. Pudlák, *Some relations between subsystems of arithmetic and the complexity of computations*, ***Logic from computer science*** (Y. N. Moschovakis, editor), Springer-Verlag, 1992, pp. 308–317.

[27] ———, *The lengths of proofs*, ***Handbook of proof theory***, Elsevier, 1998, pp. 547–637.

[28] ———, *Parallel strategies*, ***The Journal of Symbolic Logic***, vol. 68 (2003), no. 4, pp. 1242–1250.

[29] S. Salehi, ***Herbrand consistency in arithmetics with bounded induction***, Ph.D. thesis, Mathematical Institute of the Polish Academy of Sciences, 2001.

[30] G. Takeuti, ***Proof theory***, North-Holland, 1975.

[31] A. J. Wilkie and J. B. Paris, *On the schema of induction for bounded arithmetical formulas*, ***Annals of Pure and Applied Logic***, vol. 35 (1987), pp. 261–302.

[32] D. Willard, *The semantic tableaux version of the second incompleteness theorem extends almost to Robinson's arithmetic Q*, ***Automated reasoning with semantic tableaux and related methods***, LNCS 1847, Springer-Verlag, 2000, pp. 415–430.

MATHEMATICAL INSTITUTE OF THE ACADEMY OF SCIENCES
AND INSTITUTE OF THEORETICAL COMPUTER SCIENCE
PRAGUE, CZECH REPUBLIC
*E-mail*: pudlak@math.cas.cz

# REALIZABILITY FOR CONSTRUCTIVE ZERMELO-FRAENKEL SET THEORY

MICHAEL RATHJEN

**Abstract.** Constructive Zermelo-Fraenkel Set Theory, **CZF**, has emerged as a standard reference theory that relates to constructive predicative mathematics as **ZFC** relates to classical Cantorian mathematics. A hallmark of this theory is that it possesses a type-theoretic model. Aczel showed that it has a formulae-as-types interpretation in Martin-Löf's intuitionist theory of types [14, 15]. This paper, though, is concerned with a rather different interpretation. It is shown that Kleene realizability provides a self-validating semantics for **CZF**, viz. this notion of realizability can be formalized in **CZF** and demonstrably in **CZF** it can be verified that every theorem of **CZF** is realized.

This semantics, then, is put to use in establishing several equiconsistency results. Specifically, augmenting **CZF** by well-known principles germane to Russian constructivism and Brouwer's intuitionism turns out to engender theories of equal proof-theoretic strength with the same stock of provably recursive functions.

**§1. Introduction.** Realizability semantics for intuitionistic theories were first proposed by Kleene in 1945 [12]. Inspired by Kreisel's and Troelstra's [13] definition of realizability for higher order Heyting arithmetic, realizability was first applied to systems of set theory by Myhill [17] and Friedman [11]. More recently, realizability models of set theory were investigated by Beeson [6, 7] (for non-extensional set theories) and McCarty [16] (directly for extensional set theories). [16] is concerned with realizability for intuitionistic Zermelo-Fraenkel set theory, **IZF**, and employs transfinite iterations of the powerset operation through all the ordinals in defining the realizability (class) structure $V(\mathcal{A})$ over any applicative structure $\mathcal{A}$. Moreover, in addition to the powerset axiom the approach in [16] also avails itself of unfettered separation axioms. At first blush, this seems to render the approach unworkable for **CZF** as this theory lacks the powerset axiom and has only bounded separation. However, it will be shown that these obstacles can be overcome.

---

2000 *Mathematics Subject Classification.* 03F50, 03F35.

*Key words and phrases.* Constructive set theory, realizability, consistency results.

This material is based upon work supported by the National Science Foundation under Award No. DMS-0301162.

**Logic Colloquium '03**
Edited by V. Stoltenberg-Hansen and J. Väänänen
Lecture Notes in Logic, 24

Once one has demonstrated how to define $V(\mathcal{A})$ on the basis of **CZF** there still remains the task of verifying that $V(\mathcal{A})$ validates all the theorems of **CZF** when assuming just the axioms of **CZF** in the ground model. In particular the subset collection axiom poses a new challenge. Another interesting axiom that has been considered in the context of **CZF** is the regular extension axiom, **REA**. It will shown that **REA** holds in $V(\mathcal{A})$ if it holds in the background universe. The pattern propagates when it comes to forms of the axiom of choice. Taking the standard applicative structure $Kl$ based on Turing machine application, either of the axioms of countable choice, dependent choices, and the presentation axiom **PAx** propagate to $V(Kl)$ if they hold in the underlying universe. This also improves on the proof of $V(Kl) \models$ **PAx** in [16] which assumes the unrestricted axiom of choice in the ground model.

The most interesting applications of $V(Kl)$ concern principles germane to Russian constructivism and Brouwer's intuitionism that are classically refutable. For example, Church's thesis, the uniformity principle, Unzerlegbarkeit, and the assertion that every function $f : \mathbb{N}^{\mathbb{N}} \to \mathbb{N}$ is continuous hold in $V(Kl)$. As a corollary, therefore, we obtain that augmenting **CZF** by these "exotic" axioms neither increases the proof-theoretic strength nor the stock of provably recursive functions. Drawing on interpretations of **CZF** and **CZF** + **REA** in classical Kripke-Platek set theories **KP** and **KPi**, respectively, it is also shown that Markov's principle and the principle of independence of premisses may be added without changing the outcome.

The plan for the paper is as follows: Section 1.1 reviews the axioms of **CZF** while section 1.2 recalls some axioms of choice. Section 2 provides the background on applicative structures which is put to use in section 3 to define the general realizability structure, $V(\mathcal{A})$. Section 4 introduces the notion of realizability. Section 5 is devoted to showing that the axioms of **CZF** hold in the realizability structure. The validity of the regular extension in $V(\mathcal{A})$ is proved in section 6. Markov's principle and the principle of independence of premisses axioms are discussed in section 7 and are shown to not increase the proof-theoretic strength. In section 8 we briefly discuss absoluteness properties between the background universe and $V(Kl)$. Section 9 is devoted to principles germane to Russian constructivism and Brouwer's intuitionism that hold in $V(Kl)$ while section 10 is concerned with choice principles in $V(Kl)$. The last section 11 addresses Brouwerian continuity principles that hold in $V(Kl)$.

**1.1. The system CZF.** In this subsection we will summarize the language and axioms for **CZF**. The language of **CZF** is the same first order language as that of classical Zermelo-Fraenkel Set Theory, **ZF** whose only non-logical symbol is $\in$. The logic of **CZF** is intuitionistic first order logic with equality. Among its non-logical axioms are *Extensionality, Pairing* and *Union* in their usual forms. **CZF** has additionally axiom schemata which we will now proceed to summarize.

**Infinity:** $\exists x\, \forall u\, [u \in x \leftrightarrow (\emptyset = u \vee \exists v \in x\; u = v{+}1)]$ where $v{+}1 = v \cup \{v\}$.
**Set Induction:** $\forall x\, [\forall y \in x \phi(y) \rightarrow \phi(x)] \rightarrow \forall x \phi(x)$
**Bounded Separation:** $\forall a\, \exists b\, \forall x\, [x \in b \leftrightarrow x \in a \wedge \phi(x)]$
for all *bounded* formulae $\phi$. A set-theoretic formula is *bounded* or *restricted* if it is constructed from prime formulae using $\neg, \wedge, \vee, \rightarrow, \forall x \in y$ and $\exists x \in y$ only.
**Strong Collection:** For all formulae $\phi$,

$$\forall a\, [\forall x \in a\, \exists y \phi(x,y) \rightarrow \exists b\, [\forall x \in a\, \exists y \in b\, \phi(x,y) \wedge \forall y \in b\, \exists x \in a\, \phi(x,y)]].$$

**Subset Collection:** For all formulae $\psi$,

$$\forall a \forall b \exists c \forall u\, [\forall x \in a\, \exists y \in b\, \psi(x,y,u) \rightarrow \exists d \in c\, [\forall x \in a\, \exists y \in d\, \psi(x,y,u) \wedge \forall y \in d\, \exists x \in a\, \psi(x,y,u)]].$$

The Subset Collection schema easily qualifies as the most intricate axiom of **CZF**. To explain this axiom in different terms, we introduce the notion of *fullness* (cf. [1]).

DEFINITION 1.1. As per usual, we use $\langle x, y\rangle$ to denote the ordered pair of $x$ and $y$. We use $\mathbf{Fun}(g)$, $\mathbf{dom}(R)$, $\mathbf{ran}(R)$ to convey that $g$ is a function and to denote the domain and range of any relation $R$, respectively.

For sets $A, B$ let $A \times B$ be the cartesian product of $A$ and $B$, that is the set of ordered pairs $\langle x, y\rangle$ with $x \in A$ and $y \in B$. Let ${}^A B$ be the class of all functions with domain $A$ and with range contained in $B$. Let $\boldsymbol{mv}({}^A B)$ be the class of all sets $R \subseteq A \times B$ satisfying $\forall u \in A\, \exists v \in B\, \langle u, v\rangle \in R$. A set $C$ is said to be *full in* $\boldsymbol{mv}({}^A B)$ if $C \subseteq \boldsymbol{mv}({}^A B)$ and

$$\forall R \in \boldsymbol{mv}({}^A B)\; \exists S \in C\; S \subseteq R.$$

The expression $\boldsymbol{mv}({}^A B)$ should be read as the collection of *multi-valued functions* from the set $A$ to the set $B$.

Additional axioms we shall consider are:
**Exponentiation:** $\forall x \forall y \exists z\; z = {}^x y$.
**Fullness:** $\forall x \forall y \exists z\; z$ *is full in* $\boldsymbol{mv}({}^x y)$.

The next result provides an equivalent rendering of Subset Collection.

PROPOSITION 1.2. *Let* $\mathbf{CZF}^-$ *be* **CZF** *without* Subset Collection.

(i) $\mathbf{CZF}^- \vdash$ *Subset Collection* $\leftrightarrow$ *Fullness.*
(ii) $\mathbf{CZF} \vdash$ *Exponentiation.*

PROOF. [1, Proposition 2.2]. ⊣

Let **EM** be the principle of excluded third, i.e. the schema consisting of all formulae of the form $\theta \vee \neg\theta$. The first central fact to be noted about **CZF** is:

**Proposition 1.3.** **CZF + EM = ZF.**

**Proof.** Note that classically Collection implies Separation. Powerset follows classically from Exponentiation. ⊣

On the other hand, it was shown in [19, Theorem 4.14], that **CZF** has only the strength of Kripke-Platek Set Theory (with the Infinity Axiom), **KP** (see [5]), and, moreover, that **CZF** is of the same strength as its subtheory $\mathbf{CZF}^-$, i.e., **CZF** minus Subset Collection. To stay in the world of **CZF** one has to keep away from any principles that imply **EM**. Moreover, it is perhaps fair to say that **CZF** is such an interesting theory owing to the non-derivability of Powerset and Separation. Therefore one ought to avoid any principles which imply Powerset or Separation.

The first large set axiom proposed in the context of constructive set theory was the *Regular Extension Axiom*, **REA**, which Aczel introduced to accommodate inductive definitions in **CZF** (cf. [3]).

**Definition 1.4.** $A$ is inhabited if $\exists x\, x \in A$. An inhabited set $A$ is *regular* if $A$ is transitive, and for every $a \in A$ and set $R \subseteq a \times A$ if $\forall x \in a\, \exists y\, (\langle x, y\rangle \in R)$, then there is a set $b \in A$ such that

$$\forall x \in a\, \exists y \in b\, (\langle x, y\rangle \in R) \wedge \forall y \in b\, \exists x \in a \big(\langle x, y\rangle \in R\big).$$

In particular, if $R : a \to A$ is a function, then the image of $R$ is an element of $A$.

The *Regular Extension Axiom*, **REA**, is as follows: *Every set is a subset of a regular set.*

**1.2. Axioms of choice.** In many a text on constructive mathematics, axioms of countable choice and dependent choices are accepted as constructive principles. This is, for instance, the case in Bishop's constructive mathematics (cf. [8]) as well as Brouwer's intuitionistic analysis (cf. [21, Ch. 4, Sect. 2]). Myhill also incorporated these axioms in his constructive set theory [18].

The weakest constructive choice principle we shall consider is $\mathbf{AC}^{\omega,\omega}$ which asserts that whenever $\forall i \in \omega\; \exists j \in \omega\; \theta(i, j)$ then there exists a function $f : \omega \to \omega$ such that $\forall i \in \omega\; \theta(i, f(i))$.

The *Axiom of Countable Choice*, $\mathbf{AC}_\omega$, is the following scheme: whenever $\forall i \in \omega\; \exists x\; \psi(i, x)$ then there exists a function $f$ with domain $\omega$ such that $\forall i \in \omega\; \theta(i, f(i))$. Obviously $\mathbf{AC}_\omega$ implies $\mathbf{AC}^{\omega,\omega}$.

A mathematically very useful axiom to have in set theory is the *Dependent Choices Axiom*, **DC**, i.e., for all formulae $\psi$, whenever

$$(\forall x \in a)\; (\exists y \in a)\; \psi(x, y)$$

and $b_0 \in a$, then there exists a function $f : \omega \to a$ such that $f(0) = b_0$ and

$$(\forall n \in \omega)\; \psi\big(f(n), f(n+1)\big).$$

Even more useful is the *Relativized Dependent Choices Axiom*, **RDC**. It asserts that for arbitrary formulae $\phi$ and $\psi$, whenever

$$\forall x\ [\phi(x) \rightarrow \exists y(\phi(y) \wedge \psi(x,y))]$$

and $\phi(b_0)$, then there exists a function $f$ with domain $\omega$ such that $f(0) = b_0$ and

$$(\forall n \in \omega)[\phi(f(n)) \wedge \psi(f(n), f(n+1))].$$

We shall use the notation $f : X \twoheadrightarrow Y$ to convey that $f$ is a function from $X$ onto $Y$. A set $P$ is a *base* if for any $P$-indexed family $(X_a)_{a \in P}$ of inhabited sets $X_a$, there exists a function $f$ with domain $P$ such that, for all $a \in P$, $f(a) \in X_a$. The *Presentation Axiom*, **PAx**, is the statement that every set is the surjective image of a base, i.e., for all sets $A$ there exists a base $B$ and a function $f : B \twoheadrightarrow A$.

**§2. Some background on applicative structures.** In order to define a realizability interpretation we must have a notion of realizing functions on hand. A particularly general and elegant approach to realizability builds on structures which have been variably called *partial combinatory algebras*, *applicative structures*, or *Schönfinkel algebras*. These structures are best described as the models of a theory **APP**. The following presents the main features of **APP**; for full details cf. [9, 10, 7, 21]. The language of **APP** is a first-order language with a ternary relation symbol App, a unary relation symbol $N$ (for a copy of the natural numbers) and equality, $=$, as primitives. The language has an infinite collection of variables, denoted $x, y, z, \ldots$, and nine distinguished constants: $\mathbf{0}, \boldsymbol{s}_N, \boldsymbol{p}_N, \boldsymbol{k}, \boldsymbol{s}, \boldsymbol{d}, \boldsymbol{p}, \boldsymbol{p}_0, \boldsymbol{p}_1$ for, respectively, zero, successor on $N$, predecessor on $N$, the two basic combinators, definition by cases, pairing and the corresponding two projections. There is no arity associated with the various constants. The *terms* of **APP** are just the variables and constants. We write $t_1 t_2 \simeq t_3$ for App$(t_1, t_2, t_3)$.

Formulae are then generated from atomic formulae using the propositional connectives and the quantifiers.

In order to facilitate the formulation of the axioms, the language of **APP** is expanded definitionally with the symbol $\simeq$ and the auxiliary notion of an *application term* is introduced. The set of application terms is given by two clauses:

1. all terms of **APP** are application terms; and
2. if $s$ and $t$ are application terms, then $(st)$ is an application term.

For $s$ and $t$ application terms, we have auxiliary, defined formulae of the form:

$$s \simeq t := \forall y(s \simeq y \leftrightarrow t \simeq y),$$

if $t$ is not a variable. Here $s \simeq a$ (for $a$ a free variable) is inductively defined by:

$$s \simeq a \text{ is } \begin{cases} s = a, & \text{if } s \text{ is a term of } \mathbf{APP}, \\ \exists x, y[s_1 \simeq x \wedge s_2 \simeq y \wedge \mathrm{App}(x, y, a)] & \text{if } s \text{ is of the form } (s_1 s_2). \end{cases}$$

Some abbreviations are $t_1 t_2 \cdots t_n$ for $((\cdots(t_1 t_2)\cdots)t_n)$; $t\downarrow$ for $\exists y(t \simeq y)$ and $\phi(t)$ for $\exists y(t \simeq y \wedge \phi(y))$.

Some further conventions are useful. Systematic notation for *n-tuples* is introduced as follows: $(t)$ is $t$, $(s,t)$ is $\boldsymbol{p}st$, and $(t_1, \ldots, t_n)$ is defined by $((t_1, \ldots, t_{n-1}), t_n)$. In this paper, the **logic of APP** is assumed to be that of intuitionistic predicate logic with identity. **APP's non-logical axioms** are the following:

**Applicative Axioms.**

1. $\mathrm{App}(a, b, c_1) \wedge \mathrm{App}(a, b, c_2) \to c_1 = c_2$.
2. $(\boldsymbol{k}ab)\downarrow \wedge\, \boldsymbol{k}ab \simeq a$.
3. $(\boldsymbol{s}ab)\downarrow \wedge\, \boldsymbol{s}abc \simeq ac(bc)$.
4. $(\boldsymbol{p}a_0a_1)\downarrow \wedge\, (\boldsymbol{p}_0a)\downarrow \wedge\, (\boldsymbol{p}_1a)\downarrow \wedge\, \boldsymbol{p}_i(\boldsymbol{p}a_0a_1) \simeq a_i$ for $i = 0, 1$.
5. $N(c_1) \wedge N(c_2) \wedge c_1 = c_2 \to \boldsymbol{d}abc_1c_2\downarrow \wedge\, \boldsymbol{d}abc_1c_2 \simeq a$.
6. $N(c_1) \wedge N(c_2) \wedge c_1 \neq c_2 \to \boldsymbol{d}abc_1c_2\downarrow \wedge\, \boldsymbol{d}abc_1c_2 \simeq b$.
7. $\forall x\, (N(x) \to [\boldsymbol{s}_N x\downarrow \wedge\, \boldsymbol{s}_N x \neq \mathit{0} \wedge N(\boldsymbol{s}_N x)])$.
8. $N(\mathit{0}) \wedge \forall x\, (N(x) \wedge x \neq \mathit{0} \to [\boldsymbol{p}_N x\downarrow \wedge\, \boldsymbol{s}_N(\boldsymbol{p}_N x) = x])$.
9. $\forall x\, [N(x) \to \boldsymbol{p}_N(\boldsymbol{s}_N x) = x]$
10. $\varphi(\mathit{0}) \wedge \forall x[N(x) \wedge \varphi(x) \to \varphi(\boldsymbol{s}_N x)] \to \forall x[N(x) \to \varphi(x)]$.

Let $\mathit{1} := \boldsymbol{s}_N \mathit{0}$. The applicative axioms entail that $\mathit{1}$ is an application term that evaluates to an object falling under $N$ but distinct from $\mathit{0}$, i.e., $\mathit{1}\downarrow$, $N(\mathit{1})$ and $\mathit{0} \neq \mathit{1}$.

Employing the axioms for the combinators $\boldsymbol{k}$ and $\boldsymbol{s}$ one can deduce an abstraction lemma yielding $\lambda$-terms of one argument. This can be generalized using $n$–tuples and projections.

LEMMA 2.1 ((cf. [9]) **Abstraction Lemma**). *For each application term $t$ there is a new application term $t^*$ such that the parameters of $t^*$ are among the parameters of $t$ minus $x_1, \ldots, x_n$ and such that*

$$\mathbf{APP} \vdash t^*\downarrow \wedge\, t^*(x_1, \ldots, x_n) \simeq t.$$

*$\lambda(x_1, \ldots, x_n) \cdot t$ is written for $t^*$.*

The most important consequence of the Abstraction Lemma is the Recursion Theorem. It can be derived in the same way as for the $\lambda$–calculus (cf. [9], [10], [7, VI.2.7]). Actually, one can prove a uniform version of the following in **APP**.

COROLLARY 2.2 (**Recursion Theorem**).

$$\forall f \exists g \forall x_1 \cdots \forall x_n\, g(x_1, \ldots, x_n) \simeq f(g, x_1, \ldots, x_n).$$

The "standard" applicative structure is $Kl$ in which the universe $|Kl|$ is $\omega$ and $\mathrm{App}^{Kl}(x, y, z)$ is Turing machine application:

$$\mathrm{App}^{Kl}(x, y, z) \quad \text{iff } \{x\}(y) \simeq z.$$

The primitive constants of **APP** are interpreted over $|Kl|$ in the obvious way.

**§3. The general realizability structure.** The following discussion assumes that we can formalize the notion of an applicative structure in **CZF**. Moreover, for the remainder of this paper, $\mathcal{A}$ will be assumed to be a fixed but arbitrary applicative structure, which in particular is a set.

The definition of the following realizability structure is due to McCarty [16].

DEFINITION 3.1. Ordinals are transitive sets whose elements are transitive also. We use lower case Greek letters to range over ordinals. For $\mathcal{A} \models \mathbf{APP}$,

(1)
$$V(\mathcal{A})_\alpha = \bigcup_{\beta \in \alpha} \mathcal{P}(|\mathcal{A}| \times V(\mathcal{A})_\beta).$$

(2)
$$V(\mathcal{A}) = \bigcup_{\alpha} V(\mathcal{A})_\alpha.$$

As the power set operation is not available in **CZF** it is not clear whether the universe $V(\mathcal{A})$ can be formalized in **CZF**. To show this we shall review some facts showing that **CZF** accommodates inductively defined classes.

**3.1. Inductively defined classes in CZF.**

DEFINITION 3.2. An *inductive definition* is a class of ordered pairs. If $\Phi$ is an inductive definition and $\langle x, a\rangle \in \Phi$ then we write

$$\frac{x}{a}\,\Phi$$

and call $\frac{x}{a}\,\Phi$ an *(inference) step* of $\Phi$, with set $x$ of *premisses* and *conclusion* $a$. For any class $Y$, let

$$\Gamma_\Phi(Y) = \left\{a : \exists x \left(x \subseteq Y \wedge \frac{x}{a}\,\Phi\right)\right\}.$$

The class $Y$ is *$\Phi$-closed* if $\Gamma_\Phi(Y) \subseteq Y$. Note that $\Gamma_\Phi$ is monotone; i.e. for classes $Y_1$, $Y_2$, whenever $Y_1 \subseteq Y_2$, then $\Gamma_\Phi(Y_1) \subseteq \Gamma_\Phi(Y_2)$.

We define the class *inductively defined by* $\Phi$ to be the smallest $\Phi$-closed class. The main result about inductively defined classes states that this class, denoted $I(\Phi)$, always exists.

LEMMA 3.3 ((**CZF**) Class Inductive Definition Theorem). *For any inductive definition $\Phi$ there is a smallest $\Phi$-closed class $I(\Phi)$.*

*Moreover, there is a class $J \subseteq \mathbf{ON} \times V$ such that*

$$I(\Phi) = \bigcup_{\alpha} J^{\alpha},$$

*and for each $\alpha$,*

$$J^{\alpha} = \Gamma_{\Phi}\left(\bigcup_{\beta\in\alpha} J^{\beta}\right).$$

*$J$ is uniquely determined by the above, and its stages $J^{\alpha}$ will be denoted by $\Gamma_{\Phi}^{\alpha}$.*

PROOF. [2, section 4.2] or [4, Theorem 5.1]. ⊣

LEMMA 3.4. *The classes $V(\mathcal{A})_{\alpha}$ are definable in* **CZF**.

PROOF. Let $\Phi$ be the inductive definition with

$$\frac{x}{a}\Phi \quad \text{iff } \forall u \in a\ (u \in |\mathcal{A}| \times x).$$

Invoking Lemma 3.3, let $J$ be the class such that $I(\Phi) = \bigcup_{\alpha} J^{\alpha}$, and for each $\alpha$,

$$J^{\alpha} = \Gamma_{\Phi}\left(\bigcup_{\beta\in\alpha} J^{\beta}\right).$$

Now let

$$\Delta_{\alpha} := \bigcup_{\beta\in\alpha} J^{\beta}.$$

Note that $\Gamma_{\Phi}(X) = \mathcal{P}(|\mathcal{A}| \times X)$, and therefore

$$\begin{aligned}(3) \qquad \Delta_{\alpha} &= \bigcup_{\beta\in\alpha} J^{\beta} = \bigcup_{\beta\in\alpha} \Gamma_{\Phi}\left(\bigcup_{\eta\in\beta} J^{\eta}\right) \\ &= \bigcup_{\beta\in\alpha} \mathcal{P}\left(|\mathcal{A}| \times \bigcup_{\eta\in\beta} J^{\eta}\right) = \bigcup_{\beta\in\alpha} \mathcal{P}(|\mathcal{A}| \times \Delta_{\beta}).\end{aligned}$$

Letting $V(\mathcal{A})_{\alpha} := \Delta_{\alpha}$, (3) shows that the equations of definition 3.1 obtain. ⊣

LEMMA 3.5 (**CZF**). (i) *$V(\mathcal{A})$ is cumulative: for $\beta \in \alpha$, $V(\mathcal{A})_{\beta} \subseteq V(\mathcal{A})_{\alpha}$.* (ii) *If $b$ is a set such that $b \subseteq |\mathcal{A}| \times V(\mathcal{A})$ then $b \in V(\mathcal{A})$.*

PROOF. (i) is immediate by (1).

For (ii), suppose $b \subseteq |\mathcal{A}| \times V(\mathcal{A})$. Then

$$\forall x \in b\ \exists\alpha\ \exists e \in |\mathcal{A}|\ \exists z \in V(\mathcal{A})_{\alpha}\ x = \langle e, z\rangle,$$

and therefore by Strong Collection there exists a set $D$ such that

$$\forall x \in b\ \exists\alpha \in D\ \exists e \in |\mathcal{A}|\ \exists z \in V(\mathcal{A})_{\alpha}\ x = \langle e, z\rangle,$$

where $D$ is a set of ordinals. Now let $D' = \{\alpha + 1 : \alpha \in D\}$ and $\delta = \bigcup D'$ (where $\alpha + 1 := \alpha \cup \{\alpha\}$). Then $\delta$ is an ordinal as well, and $\forall \alpha \in D\ \alpha \in \delta$. Thus it follows that

$$\forall x \in b\ \exists \alpha \in \delta\ \exists e \in |\mathcal{A}|\ \exists z \in V(\mathcal{A})_\alpha\ x = \langle e, z \rangle.$$

And hence, $b \subseteq \bigcup_{\alpha \in \delta} \mathcal{P}(|\mathcal{A}| \times V(\mathcal{A})_\alpha)$, so that $b \in V(\mathcal{A})_\delta \subseteq V(\mathcal{A})$. ⊣

**§4. Defining realizability.** Having shown that the class $V(\mathcal{A})$ can be formalized in **CZF**, we now proceed to define a notion of extensional realizability over $V(\mathcal{A})$, i.e., $e \Vdash \phi$ for $e \in |\mathcal{A}|$ and sentences $\phi$ with parameters in $V(\mathcal{A})$. Except for the special treatment of bounded quantifiers, this definition is due to McCarty [16]. For $e \in |\mathcal{A}|$ we shall write $(e)_0$ and $(e)_1$ rather than $\boldsymbol{p}_0 e$ and $\boldsymbol{p}_1 e$, respectively.

DEFINITION 4.1. Bounded quantifiers will be treated as quantifiers in their own right, i.e., bounded and unbounded quantifiers are treated as syntactically different kinds of quantifiers. Let $a, b \in V(\mathcal{A})$ and $e \in |\mathcal{A}|$.

$e \Vdash a \in b$ iff $\exists c\, [\langle (e)_0, c \rangle \in b \wedge (e)_1 \Vdash a = c]$

$e \Vdash a = b$ iff $\forall f, d\, [(\langle f, d \rangle \in a \to (e)_0 f \Vdash d \in b) \wedge (\langle f, d \rangle \in b \to (e)_1 f \Vdash d \in a)]$

$e \Vdash \phi \wedge \psi$ iff $(e)_0 \Vdash \phi \wedge (e)_1 \Vdash \psi$

$e \Vdash \phi \vee \psi$ iff $[(e)_0 = \boldsymbol{0} \wedge (e)_1 \Vdash \phi] \vee [(e)_0 = \boldsymbol{1} \wedge (e)_1 \Vdash \psi]$

$e \Vdash \neg \phi$ iff $\forall f \in |\mathcal{A}| \neg f \Vdash \phi$

$e \Vdash \phi \to \psi$ iff $\forall f \in |\mathcal{A}|\, [f \Vdash \phi \to ef \Vdash \psi]$

$e \Vdash \forall x \in a\, \phi$ iff $\forall \langle f, c \rangle \in a\, ef \Vdash \phi[x/c]$

$e \Vdash \exists x \in a\, \phi$ iff $\exists c (\langle (e)_0, c \rangle \in a \wedge (e)_1 \Vdash \phi[x/c])$

$e \Vdash \forall x \phi$ iff $\forall c \in V(\mathcal{A})\ e \Vdash \phi[x/a]$

$e \Vdash \exists x \phi$ iff $\exists c \in V(\mathcal{A})\ e \Vdash \phi[x/a]$

Notice that $e \Vdash u \in v$ and $e \Vdash u = v$ can be defined for arbitrary sets $u, v$, viz., not just for $u, v \in V(\mathcal{A})$. The definitions of $e \Vdash u \in v$ and $e \Vdash u = v$ fall under the scope of definitions by transfinite recursion. More precisely, the functions

$$F_{\in}(u, v) = \{e \in |\mathcal{A}| : e \Vdash u \in v\}$$
$$G_{=}(u, v) = \{e \in |\mathcal{A}| : e \Vdash u = v\}$$

can be defined (simultaneously) on $V \times V$ by recursion on the relation

$$\langle c, d \rangle \lhd \langle a, b \rangle \quad \text{iff } (c = a \wedge d \in \boldsymbol{TC}(b)) \vee (d = b \wedge c \in \boldsymbol{TC}(a)).$$

**4.1. The soundness theorem for intuitionistic predicate logic with equality.** Except for the extra considerations concerning bounded quantifiers, the proofs of 4.2 and 4.3 are almost the same for **CZF** as the corresponding proofs for **IZF** given in [16].

LEMMA 4.2. *There are* $\mathbf{i}_r$, $\mathbf{i}_s$, $\mathbf{i}_t$, $\mathbf{i}_0$, $\mathbf{i}_1 \in |\mathcal{A}|$ *such that for all* $a, b, c \in V(\mathcal{A})$,

1. $\mathbf{i}_r \Vdash a = a$.
2. $\mathbf{i}_s \Vdash a = b \to b = a$.
3. $\mathbf{i}_t \Vdash (a = b \wedge b = c) \to a = c$.
4. $\mathbf{i}_0 \Vdash (a = b \wedge b \in c) \to a \in c$.
5. $\mathbf{i}_1 \Vdash (a = b \wedge c \in a) \to c \in b$.

*Moreover, for each formula* $\varphi(v, u_1, \ldots, u_r)$ *of* **CZF** *all of whose free variables are among* $v, u_1, \ldots, u_r$ *there exists* $\mathbf{i}_\varphi \in |\mathcal{A}|$ *such that for all* $a, b, c_1, \ldots, c_r \in V(\mathcal{A})$,

$$\mathbf{i}_\varphi \Vdash \varphi(a, \vec{c}) \wedge a = b \to \varphi(b, \vec{c}),$$

*where* $\vec{c} = c_1, \ldots, c_r$.

PROOF. Realizers for the universal closures of the above formulas can be taken from [16, chapter 2, sections 5 and 6]. Thus the above assertions follow from the "genericity" of realizers of universal statements, i.e.,

$$e \Vdash \forall v \psi(v) \quad \text{iff} \, \forall a \; e \Vdash \psi(a). \qquad \dashv$$

THEOREM 4.3. *Let* $\mathcal{D}$ *be a proof in intuitionistic predicate logic with equality of a formula* $\varphi(u_1, \ldots, u_r)$ *of* **CZF** *all of whose free variables are among* $u_1, \ldots, u_r$. *Then there is* $e_\mathcal{D} \in |\mathcal{A}|$ *such that* **CZF** *proves*

$$e_\mathcal{D} \Vdash \forall u_1 \cdots \forall u_r \; \varphi(u_1, \ldots, u_r).$$

PROOF. With the exception of the logical principles

$$(4) \qquad \forall u \in a \, \varphi(u) \leftrightarrow \forall u \; [u \in a \to \varphi(u)],$$

$$(5) \qquad \exists u \in a \, \varphi(u) \leftrightarrow \exists u \; [u \in a \wedge \varphi(u)],$$

which relate bounded to unbounded quantifiers, the proof is literally the same as in [16, chapter 2, sections 5 and 6]. Let $a \in V(\mathcal{A})$ and $\varphi$ be a formula with parameters in $V(\mathcal{A})$. We find a realizer for the formula of (4) as follows:

$$\begin{aligned}
&e \Vdash \forall v \; [v \in a \to \varphi(v)] \\
&\quad \text{iff} \; \forall x \in V(\mathcal{A}) \, \forall f \in |\mathcal{A}| [f \Vdash x \in a \text{ then } e \Vdash \varphi(x)] \\
&\quad \text{iff} \; \forall x \in V(\mathcal{A}) \, \forall f \in |\mathcal{A}| \\
&\qquad\qquad [\exists c (\langle (f)_0, c \rangle \in a \wedge (f)_1 \Vdash x = c) \text{ then } ef \Vdash \varphi(x)] \\
&\quad \text{then} \; \forall c \, \forall f \in |\mathcal{A}| [(\langle (f)_0, c \rangle \in a \wedge (f)_1 \Vdash c = c) \text{ then } ef \Vdash \varphi(c)] \\
&\quad \text{then} \; \forall \langle g, c \rangle \in ae \; (\mathbf{p} g \mathbf{i}_r) \Vdash \varphi(c) \\
&\quad \text{then} \; \lambda g \cdot e(\mathbf{p} g \mathbf{i}_r) \Vdash \forall v \in a \, \varphi(u).
\end{aligned}$$

Conversely, we have

$$\begin{aligned} & e \Vdash \forall v \in a\, \varphi(u) \\ & \quad \text{iff } \forall \langle f, c\rangle \in a\; ef \Vdash \varphi(c) \\ & \quad \text{then } \forall x \in V(\mathcal{A})\, \forall g \in |\mathcal{A}| [\exists c(\langle (g)_0, c\rangle \in a \wedge (g)_1 \Vdash x = c) \\ & \qquad\qquad \text{then } \boldsymbol{i}_\varphi(\boldsymbol{p}(g)_1(e\,(g)_0)) \Vdash \varphi(x)] \\ & \quad \text{then } \lambda g \cdot \boldsymbol{i}_\varphi(\boldsymbol{p}(g)_1(e\,(g)_0)) \Vdash \forall u[u \in v \rightarrow \varphi(u)]. \end{aligned}$$

The constants $\boldsymbol{i}_r, \boldsymbol{i}_\varphi$ are from Lemma 4.2. Letting $\boldsymbol{m}$ be

$$\boldsymbol{p}(\lambda e \cdot \lambda g \cdot e(\boldsymbol{p}g\boldsymbol{i}_r))(\lambda e \cdot \lambda g \cdot \boldsymbol{i}_\varphi(\boldsymbol{p}(g)_1(e\,(g)_0))),$$

we get

$$\boldsymbol{m} \Vdash \forall \vec{w} \forall v (\forall u \in v\, \varphi(u) \leftrightarrow \forall u\, [u \in v \rightarrow \varphi(u)]),$$

where $\forall \vec{w}$ quantifies over the remaining free variables of $\varphi$.

Similarly one finds $\bar{\boldsymbol{m}}$ such that

$$\bar{\boldsymbol{m}} \Vdash \forall \vec{w} \forall v (\exists u \in v\, \varphi(u) \leftrightarrow \exists u\, [u \in v \wedge \varphi(u)]). \quad \dashv$$

**4.2. Realizability for bounded formulae.** In the following we shall often have occasion to employ the fact that for a bounded formula $\varphi(v)$ with parameters from $V(\mathcal{A})$ and $x \subseteq V(\mathcal{A})$,

$$\{\langle e, c\rangle : e \in |\mathcal{A}| \wedge c \in x \wedge e \Vdash \varphi(c)\}$$

is a set. To prove this we shall consider an extended class of formulae.

DEFINITION 4.4. The *extended bounded formulae* are the smallest class of formulas containing the formulae of the form $x \in y$, $x = y$, $e \Vdash x \in y$, $e \Vdash x = y$, which is closed under $\wedge, \vee, \neg, \rightarrow$ and bounded quantification.

LEMMA 4.5 (**CZF**). *Separation holds for extended bounded formulae, i.e., for every extended bounded formula $\varphi(v)$ and set $x$, $\{v \in x : \varphi(v)\}$ is a set.*

PROOF. Since $F_\in$ and $G_=$ are provably total functions of **CZF**, formulas of the form $e \Vdash x \in y$ and $e \Vdash x = y$ can be treated in the context of **CZF** as though they were atomic symbols of the language. This follows from [19, Proposition 2.4] or [4, Proposition 11.12]. $\dashv$

LEMMA 4.6 (**CZF**). *Let $\varphi(v, u_1, \ldots, u_r)$ be a bounded formula of* **CZF** *all of whose free variables are among $u_1, \ldots, u_r$. Then there is an extended bounded formula $\tilde{\varphi}(v, u_1, \ldots, u_r)$ and $f_\varphi \in |\mathcal{A}|$ such that for all $a_1, \ldots, a_r \in V(\mathcal{A})$ and $e \in |\mathcal{A}|$,*

$$e \Vdash \varphi(\vec{a}) \quad \textit{iff}\; \tilde{\varphi}(f_\varphi e, \vec{a}).$$

PROOF. We proceed by induction on the generation of $\varphi$. For an atomic formula $\varphi$, the assertion follows with $\tilde{\varphi} \equiv \varphi$ and $f_\varphi$ being an index for the identity function of $\mathcal{A}$. The assertion easily follows from the respective inductive assumptions if $\varphi$ is of the form $\varphi_0 \wedge \varphi_1$ or $\varphi_0 \vee \varphi_1$.

Now suppose $\varphi$ is of the form $\forall x \in w\ \psi(x, \vec{u}, w)$. Inductively we then have for all $b, c, \vec{a} \in V(\mathcal{A})$ and $e' \in |\mathcal{A}|$,

$$e' \Vdash \psi(b, \vec{a}, c) \quad \text{iff}\ \widetilde{\psi}(f_\psi e', b, \vec{a}, c)$$

for some extended bounded formula $\widetilde{\psi}$. Hence, by the definition of realizability for bounded formulae, we can readily construct the desired extended formula $\widetilde{\varphi}$ from $\widetilde{\psi}$.

The case of a bounded existential quantifier is similar to the preceding case. ⊣

COROLLARY 4.7 (**CZF**). *Let $\varphi(v)$ be a bounded formula with parameters from $V(\mathcal{A})$ and $x \subseteq V(\mathcal{A})$. Then*

$$\{\langle e, c\rangle : e \in |\mathcal{A}| \wedge c \in x \wedge e \Vdash \varphi(c)\}$$

*is a set. Moreover, this set belongs to $V(\mathcal{A})$.*

PROOF. The above class is a set by the previous two lemmas. That the set is also an element of $V(\mathcal{A})$ follows from Lemma 3.5. ⊣

**§5. The soundness theorem for CZF.** The soundness of extensional realizability for **IZF** was shown in [16]. The proofs for the realizability of Extensionality, Pair, Infinity, and Set Induction carry over to the context of **CZF**. Union needs a little adjustment to avoid an unnecessary appeal to unbounded Separation. To establish realizability of Bounded Separation we use Separation for extended bounded formulae. Strong Collection and in particular Subset Collection are not axioms of **IZF** and therefore require new proofs.

THEOREM 5.1. *For every axiom $\theta$ of* **CZF**, *there exists a closed application term $t$ such that*

$$\mathbf{CZF} \vdash (t \Vdash \theta).$$

PROOF. We treat the axioms one after the other.

**Extensionality:** One easily checks that with

$$e = \lambda y \cdot \boldsymbol{p}(\lambda x \cdot \boldsymbol{p_0} y(\boldsymbol{p} x \boldsymbol{i_r}))(\lambda x \cdot \boldsymbol{p_1} y(\boldsymbol{p} x \boldsymbol{i_r})),$$

$$e \Vdash \forall a\ \forall b\ [\forall x\ (x \in a \leftrightarrow x \in b) \rightarrow a = b].$$

**Pair:** We need to guarantee the existence of an $e \in |\mathcal{A}|$ such that

$$\forall a, b \in V(\mathcal{A})\ e \Vdash a \in c \wedge b \in c \tag{6}$$

for some $c \in V(\mathcal{A})$. Set $e = \boldsymbol{p}(\boldsymbol{p0i_r})(\boldsymbol{p0i_r})$ and let $c = \{\langle \boldsymbol{0}, a\rangle, \langle \boldsymbol{0}, b\rangle\}$. By Lemma 3.5, $c \in V(\mathcal{A})$. One easily verifies that (6) holds for the specified $e$ and $c$.

**Union:** For each $a \in V(\mathcal{A})$, put

$$Un^{\mathcal{A}}(a) = \{\langle h, y\rangle : h \in \mathcal{A} \wedge \exists \langle f, x\rangle \in a \langle h, y\rangle \in x\},$$

so that by Lemma 3.5, $Un^{\mathcal{A}}(a) \in V(\mathcal{A})$.

Now assume $\langle f,x\rangle \in a$ and $\langle h,y\rangle \in x$. Then $\langle h,y\rangle \in Un^{\mathcal{A}}(a)$ and $\boldsymbol{phi_r} \Vdash y \in Un^{\mathcal{A}}(a)$, and hence, letting $e = \lambda u \cdot \lambda v \cdot \boldsymbol{pvi_r}$, we have $e \Vdash \forall a\, \exists w\, \forall x \in a\, \forall y \in x\; y \in w$.

**Bounded Separation:** Let $\varphi(x)$ be a bounded formula with parameters in $V(\mathcal{A})$. This time we need to find $e, e' \in |\mathcal{A}|$ such that for all $a \in V(\mathcal{A})$ there exists a $b \in V(\mathcal{A})$ such that

$$(7)\quad (e \Vdash \forall x \in b\,[x \in a \wedge \varphi(x)]) \wedge (e' \Vdash \forall x \in a\,[\varphi(x) \rightarrow x \in b]).$$

For $a \in V(\mathcal{A})$, define

$$\mathrm{Sep}^{\mathcal{A}}(a,\varphi) = \{\langle \boldsymbol{p}fg, x\rangle : f,g \in \mathcal{A} \wedge \langle g,x\rangle \in a \wedge f \Vdash \varphi(x)\}.$$

By Corollary 4.7, $\mathrm{Sep}^{\mathcal{A}}(a,\varphi) \in V(\mathcal{A})$. Put $b = \mathrm{Sep}^{\mathcal{A}}(a,\varphi)$.

To verify (7), first assume $\langle \boldsymbol{p}fg, x\rangle \in b$. Then, by definition of $b$, $\langle g,x\rangle \in a$ and $f \Vdash \varphi(x)$, so that with

$$e = \boldsymbol{p}(\boldsymbol{p}(\lambda u \cdot (u)_1)\boldsymbol{i_r})(\lambda u \cdot (u)_0),$$

$e \Vdash \forall x \in b\,[x \in a \wedge \varphi(x)]$.

Now assume $\langle g,x\rangle \in a$ and $f \Vdash \varphi(x)$. Then $\langle \boldsymbol{p}fg, x\rangle \in b$. Hence with $e' = \lambda u \cdot \lambda v \cdot \boldsymbol{p}(\boldsymbol{p}vu)\boldsymbol{i_r}$ we get $e' \Vdash \forall x \in a[\varphi(x) \rightarrow x \in b]$.

**Infinity:** The most obvious candidate to represent $\omega$ in $V(\mathcal{A})$ is $\overline{\omega}$, which is given via an injection of $\omega$ into $V(\mathcal{A})$. Recall that $\boldsymbol{0}$ denotes the zero of $\mathcal{A}$, $\boldsymbol{1} = \boldsymbol{s_N 0}$ and that 0 (the empty set) is the least element of $\omega$. Set $\underline{0} = \boldsymbol{0}$ and for $n \in \omega$, let $\underline{n+1} = \boldsymbol{s_N}\, \underline{n}$ and set $\overline{n} = \{\langle \underline{m}, \overline{m}\rangle : m \in n\}$. Then, we take

$$\overline{\omega} = \{\langle \underline{n}, \overline{n}\rangle : n \in \omega\}.$$

Clearly, by Lemma 3.5, $\overline{\omega} \in V(\mathcal{A})$. Note also that $N(\underline{n})$ holds for all $n \in \omega$. Moreover, the applicative axioms imply that if $n, m \in \omega$ and $n \neq m$, then $\underline{n} \neq \underline{m}$.

In order to show realizability of the Infinity axiom, we first have to write it out in full detail. Let $\bot_v$ be the formula $\forall u \in v\, \neg u = u$ and let $\mathrm{SC}(u,v)$ be the formula $\forall y \in v\,[y = u \vee y \in u] \wedge [u \in v \wedge \forall y \in u\; y \in v]$. Then Infinity amounts to the sentence

$$(8)\quad \exists x\, (\forall v \in x\, [\, \bot_v \vee \exists u \in x\; \mathrm{SC}(u,v)] \\ \wedge \forall v[(\, \bot_v \vee \exists u \in x\; \mathrm{SC}(u,v)) \rightarrow v \in x]).$$

Suppose $\langle f,c\rangle \in \overline{\omega}$. Then $f = \underline{n}$ and $c = \overline{n}$ for some $n \in \omega$. If $n = 0$ then $\overline{n} = 0$ and therefore $\boldsymbol{0} \Vdash \bot_{\overline{n}}$. Otherwise we have $n = k+1$ for some $k \in \omega$. If $\langle \underline{m}, \overline{m}\rangle \in \overline{n}$ then $m = k$ or $m \in k$, so that $\boldsymbol{i_r} \Vdash \overline{m} = \overline{k}$ or $\boldsymbol{p}\underline{m}\boldsymbol{i_r} \Vdash \overline{m} \in \overline{k}$, and whence $\boldsymbol{d}(\boldsymbol{p0i_r})(\boldsymbol{p1}(\boldsymbol{p}\underline{m}\boldsymbol{i_r}))\underline{m}\,\underline{k} \Vdash (\overline{m} = \overline{k} \vee \overline{m} \in \overline{k})$. As a result of the foregoing we have $\ell(\underline{k}) \Vdash \forall y \in \overline{n}\; (y = \overline{k} \vee y \in \overline{k})$, where $\ell(\underline{k}) := \lambda z \cdot \boldsymbol{d}(\boldsymbol{p0i_r})(\boldsymbol{p1}(\boldsymbol{p}z\boldsymbol{i_r}))z\,\underline{k}$. Note both that $\boldsymbol{p}\underline{k}\boldsymbol{i_r} \Vdash \overline{k} \in \overline{n}$ and

$\lambda z \cdot pzi_r \Vdash \forall y \in \overline{k}\ y \in \overline{n}$, and hence $\wp(\underline{k}) \Vdash \overline{k} \in \overline{n} \wedge \forall y \in \overline{k}\ y \in \overline{n}$, where $\wp(\underline{k}) := p(p\underline{k}i_r)(\lambda z \cdot pzi_r)$. Also note that $\underline{k} = p_N\, \underline{n}$. With

$$t(\underline{n}) := p(p_N \underline{n})(p(\ell(p_N\, \underline{n}))(\wp(p_N\, \underline{n})))$$

we thus obtain $t(\underline{n}) \Vdash \exists u \in \overline{\omega}\ \mathrm{SC}(u, \overline{n})$. In conclusion, as $n = 0$ or $n = k+1$ for some $k \in \omega$ and $\underline{n} = f$ and $\overline{n} = c$ we arrive at $d(p0 0)(p1t(f))f0 \Vdash [\bot_c \vee \exists u \in \overline{\omega}\ \mathrm{SC}(u, c)]$. Hence we have

(9) $$q^* \Vdash \forall v \in \overline{\omega}[\bot_v \vee \exists u \in \overline{\omega}\ \mathrm{SC}(u, v)]$$

where $q^* := \lambda f \cdot d(p0 0)(p1t(f))f0$.

Conversely assume $a \in V(\mathcal{A})$ and

(10) $$e \Vdash \bot_a \vee \exists u \in \overline{\omega}\ \mathrm{SC}(u, a).$$

Then either $(e)_0 = 0$ and $(e)_1 \Vdash \bot_a$ or $(e)_0 = 1$ and $(e)_1 \Vdash \exists u \in \overline{\omega}\ \mathrm{SC}(u, a)$.

The first case scenario yields $a = \emptyset = 0$. To see this assume $\langle f, c\rangle \in a$. Then $(e)_1 f \Vdash \neg c = c$, which means that $\forall g \in |\mathcal{A}|\ \neg g \Vdash c = c$. However, as $i_r \Vdash c = c$ this is absurd, showing $a = 0$. The latter yields $i_r \Vdash \overline{0} = a$ and thus

(11) $$p(e)_0 i_r \Vdash a \in \overline{\omega}.$$

The second scenario entails that $((e)_1)_0 = \underline{n}$ for some $n \in \omega$ as well as $((e)_1)_1 \Vdash \mathrm{SC}(\overline{n}, a)$. Therefore we can conclude that $t_1 \Vdash \forall y \in a\ (y = \overline{n} \vee y \in \overline{n})$, $t_2 \Vdash \overline{n} \in a$, and $t_3 \Vdash \forall y \in \overline{n}\ y \in a$ with $s := ((e)_1)_1$, $t_1 := (s)_0$, $t_2 := ((s)_1)_0$ and $t_3 := ((s)_1)_1$. Our first aim is to construct a closed application term $q^\#$ such that $q^\# \Vdash a = \overline{n+1}$. To this end assume first that $\langle f, c\rangle \in a$. Then $t_1 f \Vdash c = \overline{n} \vee c \in \overline{n}$ and $(t_1 f)_0 = 0$ or $(t_1 f)_0 = 1$. From $(t_1 f)_0 = 0$ we obtain $(t_1 f)_1 \Vdash c = \overline{n}$, and hence $p\underline{n}(t_1 f)_1 \Vdash c \in \overline{n+1}$. If, on the other hand, $(t_1 f)_0 = 1$, we conclude that $(t_1 f)_1 \Vdash c \in \overline{n}$, which entails that $((t_1 f)_1)_0 = \underline{k}$ and $((t_1 f)_1)_1 \Vdash c = \overline{k}$ for some $k \in n$, and hence $pr_0 r_1 \Vdash c \in \overline{n+1}$ where $r_i := ((t_1 f)_1)_i$. To summarize, we have

(12) $$\langle f, c\rangle \in a \rightarrow q_1(f) \Vdash c \in \overline{n+1},$$

where $q_1(f) := d(p\underline{n}(t_1 f)_1)(pr_0 r_1)(t_1 f)_0 0$.

Next assume that $\langle f, c\rangle \in \overline{n+1}$. Thus $f = \underline{k}$ and $c = \overline{k}$ for some $k \in n+1$. We then have $k = n \vee k \in n$. $k = n$ yields $t_2 \Vdash c \in a$, while $k \in n$ yields $t_3\underline{k} \Vdash \overline{k} \in a$, so that $t_3\underline{k} \Vdash c \in a$. Thus, since $f = \underline{k}$ we get $q_2(f) \Vdash c \in a$ with $q_2(f) := dt_2(t_3 f)f\underline{n}$. In conclusion,

(13) $$\langle f, c\rangle \in \overline{n+1} \rightarrow q_2(f) \Vdash c \in a.$$

With $q^\# := p(\lambda f \cdot q_1(f))(\lambda f \cdot q_2(f))$, (12) and (13) entail that $q^\# \Vdash a = \overline{n+1}$, and thus $p\underline{n+1}q^\# \Vdash a \in \overline{\omega}$.

The upshot of the foregoing is that from (10) we have concluded that (11) holds if $(e)_0 = 0$ and that (13) holds if $(e)_0 = 1$. Also note that $(e)_0 = 1$ entails $\underline{n+1} = s_N\, \underline{n} = s_N((e)_1)_0$. Thus we arrive at $\ell^{**}(e) \Vdash a \in \overline{\omega}$ with

$\ell^{**}(e) := d(p(e)_0 i_r)(p(s_N((e)_1)_0)q^{\#})(e)_0 0$. Using lambda-abstraction on $e$, it follows that

$$(14) \qquad \lambda e \cdot \ell^{**}(e) \Vdash \forall v \big[ ( \perp_v \vee \exists u \in \overline{\omega}\ \mathrm{SC}(u,v)) \to v \in \overline{\omega} \big].$$

Finally, (9) and (14) yield that $pq^*(\lambda e \cdot \ell^{**}(e))$ provides a realizer for the Infinity axiom.

**Set Induction:** Assume that for all $a \in V(\mathcal{A})$, $g \Vdash (\forall y \in a \varphi(y)) \to \varphi(a)$. We would like to construct an $e \in |\mathcal{A}|$ such that for all $b \in V(\mathcal{A})$, $eg \Vdash \varphi(b)$. To this end, suppose that $a \in V(\mathcal{A})_\alpha$ and that we have found an $e \in |\mathcal{A}|$ such that for all $b \in \bigcup_{\beta \in \alpha} V(\mathcal{A})_\beta$, $e \Vdash \varphi(b)$. Now, if $\langle h, b \rangle \in a$, then $b \in \bigcup_{\beta \in \alpha} V(\mathcal{A})_\beta$, and hence $e \Vdash \varphi(b)$, so that

$$\lambda u \cdot keu \Vdash \forall y \in a \varphi(y) \qquad \text{and} \qquad g(\lambda u \cdot keu) \Vdash \varphi(a).$$

By the recursion theorem for applicative structures (cf. [16, chap. 2, Corollary 2.7]), there is an $f^* \in |\mathcal{A}|$ that fixes $e^* = g(\lambda u.keu)$, i.e., $f^* e^* = g(\lambda u \cdot k(f^* e^*)u)$. Then, with $e = \lambda g \cdot f^* e^*$, induction on $\alpha$ yields

$$e \Vdash \forall a \left[ (\forall y \in a \varphi(y)) \to \varphi(a) \right] \to \forall a\, \varphi(a).$$

**Strong Collection:** Let $a \in V(\mathcal{A})$ and assume that $g \Vdash \forall x \in a\, \exists y\, \varphi(x,y)$. Then, for $\langle u, x \rangle \in a$ there is a $y \in V(\mathcal{A})$ such that $gu \Vdash \varphi(x,y)$. By invoking Strong Collection in the background universe, there is a set $D$ such that

$$(15) \qquad \forall \langle u, x \rangle \in a\, \exists y \in V(\mathcal{A}) \left[ \langle p(gu)u, y \rangle \in D \wedge gu \Vdash \varphi(x,y) \right],$$

and

$$(16) \qquad \forall z \in D \exists \langle u, x \rangle \in a\, \exists y \in V(\mathcal{A}) \left[ z = \langle p(gu)u, y \rangle \wedge gu \Vdash \varphi(x,y) \right].$$

In particular, $D \subseteq |\mathcal{A}| \times V(\mathcal{A})$, so that by Lemma 3.5, $D \in V(\mathcal{A})$. We need to construct $e, e' \in V(\mathcal{A})$ from $g$ such that

$$(17) \qquad e \Vdash \forall x \in a\, \exists y \in D\, \varphi(x,y),$$

$$(18) \qquad e' \Vdash \forall y \in D\, \exists x \in a\, \varphi(x,y).$$

For (17), let $\langle u, x \rangle \in a$. Then there exists a $y$ such that $\langle p(gu)u, y \rangle \in D$ and $gu \Vdash \varphi(x,y)$, and hence $p(p(gu)u)(gu) \Vdash \exists y \in D\, \varphi(x,y)$; so that, with $e = \lambda u \cdot p(p(gu)u)(gu)$, (17) obtains.

To show (18), let $\langle v, y \rangle \in D$. Then, by (16), $v = p(gu)u$ for some $u \in V(\mathcal{A})$ and there exists an $x$ such that $\langle u, x \rangle \in a \wedge gu \Vdash \varphi(x,y)$. Hence $p(v)_1(v)_0 \Vdash \exists x \in a\, \varphi(x,y)$, so that with $e' = \lambda v \cdot p(v)_1(v)_0$, (18) obtains.

**Subset Collection:** Let $a, b \in V(\mathcal{A})$ and $\varphi(x,u,y)$ be a formula with parameters in $V(\mathcal{A})$. We would like to find a realizer $r$ such that

$$(19) \qquad r \Vdash \exists q \forall u \left[ \forall x \in a\, \exists y \in b\, \varphi(x,y,u) \to \exists v \in q\, \varphi'(a,v,u) \right],$$

where $\varphi'(a, v, u)$ abbreviates the formula

$$\forall x \in a\, \exists y \in v\, \varphi(x, y, u) \wedge \forall y \in v\, \exists x \in a\, \varphi(x, y, u).$$

Set

$$b^* = \{\langle \mathbf{p}ef, d\rangle : e, f \in |\mathcal{A}| \wedge ef \downarrow \wedge \langle (ef)_0, d\rangle \in b\}.$$

Note that $b^*$ is a set. Further, let $\psi(e, f, c, u, z)$ be the formula

$$u \in V(\mathcal{A}) \wedge e, f \in |\mathcal{A}| \wedge ef \downarrow \wedge \\ \exists d\, \big[\langle \mathbf{p}ef, d\rangle = z \wedge \langle (ef)_0, d\rangle \in b \wedge (ef)_1 \Vdash \varphi(c, d, u)\big].$$

By invoking Subset Collection there exists a set $D$ such that

$$\forall u \forall e\, \big[\forall \langle f, c\rangle \in a\, \exists z \in b^* \psi(e, f, c, u, z) \rightarrow \\ \exists w \in D\, \big(\forall \langle f, c\rangle \in a \exists z \in w\, \psi(e, f, c, u, z) \wedge \\ \forall z \in w\, \exists \langle f, c\rangle \in a\, \psi(e, f, c, u, z)\big)\big].$$

Now set

$$D^* := \{w \cap b^* : w \in D\}.$$

Then $D^* \subseteq V(\mathcal{A})$, and thus

$$E := \big\{\langle 0, v\rangle : v \in D^*\big\}$$

is an element of $V(\mathcal{A})$.

Let $e \in |\mathcal{A}|$ and $u \in V(\mathcal{A})$ satisfy

(20) $$e \Vdash \forall x \in a\, \exists y \in b\, \varphi(x, y, u).$$

Then

$$\forall \langle f, c\rangle \in a\, \exists d\, \big[\langle (ef)_0, d\rangle \in b \wedge (ef)_1 \Vdash \varphi(c, d, u)\big].$$

Therefore we get $\forall \langle f, c\rangle \in a\, \exists z \in b^*\, \psi(e, f, c, u, z)$. Thus there exists $v \in D^*$ such that

$$\forall \langle f, c\rangle \in a\, \exists z \in v\, \psi(e, f, c, u, z) \wedge \forall z \in v\, \exists \langle f, c\rangle \in a\, \psi(e, f, c, u, z).$$

The latter implies the following two assertions:

$$\forall \langle f, c\rangle \in a\, \exists d\, \big[\langle \mathbf{p}ef, d\rangle \in v \wedge \langle (ef)_0, d\rangle \in b \wedge (ef)_1 \Vdash \varphi(c, d, u)\big],$$

$$\forall z \in v\, \exists \langle f, c\rangle \in a\, \exists d\, \big[z = \langle \mathbf{p}ef, d\rangle \wedge \langle (ef)_0, d\rangle \in b \wedge (ef)_1 \Vdash \varphi(c, d, u)\big].$$

With

$$m_0 := \lambda f \cdot \mathbf{p}(\mathbf{p}ef)(ef)_1,$$
$$m_1 := \lambda g \cdot \mathbf{p}\big((g)_0(g)_1\big)_0\big((g)_0(g)_1\big)_1$$

we thus obtain

$$m_0 \Vdash \forall x \in a\, \exists y \in v\, \varphi(x, y, u),$$
$$m_1 \Vdash \forall y \in v\, \exists x \in a\, \varphi(x, y, u).$$

As a result of the foregoing we have

(21) $\quad pm_0m_1 \Vdash \forall x \in a\, \exists y \in v\, \varphi(x,y,u) \wedge \forall y \in v\, \exists x \in a\, \varphi(x,y,u).$

Let $\varphi'(a,v,u)$ stand for $\forall x \in a\, \exists y \in v\, \varphi(x,y,u) \wedge \forall y \in v\, \exists x \in a\, \varphi(x,y,u)$. Thus far we have shown that (20) implies (21). Consequently,

$$\lambda e \cdot p0(pm_0m_1) \Vdash \forall x \in a\, \exists y \in b\, \varphi(x,y,u) \to \exists v \in E\, \varphi'(a,v,u),$$

and thus

$$\lambda e \cdot p0(pm_0m_1) \Vdash \exists q\, \forall u\, \big[\forall x \in a\, \exists y \in b\, \varphi(x,y,u) \to \exists v \in q\, \varphi'(a,v,u)\big],$$

verifying Subset Collection. ⊣

**§6. The soundness theorem for CZF + REA.** Next we show that the regular extension axiom holds in $V(\mathcal{A})$ if it holds in the background universe.

LEMMA 6.1 (**CZF**). 1. *If $B$ is a regular set with $2 \in B$, then $B$ is closed under unordered and ordered pairs, i.e., whenever $x, y \in B$, then $\{x,y\}, \langle x,y\rangle \in B$.*

2. *If $B$ is a regular set, then $B \cap V(\mathcal{A})$ is a set.*

PROOF. (1): Let $x, y \in B$. Then $f : 2 \to B$, where $f(0) = x$, $f(1) = y$. Hence, by regularity of $B$, the range of $f$ is in $B$, that is $\{x,y\} \in B$.

As $\langle x,y\rangle = \{\{x\},\{x,y\}\}$, $\langle x,y\rangle \in B$ follows from closure under unordered pairs.

(ii): To see this let $\kappa = \text{rank}(B)$, where the function rank is defined by $\text{rank}(x) := \bigcup\{\text{rank}(y)+1 : y \in x\}$ with $z+1 := z \cup \{z\}$. One easily shows that for all sets $x$, $\text{rank}(x)$ is an ordinal. Let $\Phi$ be the inductive definition with

$$\frac{x}{a}\Phi \quad \text{iff}\ \forall u{\in}a\ (u \in |\mathcal{A}| \times x).$$

Invoking Lemma 3.4, let $J$ be the class such that $V(\mathcal{A}) = \bigcup_\alpha J^\alpha$, and for each $\alpha$,

$$J^\alpha = \Gamma_\Phi\Bigg(\bigcup_{\beta\in\alpha} J^\beta\Bigg).$$

Moreover, define the operation $\Upsilon$ by

$$\Upsilon(X) := \{u \in B : u \subseteq |\mathcal{A}| \times X\}$$

and by recursion on $\alpha$ set

$$\Upsilon^\alpha = \Upsilon\Bigg(\bigcup_{\beta\in\alpha} \Upsilon^\beta\Bigg).$$

Let

$$\Upsilon^{<\nu} := \bigcup_{\xi\in\nu} \Upsilon^\xi.$$

Then $\Upsilon^{<\kappa}$ is a set and $\Upsilon^{<\kappa} \subseteq V(\mathcal{A})$. By induction on $\alpha$ we shall verify that

(22) $$J^{\alpha} \cap B \subseteq \Upsilon^{<\kappa}.$$

Thus assume that for all $\beta \in \alpha$, $J^{\beta} \cap B \subseteq \Upsilon^{<\kappa}$. If $c \in J^{\alpha} \cap B$, then

$$c \subseteq |\mathcal{A}| \times \bigcup_{\beta \in \alpha} (J^{\beta} \cap B)$$

since $B$ is transitive. Hence for every $x \in c$, there exists $e \in |\mathcal{A}|$ and $u \in \Upsilon^{<\kappa}$ such that $x = \langle e, u \rangle$. Thus, by definition of $\kappa$, for every $x \in c$ there exist $u, v \in B$, $e \in |\mathcal{A}|$, such that with $\eta = \mathrm{rank}(v)$, $u \in \Upsilon^{\eta}$ and $x = \langle e, u \rangle$. Using the regularity of $B$, there exists a set $d \in B$ such that for each $x \in c$ there exist $u \in B$, $v \in d$, $e \in |\mathcal{A}|$, such that with $\eta = \mathrm{rank}(v)$, $u \in \Upsilon^{\eta}$ and $x = \langle e, u \rangle$. As a result,

$$c \subseteq |\mathcal{A}| \times \Upsilon^{<\gamma},$$

where $\gamma = \mathrm{rank}(d)$. Thus $c \in \Upsilon^{\gamma} \subseteq \Upsilon^{<\kappa}$ as $\gamma \in \kappa$. Consequently, we have $J^{\alpha} \cap B \subseteq \Upsilon^{<\kappa}$.

The upshot of the above is that

$$B \cap V(\mathcal{A}) = \Upsilon^{<\kappa}$$

and therefore $B \cap V(\mathcal{A})$ is set. ⊣

THEOREM 6.2. *For every axiom $\theta$ of* **CZF + REA**, *there exists a closed application term $t$ such that*

$$\mathbf{CZF} + \mathbf{REA} \vdash (t \Vdash \theta).$$

PROOF. In view of theorem 5.1, we need only find a realizer for the axiom **REA**. Let $a \in V(\mathcal{A})$. By **REA** there exists a regular set $B$ such that $a, 2, |\mathcal{A}| \in B$. Let

$$A := B \cap V(\mathcal{A}),$$
$$C := \{\langle \mathit{0}, x \rangle : x \in A\}.$$

By Lemma 6.1, $A$ is a set; hence $C$ is a set. Moreover, as $A \subseteq V(\mathcal{A})$, it follows $C \in V(\mathcal{A})$ and

(23) $$\boldsymbol{p0i_r} \Vdash a \in C.$$

With $\tilde{\boldsymbol{m}} := \lambda x \cdot \lambda y \cdot \boldsymbol{p0i_r}$ and $\tilde{\boldsymbol{n}} := \boldsymbol{p0}(\boldsymbol{p0i_r})$ one realizes transitivity and inhabitedness of $C$, respectively, i.e.,

(24) $$\boldsymbol{p}\tilde{\boldsymbol{m}}\tilde{\boldsymbol{n}} \Vdash \forall u \in C\ \forall v \in u\ v \in C \wedge \exists x \in C\ x \in C.$$

Next, we would like to find a realizer $\boldsymbol{q}$ such that

(25) $$\boldsymbol{q} \Vdash \mathbf{Reg}(C).$$

To this end, let $b \in V(\mathcal{A})$, $f \in |\mathcal{A}|$, and $\varphi(x, y)$ be a formula with parameters in $V(\mathcal{A})$ satisfying

$$f \Vdash b \in C \wedge \forall x \in b\, \exists y \in C\, \varphi(x, y). \tag{26}$$

Then there exists $d$ such that

$$\langle f_{0,0}, d\rangle \in C \wedge f_{0,1} \Vdash b = d, \tag{27}$$

$$f_1 \Vdash \forall x \in b\, \exists y \in C\, \varphi(x, y), \tag{28}$$

where $f_{0,0} := ((f)_0)_0$, $f_{0,1} := ((f)_0)_1$, and $f_1 := (f)_1$. (27) and (28) yield

$$\boldsymbol{i}_\psi f_{0,1} f_1 \Vdash \forall x \in d\, \exists y \in C\, \varphi(x, y), \tag{29}$$

where $\boldsymbol{i}_\psi \Vdash \forall u \forall v[u = v \wedge \psi(u) \rightarrow \psi(v)]$, with $\psi(u)$ being $\forall x \in u\, \exists y \in C\, \varphi(x, y)$ (according to Lemma 4.2, $\boldsymbol{i}_\psi$ is independent of $C$ and the parameters in $\varphi$).

Since $B$ is closed under ordered pairs (Lemma 6.1) and $|\mathcal{A}| \in B$, from (29) we get

$$\begin{aligned} \forall x \forall e\, \big[&\langle e, x\rangle \in d \\ &\rightarrow \exists z \in B\, \exists y (z = \langle e, y\rangle \wedge \langle (\tilde{f} e)_0, y\rangle \in C \wedge (\tilde{f} e)_1 \Vdash \varphi(x, y))\big], \end{aligned} \tag{30}$$

where $\tilde{f} := \boldsymbol{i}_\psi f_{0,1} f_1$. Noting that $\langle v, y\rangle \in C$ entails $v = \boldsymbol{0}$, and utilizing the regularity of $B$, there exists $\boldsymbol{u} \in B$ such that

$$\forall x \forall e\, \big[\langle e, x\rangle \in d \rightarrow \exists z \in \boldsymbol{u}\, \exists y (z = \langle e, y\rangle \wedge \langle \boldsymbol{0}, y\rangle \in C \wedge (\tilde{f} e)_1 \Vdash \varphi(x, y))\big]; \tag{31}$$

$$\forall z \in \boldsymbol{u}\, \exists x, e\, \big[\langle e, x\rangle \in d \wedge \exists y (\langle \boldsymbol{0}, y\rangle \in C \wedge z = \langle e, y\rangle \wedge (\tilde{f} e)_1 \Vdash \varphi(x, y))\big]. \tag{32}$$

From (32) it follows that $\boldsymbol{u} \in A$, and thus $\langle \boldsymbol{0}, \boldsymbol{u}\rangle \in C$. So we get

$$\boldsymbol{p0i}_r \Vdash \boldsymbol{u} \in C. \tag{33}$$

Letting $s(f) := \lambda e \cdot \boldsymbol{p} e(\tilde{f} e)_1$, (31) and (32) yield

$$s(f) \Vdash \forall x \in d\, \exists y \in \boldsymbol{u}\, \varphi(x, y),$$
$$s(f) \Vdash \forall y \in \boldsymbol{u}\, \exists x \in d\, \varphi(x, y).$$

By invoking Lemma 4.2, from the latter we can effectively determine application terms $\tilde{s}(f)$ and $\hat{s}(f)$ such that

$$\tilde{s}(f) \Vdash \forall x \in b\, \exists y \in \boldsymbol{u}\, \varphi(x, y), \tag{34}$$

$$\hat{s}(f) \Vdash \forall y \in \boldsymbol{u}\, \exists x \in b\, \varphi(x, y). \tag{35}$$

Hence, letting $\tilde{q} := \lambda f \cdot p(p0i_r)(p\tilde{s}(f)\hat{s}(f))$, (33), (34), and (35) entail that

(36) $$\tilde{q} \Vdash \forall b(b \in C \wedge \forall x \in b\, \exists y \in C\, \varphi(x,y) \rightarrow \exists u \in C\,[\forall x \in b\, \exists y \in u\, \varphi(x,y) \wedge \forall y \in u\, \exists x \in b\, \varphi(x,y)]).$$

Note that the entire construction of $\tilde{q}$ shows that it neither depends on $C$ nor on the parameters of $\varphi$. Choosing $\varphi(x,y)$ to be the formula $r \subseteq b \times C \wedge \langle x,y\rangle \in r$, we deduce from (36) and (24) that

$$p(p\tilde{m}\tilde{n})\tilde{q} \Vdash \mathbf{Reg}(C).$$

Thus, in view of (23), we get

$$p(p0i_r)(p(p\tilde{m}\tilde{n})\tilde{q}) \Vdash \forall a\, \exists C\,[a \in C \wedge \mathbf{Reg}(C)]. \qquad \dashv$$

**§7. Adding Markov's principle and independence of premisses.** *Markov's Principle* (**MP**) is closely associated with the work of the school of Russian constructivists. The version of **MP** most appropriate to the set-theoretic context is the schema

$$\forall n \in \omega\,[\varphi(n) \vee \neg\varphi(n)] \rightarrow [\neg\neg\exists n \in \omega\, \varphi(n) \rightarrow \exists n \in \omega \varphi(n)].$$

The variant

$$\neg\neg\exists n \in \omega\, R(n) \rightarrow \exists n \in \omega R(n),$$

with $R$ being a primitive recursive predicate, will be denoted by $\mathbf{MP_{PR}}$. Obviously, $\mathbf{MP_{PR}}$ is implied by **MP**.

Another classically valid principle considered in connection with intuitionistic theories is the *Principle of Independence of Premisses*, **IP**, which is expressed by the schema

$$(\neg\theta \rightarrow \exists x\psi) \rightarrow \exists x(\neg\theta \rightarrow \psi),$$

where $\theta$ is assumed to be closed. A variant of **IP** is $\mathbf{IP}_\omega$:

$$(\neg\theta \rightarrow \exists n \in \omega\, \psi) \rightarrow \exists n \in \omega\,(\neg\theta \rightarrow \psi),$$

where $\theta$ is closed.

As has been shown by McCarty, both **MP** and **IP** are realized in $V(Kl)$ if one assumes classical logic in the background theory (cf. [16, Theorems 11.3 and 11.5]). In connection with **CZF** one is naturally led to ask whether these principles add any proof-theoretic strength to **CZF**?

THEOREM 7.1. (i) **CZF** *and* $\mathbf{CZF} + \mathbf{MP} + \mathbf{IP} + \mathbf{IP}_\omega$ *have the same proof-theoretic strength and the same provably recursive functions.*

(ii) $\mathbf{CZF} + \mathbf{REA}$ *and* $\mathbf{CZF} + \mathbf{REA} + \mathbf{MP} + \mathbf{IP}_\omega$ *have the same proof-theoretic strength and the same provably recursive functions.*

(i) *also obtains if one adds the axioms* **DC** (*Dependent Choices*), **PAx** (*Presentation Axiom*), *and* **ΠΣ-AC** (*for definitions see* [3]) *to* $\mathbf{CZF} + \mathbf{MP} + \mathbf{IP} + \mathbf{IP}_\omega$.

*Likewise, in* (ii) *one may add* **DC**, **PAx**, *and* **ΠΣW-AC** *to* $\mathbf{CZF} + \mathbf{REA} + \mathbf{MP} + \mathbf{IP} + \mathbf{IP}_\omega$.

PROOF. All of these results are actually corollaries of the interpretations of the systems $\mathbf{ML_1V}$ and $\mathbf{ML_{1W}V}$ of Martin-Löf type theory into the classical set theories **KP** and **KPi** of Kripke-Platek set theory, respectively, given in [19, Theorems 4.11 and 5.13]. Combining these interpretations with Aczel's formulae-as-types interpretations of set theory into Martin-Löf type theory, one obtains formulas-as-classes interpretations of **CZF** and **CZF** + **REA** in **KP** and **KPi**, respectively.

To be more precise, we shall focus on the interpretation of **CZF** in **KP** obtained in this way. The first step consists in simulating $\mathbf{ML_1V}$ in **KP** by interpreting a type $A$ as a class of natural numbers $[\![A]\!]$ and the equality relation $=_A$ on $A$ as a class $[\![=_A]\!]$ of pairs of natural numbers. Pivotally, if $A, B$ are types that have been interpreted as classes $[\![A]\!]$ and $[\![B]\!]$, then the function type $A \to B$ is interpreted as the class of indices $e$ of partial recursive functions satisfying $\forall x \in [\![A]\!]\ \{e\}(x) \in [\![B]\!]$ and $\forall (x, y) \in [\![=_A]\!]\ (\{e\}(x), \{e\}(y)) \in [\![=_B]\!]$.

Reasoning in **KP** one inductively define classes $\boldsymbol{U}$ and $\boldsymbol{V}$ such that $\boldsymbol{U}, \boldsymbol{V} \subseteq \omega$. $\boldsymbol{U}$ is the class of codes of small types while $\boldsymbol{V}$ serves as a universe for the interpretation of formulas of **CZF** (for the precise definitions of $x \in \boldsymbol{U}$ and $x \in \boldsymbol{V}$ we have to refer to [19, Definition 4.3]: $\boldsymbol{U} = \{x \in \omega : \mathbb{U} \models x\ \mathbf{set}\}$ and $\boldsymbol{V} = \{x \in \omega : \mathbb{V} \models x\ \mathbf{set}\}$). Moreover, each $\alpha \in \boldsymbol{V}$ is of the form $\sup(n, e)$ such that $n, e \in \omega$, sup is a primitive recursive pairing function on $\omega$, $n \in \boldsymbol{U}$ and $e$ is an index of a partial recursive function with

$$\{e\} : \{x : \mathbb{U} \models x \in n\} \to \boldsymbol{V}.$$

For the unique $n, e$ such that $\alpha = \sup(n, e)$ we use the shorthands $\bar{\alpha}$ and $\tilde{\alpha}$, respectively, and write $\tilde{\alpha}(x)$ for $\{e\}(x)$.

Further, the formulae-as-types interpretation associates with each formula $\varphi(u_1, \dots, u_r)$ of **CZF** and any a class-valued function

$$(\alpha_1, \dots, \alpha_r) \mapsto [\![\varphi(\alpha_1, \dots, \alpha_r)]\!]$$

(uniformly in $\alpha_1, \dots, \alpha_r$), where $\alpha_1, \dots, \alpha_r \in \boldsymbol{V}$. The interpretation is such that $[\![\bot]\!] = \emptyset$ and whenever $\mathbf{CZF} \vdash \varphi(u_1, \dots, u_r)$, then $\mathbf{KP} \vdash \forall \alpha_1, \dots, \alpha_r \in \boldsymbol{V}\ [\![\varphi(\alpha_1, \dots, \alpha_r)]\!] \neq \emptyset$. Another fact worthwhile mentioning is that this interpretation is faithful with regard to $\Pi^0_2$ statements of arithmetic, i.e. if $\psi$ is the set-theoretic rendering of a $\Pi^0_2$ statement of Peano arithmetic, then

$$\mathbf{KP} \vdash [\![\psi]\!] \neq \emptyset \leftrightarrow \psi.$$

To show that **MP** is validated under this interpretation, let $\varphi(u)$ be a set-theoretic formula with parameters from $\boldsymbol{V}$. We are to show that

$$(37)\quad [\![\forall n \in \omega\ [\varphi(n) \vee \neg\varphi(n)] \to [\neg\neg\exists n \in \omega\ \varphi(n) \to \exists n \in \omega \varphi(n)]]\!] \neq \emptyset.$$

Let $\underline{\omega}$ be the element of $V$ that serves to interpret the set $\omega$. We have $\{\tilde{\omega}\} : \omega \to V$. For $n \in \omega$ let $\bar{n} := \{\tilde{\omega}\}(n)$. Now suppose

$$e \in [\![ \forall n \in \omega \, [\varphi(n) \vee \neg\varphi(n)] ]\!] \tag{38}$$

and

$$g \in [\![ \neg\neg\exists n \in \omega \, \varphi(n) ]\!]. \tag{39}$$

(38) yields

$$\forall n \in \omega \, (\{e\}(\bar{n}) \in [\![ \varphi(\bar{n}) \vee \neg\varphi(\bar{n}) ]\!]). \tag{40}$$

It is a consequence of (40) that, for some partial recursive function $\eta$ and for each $n \in \omega$, either

$$\eta(e,n)_0 = 0 \wedge \eta(e,n)_1 \in [\![ \varphi(\bar{n}) ]\!]$$

or

$$\eta(e,n)_0 = 1 \wedge \eta(e,n)_1 \in [\![ \neg\varphi(\bar{n}) ]\!].$$

As we work in the classical theory **KP**, from (39) it follows that for some $m \in \omega$, $[\![ \varphi(\bar{m}) ]\!] \neq \emptyset$. Consequently, there exists $n$ such that $\eta(e,n)_0 = 0 \wedge \eta(e,n)_1 \in [\![ \varphi(\bar{n}) ]\!]$, and hence, with $r := \mu n \cdot \eta(e,n)_0 = 0$, $\eta(e,r)_1 \in [\![ \varphi(\bar{n}) ]\!]$, so

$$\eta(e,r) \in [\![ \exists n \in \omega \, \varphi(n) ]\!].$$

As a result, we have shown (37).

To show that **IP** is validated under this interpretation, assume that

$$e \in [\![ \neg\theta \to \exists x \psi(x) ]\!],$$

where $\theta$ is closed. Then, if $g \in [\![ \neg\theta ]\!]$, we get $0 \in [\![ \neg\theta ]\!]$, and thus $\{e\}(0) \in [\![ \exists x \psi(x) ]\!]$. Therefore, with $a := (\{e\}(0))_0$ and $e' :=:= (\{e\}(0))_1$ we get $e' \in [\![ \psi(a) ]\!]$.

Hence, if $[\![ \theta ]\!] = \emptyset$, $\{e\}(0)$ is defined and

$$\{e\}(0) \in [\![ \exists x \, (\neg\theta \to \psi(x)) ]\!].$$

On the other hand, should $[\![ \theta ]\!] \neq \emptyset$, then $[\![ \neg\theta ]\!] = \emptyset$, so

$$(\bar{0}, 0) \in [\![ \exists x \, (\neg\theta \to \psi(x)) ]\!].$$

Thus, in every case we have shown

$$[\![ (\neg\theta \to \exists x \psi(x)) \to \exists x \, (\neg\theta \to \psi(x)) ]\!] \neq \emptyset.$$

Similarly one shows that $\mathbf{IP}_\omega$ is validated under this interpretation.

The further claim that choice principles **DC**, **PAx**, and **ΠΣ-AC** (**ΠΣ$W$-AC**) may be added is a consequence of [19, Theorem 4.14 (Theorem 5.13)]. ⊣

**§8. Absoluteness properties.** The aim of this section is to show that truth in $V$ and realizability in $V(Kl)$ mean the same for almost negative arithmetic formulae. Our first task is to single out the natural candidates for representing the natural numbers in $V(Kl)$. Whenever $\exists e \in |\mathcal{A}|\ e \Vdash \varphi$, we write '$V(\mathcal{A}) \models \varphi$'.

DEFINITION 8.1. For $a, b \in V(Kl)$, set $\{a,b\}_{kl} := \{\langle 0,a\rangle, \langle 1,b\rangle\}$ and put

$$\langle a,b\rangle_{Kl} := \{\langle 0, \{a,a\}_{kl}\rangle, \langle 1, \{a,b\}_{kl}\rangle\}.$$

LEMMA 8.2. *For $a, b, x \in V(Kl)$,*

$$V(Kl) \models x \in \{a,b\}_{kl} \leftrightarrow x = a \vee x = b,$$
$$V(Kl) \models x \in \langle a,b\rangle_{Kl} \leftrightarrow x = \{a,a\}_{kl} \vee x = \{a,b\}_{kl}.$$

PROOF. See [16, Lemmas 3.2 and 3.4]. ⊣

DEFINITION 8.3. The natural internalization of $\omega$ in $V(Kl)$ is defined as follows: For each $n \in \omega$, let $\overline{n} = \{\langle m, \overline{m}\rangle : m \in n\}$ and set

$$\overline{\omega} = \{\langle n, \overline{n}\rangle : n \in \omega\}.$$

PROPOSITION 8.4 (**CZF**). *Equality and membership are realizably absolute for $\overline{\omega}$. This means that for all $n, m \in \omega$,*

$$m = n \quad \textit{iff}\ V(Kl) \models \overline{m} = \overline{n} \qquad \textit{and} \qquad m \in n \quad \textit{iff}\ V(Kl) \models \overline{m} \in \overline{n}.$$

PROOF. [16, Ch. 3, Theorem 3.11]. ⊣

Elementary recursion theory can be formalized in Heyting arithmetic (cf. [21, Vol. Ch. 3, section 6]) and a fortiori it can be formalized in **CZF**. In particular one can talk about primitive recursive relations in **CZF**. Each primitive recursive $n$-ary relation $R$ on $\omega$ is canonically represented by a formula $\varphi_R$ in the language of **CZF**. In the following we shall write $V(Kl) \models R(\overline{n_1}, \ldots, \overline{n_r})$ rather than the more accurate $V(Kl) \models \varphi_R(\overline{n_1}, \ldots, \overline{n_r})$.

PROPOSITION 8.5 (**CZF**). *When R is a primitive recursive r-ary relation R on $\omega$ and $n_1, \ldots, n_r \in \omega$, then*

$$R(n_1, \ldots, n_r) \ \textit{iff}\ V(Kl) \models R(\overline{n_1}, \ldots, \overline{n_r}).$$

PROOF. See [16, Ch. 4, Theorem 2.6]. ⊣

DEFINITION 8.6. A formula $\theta$ of the language of **CZF** which contains solely parameters from $\omega$ is said to be *almost negative arithmetic* if it is built from primitive recursive formulas $\varphi_R$, using the connectives $\wedge, \rightarrow, \neg$, bounded universal quantifiers $\forall n \in \omega$, and bounded existential quantifiers $\exists m \in \omega$ which appear only as prefixed to primitive recursive subformulae of $\theta$.

THEOREM 8.7 (**CZF**). *If $n_1, \ldots, n_r \in \omega$ and $\theta(n_1, \ldots, n_r)$ is an almost negative arithmetic formula, then there is an r-place primitive recursive function $f_\theta$ such*

$$\theta(n_1, \ldots, n_r) \quad \text{iff } f_\theta(n_1, \ldots, n_r) \Vdash \theta(\overline{n_1}, \ldots, \overline{n_r})$$
$$\text{iff } V(Kl) \models \theta(\overline{n_1}, \ldots, \overline{n_r}).$$

PROOF. This is proved by induction on the build-up of $\theta$. For the primitive recursive subformulas this follows from the proof of [16, Ch. 4, Theorem 2.6]. For the inductive steps one proceeds exactly as in the proof of [21, Sect. 4, Proposition 4.5]. ⊣

**§9. Some classical and non-classical principles that hold in $V(Kl)$.** The next definitions lists several interesting principles that are validated in $V(Kl)$.

DEFINITION 9.1. 1. *Church's Thesis*, **CT**, is formalized as

$$\forall n \in \omega\, \exists m \in \omega\, \varphi(n, m)$$
$$\rightarrow \exists e \in \omega\, \forall n \in \omega\, \exists m, p \in \omega\, [T(e, n, p) \wedge U(p, m) \wedge \varphi(n, m)]$$

for every formula $\varphi(u, v)$, where $T$ and $U$ are the set-theoretic predicates which numeralwise represent, respectively, Kleene's $T$ and result-extraction predicate $U$.

2. *Extended Church's Thesis*, **ECT**, asserts that

$$\forall n \in \omega\, [\psi(n) \rightarrow \exists m \in \omega\, \varphi(n, m)] \text{ implies}$$
$$\exists e \in \omega\, \forall n \in \omega\, [\psi(n) \rightarrow \exists m, p \in \omega\, [T(e, n, p) \wedge U(p, m) \wedge \varphi(n, m)]]$$

whenever $\psi(n)$ is an almost negative arithmetic formula and $\varphi(u, v)$ is any formula. Recall that formula $\theta$ of the language of **CZF** with quantifiers ranging over $\omega$ is said to be *almost negative arithmetic* if $\vee$ does not appear in it and instances of $\exists m \in \omega$ appear only as prefixed to primitive recursive subformulae of $\theta$.

Note that **ECT** implies **CT**, taking $\psi(n)$ to be $n = n$.

3. **UP**, the *Uniformity Principle*, is expressed by the scheme:

$$\forall x\, \exists n \in \omega\, \varphi \rightarrow \exists n \in \omega\, \forall x \varphi.$$

4. *Unzerlegbarkeit*, **UZ**, is the scheme

$$\forall x(\phi \vee \neg\phi) \rightarrow \forall x\varphi \vee \forall x \neg\varphi$$

for all formulas $\varphi$.

A set is said to be *subcountable* if it is the surjective image of a subset of $\omega$.

It is known that all the above principles hold in $V(Kl)$ if one assumes the axioms of **IZF** in the background universe $V$ (see [16]). Owing to results of sections 5 and 6, we know that Kleene realizability is self-validating for

**CZF** and **CZF + REA**. By inspection of the proofs of [16] one arrives at the following theorem:

THEOREM 9.2 (**CZF**). $V(Kl) \models \mathbf{AC}^{\omega,\omega} \wedge \mathbf{CT} \wedge \mathbf{ECT} \wedge \mathbf{UP} \wedge \mathbf{UZ}$.

PROOF. (a): The proof of $V(Kl) \models \mathbf{AC}^{\omega,\omega}$ in [16, Chap. 3, Theorem 5.1] uses intuitionistic logic and besides that just means available in **CZF**.

(b): One readily checks that the proof of Theorem 3.1, [16] of $V(Kl) \models \mathbf{CT}$ just utilizes $V(Kl) \models \mathbf{AC}^{\omega,\omega}$ and axioms from **CZF**.

(c): Inspection of [16, Theorem 8.2] shows that the proof of $V(Kl) \models \mathbf{UP}$ carries over to **CZF**.

(d): From (c) one immediately deduces that

$$V(Kl) \models y \textit{ is subcountable}$$

implies

$$V(Kl) \models \forall x\, \exists z \in y\, \varphi \rightarrow \exists z \in y\, \forall x\, \varphi.$$

Now, since any two element set is subcountable, it follows from the above that $V(Kl) \models \mathbf{UZ}$.

Ad **ECT**: Assume $d \Vdash \forall n \in \omega\, [\psi(n) \rightarrow \exists m \in \omega\, \varphi(n,m)]$, where $\psi(n)$ is an almost negative arithmetic formula. Invoking Theorem 8.7, there is a number $t_\psi(n) \in \omega$ depending primitive recursively on $n$ (and possibly further parameters of $\psi$), such that

$$\begin{aligned} V(Kl) \models \psi(n) \quad &\text{iff } t_\psi \Vdash \psi(n) \\ &\text{iff } \psi(n), \end{aligned}$$

so that with $f(n) = (\lambda u \cdot du)t_\psi(n)$ we arrive at

$$\forall n \in \omega\, \left[\psi(n) \rightarrow f(n) \Vdash \exists m \in \omega\, \varphi(\bar{n}, m)\right].$$

From the latter it follows with $h(n) = (f(n))_0$ and $l(n) = f(n)_1$ that

$$\forall n \in \omega\, \left[\psi(n) \rightarrow l(n) \Vdash \varphi(\bar{n}, \overline{h(n)})\right].$$

Taking $e$ to be an index for $h$, it is obvious that from $n$ we can effectively construct an realizer $\rho(n)$ such that

$$\forall n \in \omega\, \left[\psi(n) \rightarrow \rho(n) \Vdash \exists m, p \in \omega\, \left[T(\bar{e}, \bar{n}, p) \wedge U(p,m) \wedge \varphi(\bar{n}, m)\right]\right].$$

Hence, owing to Theorem 8.7, we can calculate effectively from $d$ a realizer $d'$ such that

$$\begin{aligned} d' \Vdash \exists e \in \omega \forall n \in \omega \big[\psi(n) \rightarrow \rho(n) \Vdash \exists m,& \\ p \in \omega\, \left[T(e,n,p) \wedge U(p,m) \wedge \varphi(n,m)\right]\big].& \quad \dashv \end{aligned}$$

The principles **MP** and **IP** are known to propagate from $V$ to $V(Kl)$. We therefore obtain the following results:

THEOREM 9.3. 1. (**CZF + MP**) $V(Kl) \models \mathbf{MP}$.

2. (**CZF + IP**) $V(Kl) \models \mathbf{IP}$.

PROOF. See [16, chap. 3, Theorem 11.3 and Theorem 11.5]. ⊣

**§10. Axioms of choice and $V(Kl)$.** While $\mathbf{AC}^{\omega,\omega}$ holds in $V(Kl)$ for free, i.e. without assuming any choice in the background universe, validity of the following choice principles in $V(Kl)$ seems to require their respective validity in $V$.

THEOREM 10.1. (i) $(\mathbf{CZF}+\mathbf{DC})\ V(Kl) \models \mathbf{DC}$.
(ii) $(\mathbf{CZF}+\mathbf{RDC})\ V(Kl) \models \mathbf{RDC}$.
(iii) $(\mathbf{CZF}+\mathbf{PAx})\ V(Kl) \models \mathbf{PAx}$.

PROOF. We shall use the abbreviations $(x, y)$, $(x)_0$, and $(x)_1$ for $\boldsymbol{p}^{Kl}xy$, $\boldsymbol{p}_0^{Kl}x$, and $\boldsymbol{p}_1^{Kl}x$, respectively.

(i): This proof is similar to the proof of [16, chap. 3, Theorem 6.1]. However, because of the special way we defined realizability for bounded quantifiers there are some differences. In particular, the set $a^{\Vdash}$ used in the proof of that Theorem will not be needed here.

Let $a, u \in V(Kl)$. Suppose

$$e \Vdash \forall x \in a \exists y \in a\ \varphi(x, y) \qquad \text{and} \qquad e^* \Vdash u \in a.$$

Then there exists $\langle k_0, u_0\rangle \in a$ such that $k_0 = (e^*)_0$ and $(e^*)_1 \Vdash u = u_0$ and

$$\forall \langle n, x\rangle \in a\ \exists y\ \left[\langle(\{e\}(n))_0, y\rangle \in a \wedge (\{e\}(n))_1 \Vdash \varphi(x, y)\right].$$

Thus, for all $n \in \omega$ and all $b \in V(Kl)$, if $\langle n, b\rangle \in a$, then $\{e\}(n)\downarrow$ and there is a $c \in V(Kl)$ such that

$$\langle(\{e\}(n))_0, c\rangle \in a \qquad \text{and} \qquad (\{e\}(n))_1 \Vdash \varphi(b, c).$$

Let $\varphi^{\Vdash}$ be such that, externally,

$$\varphi^{\Vdash}(\langle n, b\rangle, \langle m, c\rangle) \quad \text{iff } m = (\{e\}(n))_0 \wedge (\{e\}(n))_1 \Vdash \varphi(b, c).$$

By **DC** in the ground model, there is a function $F : \omega \to a$ such that

$$F(0) = \langle k_0, u_0\rangle \qquad \text{and} \qquad \forall n \in \omega \varphi^{\Vdash}(F(n), F(n+1)).$$

Next, we internalize $F$ and prove that it supplies the function required for the truth of **DC** in $V(Kl)$. If $x$ is an ordered pair $\langle u, v\rangle$, let $(x)_0^s = u$ and $(x)_1^s = v$. The appropriate internalization of $F$ is $\overline{F}$:

$$\overline{F} = \left\{\left\langle \left(n, (F(n))_0^s\right), \langle \overline{n}, (F(n))_1^s\rangle_{Kl}\right\rangle : n \in \omega\right\}.$$

Obviously, $\overline{F}$ belongs to $V(Kl)$. It remains to check that $\overline{F}$ is internally a function from $\omega$ into $a$, and that $\overline{F}$ makes the consequent of the pertinent instance of **DC** realizable as well. This part of the proof is almost the same as that of [16, chap. 3, Theorem 6.1]. However, for the readers convenience and for later reference, we shall provide the details all the same.

First, because of the properties of internal pairing in $V(Kl)$ (cf. [16, chap. 3, Lemma 3.4]), it will be shown that $V(Kl)$ believes that $\overline{F}$ is a binary relation with domain $\overline{\omega}$ and range a subset of $a$ and that this holds with a witness obtainable independently of $e$ and $e^*$. To see that $\overline{F}$ is realizably functional, assume that

$$h \Vdash \langle x, y\rangle_{Kl} \in \overline{F} \quad \text{and} \quad j \Vdash \langle x, z\rangle_{Kl} \in \overline{F}.$$

Then,

$$h_1 \Vdash \langle x, y\rangle_{Kl} = \left\langle \overline{h_{00}}, (F(h_{00}))_1^s \right\rangle_{Kl}$$

and

$$j_1 \Vdash \langle x, z\rangle_{Kl} = \left\langle \overline{j_{00}}, (F(j_{00}))_1^s \right\rangle_{Kl},$$

where $h_1 = (h)_1$, $h_{00} = ((h)_0)_0$, $j_1 = (j)_1$, and $j_{00} = ((j)_0)_0$. This holds strictly in virtue of the definition of $\overline{F}$ and the conditions on statements of membership. From the absoluteness of $\in$ and $=$ on $\omega$ (see Proposition 8.4), we know that $h_{00} = j_{00}$ and thus $(F(h_{00}))_1^s = (F(j_{00}))_1^s$. Therefore, there is a partial recursive function $\vartheta$ such that $\vartheta(h, j) \Vdash y = z$. $\vartheta$ confirms that $\overline{F}$ is realizably functional. Next, to see that $V(Kl) \models \overline{F} \subseteq \overline{\omega} \times a$, let

$$h \Vdash \langle x, y\rangle_{Kl} \in \overline{F}.$$

As above,

$$h_1 \Vdash \langle x, y\rangle_{Kl} = \left\langle \overline{h_{00}}, (F(h_{00}))_1^s \right\rangle_{Kl}.$$

Moreover, $(h_{00}, i_r) \Vdash \overline{h_{00}} \in \overline{\omega}$, and, by definition of $\overline{F}$, $(h_{01}, i_r) \Vdash (F(h_{00}))_1^s \in a$. As a result, we can effectively compute $h^*$, $h^\#$ from $h$ such that $h^* \Vdash x \in \overline{\omega}$ and $h^\# \Vdash y \in a$.

Finally, we have to check on the realizability of $\overline{F}(0) = u$ and $\forall n \in \overline{\omega}\, \varphi(\overline{F}(n), \overline{F}(n+1))$. Obviously, $((0, (e^*)_0), i_r) \Vdash \langle 0, u_0\rangle_{Kl} \in \overline{F}$; thus from $e^*$ we can effectively compute a number $e^{**}$ such that $e^{**} \Vdash \overline{F}(0) = u$. Since, for all $n \in \omega$, $\varphi^{\Vdash}(F(n), F(n+1))$, we have for all $n \in \omega$,

$$(\{e\}((F(n))_0^s))_0 = (F(n+1))_0^s$$

and

$$(\{e\}((F(n))_0^s))_1 \Vdash \varphi((F(n))_1^s, (F(n+1))_1^s).$$

Define the recursive function $\rho$ so that

$$\rho(0) = (e^*)_0 \quad \text{and} \quad \rho(n+1) = (\{e\}(\rho(n)))_0.$$

The S-m-n theorem shows that an index for $\rho$ is calculable from $e$ and $e^*$. Then, one can use induction over $\omega$ to check that, for all $n \in \omega$,

$$((n, \rho(n)), i_r) \Vdash \langle \overline{n}, (F(n))_1^s\rangle_{Kl} \in \overline{F}$$

and

$$\{e\}(\rho(n)) \Vdash \varphi((F(n))^s_1, (F(n+1))^s_1).$$

This completes the proof of (i).

(ii): **RDC** implies **DC** (see [20, Lemma 3.4]) and, on the basis of **CZF**+**DC**, **RDC** follows from the following scheme:

(41) $\forall x\, (\varphi(x) \to \exists y\, [\varphi(y) \wedge \psi(x,y)]) \wedge \varphi(b) \to$
$$\exists z\, (b \in z \wedge \forall x \in z\, \exists y \in z\, [\varphi(y) \wedge \psi(x,y)]).$$

Thus, in view of part (i) of this theorem it suffices to show that, working in **CZF** + **RDC**, $V(Kl)$ validates (41). So let $a, b \in V(Kl)$ and suppose

$$e \Vdash \forall x\, (\varphi(x) \to \exists y\, [\varphi(y) \wedge \psi(x,y)]) \quad \text{and} \quad g \Vdash \varphi(b).$$

Therefore, for all $f \in \omega$ and $x \in V(Kl)$ we have

$$f \Vdash \varphi(x) \to \exists y \in V(Kl)\, [(\{e\}(f))_0 \Vdash \varphi(y) \wedge (\{e\}(f))_1 \Vdash \psi(x,y)].$$

By applying **RDC** to the above, one can extract functions $\imath : \omega \to \omega$, $\jmath : \omega \to \omega$, and $\ell : \omega \to V(Kl)$ such that $\imath(0) = g$, $\ell(0) = b$, and for all $n \in \omega$:

$$\imath(n) \Vdash \varphi(\ell(n)) \quad \text{and} \quad \jmath(n) \Vdash \psi(\ell(n), \ell(n+1)),$$
$$\imath(n+1) = (\{e\}(\imath(n)))_0 \quad \text{and} \quad \jmath(n) = (\{e\}(\imath(n)))_1.$$

By the last line, $\imath$ and $\jmath$ are recursive functions whose indices can be effectively computed from $e$ and $g$. Thus, defining $\hbar : \omega \to \omega$ by

$$\hbar(n) = (n, (\imath(n), \jmath(n))),$$

$\hbar$ is a recursive function the index of which is calculable from $e$ and $g$. Now set

$$B = \{\langle \hbar(n), \ell(n) \rangle : n \in \omega\}.$$

Obviously, $B$ belongs to $V(Kl)$. We have

(42) $$(\hbar(0), i_r) \Vdash b \in B.$$

Moreover, if $\langle k, u \rangle \in B$, then $u = \ell((k)_0)$ and $\langle \hbar((k)_0 + 1), \ell((k)_0 + 1) \rangle \in B$ and

$$((k)_1)_0 \Vdash \varphi(u) \wedge ((k)_1)_1 \vdash \psi(u, \ell((k)_0 + 1)).$$

Thus,

(43)
$$\forall \langle k, u \rangle \in B\, \exists v \langle \hbar((k)_0 + 1), v \rangle \in B\, [((k)_1)_0 \Vdash \varphi(u) \wedge ((k)_1)_1 \vdash \psi(u, v)].$$

In view of (42) and (43) it is obvious that there is an index $e^{\#}$ calculable from $e$ and $g$ such that

$$e^{\#} \Vdash b \in B \wedge \forall x \in B\, \exists y \in B\, [\varphi(x) \wedge \psi(x,y)],$$

whence

$$e^{\#} \Vdash \exists z\ (b \in z \wedge \forall x \in z\ \exists y \in z\ [\varphi(x) \wedge \psi(x,y)]).$$

This finishes the proof of (ii)

(iii): The proof of $V(Kl) \models$ **PAx** given in [16, chap. 3, Theorem 7.6] assumes the full axiom of choice to hold in $V$ and thereby appears to be requiring nothing less than the means of **ZFC**. In consequence, we have to find an entirely different proof.

Now let $a \in V(Kl)$. We have to find a set $B^* \in V(Kl)$ such that $V(Kl)$ thinks that $B^*$ is a base that surjects onto $a$. Because **PAx** holds in the background universe, we can select a base $B$ and a surjection $\jmath : B \to a$. $a$ being a set of pairs, define $\jmath_0 : B \to \omega$ and $\jmath_1 : B \to V(Kl)$ by

$$\jmath_0(u) = \text{First}(\jmath(u)),$$
$$\jmath_1(u) = \text{Second}(\jmath(u)),$$

where for an ordered pair $z = \langle x, y\rangle$, $\text{First}(z) = x$ and $\text{Second}(z) = y$.

By transfinite recursion define

$$x^{st} = \{\langle 0, y^{st}\rangle : y \in x\}$$

for each set $x$. By $\in$-induction, $x^{st} \in V(Kl)$, and by a simultaneous $\in$-induction (see [16, chap. 3, 10.2]),

$$\text{(44)} \qquad \begin{aligned} x = y &\quad \text{iff } V(Kl) \models x^{st} = y^{st} \\ x \in y &\quad \text{iff } V(Kl) \models x^{st} \in y^{st}; \end{aligned}$$

thus $(x \mapsto x^{st})$ injects $V$ into $V(Kl)$. Now, define

$$B^* = \left\{\left\langle \jmath_0(u), \langle \overline{\jmath_0(u)}, u^{st}\rangle_{Kl}\right\rangle : u \in B\right\}.$$

First, note that $B^*$ is in one-one correspondence with $B$ via $u \mapsto \langle \jmath_0(u), \langle \overline{\jmath_0(u)}, u^{st}\rangle_{Kl}\rangle$ (owing to (44)), and hence (externally in the background universe) $B^*$ is a base as well. Let $\ell : B \to V(Kl)$ be defined by $\ell(u) = \langle \overline{\jmath_0(u)}, u^{st}\rangle_{Kl}$ and put

$$\jmath^* = \{\langle \jmath_0(u), \langle \ell(u), \jmath_1(u)\rangle_{Kl}\rangle : u \in B\}.$$

Clearly, $\jmath^* \in V(Kl)$. First, we aim at showing that

$$\text{(45)} \qquad V(Kl) \models \jmath^* \text{is a surjection from } B^* \text{ onto } a.$$

To verify $V(Kl) \models \jmath^* \subseteq B^* \times a$, suppose $e \Vdash \langle b, c\rangle_{Kl} \in \jmath^*$. Then there is a $u \in B$ such that

$$(e)_0 = \jmath_0(u) \qquad \text{and} \qquad (e)_1 \Vdash \langle b, c\rangle_{Kl} = \langle \ell(u), \jmath_1(u)\rangle_{Kl}.$$

Hence, because of

$$(\jmath_0(u), i_r) \Vdash \ell(u) \in B^* \qquad \text{and} \qquad (\jmath_0(u), i_r) \Vdash \jmath_1(u) \in a,$$

one can effectively calculate an index $e'$ from $e$ such that $e' \Vdash b \in B^* \wedge c \in a$. This shows

$$(46) \qquad V(Kl) \models \jmath^* \subseteq B^* \times a.$$

To see that $\jmath^*$ is realizably total on $B^*$, let $e \Vdash \langle c,d\rangle_{Kl} \in B^*$. Then $(e)_0 = \jmath_0(u)$ and $(e)_1 \Vdash \langle c,d\rangle_{Kl} = \ell(u)$ for some $u \in B$. Since $(\jmath_0(u), i_r) \Vdash \jmath_1(u) \in a$ and $(\jmath_0(u), i_r) \Vdash \langle \ell(u), \jmath_1(u)\rangle_{Kl} \in \jmath^*$, an index $\tilde{e}$ can be computed from $e$ such that

$$\tilde{e} \Vdash \langle c,d\rangle_{Kl} \text{ is in the domain of } \jmath^*$$

so that with (46) we can conclude that $V(Kl) \models B^*$ *is the domain of* $\jmath^*$.

To show realizable functionality of $\jmath^*$, suppose $f \Vdash \langle b,c\rangle_{Kl} \in \jmath^*$ and $h \Vdash \langle b,d\rangle_{Kl} \in \jmath^*$. Then there exist $u,v \in B$ such that $(f)_0 = \jmath_0(u)$, $(h)_0 = \jmath_0(v)$, $(f)_1 \Vdash \langle b,c\rangle_{Kl} = \langle \ell(u), \jmath_1(u)\rangle_{Kl}$, and $(h)_1 \Vdash \langle b,d\rangle_{Kl} = \langle \ell(v), \jmath_1(v)\rangle_{Kl}$. From the latter one gets $V(Kl) \models \ell(u) = \ell(v)$. Given the definition of $\ell$ and the properties of internal pairing one concludes with the aid of (44) that actually $u = v$. In consequence, an index $e'$ such that $e' \Vdash c = d$ is calculable from $f$ and $h$.

Next, to show that $\jmath^*$ realizably maps onto $a$, assume $e \Vdash x \in a$. Then $\langle (e)_0, c\rangle \in a$ and $(e)_1 \Vdash x = c$ for some $c \in V(Kl)$. As $\jmath$ maps $B$ onto $a$ there exists $u \in B$ such that $\jmath_0(u) = (e)_0$ and $\jmath_1(u) = c$. Moreover, because of $(\jmath_0(u), i_r) \Vdash \ell(u) \in B^*$ and $(\jmath_0(u), i_r) \Vdash \langle \ell(u), \jmath_1(u)\rangle_{Kl} \in \jmath^*$, one can effectively construct an index $\tilde{e}$ from $e$ such that $\tilde{e} \Vdash x$ *is in the range of* $\jmath^*$, thereby completing the proof of (45).

It remains to ensure that $V(Kl)$ thinks that $B^*$ is a base. Towards this goal, assume

$$(47) \qquad e \Vdash \forall x \in B^* \; \exists y \; \varphi(x,y)$$

for some formula $\varphi$. We are to determine an index $e'$ calculably from $e$ that satisfies

$$(48) \qquad e' \Vdash \exists G \left[\mathbf{Fun}(G) \wedge \mathbf{dom}(G) = B^* \wedge \forall x \in B^* \; \varphi(x, G(x))\right].$$

From (47) it follows that

$$\forall \langle n,c\rangle \in B^* \; \exists y \in V(Kl) \{e\}(n) \Vdash \varphi(c,y).$$

Since $B^*$ is a base in the background universe, there exists a function $F : B^* \to V(Kl)$ such that with $F(n,c) := F(\langle n,c\rangle)$,

$$(49) \qquad \forall \langle n,c\rangle \in B^* \{e\}(n) \Vdash \varphi(c, F(n,c)).$$

Next, we want to internalize $F$. The appropriate internalization of $F$ is $\tilde{F}$:

$$\tilde{F} = \{\langle (\{e\}(n), n), \langle c, F(n,c)\rangle_{Kl}\rangle : \langle n,c\rangle \in B^*\}.$$

Obviously, $\tilde{F}$ belongs to $V(Kl)$.

First we show that there exists an index $\hat{e}$ calculable from $e$ such that

(50) $$\hat{e} \Vdash \mathbf{dom}(\tilde{F}) = B^*.$$

To this end assume that $h \Vdash x \in B^*$. Then $\langle (h)_0, c\rangle \in B^*$ and $(h)_1 \Vdash x = c$ for some $c \in V(Kl)$. From $\langle (h)_0, c\rangle \in B^*$ it follows that

(51) $$((\{e\}((h)_0), (h)_0), i_r) \Vdash \langle c, F((h)_0, c)\rangle_{Kl} \in \tilde{F},$$

and hence we can compute an index $h^*$ from $h$ such that $h^* \Vdash x \in \mathbf{dom}(\tilde{F})$. Conversely, suppose $d \Vdash \langle x, y\rangle_{Kl} \in \tilde{F}$. Then there exists $\langle n, c\rangle \in B^*$ such that $\langle (d)_0, \langle c, F(n,c)\rangle_{Kl}\rangle \in \tilde{F}$ with $n = ((d)_0)_1$ and $(d)_1 \Vdash \langle x, y\rangle_{Kl} = \langle c, F(n,c)\rangle_{Kl}$. Thus $(((d)_0)_1, i_r) \Vdash c \in B^*$. In consequence, there is an index $d^*$ calculable from $d$ such that $d^* \Vdash x \in B^*$. Therefore we have all the ingredients to compose $\hat{e}$ as claimed in (50).

To show realizable functionality of $\tilde{F}$, suppose $f \Vdash \langle b, c\rangle_{Kl} \in \tilde{F}$ and $h \Vdash \langle b, d\rangle_{Kl} \in \tilde{F}$. Then there exist $\langle n, x\rangle, \langle m, y\rangle \in B^*$ such that $((f)_0)_1 = n$, $((h)_0)_1 = m$, and

(52) $$(f)_1 \Vdash \langle b, c\rangle_{Kl} = \langle x, F(n,x)\rangle_{Kl} \wedge (h)_1 \Vdash \langle b, d\rangle_{Kl} = \langle y, F(m,y)\rangle_{Kl}.$$

From (52) one gets $V(Kl) \models x = y$. Moreover, as $\langle n, x\rangle, \langle m, y\rangle \in B^*$ there are $u, v \in B$ such that

$$x = \overline{\langle \jmath_0(u), u^{st}\rangle}_{Kl} \quad \text{and} \quad n = \jmath_0(u),$$

and

$$y = \overline{\langle \jmath_0(v), v^{st}\rangle}_{Kl} \quad \text{and} \quad m = \jmath_0(v).$$

Since $V(Kl) \models x = y$, the foregoing yields $V(Kl) \models \overline{n} = \overline{m} \wedge u^{st} = v^{st}$, and so, by Proposition 8.4 and (44), we can conclude that $n = m$ and $u = v$, and hence $x = y$ and $F(n,x) = F(m,y)$. Thus, in view of (52), there is a partial recursive function $\nu$ such that $\nu(f,h) \Vdash c = d$, verifying functionality of $\tilde{F}$, i.e., $V(Kl) \models \tilde{F}$ *is a function.*

Combining the latter result with (50) and (49) allows one to construct the desired $e'$ from $e$ such that (48) holds. ⊣

**§11. Continuity principles.** Fundamental to Brouwer's development of intuitionistic mathematics are strong continuity principles incompatible with classical mathematics.

DEFINITION 11.1. Some continuity principles which pertain to Brouwer's mathematics are:[1]

1. **Cont**$(\mathbb{N}^\mathbb{N}, \mathbb{N})$: *Every function from* $\mathbb{N}^\mathbb{N}$ *to* $\mathbb{N}$ *is continuous.*

[1] For exact formalizations of the notions of complete metric space, separable metric space, and continuity in constructive set theory see [7, chap. I, section 20].

2. $\forall\mathbb{X}\forall\mathbb{Y}$ **Cont**$(\mathbb{X},\mathbb{Y})$: *For every complete separable metric space* $\mathbb{X}$ *and separable metric space* $\mathbb{Y}$, **Cont**$(\mathbb{X},\mathbb{Y})$, i.e., every function from $\mathbb{X}$ to $\mathbb{Y}$ is continuous.

Recall that $\mathbf{MP_{PR}}$ is Markov's principle for primitive recursive predicates.

THEOREM 11.2 ($\mathbf{CZF} + \mathbf{MP_{PR}}$). *$V(Kl) \models$* **Cont**$(\mathbb{N}^{\mathbb{N}},\mathbb{N})\wedge\forall\mathbb{X}\forall\mathbb{Y}$ **Cont**$(\mathbb{X},\mathbb{Y})$. *Moreover, $V(Kl)$ validates that every separable metric space is subcountable.*

PROOF. For the proof that **MP** is realized in $V(Kl)$ it suffices to have $\mathbf{MP_{PR}}$ in the background universe. As a result of this and Theorem 9.2, we have $V(Kl) \models \mathbf{MP} \wedge \mathbf{ECT}$.

By [7, IV.3.1], $\mathbf{MP_{PR}}$ proves **KLS**, where **KLS** stands for the Kreisel-Lacombe-Shoenfield's theorem asserting that every effective operation from $\mathbb{N}^{\mathbb{N}}$ to $\mathbb{N}$ is continuous. By [7, XVI.2.1.1], **ECT** implies **KLS** $\rightarrow$ **Cont**$(\mathbb{N}^{\mathbb{N}},\mathbb{N})$. In consequence, $V(Kl) \models$ **Cont**$(\mathbb{N}^{\mathbb{N}},\mathbb{N})$.

Next, let $\mathbb{X}$ be a complete separable metric space and let $\mathbb{Y}$ be a separable metric space. By [7, I.20.1], every complete separable metric space $\mathbb{X} = (X,\sigma)$ is (isometric to) a space of the form $X = \{x \in \mathbb{N}^{\mathbb{N}} : \forall n, m[\rho(x_n, x_m) < 1/n + 1/m]\}$, where $\rho$ is a metric on $\mathbb{N}$, and the metric $\sigma$ on $X$ is given by $\sigma(x,y) = \lim_{n\to\infty}\rho(x_n,y_n)$. Employing Church's thesis, $\rho$ is recursive, and $X$ is identified with a certain set of total indices. Note also that thereby the formula $x \in X$ is rendered almost negative. If $\mathbb{Y}$ is a separable metric space, then $\mathbb{Y}$ has a completion which is a complete separable metric space, and so can be identified with a subset of $\mathbb{N}^{\mathbb{N}}$ as above. Under Church's thesis, $\mathbb{N}^{\mathbb{N}}$ can be identified with a subset of $\mathbb{N}$. So every separable metric space is isometric to a subset of $\mathbb{N}$ with a recursive metric. As a result, we get **KLS**$(\mathbb{X},\mathbb{Y})$, i.e. every effective operation from $\mathbb{X}$ to $\mathbb{Y}$ is continuous (cf. [7, IV.3]). Under **ECT**, **KLS**$(\mathbb{X},\mathbb{Y})$ implies **Cont**$(\mathbb{X},\mathbb{Y})$ by [7, XVI.2.1.1]. So the upshot of the above is that the model $V(Kl)$ validates $\forall\mathbb{X}\forall\mathbb{Y}$ **Cont**$(\mathbb{X},\mathbb{Y})$ as well. In the course of the proof it was also shown that $V(Kl)$ thinks that every separable metric space is subcountable. ⊣

REFERENCES

[1] P. ACZEL, *The type theoretic interpretation of constructive set theory*, ***Logic colloquium '77*** (A. MacIntyre, L. Pacholski, and J. Paris, editors), North Holland, Amsterdam, 1978, pp. 55–66.

[2] ———, *The type theoretic interpretation of constructive set theory: Choice principles*, ***The L.E.J. Brouwer centenary symposium*** (A. S. Troelstra and D. van Dalen, editors), North Holland, Amsterdam, 1982, pp. 1–40.

[3] ———, *The type theoretic interpretation of constructive set theory: Inductive definitions*, ***Logic, methodology and philosophy of science VII*** (R. B. Marcus et al., editors), North Holland, Amsterdam, 1986, pp. 17–49.

[4] P. ACZEL and M. RATHJEN, *Notes on constructive set theory*, Technical Report 40, Institut Mittag-Leffler, The Royal Swedish Academy of Sciences, 2001, `http://www.ml.kva.se/preprints/archive2000-2001.php`.

[5] J. Barwise, ***Admissible sets and structures***, Springer-Verlag, Berlin, Heidelberg, New York, 1975.

[6] M. Beeson, *Continuity in intuitionistic set theories*, ***Logic colloquium '78*** (M. Boffa, D. van Dalen, and K. McAloon, editors), North-Holland, Amsterdam, 1979.

[7] ———, ***Foundations of constructive mathematics***, Springer-Verlag, Berlin, Heidelberg, New York, Tokyo, 1985.

[8] E. Bishop and D. Bridges, ***Constructive analysis***, Springer-Verlag, Berlin, Heidelberg, New York, Tokyo, 1985.

[9] S. Feferman, *A language and axioms for explicit mathematics*, ***Algebra and logic*** (J. N. Crossley, editor), Lecture Notes in Mathematics, vol. 450, Springer, Berlin, 1975, pp. 87–139.

[10] ———, *Constructive theories of functions and classes*, ***Logic colloquium '78*** (M. Boffa, D. van Dalen, and K. McAloon, editors), North-Holland, Amsterdam, 1979, pp. 159–224.

[11] H. Friedman, *Some applications of Kleene's method for intuitionistic systems*, ***Cambridge summer school in mathematical logic*** (A. Mathias and H. Rogers, editors), Lectures Notes in Mathematics, vol. 337, Springer, Berlin, 1973, pp. 113–170.

[12] S. C. Kleene, *On the interpretation of intuitionistic number theory*, ***The Journal of Symbolic Logic***, vol. 10 (1945), pp. 109–124.

[13] G. Kreisel and A. S. Troelstra, *Formal systems for some branches of intuitionistic analysis*, ***Annals of Mathematical Logic***, vol. 1 (1970), pp. 229–387.

[14] P. Martin-Löf, *An intuitionistic theory of types: predicative part*, ***Logic colloquium '73*** (H. E. Rose and J. Sheperdson, editors), North-Holland, Amsterdam, 1975, pp. 73–118.

[15] ———, ***Intuitionistic type theory***, Bibliopolis, Naples, 1984.

[16] D. C. McCarty, ***Realizability and recursive mathematics***, Ph.D. thesis, Oxford University, 1984.

[17] J. Myhill, *Some properties of intuitionistic Zermelo-Fraenkel set theory*, ***Cambridge summer school in mathematical logic*** (A. Mathias and H. Rogers, editors), Lectures Notes in Mathematics, vol. 337, Springer, Berlin, 1973, pp. 206–231.

[18] ———, *Constructive set theory*, ***The Journal of Symbolic Logic***, vol. 40 (1975), pp. 347–382.

[19] M. Rathjen, *The strength of some Martin-Löf type theories*, ***Archive for Mathematical Logic***, vol. 33 (1994), pp. 347–385.

[20] ———, *The anti-foundation axiom in constructive set theories*, ***Games, logic, and constructive sets*** (G. Mints and R. Muskens, editors), CSLI Publications, Stanford, 2003, pp. 87–108.

[21] A. S. Troelstra and D. van Dalen, ***Constructivism in mathematics, Volumes I, II***, North Holland, Amsterdam, 1988.

DEPARTMENT OF MATHEMATICS
OHIO STATE UNIVERSITY
COLUMBUS, OH 43210, USA
*E-mail*: rathjen@math.ohio-state.edu

# ON LONG EF-EQUIVALENCE IN NON-ISOMORPHIC MODELS

SAHARON SHELAH

**Abstract.** There has been a great deal of interest in constructing models which are non-isomorphic, of cardinality $\lambda$, but are equivalent under the Ehrefeuch-Fraissé game of length $\alpha$, even for every $\alpha < \lambda$. So under G.C.H. particularly for $\lambda$ regular we know a lot. We deal here with constructions of such pairs of models proven in ZFC, and get their existence under mild conditions.

**§1. Introduction.** There has been much work on constructing pairs of $\mathrm{EF}_{\alpha,\mu}$-equivalent non-isomorphic models of the same cardinality.

In the summer of 2003, Väänänen asked me whether we can provably in ZFC construct a pair of non-isomorphic models of cardinality $\aleph_1$ which are $\mathrm{EF}_\alpha$-equivalent even for $\alpha$ like $\omega^2$. We try to shed light on the problem for general cardinals. We construct such models for $\lambda = \mathrm{cf}(\lambda) = \lambda^{\aleph_0}$ for every $\alpha < \lambda$ simultaneously and then for singular $\lambda = \lambda^{\aleph_0}$. In subsequent work Havlin and Shelah [HvSh:866] we shall investigate further: weaken the assumption "$\lambda = \lambda^{\aleph_0}$" (e.g., $\lambda = \mathrm{cf}(\lambda) > \beth_\omega$) and generalize the results for trees with no $\lambda$-branches and investigate the case of models of a first order complete $T$ (mainly strongly dependent). We thank Chanoch Havlin and the referee for detecting some inaccuracies.

DEFINITION 1.1.

(1) We say that $M_1, M_2$ are $\mathrm{EF}_\alpha$-equivalent if $M_1, M_2$ are models (with same vocabulary) such that the isomorphism player has a winning strategy in the game $\Game_1^\alpha(M_1, M_2)$ defined below.

(1A) Replacing $\alpha$ by $< \alpha$ means: for every $\beta < \alpha$; similarly below.

(2) We say that $M_1, M_2$ are $\mathrm{EF}_{\alpha,\mu}$- equivalent <u>when</u> $M_2, M_2$ are models with the same vocabulary such that the isomorphism player has a winning strategy in the game $\Game_\mu^\alpha(M_1, M_2)$ defined below.

(3) For $M_1$, $M_2$, $\alpha$, $\mu$ as above and partial isomorphism $f$ from $M_1$ into $M_2$ we define the game $\Game_\mu^\alpha(f, M_1, M_2)$ between the player ISO and AIS as follows:

---

The author would like to thank the Israel Science Foundation for partial support of this research (Grant no. 242/03). Publication 836.

**Logic Colloquium '03**
Edited by V. Stoltenberg-Hansen and J. Väänänen
Lecture Notes in Logic, 24

(a) the play lasts $\alpha$ moves
(b) after $\beta$ moves a partial isomorphism $f_\beta$ from $M_1$ into $M_2$ is chosen increasing continuous with $\beta$
(c) in the $\beta+1$-th move, the player AIS chooses $A_{\beta,1} \subseteq M_1, A_{\beta,2} \subseteq M_2$ such that $|A_{\beta,1}| + |A_{\beta,2}| < 1 + \mu$ and then the player ISO chooses $f_{\beta+1} \supseteq f_\beta$ such that

$$A_{\beta,1} \subseteq \mathrm{Dom}(f_{\beta+1}) \text{ and } A_{\beta,2} \subseteq \mathrm{Rang}(f_{\beta+1})$$

(d) if $\beta = 0$, ISO chooses $f_0 = f$; if $\beta$ is a limit ordinal ISO chooses $f_\beta = \cup\{f_\gamma : \gamma < \beta\}$.

The ISO player loses if he had no legal move.

(4) If $f = \emptyset$ we may write $\partial^\alpha_\mu(M_1, M_2)$. If $\mu$ is 1 we may omit it. We may write $\leq \mu$ instead of $\mu^+$. The player ISO may be restricted to choose $f_{\beta+1}$ such that$(\forall a)(a \in \mathrm{Dom}(f_{\beta+1}) \wedge a \notin \mathrm{Dom}(f_\beta) \to a \in A_{\beta,1} \vee f_{\beta+1}(a) \in A_{\beta,2})$.

## §2. The case of regular $\lambda = \lambda^{\aleph_0}$.

DEFINITION 2.1.

(1) We say that $\mathfrak{x}$ is a $\lambda$-parameter <u>if</u> $\mathfrak{x}$ consists of
(a) a cardinal $\lambda$ and ordinal $\alpha^* \leq \lambda$
(b) a set $I$, and a set $S \subseteq I \times I$ (where we shall have compatibility demand)
(c) a function $\mathbf{u} : I \to \mathcal{P}(\lambda)$; we let $\mathbf{u}_s = \mathbf{u}(s)$ for $s \in I$
(d) a set $J$ and a function $\mathbf{s} : J \to I$, we let $\mathbf{s}_t = \mathbf{s}(t)$ for $t \in J$ and for $s \in I$ we let $J_s = \{t \in J : \mathbf{s}_t = s\}$
(e) a set $T \subseteq J \times J$ such that $(t_1, t_2) \in T \Rightarrow (\mathbf{s}_{t_1}, \mathbf{s}_{t_2}) \in S$.

(1A) We say $\mathfrak{x}$ is a full $\lambda$-parameter <u>if</u> in addition it consists of:
(f) a function $\mathbf{g}$ with domain $J$ such that $\mathbf{g}_t = \mathbf{g}(t)$ is a non-decreasing function from $\mathbf{u}_{\mathbf{s}(t)}$ to some $\alpha < \alpha^*$
(g) a function $\mathbf{h}$ with domain $J$ such that $\mathbf{h}_t = \mathbf{h}(t)$ is a non-decreasing function from $\mathbf{u}_{\mathbf{s}(t)}$ to $\lambda$ such that
(h) if $t_1, t_2 \in J$ and $\mathbf{s}_{t_1} = s = \mathbf{s}_{t_2}$, $\mathbf{g}_{t_1} = g = \mathbf{g}_{t_2}$ and $\mathbf{h}_{t_1} = h = \mathbf{h}_{t_2}$, $\alpha^{t_1} = \alpha = \alpha^{t_2}$ <u>then</u> $t_1 = t_2$ hence we write $t = t^\alpha_{s,g,h} = t^\alpha(s,g,h)$.

(2) We may write $\alpha^* = \alpha^*_\mathfrak{x}$, $\lambda = \lambda_\mathfrak{x}$, $I = I_\mathfrak{x}$, $J = J_\mathfrak{x}$, $J_s = J^\mathfrak{x}_s$, $t^\alpha(s,g,h) = t^{\alpha,\mathfrak{x}}(s,g,h)$, etc. Many times we omit $\mathfrak{x}$ when clear from the context.

DEFINITION 2.2. Let $\mathfrak{x}$ be a $\lambda$-parameter.

(1) For $s \in I_\mathfrak{x}$, let $\mathbb{G}^\mathfrak{x}_s$ be the group[1] generated freely by $\{x_t : t \in J_s\}$.
(2) For $(s_1, s_2) \in S_\mathfrak{x}$ let $\mathbb{G}_{s_1,s_2} = G^\mathfrak{x}_{s_1,s_2}$ by the subgroup of $\mathbb{G}^\mathfrak{x}_{s_1} \times \mathbb{G}^\mathfrak{x}_{s_2}$

[1]We also could use abelian groups satisfying $\forall x(x + x = 0)$, in this case $\mathbb{G}_s$ is the family of finite subsets of $J_2$ with the symmetric difference operation also we could use the free abelian group.

generated by

$$\{(x_{t_1}, x_{t_2}) : (t_1, t_2) \in T_{\mathfrak{x}} \text{ and } t_1 \in J^{\mathfrak{x}}_{s_1},\ t_2 \in J^{\mathfrak{x}}_{s_2}\}.$$

(3) We say $\mathfrak{x}$ is $(\lambda, \theta)$-parameter if $s \in I_{\mathfrak{x}} \Rightarrow |\mathbf{u}_s| < \theta$.

REMARK 2.3. (1) We may use $S$ a set of $n$-tuples from $I$ (or $(< \omega)$-tuples) then we have to change Definitions 2.2(2) accordingly.

DEFINITION 2.4. For a $\lambda$-parameter $\mathfrak{x}$ we define a model $M = M_{\mathfrak{x}}$ as follows (where below $I = I_{\mathfrak{x}}$, etc.).

(A) its vocabulary $\tau$ consist of

 ($\alpha$) $P_s$, a unary predicate, for $s \in I_{\mathfrak{x}}$

 ($\beta$) $Q_{s_1,s_2}$, a binary predicate for $(s_1, s_2) \in S_{\mathfrak{x}}$

 ($\gamma$) $F_{s,a}$, a unary function for $s \in I_{\mathfrak{x}}$, $a \in \mathbb{G}^{\mathfrak{x}}_s$

(B) the universe of $M$ is $\{(s, x) : s \in I_{\mathfrak{x}}, x \in \mathbb{G}^{\mathfrak{x}}_s\}$

(C) for $s \in I_{\mathfrak{x}}$ let $P^M_s = \{(s, x) : x \in \mathbb{G}^{\mathfrak{x}}_s\}$

(D) $Q^M_{s_1,s_2} = \{((s_1, x_1), (s_2, x_2)) : (x_1, x_2) \in \mathbb{G}^{\mathfrak{x}}_{s_1,s_2})\}$ for $(s_1, s_2) \in S_{\mathfrak{x}}$

(E) if $s \in I_{\mathfrak{x}}$ and $a \in \mathbb{G}^{\mathfrak{x}}_s$ then $F^M_{s,a}$ is the unary function from $P^M_s$ to $P^M_s$ defined by $F^M_{s,a}(y) = ay$, multiplication in $\mathbb{G}^{\mathfrak{x}}_s$ (for $y \in M \setminus P^M_s$ we can let $F^M_{s,a}(y)$ be $y$ or undefined).

DEFINITION 2.5.

(1) For $\mathfrak{x}$ a $\lambda$-parameter and for $I' \subseteq I_{\mathfrak{x}}$ let $M^{\mathfrak{x}}_{I'} = M_{\mathfrak{x}} \restriction \cup\{P^{M_{\mathfrak{x}}}_s : s \in I'\}$ and let $I_\gamma = I^{\mathfrak{x}}_\gamma = \{s \in I_{\mathfrak{x}} : \sup(\mathbf{u}_s) < \gamma\}$.

(2) Assume $\mathfrak{x}$ is a full $\lambda$-parameter and $\beta < \lambda$; for $\alpha < \alpha^*_{\mathfrak{x}}$ we let $\mathcal{G}^{\mathfrak{x}}_{\alpha,\beta}$ be the set of $g : \beta \to \alpha$ which are non-decreasing; <u>then</u> for $g \in \mathcal{G}^{\mathfrak{x}}_{\alpha,\beta}$

 (a) we define $h = h_g : \beta \to \lambda$ as follows: $h(\gamma) = \mathrm{Min}\{\beta' \leq \beta$: if $\beta' < \beta$ then $g(\beta') > g(\gamma)\}$

 (b) we let $I_g = I^{\mathfrak{x}}_g = \{s \in I : \mathbf{u}_s \subseteq \beta$ and $t^\alpha_{s,g\restriction \mathbf{u}_s, h_g \restriction \mathbf{u}_s}$ is well defined$\}$

 (c) we define $\bar{c}^\alpha_g = \langle c^\alpha_{g,s} : s \in I^{\mathfrak{x}}_g\rangle$ by $c^\alpha_{g,s} = x^\alpha_{t_{g,s}}$ where $t^\alpha_{g,s} = t^{\alpha,\mathfrak{x}}_{s,g\restriction \mathbf{u}_s, h_g\restriction \mathbf{u}_s}$.

(3) Let $\mathcal{G}^{\mathfrak{x}}_\alpha = \cup\{\mathcal{G}^{\mathfrak{x}}_{\alpha,\beta} : \beta < \lambda\}$ and $\mathcal{G}_{\mathfrak{x}} = \cup\{\mathcal{G}^{\mathfrak{x}}_\alpha : \alpha < \alpha^*\}$.

DEFINITION 2.6. Let $\mathfrak{x}$ be a $\lambda$-parameter.

(1) Let $\mathbf{C}_{\mathfrak{x}} = \cup\{\mathbf{C}^{\mathfrak{x}}_{I'} : I' \subseteq I_{\mathfrak{x}}\}$ where for $I' \subseteq I_{\mathfrak{x}}$ we let $\mathbf{C}^{\mathfrak{x}}_{I'} = \{\bar{c} : \bar{c} = \langle c_s : s \in I'\rangle$ satisfies $c_s \in \mathbb{G}^{\mathfrak{x}}_s$ when $s \in I'$ and $(c_{s_1}, c_{s_2}) \in \mathbb{G}_{s_1,s_2}$ when $(s_1, s_2) \in S_{\mathfrak{x}}$ and $s_1, s_2 \in I'\}$.

(2) For $\bar{c} \in \mathbf{C}^{\mathfrak{x}}_{I'}$, $I' \subseteq I_{\mathfrak{x}}$, let $f^{\mathfrak{x}}_{\bar{c}}$ be the partial function from $M_{\mathfrak{x}}$ into itself defined by $f^{\mathfrak{x}}_{\bar{c}}((s, y)) = (s, yc_s)$ for $(s, y) \in P^{M_{\mathfrak{x}}}_s$, $s \in I'$.

(3) $M_{\mathfrak{x}}$ is $P_s$-rigid <u>when</u> for every automorphism $f$ of $M_{\mathfrak{x}}$, $f \restriction P^{M_{\mathfrak{x}}}_s$ is the identity.

OBSERVATION 2.7. (1) Let $\mathfrak{x}$ be a full $\lambda$-parameter. If $g : \gamma_2 \to \alpha$ where $\alpha < \alpha^*_{\mathfrak{x}}$, $\gamma_2 < \lambda$ and the function $g$ is non-decreasing, $\gamma_1 < \gamma_2$ and $(\forall \gamma < \gamma_1)$ $(g(\gamma) < g(\gamma_1))$ <u>then</u> $I_{g\restriction\gamma_1} \subseteq I_g$ and $h_{g\restriction\gamma_1} \subseteq h_g$ and $\bar{c}^\alpha_{g\restriction\gamma_1} = \bar{c}^\alpha_g \restriction I_{g\restriction\gamma_1}$.

(2) If $g \in \mathcal{G}^\alpha_{\mathfrak{x}}$ in Definition 2.5(3), <u>then</u> $\bar{c}^\alpha_g \in \mathbf{C}^{\mathfrak{x}}_{I_g}$.

CLAIM 2.8. Assume $\mathfrak{x}$ is a full $\lambda$-parameter.

(1) For $I' \subseteq I_{\mathfrak{x}}$ and $\bar{c} \in \mathbf{C}^{\mathfrak{x}}_{I'}$, $f^{\mathfrak{x}}_{\bar{c}}$ is an automorphism of $M^{\mathfrak{x}}_{I'}$ which is the identity iff $s \in I' \Rightarrow c_s = e_{\mathbb{G}_s}$.

(2) In (1) for $s \in I'$, $f^{\mathfrak{x}}_{\bar{c}} \restriction P^{M_{\mathfrak{x}}}_s$ is not the identity iff $c_s \neq e_{\mathbb{G}_s}$.

(3) If $f$ is an automorphism of $M^{\mathfrak{x}}_{I_2}$ then $f \restriction M^{\mathfrak{x}}_{I_1}$ is an automorphism of $M^{\mathfrak{x}}_{I_1}$ for every $I_1 \subseteq I_2 \subseteq I_{\mathfrak{x}}$.

(4) If $I' \subseteq I_{\mathfrak{x}}$ and $f$ is an automorphism of $M^{\mathfrak{x}}_{I'}$, then $f = f^{\mathfrak{x}}_{\bar{c}}$ for some $\langle c_s : s \in I_{\mathfrak{x}} \rangle \in \mathbf{C}_{I'}$.

(5) If $\bar{c}_\ell \in \mathbf{C}^{\mathfrak{x}}_{I_\ell}$ for $\ell = 1, 2$ and $I_1 \subseteq I_2$ and $\bar{c}_1 = \bar{c}_2 \restriction I_1$ then $f_{\bar{c}_1} \subseteq f_{\bar{c}_2}$.

(6) The cardinality of $M_{\mathfrak{x}}$ is $|J_{\mathfrak{x}}| + \aleph_0$.

PROOF. Straight, e.g. (4) For $s \in I'$ clearly $f((s, e_{\mathbb{G}_s})) \in P^{M_{\mathfrak{x}}}_s$ so it has the form $(s, c_s), c_s \in \mathbb{G}_s$ and let $\bar{c} = \langle c_s : s \in I' \rangle$. To check that $\bar{c} \in \mathbf{C}^{\mathfrak{x}}_{I'}$ assume $(s_1, s_2) \in S_{\mathfrak{x}}$; and we have to check that $(c_{s_1}, c_{s_2}) \in \mathbb{G}_{s_1,s_2}$. This holds as $((s_1, e_{\mathbb{G}_{s_1}}), (s_2, e_{\mathbb{G}_{s_2}})) \in Q^{M_{\mathfrak{x}}}_{s_1,s_2}$ by the choice of $Q^{M_{\mathfrak{x}}}_{s_1,s_2}$ hence we have $((s_1, c_{s_1}), (s_2, c_{s_2})) = (f(s_1, e_{\mathbb{G}_{s_1}}), f(s_2, e_{\mathbb{G}_{s_2}})) \in Q^{M_{\mathfrak{x}}}_{s_1,s_2}$ hence $(c_{s_1}, c_{s_2}) \in \mathbb{G}_{s_1,s_2}$. ⊣

CLAIM 2.9. Let $\mathfrak{x}$ be a full $\lambda$-parameter $s \in I_{\mathfrak{x}}$ and $c_1, c_2 \in P^M_s$, $c^* \in \mathbb{G}_s$ and $F^{M_{\mathfrak{x}}}_{s,c^*}(c_1) = c_2$. A sufficient condition for "$(M_{\mathfrak{x}}, c_1)$, $(M_{\mathfrak{x}}, c_2)$ are $\mathrm{EF}_{\alpha,\mu}$ -equivalent" where $\alpha \leq \alpha^*_{\mathfrak{x}}$, is the existence of $R, \bar{I}, \bar{c}$ such that:

⊛ (a) $R$ is a partial order,

(b) $\bar{I} = \langle I_r : r \in R \rangle$ such that $I_r \subseteq I_{\mathfrak{x}}$ and $r_2 \leq_R r_2 \Rightarrow I_{r_1} \subseteq I_{r_2}$

(c) $R$ is the disjoint union of $\langle R_\beta : \beta < \alpha \rangle$, $R_0 \neq \emptyset$

(d) $\bar{c} = \langle \bar{c}^r : r \in R \rangle$ where $\bar{c}^r \in \mathbf{C}_{I_r}$ and $r_1 \leq r_2 = \bar{c}^{r_1} = \bar{c}^{r_2} \restriction I_{r_1}$ and $c^r_s = c^*$ so $s \in \bigcap \{I_r : r \in R\}$

(e) if $\langle r_\beta : \beta < \beta^* \rangle$ is $\leq_R$-increasing, $\beta < \beta^* \Rightarrow r_\beta \in R_\beta$ and $\beta^* < \alpha$ then it has an $\leq_R$-ub from $R_{\beta^*}$

(f) if $r_1 \in R_\beta$, $\beta + 1 < \alpha$ and $I' \subseteq I$, $|I'| < \mu$ then $(\exists r_2)(r_1 \leq r_2 \in R_{\beta+1} \wedge I' \subseteq I_{r_2})$.

PROOF. Easy. Using Claim 2.8(1), (5). ⊣

CLAIM 2.10.

(1) Let $\mathfrak{x}$ be a $\lambda$-parameter and $I' \subseteq I_{\mathfrak{x}}$. A necessary and sufficient condition for "$M^{\mathfrak{x}}_{I'}$ is $P_s$-rigid" is:

⊛$_1$ there is no $\bar{c} \in \mathbf{C}^{\mathfrak{x}}_{I'}$ with $c_s \neq e_{\mathbb{G}_s}$.

(2) Let $\mathfrak{x}$ be a full $\lambda$-parameter and assume that $s(*) \in I_{\mathfrak{x}}$, $\alpha < \alpha^*_{\mathfrak{x}}$, $\alpha \geq \omega$ for notational simplicity and $t^* \in J^{\mathfrak{x}}_{s(*)}$. The models $M_1 = (M, (s, e_{\mathbb{G}_s}))$, $M_2 = (M, (s, x_{t^*}))$ are $\mathrm{EF}_{\alpha,\lambda}$-equivalent when:

⊛$_{2,\alpha}$ (i) $\lambda$ is regular, $s \in I_{\mathfrak{x}} \Rightarrow |\mathbf{u}^{\mathfrak{x}}_s| < \lambda$

(ii) if $s \in I_{\mathfrak{x}}$ and $g \in \mathcal{G}_{\mathfrak{x}}$ and $\mathbf{u}^{\mathfrak{x}}_s \subseteq \mathrm{Dom}\,(g)$ then $t^{\alpha,\mathfrak{x}}_{s,g\restriction \mathbf{u}_s, h_g \restriction \mathbf{u}_s}$ is well defined

(iii) if $(s_1, s_2) \in S_{\mathfrak{r}}$ and $t_1 = t^{\alpha}_{s_1,g_1,h_1}$, $t_2 = t^{\alpha}_{s_2,g_2,h_2}$ are well defined then $(t_1, t_2) \in T_{\mathfrak{r}}$ when for some $g \in \mathcal{G}_{\mathfrak{r}}$ we have $g_{t_1} \cup g_{t_2} \subseteq g$ and $h_1 \cup h_2 \subseteq h_g$

(iv) $t^* = t^{\alpha,\mathfrak{r}}_{s(*),g,h_g}$ where $g : \mathbf{u}_{s(*)} \to \{0\}$ and $h_g$ is constantly $\gamma^* = \cup\{\gamma + 1 : \gamma \in \mathbf{u}_{s(*)}\}$.

PROOF. (1) Toward contradiction assume that $f$ is an automorphism of $M^{\mathfrak{r}}_{I}$, such that $f \restriction P^{M_{\mathfrak{r}}}_{s}$ is not the identity. By Claim 2.8(4) for some $\bar{c} \in \mathbf{C}^{\mathfrak{r}}_{I}$, we have $f = f_{\bar{c}}$. So $f_{\bar{c}} \restriction P^{M_{\mathfrak{r}}}_{s} = f \restriction P^{M_{\mathfrak{r}}}_{s} \neq \text{id}$ hence by Claim 2.8(1) we have $c_s \neq e_{\mathbf{G}_s}$, contradicting the assumption $\circledast_1$.

(2) We apply Claim 2.9. For every $i < \alpha$ and non-decreasing function $g \in \mathcal{G}^{\mathfrak{r}}_{\alpha}$ from some ordinal $\gamma = \gamma_g$ into $i$ we define $\bar{c}^{\alpha}_{g} = \langle c^{\alpha}_{g,s} : s \in I_{g_p}\rangle$, $c^{\alpha}_{g,s} = (s, x_{t^{\alpha}_{g,s}})$, $t^{\alpha}_{g,s} = t^{\alpha}_{s,g\restriction \mathbf{u}_s, h_g \restriction \mathbf{u}_s}$. Let $R_i = \{g : g$ a non-decreasing function from some $\gamma < \lambda$ to $1 + i$ such that $\gamma^* \leq \gamma$, $g \restriction \gamma^*$ is constantly zero, $\gamma^* < \gamma \Rightarrow g(\gamma^*) = 1\}$ and let $R = \cup\{R_i : i < \alpha\}$ ordered by inclusion. Let $\bar{I} = \langle I_g : g \in R\rangle$ and $\bar{c} = \langle \bar{c}^{\alpha}_{g} : g \in R\rangle$. It is easy to check that $(R, \bar{I}, \bar{c})$ is as required. ⊣

CLAIM 2.11.

(1) Assume $\alpha^* \leq \lambda = \text{cf}(\lambda) = \lambda^{\aleph_0}$. Then for some full $(\lambda, \aleph_1)$-parameter $\mathfrak{r}$ we have $|I| = \lambda = |J|$, $\alpha^*_{\mathfrak{r}} = \alpha^*$ and condition $\circledast_1$ of Claim 2.10(1) holds and for every $s(*) \in I_{\mathfrak{r}} \setminus \{\emptyset\}$ condition $\circledast_{2.\alpha}$ of Claim 2.10(2) holds whenever $\alpha < \alpha^*$.

(2) Moreover, if $s \in I_{\mathfrak{r}} \setminus \{\emptyset\}$ then for some $c_1 \neq c_2 \in P^{M_{\mathfrak{r}}}_{s}$ and $(M, c_1)$, $(M, c_2)$ are $\text{EF}_{\alpha,\lambda}$-equivalent for every $\alpha < \alpha^*_{\mathfrak{r}}$ but not $\text{EF}_{\alpha^*_{\mathfrak{r}},\lambda}$-equivalent.

Claim 2.11(1) clearly implies

CONCLUSION 2.12.

(1) If $\lambda = \text{cf}(\lambda) = \lambda^{\aleph_0}$, $\alpha^* \leq \lambda$ then for some model $M$ of cardinality $\lambda$ we have:
 (a) $M$ has no non-trivial automorphism
 (b) for every $\alpha < \lambda$ for some $c_1 \neq c_2 \in M$, the model $(M, c_1)$, $(M, c_2)$ are $\text{EF}_{\alpha}$-equivalent and even $\text{EF}_{\alpha,\lambda}$-equivalent.

(2) We can strengthen clause (b) to: for some $c_1 \neq c_2$ for every $\alpha < \lambda$ the models $(M, c_1)$, $(M, c_2)$ are $\text{EF}_{\alpha,\lambda}$-equivalent.

PROOF OF CLAIM 2.11. (1) Assume $\alpha_* > \omega$ for notational simplicity. We define $\mathfrak{r}$ by ($\lambda_{\mathfrak{r}} = \lambda$ and):

⊠ (a) ($\alpha$) $I = \{u : u \in [\lambda]^{\leq \aleph_0}\}$
 ($\beta$) the function $\mathbf{u}$ is the identity on $I$
 ($\gamma$) $S = \{(u_1, u_2) : u_1 \subseteq u_2 \in I\}$
 ($\delta$) $\alpha^*_{\mathfrak{r}} = \alpha^*$

(b) ($\alpha$) $J$ is the set of quadruple $(u,\alpha,g,h)$ satisfying

(i) $u \in I, \alpha < \alpha^*$

(ii) $h$ is a non-decreasing function from $u$ to $\lambda$

(iii) $g$ is a non-decreasing function from $u$ to $\alpha$

(iv) if $\beta_1, \beta_2 \in u$ and $g(\beta_1) = g(\beta_2)$ then $h(\beta_1) = h(\beta_2)$

(v) $h(\beta) > \beta$

($\beta$) let $t = (u^t, \alpha^t, g^t, h^t)$ for $t \in J$ so naturally $\mathbf{s}_t = u$, $\mathbf{g}_t = g^t, \mathbf{h}_t = h^t$

($\gamma$) $T = \{(t_1,t_2) \in J \times J : \alpha^{t_1} = \alpha^{t_2}, u^{t_1} \subseteq u^{t_2}, h^{t_1} \subseteq h^{t_2} \text{ and } g^{t_1} \subseteq g^{t_2}\}$.

Now

$(*)_0$ $\mathfrak{x}$ is a full $(\lambda, \aleph_1)$-parameter
[Why? Just read Definitions 2.1 and 2.2(3).]

$(*)_1$ for any $s(*) \in I \backslash \{\emptyset\}$, $\mathfrak{x}$ satisfies the demands for $\circledast_{2,\alpha}(i), (ii), (iii), (iv)$ from Claim 2.10(2) for every $\alpha < \alpha^*$
[Why? Just check.]

$(*)_2$ if $u_1 \subseteq u_2 \in I$, we define the function $\pi_{u_1,u_2} : J_{u_2} \to J_{u_1}$ by $\pi_{u_1,u_2}(t) = (u_1, \alpha^t, g^t \restriction u_1, h^t \restriction u_1)$ for $t \in J_{u_2}$,
[Why is $\pi_{u_1,u_2}$ a function from $J_{u_2}$ into $J_{u_1}$? Just check.]

$(*)_3$ for $u_1 \subseteq u_2$ we have

($\alpha$) $T \cap (J_{u_1} \times J_{u_2}) = \{(\pi_{u_1,u_2}(t_2), t_2) : t_2 \in J_{u_2}\}$ hence

($\beta$) $\mathbb{G}_{u_1,u_2} = \{(\hat{\pi}_{u_1,u_2}(c_2), c_2) : c_2 \in \mathbb{G}_{u_2}\}$ where $\hat{\pi}_{u_1,u_2} \in \mathrm{Hom}(\mathbb{G}^{\mathfrak{x}}_{u_2}, \mathbb{G}^{\mathfrak{x}}_{u_1})$ is the unique homomorphism from $\mathbb{G}^{\mathfrak{x}}_{u_2}$ into $\mathbb{G}^{\mathfrak{x}}_{u_1}$ mapping $x_{t_2}$ to $x_{t_1}$ whenever $\pi_{u_1,u_2}(t_2) = t_1$
[Why? Check.]

$(*)_4$ if $u_1 \cup u_2 \subseteq u_3 \in I$, $t_3 \in J_{u_3}$ and $t_\ell = \pi_{u_\ell,u_3}(t_3)$ for $\ell = 1,2$ then $\mathbf{g}_{t_1}$, $\mathbf{g}_{t_2}$ are compatible functions as well as $\mathbf{h}_{t_1}$, $\mathbf{h}_{t_2}$ and $\alpha^{t_1} = \alpha^{t_2}$ moreover $\mathbf{g}_{t_1} \cup \mathbf{g}_{t_2}$ is non-decreasing, $\mathbf{h}_{t_1} \cup \mathbf{h}_{t_2}$ is non-decreasing
[Why? Just check.]

$(*)_5$ clause $\circledast_1$ of Claim 2.10(1) holds for $I' = I, s(*) \in I \setminus \{\emptyset\}$.

[Why? Assume $\bar{c} \in C^{\mathfrak{x}}_I$ is such that $c_{s(*)} \neq e_{\mathbb{G}_{s(*)}}$. For each $u \in I$, $c_u$ is a word in the generators $\{x_t : t \in J_u\}$ of $\mathbb{G}_u$ and let $\mathbf{n}(u)$ be the length of this word and $\mathbf{m}(u)$ the number of generators appearing in it.

Now by $(*)_3$ we have $u_1 \subseteq u_2 \Rightarrow \mathbf{n}(u_1) \leq \mathbf{n}(u_2) \wedge \mathbf{m}(u_1) \leq \mathbf{m}(u_2)$. As $(I, \subseteq)$ is $\aleph_1$-directed, for some $u_* \in I$ we have $u_* \subseteq u \in I \Rightarrow \mathbf{n}(u) = n_* \wedge \mathbf{m}(u) = m_*$ and let $c_u = (\ldots, x^{i(\ell)}_{t(u,\ell)}, \ldots)_{\ell < n_*}$ where $i(\ell) \in \{1,-1\}$ and $t(u,\ell) \in J^{\mathfrak{x}}_u$ and $t(u,\ell) = t(u,\ell+1) \Rightarrow i(\ell) = i(\ell+1)$. Clearly $u_* \subseteq u_1 \subseteq u_2 \in I \;\&\; \ell < n_* \Rightarrow \pi_{u_1,u_2}(t(u_2,\ell)) = t(u_1,\ell) \wedge \alpha^{t(u_2,\ell)} = \alpha^{t(u_*,\ell)}$. By our assumption toward contradiction necessarily $n_* > 0$.

As $\{u : u_* \subseteq u \in I\}$ is directed, by $(*)_4$ above, for each $\ell < n_*$ any two of the functions $\{g^{t(u,\ell)} : u_* \subseteq u \in I\}$ are compatible so $g_\ell =: \cup\{g^{t(u,\ell)} : u \in I\}$ is a non-decreasing function from $\lambda = \cup\{u : u \in I\}$ to $\alpha^*$ and $h_\ell =: \cup\{h^{t(u,\ell)} : u_* \subseteq u \in I\}$ is similarly a non-decreasing function from $\lambda$ to $\lambda$. It also follows that for some $\alpha^*_\ell$ we have $\alpha^*_\ell =: \alpha^{t(u,\ell)}$ whenever $u_* \subseteq u \in I$ in fact $\alpha^*_\ell = \alpha^{t(u_*,\ell)}$ is O.K. For each $i \in \text{Rang}(g_\ell) \subseteq \alpha^*_\ell$ choose $\beta_{\ell,i} < \lambda$ such that $g_\ell(\beta_{\ell,i}) = i$ and let $E = \{\delta < \lambda : \delta$ a limit ordinal $> \sup(u_*)$ such that $i < \alpha^*_\ell \;\&\; \ell < n_* \;\&\; i \in \text{Rang}(g_\ell) \Rightarrow \beta_{\ell,i} < \delta$ and $\beta < \delta \;\&\; \ell < n \Rightarrow h_\ell(\beta) < \delta\}$, it is a club of $\lambda$. Choose $u$ such that $u_* \subseteq u$ and $\text{Min}(u \backslash u_*) = \delta^* \in E$.

Now what can $g_\ell(\text{Min}(u\backslash u_*))$ be? It has to be $i$ for some $i < \alpha^*_\ell < \alpha^*$ hence $i \in \text{Rang}(g_\ell)$ so for some $u_1, u_* \subseteq u_1 \subseteq \delta^*$ and $\beta_{\ell,i} \in u_1$ so $h_\ell(\beta_{\ell,i}) < \delta^*$ hence considering $u \cup u_1$ and recalling clause $(\alpha)(vi)$ of (b) from definition of $\mathfrak{r}$ in the beginning of the proof we have $h_\ell(\beta_{\ell,i}) < h_\ell(\delta^*)$ hence by (clause $(b)(\alpha)(v)$) we have $i = g_\ell(\beta_{\ell,i}) < g_\ell(\delta^*)$, contradiction.]

(2) A minor change is needed in the choice of $T^{\mathfrak{r}}$

$$T^{\mathfrak{r}} = \{(t_1, t_2) : (t_1, t_2) \in J \times J \text{ and } u^{t_1} \subseteq u^{t_2},\ h^{t_1} \subseteq h^{t_2},\ g^{t_1} \subseteq g^{t_2},$$
$$\gamma^{t_1} \le \gamma^{t_2} \text{ and if } \text{Rang}(g^{t_1}) \nsubseteq \{0\} \text{ then } \alpha^{t_1} = \alpha^{t_2}\}. \qquad \dashv$$

**§3. The singular case.** We deal here with singular $\lambda = \lambda^{\aleph_0}$ and our aim is the parallel of Conclusion 2.12 constructing a pair of $\text{EF}_\alpha$-equivalent for every $\alpha < \lambda$ non-isomorphic models of cardinality $\lambda$. But it is natural to try to construct a stronger example: This is done here:

⊛ for each $\gamma < \kappa = \text{cf}(\lambda)$, in the following game the ISO player wins.

Definition 3.1.

(1) For models $M_1, M_2, \lambda$ and partial isomorphism $f$ from $M_1$ to $M_2$ and $\gamma < \text{cf}(\lambda)$ we define a game $\partial^*_{\gamma,\lambda}(f, M_1, M_2)$. A play lasts $\gamma$ moves, in the $\beta < \gamma$ move a partial isomorphism $f_\beta$ was formed increasing with $\beta$, extending $f$, satisfying $|\text{Dom}(f_\beta)| < \lambda$. In the $\beta$-th move if $\beta = 0$, the player ISO choose $f_0 = f$, if $\beta$ is a limit ordinal the ISO player chooses $f_\beta = \cup\{f_\varepsilon : \varepsilon < \beta\}$. In the $\beta + 1 < \gamma$ move the player AIS chooses $\alpha_\beta < \lambda$ and then they play a sub-game $\partial^{\alpha_\beta}_1(f_\beta, M_1, M_2)$ from Definition 1.1(3) producing an increasing sequence of partial isomorphisms $\langle f^\beta_i : i < \alpha_\beta\rangle$ and let their union be $f_{\beta+1}$. ISO wins if he always has a legal move.

(2) If ISO wins the game (i.e. has a winning strategy) then we say $M_1, M_2$ are $\text{EF}^*_{\gamma,\lambda}$-equivalent, we omit $\lambda$ if clear from the context. If $f = \emptyset$ we may write $\partial^*_{\gamma,\lambda}(M_1, M_2)$

Remark. For $(M, c_1), (M, c_2)$ to be $\text{EF}^*_{<\alpha,\lambda}$-equivalent not $\text{EF}^*_{\alpha,\lambda}$-equivalent not just $\text{EF}^*_\alpha$-equivalent not $\text{EF}^*_{\alpha+1}$-equivalent we may need a minor change.

Hypothesis 3.2. $j_* \le \kappa = \text{cf}(\lambda) < \lambda, \kappa > \aleph_0, \bar{\mu} = \langle \mu_i : i < \kappa\rangle$ is increasing continuous with limit $\lambda$, $\mu_0 = 0$, $\mu_1 = \kappa (= \text{cf}(\lambda))$, $\mu_{i+1}$ is regular $> \mu_i^+$ and let $\mu_\kappa = \lambda$ and for $\alpha < \lambda$ let $\mathbf{i}(\alpha) = \text{Min}\{i : \mu_i \le \alpha < \mu_{i+1}\}$.

DEFINITION 3.3. Under the Hypothesis 3.2 we define a $\lambda$-parameter $\mathfrak{x} = \mathfrak{x}_{j_*,\bar\mu}$ as follows:

(a) ($\alpha$) $I$ is the set of $u \in [\lambda \setminus \kappa]^{\leq \aleph_0}$

($\beta$) $\mathbf{u} : I \to \mathcal{P}(\lambda \setminus \kappa)$ is the identity,

($\gamma$) $S = \{(u_1, u_2) : u_1 \subseteq u_2 \in [\lambda \setminus \kappa]^{\leq \aleph_0}\}$

($\delta$) $\alpha^*_{\mathfrak{x}} = j_*$

(b) $J$ is the set of tuples $t = (u, j, g, h) = (u^t, j^t, g^t, h^t)$ such that

($\alpha$) $u \in I$

($\beta$) $j < j_*$

($\gamma$) (i) $g$ is a non-decreasing function from $u_g = u \cup v_g$ to $\lambda$ where $v_g = \{\mathbf{i}(\alpha) : \alpha \in u \text{ and } g(\alpha) = \mu^+_{i(\alpha)}\}$

(ii) $\alpha \in u \Rightarrow g(\alpha) \in [\mu_{\mathbf{i}(\alpha)}, \mu^+_{i(\alpha)}]$

(iii) if $i \in v_g$ then $g(i) < j^t (< \kappa = \mu_1)$

(iv) $v_g$ is an initial segment of $\{\mathbf{i}(\alpha) : \alpha \in u\}$

($\delta$) (i) $h$ is a non-decreasing function with domain $u_g \cup v_g$

(ii) $\alpha \in u \Rightarrow h(\alpha) \in [\mu_{\mathbf{i}(\alpha)}, \mu_{\mathbf{i}(\alpha)+1}]$ and if $i \in v_g$ then $h(i) < \kappa$

(iii) if $\beta_1 < \beta_2$ are from $u_g \cup v_g$ and $\mathbf{i}(\beta_1) = \mathbf{i}(\beta_2)$ then $g(\beta_1) = g(\beta_2) \Leftrightarrow h(\beta_1) = h(\beta_2)$

(iv) $\alpha < h(\alpha)$ for $\alpha \in u_g \cup v_g$ and $g(\alpha) = \mu^+_{i(\alpha)} \Leftrightarrow h(\alpha) = \mu^+_{i(\alpha)}$ for $\alpha \in u$

(c) $T$ is the set of pairs $(t_1, t_2) \in J \times J$ satisfying

(i) $u^{t_1} \subseteq u^{t_2} \in I$ and

(ii) $g^{t_1} \subseteq g^{t_2}$, $h^{t_1} \subseteq h^{t_2}$, $j^{t_1} = j^{t_2}$

OBSERVATION 3.4. $\mathfrak{x}_\lambda = \mathfrak{x}_{j_*,\bar\mu}$ is a full $\lambda$-parameter.

PROOF. Read the Definition 2.1(1) + 2.1(1A). ⊣

CLAIM 3.5. Assume $s \in I_{\mathfrak{x}}$, $c_1 = (s, e_{G_s})$, $c_2 = (s, x_t)$, $t \in J_s$, and for simplicity $\mathrm{Rang}(g^t \restriction [\mu_{1+i}, \mu_{1+i+1})) \subseteq \{\mu_{1+i}\}$, $\mathrm{Rang}(g^t \restriction \kappa) = \{0\}$ and $\omega < j^t < j_*$. <u>Then</u> $(M_{\mathfrak{x}}, c_1)$, $(M_{\mathfrak{x}}, c_2)$ are $\mathrm{EF}^*_{\lambda,j^t}$-equivalent.

PROOF. So $t$, $j^t$ are fixed. For $i_* < \kappa$, $j < j_*$ let

(a) $B_{i_*} = \{\bar\beta : \bar\beta = \langle \beta_i : i < \kappa \rangle$ and $\mu_i \leq \beta_i \leq \mu_{i+1}$ and $\beta_0 = i_*$ and $(\beta_{1+i} = \mu_{1+i+1} \equiv 1 + i < i_*)\}$

(b) for $\bar\beta \in B_{i_*}$ let $A_{\bar\beta} = \cup\{[\mu_i, \beta_i) : i < \kappa\}$ which by our conventions is equal to $i_* \cup \bigcup\{[\mu_j, \mu_{j+1}) : 1 \leq j < i_*\} \cup \bigcup\{[\mu_i, \beta_i) : i \in [i_*, \kappa)\}$

(c) for $\bar\beta \in B_{i_*}$ let $\mathcal{G}_{j,i_*,\bar\beta} = \{g : g$ is a function from $A_{\bar\beta}$ to $\lambda$, non-decreasing and the function $g \restriction \kappa$ is into $j$ and the function $g \restriction [\mu_{1+i}, \mu_{1+i+1})$ is into $[\mu_i, \mu^+_i]$ and $1 \leq i < i_* \Leftrightarrow (\exists \alpha)(\mu_i \leq \alpha < \mu_{i+1} \wedge g(\alpha) = \mu^+_i)\}$

(d) for $g \in \mathcal{G}_{j,i_*\bar\beta}$, $\bar\beta \in B_{i_*}$ we define $h_g : A_{\bar\beta} \to \lambda$ as follows: if $\gamma \in A_{\bar\beta}$ then $h(\gamma) = \mathrm{Min}\{\beta' \leq \beta_{\mathbf{i}(\gamma)}$: $i(\gamma) > 0 \wedge g(\gamma) = \mu^+_{i(\gamma)}$ then $\beta' = \mu_{i(\gamma)+1}$, otherwise $\beta' \in [\mu_{i(\gamma)}, \beta_{i(\gamma)}]$ and $\beta' \neq \beta_{i(\gamma)} \Rightarrow g(\gamma) < g(\beta')\}$

(e) $\mathcal{G}_{j,i_*} = \cup\{\mathcal{G}_{j,i_*,\bar\beta} : \bar\beta \in B_{i_*}\}$ and $\mathcal{G}_j = \cup\{\mathcal{G}_{j,i_*} : i_* < \kappa\}$.

Let $R = \mathcal{G}_{j'}$ and for $g \in R$ let $i_*(g)$ be the unique $i_* < \kappa$ such that $g \in \mathcal{G}_{j',i_*}$ and $\bar{\beta}_g$ the unique $\bar{\beta} \in B_{i_*}$ such that $g \in \mathcal{G}_{j',i_*(g),\bar{\beta}}$ and $\bar{\beta} = \langle \beta_i(g) : i < \kappa \rangle$

On $R$ we define a partial order $g_1 \leq g_2 \Leftrightarrow g_1 \subseteq g_2 \wedge h_{g_1} \subseteq h_{g_2}$.

For $g \in R$ we define $I_g$, $\bar{c}_g$ as follows

⊛ (a) $I_g = \{u \in I : u \subseteq \mathrm{Dom}(g) \setminus \kappa\}$

(b) $\bar{c}_g = \langle c_{g,s} : s \in I_g \rangle$

(c) $c_{g,s} = x_{t_g(s)}$ where $t_g(s) = (s, j, g{\restriction}u_{g,s}, h_g{\restriction}u_{g,s})$ where $u_{g,s} = u \cup \{\mathrm{i}(\alpha) : \alpha \in u \text{ and } g(\alpha) = \mu^+_{\mathrm{i}(\alpha)}\}$.

Let $g_* \in \mathcal{G}_1$ be chosen such that for $i > 0$, $\beta_i(g_*) = \sup(\{g'(\alpha) : \alpha \in u' \cap [\mu_i, \mu_{i+1})\} \cup \{\mu_i\})$ and $\beta_0(g_*) = \cup\{\mathrm{i}(\alpha)+1 : \alpha \in u' \text{ and } g'(\alpha) = \mu^+_{\mathrm{i}(\alpha)}\} \cup \{1\}$. Let $\bar{c}_* = \bar{c}_{g_*}$ and $f_* = f^{\mathfrak{x}}_{\bar{c}_*}$ is the partial automorphism of $M_\gamma$ with domain $\cup\{P_u^{M_{\mathfrak{x}}} : u \in I_{g_*}\}$ from Definition 2.6. We prove that the player ISO wins in the game $\Game^*_{\lambda,j}(f_*, M_1, M_1)$, as $f_*(c_1) = c_2 (\in P_{u'}^{M_{\mathfrak{x}}})$ this is enough. Recall that a play last $j$ moves; now the player ISO commit himself to choose in the $\beta < j$ move on the side a function $g_\beta \in \mathcal{G}_{1+\beta}$, increasing with $\beta$, $g_0 = g_*$ and his actual move $f_\beta$ is $f^{\mathfrak{x}}_{\bar{c}_\beta}$ where $\bar{c}_\beta = \bar{c}_{g_\beta}$. For the $\beta$-th move if $\beta = 0$ or $\beta$ limit let $g_\beta = \cup\{g_\varepsilon : \varepsilon < \beta\} \cup g_* \in \mathcal{G}_{1+\beta}$. In the $(\beta+1)$-th move let the AIS player choose $\alpha_\beta < \lambda$. Now the player ISO, on the side, first choose $i_\beta < \kappa$ such that $i_*(g_\beta) < i_\beta$, and $\mu_{i_\beta} > \alpha_\beta$, second he chooses $g^+_\beta \in \mathcal{G}_{i_\beta 1+\beta+1}, i_\beta$ satisfying:

⊛ (a) $g^+_\beta$ extends $g_\beta$,

(b) $\mathrm{Dom}(g^+_\beta) \cap \kappa = i_\beta$

(c) $g^+_\beta{\restriction}(i_\beta \setminus \mathrm{Dom}(g_\beta))$ is constantly $1 + \beta$

(d) if $0 < i \in \mathrm{Dom}(g_\beta) \cap \kappa$ then $g^+_\beta{\restriction}[\mu_i, \mu_{i+1}) = g_\beta{\restriction}[\mu_i, \mu_{i+1})$

(e) if $i \notin \mathrm{Dom}(g_\beta) \cap \kappa$ and $i\ \mathrm{Dom}(g^+_\beta) \cap \kappa$ then $\mathrm{Dom}(g^+_\beta{\restriction}[\mu_i, \mu_{i+1})) = [\mu_i, \mu_{i+1})$ and $\varepsilon \in [\mu_i, \mu_{i+1}) \setminus \mathrm{Dom}(g_\beta) \Rightarrow g^+_\beta(\varepsilon) = \mu^+_i$

(f) if $i < \kappa, i \notin \mathrm{Dom}(g^+_\beta)$ then $g^+_\beta{\restriction}[\mu_i, \mu_{i+1}) = g_\beta{\restriction}[\mu_i, \mu_{i+1})$

Now ISO and AIS has to play the sub-game $\Game_1^{\alpha_\beta}(f_\beta, M_1, M_2)$. The player ISO has to play $f_{\beta,\alpha}$ in the $\alpha$-th move for $\alpha \leq \alpha_\beta$ and on the side he chooses $g_{\beta,\alpha} \in \mathcal{G}_{1+\beta+1}$ with large enough domain and range, to make it a legal move, increasing with $\alpha$, and $g_{\beta,0} = g^+_\beta$ and $g_{\beta,\alpha}{\restriction}\mu_{i_\beta} = g^+_\beta{\restriction}\mu_{i_\beta}$. Now obviously $\{g : g \in \mathcal{G}_{1+\beta+1}, g^+_\beta \subseteq g\}$ is closed under increasing union of length $< \mu_{i_\beta}$, it is enough to show that he can make the $(\alpha+1)$-th move which is trivial so we are done. ⊣

Claim 3.6. $M_{\mathfrak{x}}$ is $P_s$-rigid for $s \in I^*$.

Proof. We imitate the proof of Claim 2.11.

$(*)_0$ $\mathfrak{x}$ is a full $(\lambda, \aleph_1)$-parameter

$(*)_1$ if $u_1 \subseteq u_2 \in I$, we define the function $\pi_{u_1,u_2} : J_{u_2} \to J_{u_1}$ by $F_{u_1,u_2}(t) = (u_1, j^t, g^t{\restriction}u_1, h^t{\restriction}u_1)$ for $t \in J_{u_2}$,

$(*)_2$ if $u_1 \subseteq u_2 \subseteq u_3$ are from $I$ then $\pi_{u_1,u_3} = \pi_{u_1,u_2} \circ \pi_{u_2,u_3}$ that is $\pi_{u_1,u_2}(t) = \pi_{u_1,u_2}(\pi_{u_2,u_3}(t))$

$(*)_3$ for $u_1 \subseteq u_2$ we have

$(\alpha)$ $T \cap (J_{u_1} \times J_{u_2}) = \{(\pi_{u_1,u_2}(t_2), t_2) : t_2 \in J_{u_2}\}$

$(\beta)$ $\mathbb{G}_{u_1,u_2} = \{(\hat{\pi}_{u_1,u_2}(c_2), c_2) : c_2 \in \mathbb{G}_{u_2}\}$ where $\hat{\pi}_{u_1,u_2} \in \mathrm{Hom}(\mathbb{G}^{\mathfrak{x}}_{u_2}, \mathbb{G}^{\mathfrak{x}}_{u_1})$ is the unique homomorphism from $\mathbb{G}^{\mathfrak{x}}_{u_2}$ into $\mathbb{G}^{\mathfrak{x}}_{u_1}$ mapping $x_{t_2}$ to $x_{t_1}$ whenever $\pi_{u_1,u_2}(t_2) = t_1$

[Why? Check.]

$(*)_4$ if $u_1 \cup u_2 \subseteq u_3 \in I$, $t_3 \in J_{u_3}$ and $t_\ell = \pi_{u_\ell,u_3}(t_3)$ for $\ell = 1, 2$ then, recalling Definition 2.1 (1A)(h), $g^{t_1}, g^{t_2}$ are compatible functions as well as $h^{t_1}, h^{t_2}$ and $j^{t_1} = j^{t_2}$ moreover $g^{t_1} \cup g^{t_2}$ is non-decreasing, $h^{t_1} \cup h^{t_2}$ is non-decreasing

[Why? Just check.]

$(*)_5$ clause $\circledast_1$ of Claim 2.10(1) holds for $I' = I (= I_{\mathfrak{x}})$.

Why? Assume $\bar{c} \in C^{\mathfrak{x}}_I$ is such that $c_{s(*)} \neq e_{\mathbb{G}_{s(*)}}$ for some $s(*) \in I$. For each $u \in I$, $c_u$ is a word in the generators $\{x_t : t \in J_u\}$ of $\mathbb{G}_u$ and let $\mathbf{n}(u)$ be the length of this word and $\mathbf{m}(u)$ the number of generators appearing in it.

Now by clause $(\beta)$ of $(*)_3$ we have $u_1 \subseteq u_2 \Rightarrow \mathbf{n}(u_1) \leq \mathbf{n}(u_2) \wedge \mathbf{m}(u_1) \leq \mathbf{m}(u_2)$. As $(I, \subseteq)$ is $\aleph_1$-directed, for some $u_* \in I$, $n_* < \omega$ and $m_* < \omega$ we have $u_* \subseteq u \in I \Rightarrow \mathbf{n}(u) = n_* \wedge \mathbf{m}(u) = m_*$ and let $c_u = (\dots, x^{i(u,\ell)}_{t(u,\ell)}, \dots)_{\ell < n_*}$ where $k(u,\ell) \in \{1, -1\}$ and $t(u,\ell) \in J^{\mathfrak{x}}_u$ and $t(u,\ell) = t(u,\ell+1) \Rightarrow k(\ell) = k(u_1+1)$. Clearly $u_* \subseteq u_1 \subseteq u_2 \in I \mathbin{\&} \ell < n_* \Rightarrow \pi_{u_1,u_2}(t(u_2,\ell)) = t(u_1,\ell) \wedge k(u_1,\ell) = k(u_2,\ell) = k(u_*,\ell)$ hence $j^{t(u_2,\ell)} = j^{t(u_*,\ell)} \wedge j^{t(u_2,\ell)} = j^{t(u_*,\ell)}$. By our assumption toward contradiction necessarily $n_* > 0$ and let $k(\ell) = k(u_*,\ell)$.

As $\{u : u_* \subseteq u \in I\}$ is directed, by $(*)_4$ above, for each $\ell < n_*$ any two of the functions $\{g^{t(u,\ell)} : u_* \subseteq u \in I\}$ are compatible so $g_\ell =: \cup\{g^{t(u,\ell)} : u \in I\}$ is a non-decreasing function from $Y_{i_\ell(*)}$ to $\lambda$ where $Y_{i_\ell(*)} = (\lambda \setminus \kappa) \cup i_\ell(*)$ for some $i_\ell(*) \leq \kappa$ and $h_\ell =: \cup\{h^{t(u,\ell)} : u_* \subseteq u \in I\}$ is similarly a non-decreasing function from $Y_{i_\ell(*)}$ to $\lambda$. Also $g_\ell$ maps $[\mu_i, \mu_{i+1})$ into $[\mu_i, \mu_i^+]$ for $i < \kappa$ and maps $\kappa$ to $\kappa$.

CASE 1. $i_\ell(*) = \kappa$.

It also follows that for some $j^*_\ell$ we have $j^*_\ell =: j^{t(u,\ell)}$ whenever $u_* \subseteq u \in I$ in fact $j^*_\ell = j^{t(u_*,\ell)}$ is O.K. and $j^*_\ell < j_* \leq \kappa$. For each $i \in \mathrm{Rang}(g_\ell \restriction \kappa)$ choose $\beta_{\ell,i} < \kappa$ such that $g_\ell(\beta_{\ell,i}) = i$ and let $E = \{\delta < \kappa : \delta$ a limit ordinal $> \sup(u_* \cap \kappa)$ such that $i < j^*_\ell \mathbin{\&} \ell < n_* \mathbin{\&} i \in \mathrm{Rang}(g_\ell) \Rightarrow \beta_{\ell,i} < \delta$ and $\beta < \delta \mathbin{\&} \ell < n \Rightarrow h_\ell(\beta) < \delta\}$, it is a club of $\kappa$. Choose $u$ such that $u_* \subseteq u$ and $\mathrm{Min}(u \cap \kappa \setminus u_*) = \delta^* \in E$.

Now what can $g^{t(u,\ell)}(\mathrm{Min}(u \setminus u_*))$ be?

It has to be $i$ for some $i < j^*_\ell < j^*$ hence $i \in \mathrm{Rang}(g_\ell)$ so for some $u_1, u_* \subseteq u_1 \subseteq \delta^*$ and $\beta_{\ell,i} \in u_1$ so $h_\ell(\beta_{\ell,i}) < \delta^*$ hence considering $u \cup u_1$ and recalling clause $(\delta)(iv)$ of (b) from Definition 3.3 of $\mathfrak{x}$ we have $h_\ell(\beta_{\ell,i}) < h_\ell(\delta^*)$ hence by (clause $(b)(\alpha)(iii)$) we have $i = g_\ell(\beta_{\ell,i}) < g_\ell(\delta^*)$, contradiction.

CASE 2. $i_\ell(*) \neq \kappa$ so $i_\ell(*) < \kappa$.

Clearly if $i \in (i_\ell(*), \kappa)$ and $\alpha \in [\mu_i, \mu_{i+1})$ then $g_\ell(\alpha) \neq \mu_i^+$ (see clause (b)$(\gamma)$(iii) of Definition 3.3) hence $g_\ell \restriction [\mu_i, \mu_{i+1})$ is a non-decreasing function from $[\mu_i, \mu_{i+1})$ to $\mu_i^+$, but $\mu_{i+1}$ is regular $> \mu_i^+$ (see Hypothesis 3.2) hence $g_\ell \restriction [\mu_i, \mu_{i+1})$ is eventually constant say $\gamma_i \in [\mu_i, \mu_{i+1})$ and $g_\ell \restriction [\gamma_i, \mu_{i+1})$ is constantly $\varepsilon_i \in [\mu_i, \mu_i^+)$. So also $h_\ell \restriction [\gamma_i, \mu_{i+1}^+)$ is constant and its value is $< \mu_{i+1}$, and we get contradiction as in Case 1. ⊣

CONCLUSION 3.7. If $\lambda = \lambda^{\aleph_0} > \mathrm{cf}(\lambda) > \aleph_0$ then for every $\alpha < \mathrm{cf}(\lambda)$ there are non-isomorphic models $M_1, M_2$ of cardinality $\lambda$ which are $\mathrm{EF}^*_{\alpha,\lambda}$-equivalent.

PROOF. By Claim 3.5 + 3.6 as the cardinality of $M_{\mathfrak{x}}$ is $\lambda$. ⊣

REMARK 3.8. By minor changes, for some $t \in P_u^M$, $u = \emptyset$ letting $c_1 = e_{G_u}$, $c_2 = x_t$ we have: $(M_{\mathfrak{x}}, c_1), (M_{\mathfrak{x}}, c_2)$ are non-isomorphism but $\mathrm{EF}^*_{\lambda,j}$-equivalent for every $j < \kappa = \mathrm{cf}(\lambda)$. This is similar to the parallel remark in the end of §1.

## REFERENCES

[HvSh:866] CHANOCH HAVLIN and SAHARON SHELAH, *More on* $\mathrm{EF}_{\alpha,\lambda}$*-equivalence in non-isomorphic models*, in preparation.

INSTITUTE OF MATHEMATICS
THE HEBREW UNIVERSITY OF JERUSALEM
JERUSALEM 91904, ISRAEL
and
DEPARTMENT OF MATHEMATICS
RUTGERS UNIVERSITY
NEW BRUNSWICK, NJ 08854, USA
*E-mail*: shelah@math.huji.ac.il

# THE ∀∃ THEORY OF $\mathcal{D}(\leq, \vee, ')$ IS UNDECIDABLE

RICHARD A. SHORE[†] AND THEODORE A. SLAMAN[‡]

**Abstract.** We prove that the two quantifier theory of the Turing degrees with order, join and jump is undecidable.

**§1. Introduction.** Our interest here is in the structure $\mathcal{D}$ of all the Turing degrees but much of the motivation and setting is shared by work on other degree structures as well. In particular, Miller, Nies, and Shore [2004] provides an undecidability result for the r.e. degrees $\mathcal{R}$ at the two quantifier level with added function symbols as we do here for $\mathcal{D}$. To set the stage, we begin then with an adaptation of the introduction from that paper including statements of some of its results and so discuss $\mathcal{R}$, the r.e. degrees and $\mathcal{D}(\leq \mathbf{0}')$, the degrees below $\mathbf{0}'$ as well.

A major theme in the study of degree structures of all types has been the question of the decidability or undecidability of their theories. This is a natural and fundamental question that is an important goal in the analysis of these structures. It also serves as a guide and organizational principle for the development of construction techniques and algebraic information about the structures. A decision procedure implies (and requires) a full understanding and control of the first order properties of a structure. Undecidability results typically require and imply some measure of complexity and coding in the structure. Once a structure has been proven undecidable, it is natural to try to determine both the extent and source of the complexity. On the one hand, one wants to determine the degree of the theory. On the other hand, one strives to find the dividing line between decidability and undecidability in terms of fragments of the theory. The first has frequently brought with it considerable information about second order properties such as definability and automorphisms. The second requires the most algebraic information and development of construction techniques.

[†]Partially supported by NSF Grant DMS-0100035. Thanks also to Andre Nies for some helpful conversations about this work.

[‡]Partially supported by NSF Grant DMS-9988644.

Logic Colloquium '03
Edited by V. Stoltenberg-Hansen and J. Väänänen
Lecture Notes in Logic, 24

For $\mathcal{D}$ and $\mathcal{D}(\leq \mathbf{0}')$ the results came fairly early. The first paper on the structure $\mathcal{D}$ of the Turing degrees as a whole, Kleene and Post [1954], developed the finite extension method (essentially Cohen forcing for one quantifier formulas of arithmetic) and proved that all finite partial orderings can be embedded in both $\mathcal{D}$ and $\mathcal{D}(\leq \mathbf{0}')$. As these structures are partial orderings, this suffices to show that the one quantifier ($\exists$) theories are decidable. (An existential sentence is true in either structure if and only if it is consistent with the theory of partial orders, or equivalently, if there is a partial order with a domain of size the number of variables in the formula which satisfies the formula.)

Once the embedding problem is settled, the next level of algebraic questions about the structures concern extension of embeddings. The first example here is density (or, from the other side minimal covers). A long development of construction techniques building on Spector's original construction 1956 of a minimal degree, essentially by forcing with recursive trees, lead to Lachlan's [1968] result that every countable distributive lattice is isomorphic to an initial segment of $\mathcal{D}$. This coding of distributive lattices is sufficient to get the undecidability of the theory as Lachlan [1968] notes. Combining these initial segment techniques with simultaneous control of the join and Spector's [1956] exact pair theorem, Simpson [1977] showed that the theory of $\mathcal{D}$ is recursively isomorphic to $Th^2(\mathbb{N})$, true second order arithmetic.

Finding the dividing line between decidability and undecidability required Lerman's [1971] result that every finite lattice (not just the distributive ones) is isomorphic to an initial segment of $\mathcal{D}$. On one hand, combining this with the finite extension method solved the extension of embedding problem in such a way that it gave the decidability of the two quantifier ($\forall\exists$) theory of $\mathcal{D}$ (Shore [1978] and Lerman [1983, Appendix A]). (By the extension of embedding problem we mean determining for which partial orders $\mathcal{X} \subseteq \mathcal{Y}$ does every embedding of $\mathcal{X}$ into $\mathcal{D}$ have an extension to one of $\mathcal{Y}$.) The ability to code all finite lattices also sufficed for Schmerl (see Lerman [1983, Appendix A]) to prove that the three quantifier ($\forall\exists\forall$) theory of $\mathcal{D}$ is undecidable.

A similar analysis of $\mathcal{D}(\leq 0')$ was then carried out. First came a significant elaboration of the construction techniques to get enough initial segments results below $0'$ to give undecidability (Epstein [1979] and Lerman). Lerman then proved the full analog that every finite (even recursive) lattice is isomorphic to an initial segment of $\mathcal{D}(\leq 0')$ (Lerman [1983, XII]). This immediately gives the undecidability of the three quantifier theory. Then these results were extended and combined with extension of embedding results below an arbitrary r.e. degree (Lerman and Shore [1988]) to get the decidability of the two quantifier theory. They were also used to show (Shore [1981]) that the theory of $\mathcal{D}(\leq 0')$ is recursively isomorphic to true first order arithmetic.

The road has been much longer for the analysis of the r.e. degrees, $\mathcal{R}$. It began with the finite injury (or $0'$) priority method of Friedberg [1957] and Muchnik [1956] that produced incomparable r.e. degrees and so an embedding of the simplest partial (nonlinear) order. This method sufficed to embed all finite (even countable) partial orderings (Sacks [1963]) and so decide the one quantifier theory of $\mathcal{R}$ in the same way that Kleene and Post's work decided that of $\mathcal{D}$ and $\mathcal{D}(\leq 0')$. As the r.e. degrees are dense (by the infinite injury (or $0''$) methods of Sacks [1964]), the next steps in the analysis could not follow the path laid out for $\mathcal{D}$. Many years of development of construction techniques and algebraic information ensued. Lachlan's monster (or $0'''$ injury) methods were eventually used by Harrington and Shelah [1982] to prove that $\mathcal{R}$ is undecidable. The degree of its theory, as by now one should expect, is also that of true first order arithmetic (Harrington and Slaman; Slaman and Woodin; Nies, Shore, and Slaman [1998]).

This leaves us with determining the boundary line between decidability and undecidability for $\mathcal{R}$. Once again, a long hiatus and much work on other developments led to the undecidability of the three quantifier theory by Lempp, Nies, and Slaman [1998]. The extension of embedding problem was solved by Slaman and Soare [2001] but the question of the decidability of the two quantifier theory of $\mathcal{R}$ remains open. A major obstacle is the lattice embedding problem of determining which finite lattices can be embedded in $\mathcal{R}$. Despite some forty years of effort by many researchers on both embedding and nonembedding results, this question is still unsolved. The best result to date is Lerman [2000] which shows that the question for an important class of lattices is of degree at most $0''$. Even if the lattice embedding problem is shown to be decidable, there are further difficulties related to Lachlan's [1966] nondiamond result that there is no embedding of the four element Boolean algebra into $\mathcal{R}$ that preserves both 0 and 1.

The situation for these three degree structures is summarized in the following table:

| | $\mathcal{R}$ | $\mathcal{D}$ | $\mathcal{D}(\leq 0')$ |
|---|---|---|---|
| $\exists(\leq)$ | Dec | Dec | Dec |
| $\forall\exists(\leq)$ | ? | Dec | Dec |
| $\forall\exists\forall(\leq)$ | Undec | Undec | Undec |
| $Th(\leq)$ | $Th(N)$ | $Th^2(N)$ | $Th(N)$ |

Thus we remain a long way from the decidability of the two quantifier theory of $\mathcal{R}$. On the other hand, the methods used to prove undecidability of other degree structures, interpretation of theories with simple fragments known to be undecidable, cannot work for the two quantifier theory of $\mathcal{R}$ with just $\leq_T$, or even any extension by relation symbols, since the most we can code into this fragment is the validity (perhaps in all finite models) of an $\forall\exists$ sentence in a finite relational language but this problem is always decidable. (The point here is that, since the language is relational, any such sentence with $n$ variables is satisfiable if and only if it is satisfiable in some structure of size at most $n$. As there are only finitely many such structures, this question is decidable. The basic result is classical (Bernays and Schönfinkel [1928] and Ramsey [1930]). Its application to interpretations in structures such as $\mathcal{R}$ is pointed out in Shore [1999, p. 179].)

The only hope for an undecidability result at the two quantifier level for $\mathcal{R}$ then is to add function symbols. One would then try to interpret some theory with function symbols or, more directly, to code register machines. (The coding of register machines is at the base of much of the work on undecidability of various severely restricted quantification classes of formulas as in Börger, Grädel, and Gurevich [1997].) This raises the natural question about the boundary between decidability and undecidability for all these degree structures: What happens when we add additional function symbols to the language?

In all these settings the most natural one to be considered is the join operator $\vee$. As the structures remain uniformly locally finite the arguments for the unlikeliness of interpretations of theories providing undecidability remain in place. (The closure of any finite set is finite with size bounded by a fixed recursive function of the cardinality of the starting set and so cannot, on its own, be used to generate the infinite (or at least unbounded) structures need for coding even register machines.) Indeed, for all the degrees, the ∀∃ theory of this structure, $\mathcal{D}(\leq, \vee)$, is decidable by Jockusch and Slaman [1993].

The next thing to try in terms of the known structural work on $\mathcal{R}$ is the infimum operator $\wedge$. This has the advantage that finitely generated substructures can be infinite (Lerman, Shore, and Soare [1984]). The obvious problem with this approach is that not every pair of r.e. degrees has an infimum and so $\wedge$ is not a total function on $\mathcal{R}$ as is required. We can, of course, consider total extensions of the partial infimum relation but would not want the undecidability to be an artifact of our (perhaps perverse) choice of extension. The solution of Miller, Nies, and Shore is to prove undecidability in a sufficiently uniform way so that the proof is independent of the choice of extension.

Theorem 1.1 (Miller, Nies, and Shore [2004]). *For any total extension $\wedge$ of the partial infimum relation on $\mathcal{R}$, the two quantifier (∀∃) theory of $\mathcal{R}(\leq, \vee, \wedge)$ is undecidable.*

They noted that the coding methods used for this result can be applied to both $\mathcal{D}$ and $\mathcal{D}(\leq \mathbf{0}')$ along with known initial segment constructions to get similar results.

Corollary 1.2 (Miller, Nies, and Shore [2004]). *For any total extension $\wedge$ of the partial infimum relation on $\mathcal{D}$ ($\mathcal{D}(\leq 0')$), the two quantifier (∀∃) theory of $\mathcal{D}$ ($\mathcal{D}(\leq 0')$) with $\leq$, $\vee$ and $\wedge$ is undecidable.*

In the setting of the degrees as a whole, however, there is a second natural operator to consider adding to our language, the jump operator. The jump is definable in $\mathcal{D}$ by Shore and Slaman [1999] but the definition involves coding models of arithmetic and discussing automorphisms. Its complexity is very high and so sheds no light on the boundary between decidability and undecidability in $\mathcal{D}(\leq,')$, the degrees with jump which is our topic here.

Decidability results to date on the theory with the jump operator include the following:

THEOREM 1.3 (Jockusch and Soare, see Lerman [1983, III.4.21]). *The theory of $\mathcal{D}$ with just $'$ (no $\leq$) is decidable.*

THEOREM 1.4 (Hinman and Slaman [1991]). *The $\exists$-theory of $\mathcal{D}(\leq,')$ is decidable.*

THEOREM 1.5 (Montalban [2003]). *The $\exists$-theory of $\mathcal{D}(\leq,\vee,')$ is decidable.*

We show that Montalban's decidability result is the best possible (actually, our result was proven first so he provided the proof of sharpness for our result) and so solve problem IV.7 (attributed to Jockusch) of Arslanov and Lempp [1999]:

THEOREM 1.6. *The $\forall\exists$-theory of $\mathcal{D}(\leq,\vee,')$ is undecidable.*

Our route to undecidability is via the coding of register machines as in Miller, Nies, and Shore [2004]. We describe the machines and their coding in predicate logic in the next section. Once we see how they are interpreted in predicate logic it will be clear that our undecidability result would follow immediately from the existence of a $\Delta_0$ formula with $x$ free and additional free variables such that, as the additional variables range over the degrees, the sets defined by the formula range over all countable subsets of $\mathcal{D}$. This result is provided by our main technical theorem.

THEOREM 1.7. *Given $\{\mathbf{y}_i \mid i \in \omega\}$ and $\{\mathbf{z}_j \mid j \in \omega\}$ disjoint countable sets of degrees, there are $\mathbf{g}$ and $\mathbf{h}$ such that $\forall i, j \in \omega((\mathbf{y}_i \oplus \mathbf{g})' \leq_T (\mathbf{y}_i \oplus \mathbf{h})'$ & $(\mathbf{z}_j \oplus \mathbf{g})' \nleq_T (\mathbf{z}_j \oplus \mathbf{h})')$.*

COROLLARY 1.8. *If $\mathbf{C}$ is any countable set of degrees then it is uniformly $\Delta_0$-definable in parameters, i.e. there is a single formula which defines every such $\mathbf{C}$ as the parameters vary.*

PROOF. Choose any strict upper bound $\mathbf{z}$ for $\mathbf{C}$. Let $\{\mathbf{y}_i\} = \{\mathbf{x} < \mathbf{z} : \mathbf{x} \in \mathbf{C}\}$ and $\{\mathbf{z}_j\} = \{\mathbf{x} < \mathbf{z} : \mathbf{x} \notin \mathbf{C}\}$. Apply the theorem to get the degrees $\mathbf{g}$ and $\mathbf{h}$. $\mathbf{C}$ is then $\{\mathbf{x} < \mathbf{z} : (\mathbf{x} \vee \mathbf{g})' \leq_T (\mathbf{x} \vee \mathbf{h})'\}$ and so our desired formula is $x < w \,\&\, (x \vee u)' \leq_T (x \vee v)'$. ⊣

We prove this theorem in §3. In the final section we summarize the state of affairs and point out some new problems suggested by our view of these matters.

**§2. Coding register machines.** In this section we will explain the algebraic aspects of our codings and derive the main theorem, assuming these codings can be interpreted in $\mathcal{D}(\leq,')$. The next section will provide the recursion theoretic arguments to show that the structures described here can be realized in $\mathcal{D}(\leq,')$. For completeness, we reprise (with some small simplification allowed by our construction here) the presentation of coding register machines in Miller, Nies, and Shore [2004] beginning with a standard description of the $k$-register machines of Shepherdson and Sturgis [1963] and Minsky [1961] and their representation in predicate logic as in Nerode and Shore [1997, III.8] or Börger, Grädel, and Gurevich [1997, 2.1].

A *k-register machine* consists of $k$ many storage locations called registers. Each register contains a natural number. There are only two types of operations that these machines can perform in implementing a program. First, they can increase the content of any register by one and then proceed to the next instruction. Second, they can check if any given register contains the number 0 or not. If so, they go on to the next instruction. If not, they decrease the given register by one and can be told to proceed to any instruction in the program. Formally, we define register machine programs and their execution as follows:

A *k-register machine program I* is a finite sequence $I_1, \ldots, I_t, I_{t+1}$ of instructions operating on a sequence of numbers $x_1, \ldots, x_k$, where each instruction $I_m$, for $m \leq t$, is of one of the following two forms:

(i) $x_i := x_i + 1$ (replace $x_i$ by $x_i + 1$)

(ii) If $x_i \neq 0$, then $x_i := x_i - 1$ and go to $j$. (If $x_i \neq 0$, replace it by $x_i - 1$ and proceed to instruction $I_j$.)

It is assumed that after executing some instruction $I_m$, the execution proceeds to $I_{m+1}$, the next instruction on the list, unless $I_m$ directs otherwise. The execution of such a program proceeds in the obvious way on any input of values for $x_1, \ldots, x_k$ (the initial content of the registers) to change the values of the $x_i$ and progress through the list of instructions. The final instruction, $I_{t+1}$, is always a halt instruction. Thus, if $I_{t+1}$ is ever reached, the execution terminates with the current values of the $x_i$. In general, we denote the assertion that an execution of the program $I$ is at instruction $I_m$ with values $n_1, \ldots, n_k$ of the variables by $I_m(n_1, \ldots, n_k)$.

The standard translation of a register machine $M$ describes the action of $M$ by a system of universal axioms in the language of one unary function $s$ thought of as the successor function on ℕ. For technical reasons peculiar to our later coding, we want to use distinct domains $D_i$ with least elements $0_i$. In our application here, these sets will be of the form $\{\mathbf{q}_i^{(n)} \mid n \in \omega\}$ for arithmetically independent degrees $\mathbf{q}_i$. The successor operators $s_i$ will all be the jump operator. For now, we describe the axioms needed in predicate logic with additional $k$-ary relations $P_m$ corresponding to the instructions $I_m$.

For each instruction $I_m$, $1 \leq m \leq t$, include an axiom of the appropriate form:

(i) $P_m(x_1, \ldots, x_k) \to P_{m+1}(x_1, \ldots, x_{i-1}, s_i(x_i), x_{i+1}, \ldots, x_k)$.

(ii) $P_m(x_1, \ldots, x_{i-1}, 0, x_{i+1}, \ldots, x_k) \to P_{m+1}(x_1, \ldots, x_{i-1}, 0, x_{i+1}, \ldots, x_k) \wedge$
$P_m(x_1, \ldots, x_{i-1}, s_i(y), x_{i+1}, \ldots x_k) \to P_j(x_1, \ldots, x_{i-1}, y, x_{i+1}, \ldots, x_k)$.

(Note that being a successor is equivalent to being nonzero, i.e. not equal to $\mathbf{q}_i$ which is the $0_i$ of $D_i$.)

Let $P(I)$ be the finite set of universal axioms corresponding in this translation to register program $I$. It is easy to prove that, program $I$ halts on input $(n_1, \dots, n_k)$ if and only if the sentence $F_k(n_1, \dots, n_k) \equiv P_1(s^{n_1}(0), \dots s^{n_k}(0)) \rightarrow \exists x_1, \dots, \exists x_k [P_{t+1}(x_1, \dots, x_k)]$ is a logical consequence of $P(I)$. As it is a classical fact (Shepherdson and Sturgis [1963]; Minsky [1961]) that the halting problem for 2-register machine programs is r.e. complete, it suffices to code all such models with binary predicates to get undecidability.

As usual for interpretations, we now want to provide formulas $\Delta_i(\vec{q}, x)$, $\Pi_m(\vec{q}, x, y)$ defining, for each choice of parameters $\vec{q}$, sets $D_i$ $(i = 1, 2)$ and binary relations $P_m$ on $D_1 \times D_2$ $(1 \leq m \leq t+1)$. (Remember $s_i$ will be interpreted as the jump operator.) We take $q_1$ and $q_2$ to be the interpretations of 0 in $D_1$ and $D_2$ respectively. We now interpret our formulas $P(I) \rightarrow F(n_1, \dots, n_k)$ in the usual way. We relativize the quantifiers to the appropriate domain, i.e. $\exists x_i(\dots)$ becomes $\exists x_i(\Delta_i(\vec{q}, x_i) \wedge \cdots)$ and $\forall x_i(\cdots)$ becomes $\forall x_i(\Delta_i(\vec{q}, x_i) \rightarrow \cdots)$. We then replace occurrences of $s_i(x_i)$ by $x_i'$ and ones of $P_m(x_1, x_2)$ by $\Pi_m(\vec{q}, x_1, x_2)$. We indicate this translation by $*$. We also need a correctness condition $\Theta$ that says that $q_i \in D_i$ and the jump is a function on the $D_i$ : $\Delta_1(\vec{q}, q_1) \wedge \Delta_2(\vec{q}, q_2) \wedge \forall x_1(\Delta_1(\vec{q}, x_1) \rightarrow \Delta_1(\vec{q}, x_1')) \wedge \forall x_2(\Delta_2(\vec{q}, x_2) \rightarrow \Delta_2(\vec{q}, x_2'))$. The class of sentences of $\mathcal{R}(\leq, \vee, \wedge)$ that we want will then be those of the form $\forall \vec{q}\,[\Theta \rightarrow (P(I)^* \rightarrow F_2^*)]$ where $I$ ranges over programs for 2-register machines.

It is clear that to get these sentences to be $\forall\exists$ ones it is sufficient to get quantifier free definitions ($\Delta_i$ and $\Pi_m$) of the domains and relations (and the worst that would work would be equivalent $\Sigma_1$ and $\Pi_1$ definitions). For the undecidability it suffices, of course, for the $D_i$ to include ones isomorphic to $\{\mathbf{q}_i^{(n)}\}$ and the relations to include all binary relations on the $D_1 \times D_2$ as the parameters vary. If we take $\mathbf{q}_1$ and $\mathbf{q}_2$ to be arithmetically independent, i.e. $\mathbf{q}_i^{(m)} \leq_T \mathbf{q}_1^{(n_1)} \vee \mathbf{q}_2^{(n_2)} \Leftrightarrow m \leq n_i$ then we can code any relation $R_m \subseteq D_1 \times D_2$ in a $\Delta_0$ way from the $D_j$ and one additional countable set of degrees $T_m = \{\mathbf{q}_1^{(n_1)} \vee \mathbf{q}_2^{(n_2)} : \langle n_1, n_2 \rangle \in R_m\}$ by $R_m(x_1, x_2) \Leftrightarrow x_1 \in D_1 \,\&\, x_2 \in D_2 \,\&\, x_1^{(n_1)} \vee x_2^{(n_2)} \in T_m$.

So to prove our $\forall\exists$ undecidability result for $\mathcal{D}(\leq, ')$ it suffices to be able to define arbitrary countable subsets of $\mathcal{D}$ by a $\Delta_0$ formula in parameters. As we pointed out at the end of the introduction, Theorem 1.7 provides this $\Delta_0$ formula.

**§3. The coding theorem.** We now prove our main technical result that says that by joining arbitrary given degrees with degrees $\mathbf{g}$ and $\mathbf{h}$ of our choosing we can make the jumps of the results be comparable or not as desired.

THEOREM 1.7. *Given* $\{\mathbf{y}_i \mid i \in \omega\}$ *and* $\{\mathbf{z}_j \mid j \in \omega\}$ *disjoint countable sets of degrees, there are* $\mathbf{g}$ *and* $\mathbf{h}$ *such that* $\forall i, j \in \omega\; ((\mathbf{y}_i \oplus \mathbf{g})' \leq_T (\mathbf{y}_i \oplus \mathbf{h})' \;\&\; (\mathbf{z}_j \oplus \mathbf{g})' \not\leq_T (\mathbf{z}_j \oplus \mathbf{h})')$.

We fix representatives $Y_i$ and $Z_j$ $(i, j \in \omega)$ of the degrees $\mathbf{y}_i$ and $\mathbf{z}_j$, respectively and construct sets $G$ and $H$ such that for every $i, j \in \omega$ the following requirements are satisfied:

- $C_i : (Y_i \oplus G)' \leq_T (Y_i \oplus H)'$.
- $D_j : (Z_j \oplus G)' \not\leq_T (Z_j \oplus H)'$.

Our construction will be a forcing argument. Our coding procedure for the $C_i$ requirements relies on the fact that, for any $A$ and $B$, $A' \in \Pi_2^B \Rightarrow A' \leq_T B'$ (Theorem 4.3 of Soare [1987, IV] relativized to $B$). Thus our plan for satisfying $C_i$ is to make sure that, for almost every $n$, if (we force) $n \in (Y_i \oplus G)'$ then we immediately force some canonically chosen $\Pi_2^0$ fact (the $h(n)$th one for some recursive $h$) about $Y_i \oplus H$ to be true. On the other hand, if (we force) $n \notin (Y_i \oplus G)$ we want to immediately force this fact to be false. Thus $h$ provides a one-one reduction from $A'$ to the complete $\Pi_2^B$ set and so shows that $A' \in \Pi_2^B$. We make a list of the corresponding subrequirements $C_{i,n}$ where $h$ is some specific recursive function that we will define later:

- $C_{i,n} : n \in (Y_i \oplus G)' \Leftrightarrow h(n) \notin (Y_i \oplus H)''$.

To satisfy $C_i$ we must satisfy $C_{i,n}$ for almost every $n$.

Our diagonalization strategy to satisfy the requirements $D_j$ is based on the fact that if $A' \leq_T B'$ then $\Sigma_1^A \in \Delta_2^B$ and so $\Pi_2^A \in \Pi_2^B$. Thus there is a $\Delta_0$ formula $\theta$ such that if $(Z_j \oplus G)' \leq_T (Z_j \oplus H)'$ then there is a recursive function $f_k$ such that $n \notin (Z_j \oplus G)'' \Leftrightarrow \forall u \exists v \theta(f_k(n), u, v, Z_j, H)$. (Here we have listed all the recursive functions as $f_k$, $k \in \omega$.) Our plan is then for each $k$ (i.e. for each recursive function) to choose an $n$ and, at some stage $s$ of the construction, meet the following subrequirement:

- $D_{j,k} : \exists n \neg[n \notin (Z_j \oplus G)'' \Leftrightarrow \forall u \exists v \theta(f_k(n), u, v, Z_j, H)]$.

We will accomplish this by arranging that either we force

$$\exists u \forall v \neg\theta(f_k(n), u, v, Z_j, H)$$

at stage $s$ while immediately forcing the $\Pi_2^{Z_j \oplus G}$ fact that $n \notin (Z_j \oplus G)''$ to be true or, at stage $s$ we force $\forall u \exists v \theta(f_k(n), u, v, Z_j, H)$ while immediately making the $\Sigma_2^{Z_j \oplus G}$ fact $n \in (Z_j \oplus G)''$ true. In either case, we violate the biconditional equivalent to $(Z_j \oplus G)' \leq_T (Z_j \oplus H)$ and meeting $D_{j,k}$ for all $k$ guarantees that $(Z_j \oplus G)' \not\leq_T (Z_j \oplus H)$ as required to satisfy $D_j$.

To give us immediate control (both positively and negatively) over a canonical list of $\Pi_2$ facts, we employ a variation of Kumabe-Slaman forcing (See Shore and Slaman [1999]). That forcing builds a (not necessarily recursive) Turing functional $\Phi$ and allows one to immediately guarantee that, for any set $X$, $\Phi(X)$ is a finite function (a canonical $\Sigma_2^{X \oplus \Phi}$ fact). We wish to extend this to allow such control over a recursive set of such facts and to also allow "immediate forcing" of their negations. The first goal is achieved by considering the action of $\Phi(X)$ on individual columns (of inputs) rather than all possible inputs. The second is met by adding on an additional second set of restraints on extensions that prevent the $\Sigma_2$ fact of interest from ever being immediately forced (for any witness). General genericity arguments will then guarantee that once we have done this, the $\Pi_2$ fact will immediately be forced.

We now supply the formal definitions of our forcing relation.

DEFINITION 3.1. A *Turing functional* $\Phi$ is a set of sequences $(x, y, \sigma)$ (the axioms) such that $x$ is a natural number (the input), $y$ is either 0 or 1 (the output), and $\sigma$ is a finite binary sequence (the use). Furthermore, for all $x$, for all $y_1$ and $y_2$, and for all compatible $\sigma_1$ and $\sigma_2$, if $(x, y_1, \sigma_1) \in \Phi$ and $(x, y_2, \sigma_2) \in \Phi$, then $y_1 = y_2$ and $\sigma_1 = \sigma_2$. (This says that only one axiom in $\Phi$ applies to a particular oracle on any particular input. It is technically useful.)

We write $\Phi(x, \sigma) = y$ to indicate that there is an initial segment $\tau$ of $\sigma$, possibly equal to $\sigma$, such that $(x, y, \tau) \in \Phi$. If $X \subseteq \omega$, we write $\Phi(x, X) = y$ to indicate that there is an $\ell$ such that $\Phi(x, X \upharpoonright \ell) = y$, and write $\Phi(X)$ for the function evaluated in this way. (Note that the set of axioms need not be recursively enumerable so the functions computed are partial recursive in the functional $\Phi$ plus the input set $X$.)

DEFINITION 3.2. Conditions $p$ in our forcing notion $\mathcal{P}$ are triples $\langle \Phi_p, \mathbf{X}_p, \mathbf{W}_p \rangle$ where $\Phi_p$ is a finite Turing functional and both $\mathbf{X}_p$ and $\mathbf{W}_p$ are finite collections of pairs $\langle X, n \rangle$ (or $\langle W, n \rangle$) consisting of one set and one natural number such that $\mathbf{X}_p \cap \mathbf{W}_p = \emptyset$. We say that $q$ extends $p, q \leq p$, if

1. $\Phi_p \subseteq \Phi_q \,\&\, ([(x, y, \sigma) \in \Phi_p \,\&\, (x', y', \sigma') \in \Phi_q - \Phi_p] \rightarrow |\sigma| < |\sigma'|)$.
2. $\mathbf{X}_p \subseteq \mathbf{X}_q \,\&\, \mathbf{W}_p \subseteq \mathbf{W}_q$.
3. $(\forall x, y, \langle X, n \rangle \in \mathbf{X}_p)(\Phi_q(\langle n, x \rangle, X) = y \rightarrow \Phi_p(\langle n, x \rangle, X) = y)$.

If $K$ is a filter on $\mathcal{P}$ then $\Phi_K = \cup\{\Phi_p \mid p \in K\}$ is a Turing functional. It is this functional that will be our generic object $\Phi$ and about which we will be able to speak in our forcing language. The second clause of (1) says that only longer axioms than the ones we already have can be added on by an extension. Clause (3) says that no new axioms can be added which apply to the oracle $X$ on any input in column $n$.

One can now give a common definition (in the setting of forcing in arithmetic) of $p \Vdash \psi$ for $p \in \mathcal{P}$ and $\psi$ a sentence of arithmetic with bounded as well as unbounded quantifiers and an added unary relation symbol $\Phi$ (for the generic Turing functional) as well as ones $Y_i$ and $Z_j$ for our fixed choice of representatives of the degrees mentioned in Theorem 1.7.

DEFINITION 3.3. We define $p \Vdash \psi$ by induction on formulas where we view $\forall x$ as an abbreviation for $\neg\exists x \neg$ and consider sentences with leading (unbounded) quantifiers ($\exists$ and $\forall$) and negation symbols followed by a bounded formula.

1. If $\psi$ is $\Delta_0$ (i.e. has only bounded quantifiers) then $p \Vdash \psi$ if and only if $\psi(\Phi_K)$ is true for every filter $K$ on $\mathcal{P}$ containing $p$. (Thus if $\psi$ and $\psi'$ are logically equivalent $\Delta_0$ formulas then $p \Vdash \psi \Leftrightarrow p \Vdash \psi'$.)

2. $p \Vdash \exists n\psi(n)$ if and only if there is an $n \in \omega$ such that $p \Vdash \psi(n)$.
3. $p \Vdash \neg\psi$ if and only if no $q \leq p$ forces $\psi$.

We say that a filter $K$ on $\mathcal{P}$ is 1-*generic* if for every $\Delta_0$ formula $\psi(x)$ there is a $p \in K$ such that either $p \Vdash \exists x\psi(x)$ or $p \Vdash \neg\exists x\psi(x)$ (or equivalently, $p \Vdash \forall x \neg\psi(x)$).

Clause (3) of Definition 3.2 is the key to making our canonical $\Pi_2$ facts false as putting $\langle X, n\rangle$ into $\mathbf{X}_p$ immediately forces $\Phi(X) \upharpoonright \omega^{[n]}$ to be finite. The restriction that $\mathbf{X}_p \cap \mathbf{W}_p = \emptyset$ means that once a pair $\langle W, n\rangle$ has gone into $\mathbf{W}_p$ then it can never go into $\mathbf{X}_p$ and so, if $\Phi$ is sufficiently generic (even 1-generic) $\Phi(W)$ will be total on $\omega^{[n]}$. Thus we see that our canonical $\Pi_2^{X\oplus\Phi}$ facts can be taken to be that $\Phi(X)$ is total on $\omega^{[n]}$. Putting $\langle X, n\rangle$ into $\mathbf{X}_p$ will immediately force this sentence to be false while putting it into $\mathbf{W}_p$ will make it true as long as $K$ is at least 1-generic. In the terminology of the subrequirements $C_{i,n}$ this corresponds to choosing a recursive $h$ such that $h(n) \notin (X \oplus \Phi)'' \Leftrightarrow \forall x \exists y(\Phi(\langle n, x\rangle, X) = y)$ for every $\Phi$ and $X$.

We now list some of the basic facts about this forcing relation. Key among them are that forcing for bounded sentences is recursive (in the set parameters) and that, if $p$ belongs to a 1-generic filter $K$ and $p \Vdash \Theta(\Phi)$, where $\Theta$ is $\Sigma_2$ or $\Pi_2$, then $\Theta(\Phi_K)$ holds. These facts allow us to control the forcing relation when needed and guarantee that forcing the sentences relevant to our requirements makes them true of the functionals associated with the 1-genreic filters that we construct.

REMARK 3.4. If $\psi$ is $\Delta_0$ and $K$ is a filter on $\mathcal{P}$ then we have the following facts:

1. $p \Vdash \exists u\psi(u, \Phi) \Rightarrow \exists q \leq p \exists u[(\Phi_q, \emptyset, \emptyset) \Vdash \psi(u, \Phi)]$ since some initial segment $\Phi_q$ of any $\Phi_K$ with $p \in K$ suffices to guarantee the truth of the $\Delta_0$ sentence. Of course, if $p \in K$ then $\exists x\psi(x, \Phi_K)$ is true. Conversely, if $\exists x\psi(x, \Phi_K)$ is true then $\exists p \in \mathcal{P}(p \Vdash \exists x\psi(x, \Phi))$ for the same reason and the fact that the definition of the forcing order guarantees that once a use $\sigma$ is included in an axiom of $\Phi$ then no axioms (with codes) smaller than $\sigma$ can ever be inserted.
2. $(\Phi_q, \emptyset, \emptyset) \Vdash \psi(u, \Phi)$ is a uniformly recursive relation in $q, u, \psi, \Phi$ and the $Y_i$ and $Z_j$ appearing in $\psi$ as the truth of $\psi(u, \Phi_K)$ depends only on initial segments of $\Phi_K$ (and the $Y_i$ and $Z_j$). We abbreviate this relation as $\Phi_q \Vdash \psi(u, \Phi)$.
3. If $p \Vdash \forall u\psi(u, \Phi)$ then $[(\Phi_p, \mathbf{X}_p, \emptyset) \Vdash \forall u\psi(u, \Phi)]$ as otherwise there would be a $u \in \omega$ and $q \leq (\Phi_p, \mathbf{X}_p, \emptyset)$ such that $q \Vdash \neg\psi(u, \Phi)$ and so a $q' \leq q$ such that $(\Phi_{q'}, \emptyset, \emptyset) \Vdash \neg\psi(u, \Phi)$ but then $(\Phi_{q'}, \mathbf{X}_p, \mathbf{W}_p)$ extends $p$ but also forces $\neg\psi(u, \Phi)$ for a contradiction.
4. If $p \Vdash \forall u\psi(u, \Phi)$ and $p \in K$ then $\forall u\psi(u, \Phi_K)$ as otherwise there would be a $u$ such that $\neg\psi(u, \Phi_K)$ and so a $q \in K$ such that $q \Vdash \neg\psi(u, \Phi)$ contradicting the compatibility required by $p, q \in K$ and the definition

of forcing. Conversely, if $K$ is 1-generic and $\forall u\psi(u, \Phi_K)$ then $\exists p \in K(p \Vdash \forall u\psi(u, \Phi))$ as otherwise there would be a $q \in K$ such that $q \Vdash \exists u\neg\psi(u, \Phi)$ and so $\exists u\neg\psi(u, \Phi_K)$ would hold for a contradiction.

5. If $p \Vdash \exists x\forall y\psi(x, y, \Phi)$ and $p \in K$ then $\exists x\forall y\psi(x, y, \Phi_K)$ by the definition of forcing and (4). Similarly, if $K$ is 1-generic, $p \in K$ and $p \Vdash \forall x\exists y\psi(x, y, \Phi)$ then $\forall x\exists y\psi(x, y, \Phi_K)$ as otherwise there would be an $x$ such that $\neg\exists y\psi(x, y, \Phi_K)$, and so (by 4) a $q \in K$ such that $q \Vdash \neg\exists y\psi(x, y, \Phi)$. Again the compatibility of $p$ and $q$ gives us an $r$ extending both for a contradiction (as $r \leq q$, $r \Vdash \exists x\neg\exists y\psi(x, y, \Phi)$ but as $r \leq p$, this contradicts the assumption that $p \Vdash \forall x\exists y\psi(x, y, \Phi)$.

We will build $G$ and $H$ to be the Turing functionals corresponding to two 1-generic filters $K$ and $L$ for this forcing and to satisfy a specific list of other conditions by building two sequences $p_s, q_s$ such that $\cup\Phi_{p_s} = \Phi_K = G$ and $\cup\Phi_{q_s} = \Phi_L = H$. The requirements that we have to satisfy in addition to those for 1-genericity are ones to guarantee that $C_i$ and $D_j$ are true of $G$ and $H$.

As described above, our plans with our now known list of canonical $\Pi_2$ facts are as follows. To satisfy $C_i$ we make sure that, for almost every $n$, if $n \in (Y_i \oplus G)'$ then we put $\langle Y_i, n\rangle$ into $\mathbf{W}_{q_m}$ for some $m$ while if $n \notin (Y_i \oplus G)'$ we put it into $\mathbf{X}_{q_m}$ for some $m$. In the first case, this makes $h(n) \notin (Y_i \oplus H)''$, indeed $\operatorname{dom}\Phi_L(Y_i) \upharpoonright \omega^{[n]} = \omega^{[n]}$ for any 1-generic $L$ containing $q$. In the second case, this makes $\operatorname{dom}\Phi_L(Y_i) \upharpoonright \omega^{[n]}$ finite and so $h(n) \in (Y_i \oplus H)''$. Thus, if for almost all $n$ we meet the requirements $C_{i,n}$, then $(Y_i \oplus G)' \in \Pi_2^{Y_i \oplus H}$ and $(Y_i \oplus G)' \leq_T (Y_i \oplus H)'$ as required.

To satisfy $D_j$, our plan is, for each $k$, to choose an $n$ such that $n \notin (Z_j \oplus G)'' \Leftrightarrow (G(Z_j) \upharpoonright \omega^{[n]}$ is total) and, at some stage $s$ of the construction, meet the subrequirement $D_{j,k}$ that $\neg[n \notin (Z_j \oplus G)'' \Leftrightarrow \forall u\exists v\theta(f_k(n), u, v, Z_j, H)]$, i.e. $\neg[G(Z_j) \upharpoonright \omega^{[n]}$ is total $\Leftrightarrow \forall u\exists v\theta(f_k(n), u, v, Z_j, H)]$ by arranging that either $q_s \Vdash \exists u\forall v\neg\theta(f_k(n), u, v, Z_j, \Phi)$ while $\langle Z_j, n\rangle \in \mathbf{W}_{p_s}$ which guarantees that $G(Z_j) \upharpoonright \omega^{[n]}$ is total; or $q_s \Vdash \forall u\exists v\theta(f_k(n), u, v, Z_j, \Phi)$ while $\langle Z_j, n\rangle \in \mathbf{X}_{p_s}$ which guarantees that $G(Z_j) \upharpoonright \omega^{[n]}$ is finite. In either case we violate the biconditional equivalent to $(Z_j \oplus G)' \leq_T (Z_j \oplus H)'$. Meeting $D_{j,k}$ for all $k$ guarantees that $(Z_j \oplus G)' \not\leq_T (Z_j \oplus H)$ as required. We now give the details of the construction.

**Construction:** We make an $\omega$-list of the subrequirements $C_{i,n}, D_{j,k}$ described above as well as others $P_m$ to decide the one quantifier sentences $\exists w\psi_m$ about $G$ and $H$. We begin at stage 0 with the empty conditions $p_0 = q_0 = \langle\emptyset, \emptyset, \emptyset\rangle$. Suppose we are at stage $s+1$ with $p_s, q_s$ defined. We let $t_i$ be the least $s$ at which $Y_i$ is mentioned in one of the conditions, i.e. $\langle Y_i, m\rangle \in \mathbf{X}_{p_s} \cup \mathbf{W}_{p_s} \cup \mathbf{X}_{q_s} \cup \mathbf{W}_{q_s}$ for some $m$. Our action depends on the requirement assigned to $s$, i.e. the $s$th one on our list.

$P_m$: If there is a finite $\Phi' \supseteq \Phi_{p_s}$ (consistent with $p_s$, i.e. $(\Phi', \mathbf{X}_{p_s}, \mathbf{W}_{p_s}) \leq p_s$) and a $w$ such that $\langle\Phi', \mathbf{X}_{p_s}, \mathbf{W}_{p_s}\rangle \Vdash \psi_m(w, \Phi)$ then we choose one and set

$p_{s+1} = \langle \Phi', \mathbf{X}_{p_s}, \mathbf{W}_{p_s} \rangle$. If not we let $p_{s+1} = p_s$. We then do the same for $H$ to define $q_{s+1}$ from $q_s$.

$C_{i,n}$: If $\langle Y_i, n \rangle$ is already on either list in $q_s$ let $p_{s+1} = p_s$ and $q_{s+1} = q_s$. Otherwise, ask if there is a finite $\Phi' \supseteq \Phi_{p_s}$ (consistent with $p_s$) such that $\langle \Phi', \mathbf{X}_{p_s}, \mathbf{W}_{p_s} \rangle \Vdash n \in (Y_i \oplus \Phi)'$. If so, choose one and let $p_{s+1} = \langle \Phi', \mathbf{X}_{p_s}, \mathbf{W}_{p_s} \rangle$ and $q_{s+1} = \langle q_s, \mathbf{X}_{q_s}, \mathbf{W}_{q_s} \cup \{\langle Y_i, n \rangle\} \rangle$. If not, let $p_{s+1} = p_s$ and $q_{s+1} = \langle \Phi_{q_s}, \mathbf{X}_{q_s} \cup \{\langle Y_i, n \rangle\}, \mathbf{W}_{q_s} \rangle$.

$D_{j,k}$: Choose an $n$ such that $\langle Z_j, n \rangle$ does not appear on either list in $p_s$. Ask if there is a $u$ and $q' \leq q_s$ (which by Remark 3.4 can be to assumed to have $\mathbf{W}_{q'} = \mathbf{W}_{q_s}$) such that $q' \Vdash \forall v \neg \theta(f_k(n), u, v, Z_j, \Phi)$ and a $p' \leq p_s$ (which again by Remark 3.4 can be assumed to have $\mathbf{W}_{p'} = \mathbf{W}_{p_s}$) so that for any $m$ and any $t_i \leq s$ such that $Y_i \leq_T Z_j$ and $\langle Y_i, m \rangle \in \mathbf{X}_{q'} - \mathbf{X}_{q_{t_i}}$ we have that $p' \Vdash m \notin (Y_i \oplus \Phi)'$.

If not ($\Pi_2$ outcome), then we let $p_{s+1} = \langle \Phi_{p_s}, \mathbf{X}_{p_s} \cup \{\langle Z_j, n \rangle\}, \mathbf{W}_{p_s} \rangle$ and $q_{s+1} = q_s$.

If there are such $u, q'$ and $p'$ ($\Sigma_2$ outcome) then we claim (and will verify in Lemma 3.8 below) that we can choose such a $q'$ with no $\langle Y_i, l \rangle \in \mathbf{X}_{q'} - \mathbf{X}_{q_s}$ with $Y_i \not\leq_T Z_j$ and a $p'$ for which $\langle Z_j, n \rangle \notin \mathbf{X}_{p'}$. We choose such and set $p_{s+1} = \langle \Phi_{p'}, \mathbf{X}_{p'}, \mathbf{W}_{p'} \cup \{\langle Z_j, n \rangle\} \rangle$ and $q_{s+1} = q'$.

**Verifications:** It is clear that there are no difficulties carrying out these instructions when the requirement being considered is $P_m$ or $C_{i,n}$. Moreover, it is also clear that, in the case of $P_m$, we have guaranteed that the $\Sigma_1$ formula is forced at $s + 1$ if possible and otherwise no extension forces it. Thus $K$ and $L$ are 1-generic. Similarly, in the case of $C_{i,n}$, as long as from some stage $t$ onward no actions other than for $C_{i,m}$ put any $\langle Y_i, m \rangle$ on any list in $q_s$ without our forcing the corresponding outcome for $m \in (Y_i \oplus G)'$, then we satisfy $C_i$. In fact, our actions guarantee a bit more. If $\langle Y_i, m \rangle$ is put into $\mathbf{X}_{q_s}$ at $s \geq t_i$ (by $C_{i,m}$ or some $D_{j,k}$ acting for its $\Sigma_2$ outcome as described above) then $p_{s+1} \Vdash m \notin (Y_i \oplus \Phi)'$ as required; and no action other than for $C_{i,m}$ can put $\langle Y_i, m \rangle$ into $\mathbf{W}_{q_s}$ for $s \geq t_i$. Thus we are left with analyzing the way we decide the $\Pi_2$ questions for $D_{j,k}$ at $s_{j,k}$.

The simpler case is the $\Pi_2$ outcome. Here we only have to show that, for each $u$, when we consider the requirement $P_m$ for the formula $\exists v \theta(f_k(n), u, v, Z_j, \Phi)$ for $H$ at $s > s_{j,k}$ that we force the $\Sigma_1$ outcome. If not, then by our construction, $q_s \Vdash \forall v \neg \theta(f_k(n), u, v, Z_j, \Phi)$ and $q_s \leq q_{s_{j,k}}$. We now claim that if $\langle Y_i, m \rangle \in \mathbf{X}_{q_s} - \mathbf{X}_{q_{t_i}}$ then $p_s \Vdash m \notin (Y_i \oplus \Phi)'$ which would put us in the $\Sigma_2$ outcome for a contradiction. To verify the claim consider how $\langle Y_i, m \rangle$ could have entered $\mathbf{X}_{q_s}$. No action for $P_m$ adds anything to either list. Action for $C_{h,l}$ adds at most $\langle Y_h, l \rangle$ to the $\mathbf{X}$ list but only when it already forces $l \notin (Y_h \oplus \Phi)'$ (for $G$) as required. Finally, by the claim in the $\Sigma_2$ case of the construction (Lemma 3.8 below) action for some $D_{h,l}$ at a stage $t$ after $t_i$ (and before $s$) puts $\langle Y_i, m \rangle$ into $\mathbf{X}_{q_s}$ only when it also makes $p_{t+1} \Vdash m \notin (Y_i \oplus \Phi)'$ as required.

The $\Sigma_2$ case requires a deeper analysis of the forcing relation along the lines of Shore and Slaman [1999] albeit somewhat more elaborate combinatorially. The idea here (as in Shore and Slaman [1999]) is that if there is a $q' \leq q$ such that $q' \Vdash \forall v \neg\theta(f_k(n), u, v, Z_j, \Phi)$ then, first, it can be taken to be of the form $(\Phi', \mathbf{X}', \mathbf{W}_q)$. Next, by the basic properties of forcing, no extension $q'' = (\Phi'', \mathbf{X}'', \mathbf{W}_q)$ of $q'$ forces $\exists v\theta(f_k(n), u, v, Z_j, \Phi)$ so, if for some $m$, some $\Phi'' \supseteq \Phi'$ makes $\theta(f_k(n), u, m, Z_j, H)$ true then it must be incompatible with the conditions imposed by $\mathbf{X}'$, i.e. some axiom in $\Phi'' - \Phi'$ with input in column $e$ applies to an $X_i$ with $\langle X_i, e\rangle \in \mathbf{X}'$. We can form a tree of approximations to such an $\mathbf{X}'$ by considering those sequences $\langle \sigma_l, s_l\rangle$ such that any such $\Phi''$ has an axiom compatible with one of the $\langle \sigma_l, s_l\rangle$. Now we must also associate with any such $q'$ a condition $p' \leq p$ that forces $m \notin (Y_i \oplus \Phi)'$ for the $i$ and $m$ that we worry about in the definition of the $\Sigma_2$ outcome. This adds one more layer of approximations to produce $p'$ from a path in the appropriate tree along with $q'$. We are thus led to the following definitions.

DEFINITION 3.5. Suppose we are given conditions $p$ and $q$ and an instance of one of our formulas $\forall v \neg\theta(a, u, v, Z, \Phi)$ where we are thinking about $H$ as the interpretation of $\Phi$ and have written $a$ for $f_k(n)$ and $Z$ for $Z_j$. We say that $i$ *is crucial* (*for $p$ and $q$*) if $Y_i \leq_T Z$ and $Y_i$ is on some list in $p$ or $q$. Let $t$ be larger than all numbers mentioned in $p$ or $q$ (either in one of the lists or one of the Turing functionals) and such that $U \upharpoonright t \neq V \upharpoonright t$ for any sets $U \neq V$ appearing on any list in $p$ or $q$. For each $m \geq t$ and finite Turing functionals $\Gamma$ and $\Theta$ such that $\langle \Gamma, \mathbf{X}_p, \mathbf{W}_p\rangle \leq p$ and $\langle \Theta, \mathbf{X}_q, \mathbf{W}_q\rangle \leq q$ we define the tree $T_m(\Gamma, \Theta, p, q) = T$ whose nodes are sequences $\langle\langle \sigma_l, s_l\rangle, \vec{\tau}_l\rangle_{l \leq m}$ (coded as numbers in some fixed way as are all finite sets and sequences) where each $\vec{\tau}_l$ is itself a sequence $\langle \tau_{l,e}, t_{l,e}\rangle_{e \leq m}$ with the following properties:

1. $|\sigma_l|, |\tau_{l,e}| \geq m$ and all of these lengths are the same; $\langle W, s_l\rangle \in \mathbf{W}_q \Rightarrow \sigma_l \upharpoonright m \not\subseteq W \upharpoonright m$; $\langle W, t_{l,e}\rangle \in \mathbf{W}_p \Rightarrow \tau_{l,e} \upharpoonright m \not\subseteq W \upharpoonright m$.
2. If $\Theta' \supseteq \Theta$ with the code for $\Theta'$ less than the one for the node and $\Theta' \Vdash \theta(a, u, v, Z, \Phi)$ for some $v$ also less than the code for the node, then $\Theta' - \Theta$ contains an axiom $(\langle s, z\rangle, y, \sigma)$ such that, for some $l \leq m, s = s_l$ and $\sigma$ is compatible with $\sigma_l$.
3. If $\sigma_l \subseteq Y_i$ for some crucial $i$, $\Gamma' \supseteq \Gamma$ with the code for $\Gamma'$ less than the one for the node and $\Gamma' \Vdash s_l \in (Y_i \oplus \Phi)'$ with a witness for convergence also less than the code for the node then $\Gamma' - \Gamma$ contains an axiom $(\langle t, z\rangle, y, \tau)$ such that, for some $e \leq m, t = t_{l,e}$ and $\tau$ is compatible with $\tau_{l,e}$.

We order the nodes of $T$ in the expected way: $\langle\langle \sigma'_l, s'_l\rangle, \vec{\tau}'_l\rangle \preceq_T \langle\langle \sigma_l, s_l\rangle, \vec{\tau}_l\rangle \Leftrightarrow (\forall l \leq m)(\sigma'_l \subseteq \sigma_l \,\&\, s_l = s'_l \,\&\, (\forall e \leq m)(\tau'_{l,e} \subseteq \tau_{l,e} \,\&\, t_{l,e} = t'_{l,e}))$.

LEMMA 3.6. *$T$ is a finitely branching tree recursive in $Z$.*

PROOF. $T$ is obviously finitely branching since the $\sigma_l$ and $\tau_{l,e}$ are binary sequences. As $m$ is fixed the information needed in (1) is finite so it is a recursive condition. Since forcing a $\Delta_0$ formula is recursive (in the set parameters), (2) is recursive in $Z$. The list of $Y_i$ mentioned in $p$ or $q$ and recursive in $Z$ is finite and $Z$ can tell if $\sigma_l$ is an initial segment of one of these $Y_i$ while the rest of the condition is recursive given our bound on the witness for convergence in $s_l \in (Y_i \oplus \Phi)'$. ⊣

LEMMA 3.7. *If there are $q' = (\Theta, \mathbf{X}_{q'}, \mathbf{W}_q)$ and $p' = (\Gamma, \mathbf{X}_{p'}, \mathbf{W}_p)$ as required in the definition of the $\Sigma_2$ outcome for $D_{j,k}$ then for some (large enough) $m$, $T_m(\Gamma, \Theta, p, q) = T$ has an infinite path.*

PROOF. We claim that if $\mathbf{X}_{q'} = \{\langle S_l, s_l\rangle \mid l \leq m\}$ and $\mathbf{X}_{p'} = \{\langle T_e, t_e\rangle \mid e \leq m\}$ (we allow duplications to keep the indexing the same and $m \geq t$) then $\{\langle\langle S_l \restriction n, s_l\rangle_{l\leq m}, \vec{\tau}_{l,n}\rangle \mid n > m\}$ is a path in $T$ where $\vec{\tau}_{l,n} = \langle T_e \restriction n, t_e\rangle_{e\leq m}$. Here we have chosen $m$ larger than $t$ and the cardinalities of $\mathbf{X}_{q'}$ and $\mathbf{X}_{p'}$ as well as to insure that if $U \neq V$ are mentioned in $p'$ or $q'$ then $U \restriction m \neq V \restriction m$. Note that if $\langle X, n\rangle \in \mathbf{X}_{q'} - \mathbf{X}_q$ and $\langle W, n\rangle \in \mathbf{W}_q$ then $X \neq W$ by the definition of the forcing ordering and so $X \restriction m \neq W \restriction m$ and similarly for $p$. To see that these nodes are on $T$ check that each condition is satisfied for $\langle\langle S_l \restriction n, s_l\rangle_{l\leq m}, \vec{\tau}_{l,n}\rangle$:

1. The lengths are all larger than $m$ by definition. The restrictions associated with $\mathbf{W}_q$ and $\mathbf{W}_p$ are satisfied by our choice of $m$ and the note above.

2. If $\Theta' \supseteq \Theta$ and $\Theta' \Vdash \theta(a, u, v, Z, \Phi)$ for some $v$ then $\Theta'$ is not consistent with $q'$ and so $\Theta' - \Theta$ must contain an axiom $(\langle s, z\rangle, y, \sigma)$ such that there is an $l \leq m$ such that $s_l = s$ and $\sigma \subseteq S_l$. Clearly this $\sigma$ is compatible with $S_l \restriction n$.

3. If $S_l \restriction n \subseteq Y_i$ and $i$ is crucial (and so, in particular, $S_l = Y_i$ by our choice of $m$), $Y_i \leq_T Z$, $\Gamma' \supseteq \Gamma$ and $\Gamma' \Vdash s_l \in (Y_i \oplus \Phi)'$ then $\Gamma'$ is not consistent with $p'$ since by our requirements when we acted for $D_{j,k}$, $p' \Vdash s_l \notin (S_l \oplus \Phi)'$ and so $\Gamma' - \Gamma$ must contain an axiom $(\langle t, z\rangle, y, \tau)$ such that there is an $e \leq m$ such that $t = t_e$ and $\tau \subseteq T_e$. Clearly this $\tau$ is compatible with $T_e \restriction n$. ⊣

LEMMA 3.8. *If there is an infinite path $P = \{\langle\langle\sigma_{n,l}, s_l\rangle, \vec{\tau}_{n,l}\rangle\rangle_{l\leq m} \mid n \in \omega\}$ in some $T_m(\Gamma, \Theta, p, q) = T$ then there are $q'$ and $p'$ as required by the claim in the implementation of the $\Sigma_2$ outcome for $D_{j,k}$.*

PROOF. Any infinite path clearly produces sets $S_l = \bigcup\{\sigma_{n,l} \mid n \in \omega\}$ for $l \leq m$ and $R_{l,e} = \bigcup\{\tau_{n,l,e} \mid n \in \omega\}$ for $e \leq m$ if $S_l = Y_i$ some crucial $i$. As $T$ is recursive in $Z$ and has an infinite path we can choose one that does not compute any $Y_i \not\leq_T Z$ by Jockusch and Soare [1972] and so we may assume that no $S_l = Y_i$ for any crucial $i$. We let $q' = (\Theta, \mathbf{X}_q \cup \{\langle S_l, s_l\rangle \mid l \leq m\}, \mathbf{W}_q)$ which is clearly a condition since no $\langle S_l, s_l\rangle \in \mathbf{W}_q$ by clause (1) of the definition of $T$. We first argue that $q' \Vdash \forall v \neg\theta(a, u, v, Z, H)$. If not, then by Remark 3.4, there would be a $(\Theta', \mathbf{X}_{q'}, \mathbf{W}_{q'}) \leq q'$ and a $v$ such that $\Theta' \Vdash \theta(a, u, v, Z, \Phi)$. By clause (2) of the definition of $T$ there would then be an axiom $(\langle s, z\rangle, y, \sigma) \in \Theta' - \Theta$ and an $l \leq m$ such that $s = s_l$ and $\sigma$ is compatible with $\sigma_{n,l} \subseteq S_l$ for infinitely many $n$ and so $\sigma \subseteq S_l$. As this contradicts the definition of $(\Theta', \mathbf{X}_{q'}, \mathbf{W}_{q'}) \leq q'$, we have that $q' \Vdash \forall v \neg\theta(a, u, v, Z, \Phi)$ and there is no $\langle Y_i, h\rangle \in \mathbf{X}_{q'} - \mathbf{X}_q$ with $Y_i \leq_T Z$ as required.

We next argue that $\bar{p} = (\Gamma, \mathbf{X}_p \cup \{R_{l,e} \mid S_l = Y_i \text{ some crucial } i \text{ and } e \leq m\}, \mathbf{W}_p) \Vdash s_l \notin (S_l \oplus \Phi)'$ for every $\langle S_l, s_l\rangle \in \mathbf{X}_{q'} - \mathbf{X}_{q_{t_i}}$ with $S_l = Y_i$ for some crucial $i$. Suppose $S_l = Y_i$ for some crucial $i$ but $\bar{p} \nVdash s_l \notin (S_l \oplus \Phi)'$. In this case, there is a $\Gamma' \supseteq \Gamma$ consistent with $\bar{p}$ such that $\Gamma' \Vdash s_l \in (S_l \oplus \Phi)'$. By clause (3) of the definition of $T$, there must be a $(\langle t, z\rangle, y, \tau) \in \Gamma' - \Gamma$ such that, for some $e \leq m$, $t = t_{l,e}$ and $\tau$ is compatible with $\tau_{n,l,e}$ for infinitely many $n$. Thus $\tau \subseteq R_{l,e} \in \mathbf{X}_{\bar{p}}$ and so $\Gamma'$ is not consistent with $\bar{p}$ for the desired contradiction.

Finally, we argue that we can find a $p' \leq p$ such that $\langle Z, n\rangle \notin \mathbf{X}_{p'}$ and $p' \Vdash s_l \notin (S_l \oplus \Phi)'$ for all the $l$'s for which $S_l = Y_i$ for a crucial $i$ and $\langle S_l, s_l\rangle \notin \mathbf{X}_{q_{t_i}}$ which will complete the proof. For each such $l$ we consider the tree $T_{m,l}$ whose nodes are sequences $\langle \tau_e, t_e\rangle_{e \leq m}$ such that $(\forall e, e' \leq m)$ $(|\tau_e| = |\tau_{e'}| > m \,\&\, (\langle W, t_e\rangle \in \mathbf{W}_p \rightarrow \tau_e \restriction m \neq W \restriction m))$ and $(\forall \Gamma' \supseteq \Gamma)$ (if the code for $\Gamma'$ is less than the one for $\vec{\tau}$ and $\Gamma' \Vdash s_l \in (Y_i \oplus \Phi)'$ with a witness for convergence also less than the code for $\vec{\tau}$ then $\Gamma' - \Gamma$ contains an axiom $(\langle t, z\rangle, y, \tau)$ such that, for some $e \leq m$, $t = t_e$ and $\tau$ is compatible with $\tau_e$). We order $T_{m\cdot l}$ by $\langle \tau'_e, t'_e\rangle_{e \leq m} \preceq_{T_{m,l}} \langle \tau_e, t_e\rangle_{e \leq m} \Leftrightarrow (\forall e \leq m)(t'_e = t_e \,\&\, \tau'_e \subseteq \tau_e)$.

This tree is finitely branching and recursive in $Y_i$. Our original sequence $\langle \vec{\tau}_{n,l} \mid n \in \omega\rangle$ is a path in $T_{m,l}$ and so there is one that does not compute $Z \nleq_T Y_i$. (Note here that as $i$ is crucial $Y_i \leq_T Z$ but by the assumption of the Theorem they have different degree.) Let $\langle \vec{\tau}'_{n,l} \mid n \in \omega\rangle$ be such a path for each $l$ as required and let $R'_{l,e} = \cup\{\tau'_{n,l,e} \mid n \in \omega\}$. As before we know that $p_l = (\Gamma, \mathbf{X}_p \cup \{\langle R_{l,e}, t'_{l,e}\rangle \mid e \leq m\}, \mathbf{W}_p) \Vdash s_l \notin (S_l \oplus \Phi)'$ while no $R'_{l,e}$ is $Z$. We can now get the required $p'$ as $(\Gamma, \mathbf{X}_p \cup \{\langle R'_{l,e}, t'_{l,e}\rangle \mid S_l = Y_i$ for some crucial $i$ and $\langle S_l, s_l\rangle \notin \mathbf{X}_{q_{t_i}} \,\&\, e \leq m\}, \mathbf{W}_p)$. As $p'$ extends every $p_l$ for $l$ on our list, $p' \Vdash s_l \notin (S_l \oplus \Phi)'$ for all these $l$'s while $\langle Z, n\rangle \notin \mathbf{X}_p$ by our choice of $n$ and no $R_{l,e} = Z$, $p'$ is as required to implement the $\Sigma_2$ outcome for $D_{j,k}$. ⊣

**§4. Conclusions and questions.** We summarize much of the current state of affairs about the dividing line between decidability and undecidability in the structures $\mathcal{R}$, $\mathcal{D}$ and $\mathcal{D}(\leq 0')$ in various languages in the following tables.

| | $\mathcal{R}$ | $\mathcal{D}$ | $\mathcal{D}(\leq \mathbf{0}')$ |
|---|---|---|---|
| $\exists(\leq, \vee)$ | Dec | Dec | Dec |
| $\forall\exists(\leq, \vee)$ | ? | Dec | ? |
| $\forall\exists\forall(\leq, \vee)$ | Undec | Undec | Undec |

| | $\mathcal{R}$ | $\mathcal{D}$ | $\mathcal{D}(\leq \mathbf{0}')$ |
|---|---|---|---|
| $\exists(\leq, \vee, \wedge)$ | ? | Dec | Dec |
| $\forall\exists(\leq, \vee, \wedge)$ | Undec | Undec | Undec |

| | $\mathcal{D}$ |
|---|---|
| $\exists(\leq, \vee, ')$ | Dec |
| $\forall\exists(\leq, \vee, ')$ | Undec |

The first and third tables are straightforward. The second requires some explanation of the use of $\wedge$. We understand decidability in the $\exists$ line to mean that every finite lattice is embeddable, indeed this is true even if we preserve 0 and 1 (if it exists). (For $\mathcal{R}$ the question of which lattices are embeddable even without preserving 0 or 1 is, of course, still open.) Undecidability in the $\forall\exists$ line means that the theory is undecidable (via the same set of sentences) for every total extension of the partial infimum relation.

In addition to the long standing and well known open questions about $\mathcal{R}$, these tables suggest two other natural areas for investigation on the boundary between decidability and undecidability. The first is the $\forall\exists$ theory of $\mathcal{D}(\leq, ')$. This is a very rich theory that includes many interesting and difficult subproblems and a number of partial results. For example, it includes the $\exists$ theory of $\mathcal{D}(\leq, \vee, ')$ shown decidable by Montalban [2003]. It is also possible to define 0 at this level without any added complexity and so it is equivalent to the $\forall\exists$ theory of $\mathcal{D}(0, \leq, ')$ which includes the $\exists$ theory of $\mathcal{D}(0, \leq, \vee, ')$. Now, for all the other decidability results in these tables which do not involve the jump operator, adding 0 to the language presents no serious extra difficulties. This is far from true once we allow the jump. Indeed even the $\exists$ theory of $\mathcal{D}(0, \leq, ')$ is a complex problem about which there are many difficult and interesting partial results. The crucial point is that the addition of **0** requires very fine control over the complexity of the degrees realizing some given $\exists$ statement in the language with just $\leq$ and $'$. The formula can now specify that witnesses lie in precise intervals with endpoints of the form $\mathbf{0}^{(n)}$. In contrast, the realizations in the decidability results for the $\exists$ theory of $\mathcal{D}(\leq, ')$ and $\mathcal{D}(\leq, \vee, ')$ (Hinman and Slaman [1991]; Montalban [2003]) produce witnesses which are not even hyperarithmetic.

Among the partial results on the $\exists$ theory with 0 we mention Lerman [1985] which proves the decidability of the $\exists$ theory of $\mathcal{D}$ with $\leq$ and additional predicates for the classes in the High/Low hierarchy. Montalban [2004] (answering a question of Lerman [1985]) proves the decidability of $\exists$ theory of $\mathcal{D}$ with $\leq, 0$ and additional predicates for the classes $\mathrm{GL}_n$, $\mathrm{GH}_n$ and GI (for $n \geq 0$) in the generalized high/low hierarchies (which also falls within the $\exists$ theory of $\mathcal{D}(0, \leq, \vee, '))$. Lempp and Lerman [1996] prove the $\exists$ theory of $\mathcal{R}$ with $0, 1, \leq$ and predicates for $\mathbf{x}^{(n)} \leq_T \mathbf{y}^{(n)}$ for every $n$ is decidable. And so, a fortiori, also the $\exists$ theory of $\mathcal{D}$ with $0, \leq$ and all of these predicates. Indeed, they can even add $\vee$ to the language as long as 1 ($0'$) is omitted. It seems plausible that their methods (which are very complex) may suffice to prove the decidability of the $\exists$ theory of $\mathcal{D}(0, \leq, ')$ and perhaps even with $\vee$ added to the language[1].

As for the full $\forall\exists$ theory of $\mathcal{D}(\leq, ')$, there are many difficult open problems along the road to decidability (including, for example, controlling the jumps

[1]Lerman (personal communication) has recently announced a proof of this result (with $\vee$) along these lines.

of initial segments). On the other hand, the path to undecidability seems quite dark. It is not immediately blocked by our general comments in the introduction on proving undecidability of $\forall\exists$ theories since we do have a function symbol $'$ that generates $\omega$-sequences. The problem is how to define them (and relations on them) in a $\Delta_0$ way using only $\leq$ and $'$.

The new area suggested by these charts is the decision problem for the $\forall\exists$ theory of $\mathcal{D}(\leq \mathbf{0}')$ with $\leq$ and $\vee$. This problem turns out to be far from straightforward. It would seem that our only hope is to prove decidability. We have the required initial segment results for decidability from the proof of decidability without $\vee$ (Lerman and Shore [1988]). The extension of embeddings part, however, has, surprisingly, something of the flavor of the $\forall\exists$ theory of $\mathcal{R}$.

Montalban has shown that we cannot reduce the full $\forall\exists$ decision problem to the extension of embedding problem as has been done in all the other successful proofs of decidability at the $\forall\exists$ level. (The extension of embedding problem for a structure $\mathcal{M}$ asks for a characterization of the partial orders $\mathcal{X} \subseteq \mathcal{Y}$ such that every embedding $f : \mathcal{X} \to \mathcal{M}$ can be extended to one $g : \mathcal{Y} \to \mathcal{M}$. The decision problem for $\forall\exists$ sentences about one of the degree structures is equivalent to deciding all problems of the form "every embedding of $\mathcal{X}$ can be extended to an embedding of some $\mathcal{Y}_i$ from a specified list." A reduction argument shows that the list can always be taken to have only one element.)

Proposition 4.1 (Montalban). *For every* $\mathbf{x}_1 < \mathbf{x}_2$ *in* $(\mathbf{0}, \mathbf{0}')$ *there is either a* $\mathbf{y}$ *such that* $\mathbf{0} < \mathbf{y} < \mathbf{x}_1$ *or one such that* $\mathbf{x}_1 < \mathbf{y} < \mathbf{1}$ *and* $\mathbf{x}_2 \vee \mathbf{y} = \mathbf{1}$ *but neither disjunct holds for every* $\mathbf{x}_1 < \mathbf{x}_2$ *in* $(\mathbf{0}, \mathbf{0}')$.

Proof. If $\mathbf{x}_1 \notin \mathbf{L}_2$ then it is not minimal and so there is a nonrecursive $\mathbf{y} < \mathbf{x}_1$. On the other hand, if $\mathbf{x}_1 \in \mathbf{L}_2$ then $\mathbf{0}'$ is high over $\mathbf{x}_1$ and so by the join theorem for high degrees (Posner [1977] or see Lerman [1983, IV.9]) there is a $\mathbf{y}$ which joins $\mathbf{x}_2$ up to $\mathbf{0}'$. However, there are counterexamples to each disjunct.

Of course, any minimal degree $\mathbf{x}_1 < \mathbf{0}'$ and any $\mathbf{x}_2$ between $\mathbf{x}_1$ and $0'$ supply a counterexample to the first disjunct. On the other hand, if we consider the construction of Slaman and Steel [1989] of nonrecursive r.e. degrees $\mathbf{a} < \mathbf{b}$ such that no degree $\mathbf{c} < \mathbf{b}$ joins $\mathbf{a}$ to $\mathbf{b}$ and apply the pseudojump inversion theorem of Jockusch and Shore [1983], we get $\mathbf{x}_1 < \mathbf{x}_2 < \mathbf{0}'$ (replacing $\mathbf{0} < \mathbf{a} < \mathbf{b}$) such that no $\mathbf{y}$ below $\mathbf{0}'$ and above $\mathbf{x}_1$ joins $\mathbf{x}_2$ up to $\mathbf{0}'$ as required. ⊣

This is reminiscent of the nondiamond (and other) phenomenon in $\mathcal{R}$. (The nondiamond theorem of Lachlan [1966] says that for every pair of incomparable r.e. degrees $\mathbf{x}_1, \mathbf{x}_2$ there is either a $\mathbf{y}$ with $\mathbf{x}_1, \mathbf{x}_2 \leq_T \mathbf{y} <_T \mathbf{0}'$ or one with $0 <_T \mathbf{y} \leq_T \mathbf{x}_1, \mathbf{x}_2$. On the other hand, there are $\mathbf{x}_1, \mathbf{x}_2$ such that $\mathbf{x}_1 \vee \mathbf{x}_2 = \mathbf{0}'$ and ones such that $\mathbf{x}_1 \wedge \mathbf{x}_2 = \mathbf{0}$.) It suggests that there are many problems to consider here.

## REFERENCES


M. M. Arslanov and S. Lempp (editors) [1999], ***Recursion theory and complexity, proceedings of the Kazan '97 workshop***, Walter de Gruyter, Berlin.

P. Bernays and M. Schönfinkel [1928], *Zum Entscheidungsproblem der mathematischen Logik*, ***Mathematische Annalen***, vol. 99, pp. 342–372.

E. Börger, E. Grädel, and Y. Gurevich [1997], ***The classical decision problem***, Springer, Berlin.

R. L. Epstein [1979], ***Degrees of unsolvability: Structure and theory***, Lecture Notes in Mathematics, vol. 759, Springer-Verlag, Berlin.

R. M. Friedberg [1957], *Two recursively enumerable sets of incomparable degrees of unsolvability*, ***Proceedings of the National Academy of Sciences***, vol. 43, pp. 236–238.

L. Harrington and S. Shelah [1982], *The undecidability of the recursively enumerable degrees (research announcement)*, ***American Mathematical Society. Bulletin. New Series***, vol. 6, pp. 79–80.

P. G. Hinman and T. A. Slaman [1991], *Jump embeddings in the Turing degrees*, ***The Journal of Symbolic Logic***, vol. 56, pp. 563–591.

C. G. Jockusch, Jr. and R. A. Shore [1983], *Pseudojump operators. I. The r.e. case*, ***Transactions of the American Mathematical Society***, vol. 275, pp. 599–609.

C. G. Jockusch, Jr. and T. A. Slaman [1993], *On the $\Sigma_2$-theory of the upper semilattice of Turing degrees*, ***The Journal of Symbolic Logic***, vol. 58, pp. 193–204.

C. G. Jockusch, Jr. and R. I. Soare [1972], *$\Pi^0_1$ classes and degrees of theories*, ***Transactions of the American Mathematical Society***, vol. 173, pp. 33–56.

S. C. Kleene and E. L. Post [1954], *The upper semi-lattice of degrees of recursive unsolvability*, ***Annals of Mathematics. Second Series***, vol. 59, pp. 379–407.

A. H. Lachlan [1966], *The impossibility of finding relative complements for recursively enumerable degrees*, ***The Journal of Symbolic Logic***, vol. 31, pp. 434–454.

A. H. Lachlan [1968], *Distributive initial segments of the degrees of unsolvability*, ***Zeitschrift für Mathematische Logik und Grundlagen der Mathematik***, vol. 14, pp. 457–472.

S. Lempp and M. Lerman [1996], *The decidability of the existential theory of the poset of recursively enumerable degrees with jump relations*, ***Advances in Mathematics***, vol. 120, pp. 1–142.

S. Lempp, A. Nies, and T. A. Slaman [1998], *The $\Pi_3$-theory of the computably enumerable Turing degrees is undecidable*, ***Transactions of the American Mathematical Society***, vol. 350, pp. 2719–2736.

M. Lerman [1971], *Initial segments of the degrees of unsolvability*, ***Annals of Mathematics. Second Series***, vol. 93, pp. 365–389.

M. Lerman [1983], ***Degrees of unsolvability***, Springer-Verlag, Berlin.

M. Lerman [1985], *On the ordering of classes in high/low hierarchies*, ***Recursion theory week*** *(Oberwolfach, 1984)* (H.-D. Ebbinghaus, G. H. Müller, and G. E. Sacks, editors), Lecture Notes in Mathematics, vol. 1141, Springer, Berlin, pp. 260–270.

M. Lerman [2000], *A necessary and sufficient condition for embedding principally decomposable finite lattices into the computably enumerable degrees*, ***Annals of Pure and Applied Logic***, vol. 101, pp. 275–297.

M. Lerman and R. A. Shore [1988], *Decidability and invariant classes for degree structures*, ***Transactions of the American Mathematical Society***, vol. 310, pp. 669–692.

M. Lerman, R. A. Shore, and R. I. Soare [1984], *The elementary theory of the recursively enumerable degrees is not $\aleph_0$-categorical*, ***Advances in Mathematics***, vol. 53, pp. 301–320.

R. G. Miller, A. O. Nies, and R. A. Shore [2004], *The ∀∃-theory of $\mathcal{R}(\leq, \vee, \wedge)$ is undecidable*, ***Transactions of the American Mathematical Society***, vol. 356, pp. 3025–3067.

M. L. Minsky [1961], *Recursive unsolvability of Post's problem of tag and other topics in the theory of Turing machines*, ***Annals of Mathematics***, vol. 74, pp. 437–454.

A. MONTALBAN [2003], *Embedding jump upper semilattices into the Turing degrees*, ***The Journal of Symbolic Logic***, vol. 68, pp. 989–1014.

A. MONTALBAN [2004], *There is no ordering on the classes in the generalized high/low hierarchy*, ***Archive for Mathematical Logic***, to appear.

A. A. MUCHNIK [1956], *On the unsolvability of the problem of reducibility in the theory of algorithms*, ***Doklady Akademii Nauk SSSR. New Series***, vol. 108, pp. 29–32.

A. NERODE AND R. A. SHORE [1997], ***Logic for applications***, 2nd ed., Springer, New York.

A. NIES, R. A. SHORE, AND T. A. SLAMAN [1998], *Interpretability and definability in the recursively enumerable degrees*, ***Proceedings of the London Mathematical Society. Third Series***, vol. 77, pp. 241–291.

D. POSNER [1977], ***High degrees***, Ph.D. thesis, Univeristy of California, Berkeley.

F. P. RAMSEY [1930], *On a problem of formal logic*, ***Proceedings of the London Mathematical Society. Second Series***, vol. 30, pp. 338–384.

G. E. SACKS [1963], *Recursive enumerability and the jump operator*, ***Transactions of the American Mathematical Society***, vol. 108, pp. 223–239.

G. E. SACKS [1964], *The recursively enumerable degrees are dense*, ***Annals of Mathematics. Second Series***, vol. 80, pp. 300–312.

J. SHEPHERDSON AND H. STURGIS [1963], *Computability of recursive functions*, ***Journal of the Association for Computing Machinery***, vol. 10, pp. 217–255.

R. A. SHORE [1978], *On the $\forall\exists$-sentences of $\alpha$-recursion theory*, ***Generalized recursion theory II*** (J. E. Fenstad, R. O. Gandy, and G. E. Sacks, editors), Studies in Logic and the Foundations of Mathematics, vol. 94, North-Holland, Amsterdam, pp. 331–354.

R. A. SHORE [1981], *The theory of the degrees below* $\mathbf{0}'$, ***Journal of the London Mathematical Society***, vol. 24, pp. 1–14.

R. A. SHORE [1999], *The recursively enumerable degrees*, ***Handbook of computability theory*** (E. R. Griffor, editor), Studies in Logic and the Foundations of Mathematics, vol. 140, North-Holland, Amsterdam, pp. 169–197.

R. A. SHORE AND T. A. SLAMAN [1999], *Defining the Turing jump*, ***Mathematical Research Letters***, vol. 6, pp. 711–722.

S. G. SIMPSON [1977], *First order theory of the degrees of recursive unsolvability*, ***Annals of Mathematics. Second Series***, vol. 105, pp. 121–139.

T. A. SLAMAN AND R. I. SOARE [2001], *Extension of embeddings in the computably enumerable degrees*, ***Annals of Mathematics***, vol. 153, pp. 1–43.

T. A SLAMAN AND J. R. STEEL [1989], *Complementation in the Turing degrees*, ***The Journal of Symbolic Logic***, vol. 54, pp. 160–176.

R. I. SOARE [1987], ***Recursively enumerable sets and degrees***, Springer-Verlag, Berlin.

C. SPECTOR [1956], *On degrees of recursive unsolvability*, ***Annals of Mathematics. Second Series***, vol. 64, pp. 581–592.

DEPARTMENT OF MATHEMATICS
CORNELL UNIVERSITY
ITHACA, NY 14853, USA
*E-mail*: shore@math.cornell.edu

DEPARTMENT OF MATHEMATICS
UNIVERSITY OF CALIFORNIA, BERKELEY
BERKELEY, CA 94720, USA
*E-mail*: slaman@math.berkeley.edu

# COCOVERING AND SET FORCING

M. C. STANLEY

**Abstract.** Let $\kappa$ be an uncountable regular cardinal and let $\mu$ be a cardinal of cofinality greater than $\kappa$ in the model $V$ of ZFC. It is shown that if certain combinatorial properties hold between $V$ and an outer model $W$, then every subset of $\mu$ in $W$ is set generic for a forcing of $V$-cardinality less than $\kappa$. This leads to a combinatorial characterization of those outer models $W$ that are set generic extensions of $V$.

**§1. Introduction.** Let $V[G]$ be the result of adding a Cohen real and a Cohen subset of $\omega_2$ to a model $V$ of the GCH. Every infinite cardinal gets a new subset in $V[G]$, namely, the Cohen real. Yet, in some sense, only $\omega$ and $\omega_2$ get new subsets. One way to capture this is to say that a sort of dual to Covering holds between $V$ and $V[G]$, namely, "$\kappa$-cocovering" for $\kappa$ other than $\omega$ and $\omega_2$.

We need some definitions.

If $\kappa$ is a regular cardinal in $V$, say that *$\kappa$-cocovering* holds between $V$ and the outer model $W$ if, given any unbounded $a \subseteq \kappa$ that lies in $W$, there exists $b \in V$ such that $b \subseteq a$ and $b$ is unbounded in $\kappa$.

If $V$ is a standard transitive model of ZFC, say that $W \supseteq V$ is an *outer model* of $V$ if $W$ is also a standard transitive model of ZFC and $V \cap \mathrm{OR} = W \cap \mathrm{OR}$. In this paper we are only concerned with outer models such that $(W; V)$ satisfies ZFC (in a language with a predicate symbol for $V$).

Ultimately, ZFC is our metatheory; talk of models can be understood as talk of *set*, or even *countable set* models. The reader will note that our results usually can be paraphrased in more traditional terminology, perhaps at the cost of some generality.

This paper proves a sort of converse to the observation with which we began. The general case is not quite so simple as this example suggests. We show that if $W$ is any outer model of $V$, and sufficient cocovering (and maybe a little covering) hold around $\kappa$, then every subset of $\kappa$ that lies in $W$ is set generic over $V$ for a forcing of $V$-cardinality less than $\kappa$.

Research supported by N.S.F. Grant DMS 9505157.

**Logic Colloquium '03**
Edited by V. Stoltenberg-Hansen and J. Väänänen
Lecture Notes in Logic, 24

To be precise, let us begin with a special case in which we assume a little of the GCH in the inner model $V$. Corollary 2.12 draws the following corollary of our main theorem.

THEOREM. *Assume that*

- *in* $V$: $\kappa$ *is regular,* $\kappa^{<\kappa} = \kappa$, *and* $2^\kappa = \kappa^+$. *Also assume that*
- *both* $\kappa$ *and* $\kappa^+$*-cocovering hold between* $V$ *and* $W$.

*Then every subset of* $\kappa^+$ *that lies in* $W$ *is set generic over* $V$ *for a forcing of* $V$*-cardinality less than* $\kappa$.

In light of the example with which we began, requiring both $\kappa$ and $\kappa^+$-cocovering is a surprise. Example 2.15 shows that this is necessary in general.

The author's interest in this topic was sparked by considering alternative strategies for coding the universe. All known proofs of Jensen's Coding Theorem use variants of almost disjoint coding. A generic real codes a subset of $\omega_1$; that subset of $\omega_1$ codes a subset of $\omega_2$; and so forth. Coding requirements (at least) at limits eventually overlap. Substantial technical machinery is necessary to handle the project. Could there be an easier way?

One informal idea is this: Rather than proceeding cardinal-by-cardinal, code across bigger gaps in the ordinals, say using homogeneous sets for partitions of some sort in place of almost disjoint coding. The author's opinion is that the theorem stated above implies that such a program is unlikely succeed. Over a model of the GCH (for example, $V = L$), a real that does not add "genuinely new" subsets to some $V$-regular cardinal and its $V$-successor is set-generic.

Something like its cocovering hypotheses is necessary for this theorem. On the other hand, the cardinal arithmetic assumptions on $V$ appear spurious. To eliminate these, we need a more devilish form of cocovering:

$(\pitchfork_{\lambda,<\kappa})$ For each $F : [\lambda]^{<\kappa} \to \lambda$ in $W$, there exists $f \in V$ such that $f \subseteq F$ and $\bigcup \mathrm{dom}(f) = \lambda$.

We might remark that $\pitchfork_{\lambda,<\kappa}$ is just a two-variable generalization of $\lambda$-cocovering. Another way to state $\lambda$-cocovering is, *given* $F : \lambda \to \lambda$ *that lies in* $W$, *there exists* $f \in V$ *such that* $f \subseteq F$ *and* $\bigcup \mathrm{dom}(f) = \lambda$. Indeed, for $V$-regular $\kappa$, $\pitchfork_{\kappa,<\kappa}$ is equivalent to $\kappa$-cocovering (see Lemma 3.4).

Like $\lambda$-cocovering, if $G$ is generic for a forcing of $V$-cardinality less than $\kappa$, then $\pitchfork_{\lambda,<\kappa}$ holds between $V$ and $W = V[G]$, for all $\lambda \geq \kappa$. (See Lemma 3.2.)

Using $\pitchfork_{\lambda,<\kappa}$ in place of cocovering, we get for example (Corollary 3.7):

THEOREM. *In* $V$, *let* $\kappa$ *be a regular uncountable cardinal and set* $\lambda = 2^\kappa$. *Assume that there are no limit cardinals (of* $V$*) between* $\kappa$ *and* $\lambda$. *If* $\pitchfork_{\lambda,<\kappa}$ *holds, then every subset of* $\kappa^+$ *in* $W$ *is set generic over* $V$ *for a forcing of* $V$*-cardinality less than* $\kappa$.

If we allow more characters in our story, we can eliminate the (again spurious) hypothesis that $(2^\kappa)^V$ is not much larger than $\kappa$. In full generality, the main theorem of this paper (Theorem 8.1) is

SET FORCING THEOREM. *In $V$, let $\kappa$, $\mu$, and $\lambda$ be cardinals such that*

- $\omega_1 \leq \kappa < \mu \leq \lambda$,
- *$\kappa$ is regular,*
- $\mathrm{cf}(\mu) > \kappa$, *and*
- $\lambda^{<\mu} = \lambda$.

*Assume that*

- *$<\kappa$-covering for subsets of $\lambda$ and*
- $\pitchfork_{\lambda,<\kappa}$

*hold between $V$ and $W$. Then every subset of $\mu$ in $W$ is set generic over $V$ for a forcing of $V$-cardinality less than $\kappa$.*

*$<\kappa$-covering for subsets of $\lambda$* is the requirement that, given any $a \subseteq \lambda$ that has $W$-cardinality less than $\kappa$, there exists $b \in V$ such that $b \supseteq a$ and $b$ has $V$-cardinality less than $\kappa$. If there are no limit cardinals between $\kappa$ and $\lambda$, then $<\kappa$-covering for subsets of $\lambda$ is a consequence of $\pitchfork_{\lambda,<\kappa}$. (See Lemmas 3.3 and 3.8.)

Like $\lambda$-cocovering and $\pitchfork_{\lambda,<\kappa}$, for $\lambda \geq \kappa$, $<\kappa$-covering is preserved by forcing of size less than $\kappa$.

The Set Forcing Theorem is a corollary of a factoring lemma for class forcing (Lemma 7.1). But the Set Forcing Theorem applies to arbitrary outer models of $V$. How is this possible?

If $\mathbb{P}$ is a $V$-amenable partial ordering, say that a filter $G \subseteq \mathbb{P}$ is *internally generic* if $G$ *decides* every set $d \in V$ of conditions. That is, there exists $p \in G$ such that either $p \in d$ or $p$ is incompatible with every element of $d$. If $\mathbb{P} \in V$, then this is equivalent to ordinary (set) genericity. However in the case of class forcing, this sort of genericity is too weak to be of much interest. (See [S] for a discussion of levels of class genericity.)

It is an essentially trivial observation that any outer model of $V$ is an internally generic extension of $V$ (see Lemma 8.2). A substantial fraction of the effort that goes into proving the Factor Lemma goes into seeing that it applies to forcing even in the weak sense of internal genericity. The point is that even internal genericity over a set-sized factor amounts to ordinary set genericity.

If $V$ is "sufficiently non-minimal" as an inner model of $W$, then $W$ is a generic extension of $V$ in a better sense: There exists a $V$-amenable class partial ordering $\mathbb{P}$ and a filter $G$ such that $W = V[G]$ and $G$ is "generic with respect to the language of set theory." This is the main theorem of [S]. No use is made of it in this paper, but the Factor Lemma is proved in a form that applies to this forcing, as well as to forcing in larger languages and to set forcing.

In §4 we pursue two sorts of applications of the Set Forcing Theorem. One is to give a combinatorial characterization of set-forcing extensions. This was first done in the early 1970s by Vopěnka and Hájek and, in a different way, later by Bukovský (see [B]). These are discussed in §4.1. The former result is a corollary of our characterization (see Lemma 4.7); the latter is of a different nature.

Let *Jensen $\kappa$-covering* be the following principle: If $a \in W$ is a set of ordinals, then there exists $b \in V$ such that $b \supseteq a$ and $|b|^W \leq \max(\kappa, |a|^W)$.

In this terminology, Jensen's Covering Lemma is that if $W$ satisfies "$0^\#$ does not exist," then Jensen $\omega_1^W$-covering holds.

Theorem 4.3 characterizes set-forcing extensions, assuming some of the GCH in $V$.

THEOREM. *Assume that in $V$ there exist unboundedly many regular $\kappa$ such that $\kappa^{<\kappa} = \kappa$ and $2^\kappa = \kappa^+$. The following are equivalent:*

(a) *$W$ is a set generic extension of $V$.*
(b) *There exists a $W$-cardinal $\nu$ such that*
   - *Jensen $\nu$-covering and*
   - *$\lambda$-cocovering, for $\lambda \geq \nu$,*

   *hold between $V$ and $W$.*

Using $\pitchfork_{\lambda,<\nu}$, Theorem 4.2 characterizes set forcing extensions without the assumptions on cardinal arithmetic in $V$.

The other application in this paper concerns supercompact cardinals. If $D$ is a normal fine $\kappa$-complete ultrafilter on $\mathcal{P}_\kappa\lambda$, let $j_D$ be the associated ultrapower embedding. Let $D$ be such an ultrafilter in $V$. In $(W; V)$ define

$$\widehat{D} = \{A \subseteq \mathcal{P}_\kappa\lambda : \exists B \in D\; B \subseteq A\}.$$

Say that $D$ *automatically extends* in $W$ if,

- in $W$, $\widehat{D}$ is a normal fine $\kappa$-complete ultrafilter on $\mathcal{P}_\kappa\lambda$ and
- $j_{\widehat{D}} \restriction V = j_D$.

If $W = V[G]$, where $G$ is $\mathbb{P}$-generic over $V$ and $\mathbb{P} \in V_\kappa$, then $D$ automatically extends in $W$. Theorem 4.10 proves a converse:

THEOREM. *In $V$, let $\lambda \geq \mu > \kappa$ be uncountable cardinals such that $\kappa$ is regular,* $\mathrm{cf}(\mu) > \kappa$, *and $\lambda^{<\mu} = \lambda$. If there exists a normal ultrafilter $D \in V$ on $(\mathcal{P}_\kappa\lambda)^V$ that automatically extends in $W$, then each subset of $\mu$ that lies in $W$ is $\mathbb{P}$-generic over $V$, for some forcing $\mathbb{P} \in V_\kappa$.*

If this happens at every $\lambda \geq \kappa$, then $W$ is a set-generic extension of $V$ (Theorem 4.11):

THEOREM. *Suppose that for every $\lambda \geq \kappa$ there exists a normal ultrafilter $D \in V$ on $(\mathcal{P}_\kappa\lambda)^V$ that automatically extends in $W$. Then there exists a forcing $\mathbb{P} \in V_\kappa$ and a filter $G$ such that $G$ is $\mathbb{P}$-generic over $V$ and $W = V[G]$.*

Proposition 4.13 shows that the analog of Theorem 4.10 for measures on $\kappa$ is false.

This paper has eight sections. Section 2 develops some useful facts about $\kappa$-covering and $\kappa$-cocovering. Section 3 does the same for $\pitchfork_{\lambda,<\kappa}$. Using this work, Section 4 makes the applications mentioned above. Sections 5–8 prove the Set Forcing Theorem. Section 5 establishes the basic consequence of $\pitchfork_{\lambda,<\kappa}$ used in the proof, namely, "superdensity reduction". Section 6 applies it repeatedly to get a technical Iterated Reduction Lemma aimed at "closing off" in set cardinality the set-complete subalgebra generated by a set of conditions. Section 7 applies this apparatus to prove the Factor Lemma. Section 8 then easily concludes the Set Forcing Lemma from the Factor Lemma.

So far as the author knows, with one exception, the notation in this paper is consistent with [J]. That exception is that $\langle\langle\alpha, \beta\rangle\rangle$ is used to indicate Gödel paring.

**§2. Cocovering, covering, and set genericity with GCH assumptions.** In this section, adopt the convention that $V$ is a standard transitive model of ZFC and that $W$ is an outer model of $V$ such that $(W; V) \models$ ZFC.

**2.1. $\kappa$-cocovering.** Recall from §1 the following

DEFINITION. Suppose that $\kappa$ is an infinite regular cardinal in the inner model $V$. Say that *$\kappa$-cocovering* holds between $V$ and $W$ if, given any unbounded $a \subseteq \kappa$ that lies in $W$, there exists an unbounded $b \subseteq a$ that lies in $V$.

The following observation is universally known.

PROPOSITION 2.1. *In $V$, let $\kappa$ be regular and let $\mathbb{P}$ be a partial ordering of cardinality less than $\kappa$. If $G$ is $\mathbb{P}$-generic over $V$, then $\kappa$-cocovering holds between $V$ and $V[G]$.*

PROOF. This is easy to prove directly. Alternatively, it follows from Lemmas 3.2 and 3.4. ⊣

Our definition of "$\kappa$-cocovering" presupposes that $\kappa$ is regular in the inner model $V$. Trivially, we have

LEMMA 2.2. *If $\kappa$-cocovering holds between $V$ and $W$, then $\kappa$ is also regular in the outer model $W$.*

Further evidence that $\kappa$-cocovering is a strong hypothesis is provided by the following two easy lemmas that will be useful later.

LEMMA 2.3. *If $\omega$-cocovering holds between $V$ and $W$, then $\mathcal{P}^W(\omega) = \mathcal{P}^V(\omega)$.*

PROOF. In $V$, fix a bijection $f : [\omega]^{<\omega} \to \omega$. Let $a \in W$ be a subset of $\omega$. If $b \in V$ is an unbounded subset of $f''\{a \cap n : n \in \omega\}$, then $a = \bigcup_{i \in b} f^{-1}(i)$ lies in $V$. ⊣

LEMMA 2.4. *If $\kappa$-cocovering holds between $V$ and $W$, and then $(\kappa^+)^W \leq (2^\kappa)^V$.*

PROOF. If $\kappa = \omega$, this follows from Lemma 2.3. Assume that $\kappa$ is uncountable in $V$ (hence in $W$). For a contradiction, suppose that $(2^\kappa)^V < (\kappa^+)^W$. In $(W; V)$, let $C$ be a diagonal intersection of all of the club subsets of $\kappa$ that lie in $V$. Working in $V$, let $D$ be the closure of an unbounded subset of $C$. Then $D$ is eventually contained in every club subset of $\kappa$, which is absurd. ⊣

**2.2. $\kappa$-covering.**

DEFINITION. If $\kappa$ is a cardinal in the outer model $W$ and $\lambda \geq \kappa$, say that *$\kappa$-covering for subsets of $\lambda$* holds between $V$ and $W$ if, given any $a \subseteq \lambda$ such that $a \in W$ and $|a|^W = \kappa$, there exists $b \in V$ such that $b \supseteq a$ and $|b|^V = \kappa$. Say that *$\kappa$-covering* holds between $V$ and $W$ if $\kappa$-covering for subsets of $\lambda$ holds for all $\lambda \geq \kappa$. Let *$<\kappa$-covering for subsets of $\lambda$* and *$<\kappa$-covering* have the same definitions with "$<$" in place of "$=$".

In general, the requirement that the covering set $b$ has cardinality $\kappa$ *in the inner model* $V$ is strong. The context of this paper typically softens this in the sense that usually we also assume some cocovering, which implies that relevant $V$-cardinals are $W$-cardinals.

Like Proposition 2.1, another familiar observation is

PROPOSITION 2.5. *Let $\kappa$ be regular in $V$ and let $\mathbb{P} \in V$ be a partial ordering of $V$-cardinality less than $\kappa$. If $G$ is $\mathbb{P}$-generic over $V$, then $<\kappa$-covering and $\lambda$-covering, for $\lambda \geq \kappa$, hold between $V$ and $V[G]$.*

The following two examples show that covering and cocovering are independent phenomena.

EXAMPLE 2.6. A pair of models such that

(a) $\kappa$-covering holds for all $W$-cardinals $\kappa$, but
(b) $\kappa$-cocovering fails for all $V$-regular $\kappa$.

Let $V$ satisfy the GCH and let $W = V[G]$ be the result of adding a Cohen subset to each $V$-regular $\kappa$ using an Easton support product.

EXAMPLE 2.7. A pair of models such that

(a) there exists an uncountable $W$-cardinal $\lambda$ such that $\kappa$-covering for subsets of $\lambda$ fails for all $W$-cardinals $\kappa \leq \lambda$, but
(b) $\kappa$-cocovering holds for all $V$-regular $\kappa$.

Let $\lambda$ be measurable in $\bar{V}$. Let $y$ be a $\bar{V}$-generic Prikry sequence, and let $x$ consist of every other element of $y$. Set $V = \bar{V}[x]$ and $W = \bar{V}[y] = V[y]$.

Together covering and cocovering are strong indeed. The following lemma will be useful in applications.

LEMMA 2.8. *Assume that*

- *$\kappa$-cocovering holds, for all infinite $V$-regular $\kappa$, and*
- *$\kappa$-covering holds, for all uncountable $W$-cardinals $\kappa$.*

*Then $V = W$.*

The models $V$ and $W$ have the same cardinals and cofinalities under the hypothesis that $\kappa$-cocovering holds for all infinite $V$-regular $\kappa$. Thus, in the presence of the first hypothesis, the second is just that Jensen's Covering Lemma holds between $V$ and $W$.

PROOF OF LEMMA 2.8. Proceed by induction on infinite $\kappa$ to see that $\mathcal{P}^V(\kappa) = \mathcal{P}^W(\kappa)$. This suffices, since $V$ and $W$ are standard transitive models of ZFC with the same ordinals.

If $\kappa = \omega$, then we have already seen that $\omega$-cocovering implies $\mathcal{P}^V(\omega) = \mathcal{P}^W(\omega)$.

Now suppose that $\kappa > \omega$ is a cardinal and that any bounded subset of $\kappa$ in $W$ lies in $V$. In $V$, fix a one-to-one enumeration $\langle x_i : i < \lambda \rangle$ of all bounded subsets of $\kappa$. Also fix in $V$ an increasing sequence $\langle \alpha_\delta : \delta < \mathrm{cf}(\kappa) \rangle$ cofinal in $\kappa$. In $W$, fix $x \subseteq \kappa$. We may assume that $x$ is unbounded in $\kappa$. Set

$$I = \{ i < \lambda : x_i = x \cap \alpha_\delta, \text{ for some } \delta < \mathrm{cf}(\kappa) \}.$$

Let $\alpha$ be least such that $|I \cap \alpha| = \mathrm{cf}(\kappa)$. Then $\mathrm{cf}(\alpha) = \mathrm{cf}(\kappa)$ and $\bigcup_{i \in I \cap \alpha} x_i = x$. Let $J \in V$ be such that $J \supseteq I \cap \alpha$ and $|J| \leq \max(\mathrm{cf}(\kappa), \omega_1) \leq \kappa$. Let $f \in V$ be a bijection $f : |J| \to J$ and let $z$ be the inverse image of $I$ under $f$. If $|J| < \kappa$, then $z \in V$ by induction, hence $x = \bigcup_{i \in f''z} x_i \in V$. So suppose $|J| = \kappa$. Then $\kappa = \max(\mathrm{cf}(\kappa), \omega_1)$. It follows that $\kappa = \mathrm{cf}(\kappa)$, since otherwise $\kappa = \omega_1 > \mathrm{cf}(\kappa)$, which is absurd. By $\kappa$-cocovering, let $y \in V$ be a subset of $z$ that is cofinal in $\kappa$. Then $f''y$ is cofinal in $I \cap \alpha$. It follows that $x = \bigcup_{i \in f''y} x_i \in V$. ⊣

The following easy observation is used in the proof of Corollary 2.12.

LEMMA 2.9. *Assume that $\kappa$ is regular in $W$ Then $\kappa$-covering for subsets of $\lambda$ implies $<\kappa$-covering for subsets of $\lambda$.*

PROOF. Let $b \subset \lambda$ be a set of $W$-cardinality less than $\kappa$. In $W$, extend $b$ to $b'$ having cardinality $\kappa$. Let $a' \in V$ have $V$-cardinality $\kappa$ and cover $b'$. Let $f : a' \to \kappa$ be a bijection in $V$. Because $\kappa$ is regular in $W$, there exists $\delta < \kappa$ such that $f''b \subseteq \delta$. The inverse image of $\delta$ under $f$ is a set in $V$ of $V$-cardinality less than $\kappa$ that covers $b$. ⊣

**2.3. Set forcing, covering, and cocovering.** The next two lemmas concern two ways in which forcing can kill cocovering. If $\kappa^{<\kappa} = \kappa$ in $V$, then $\kappa$-covering for subsets of $(\kappa^{<\kappa})^V$ is trivial. It is a consequence of the first of these lemmas that, in this case, if $\kappa$ is (uniformly) least such that forcing with $\mathbb{P}$ adds a subset to $\kappa$, then $\kappa$-cocovering fails between $V$ and any $\mathbb{P}$-generic extension.

The second lemma is that if $\kappa$ is the uniform density of $\mathbb{P}$, then forcing with $\mathbb{P}$ kills $\kappa$-cocovering.

An example of both is provided by Sacks forcing over a model of CH. The least ordinal (uniformly) that gets a new subset in a Sacks extension is $\omega$; assuming CH, Sacks forcing has uniform density $\omega_1$. Neither $\omega$ nor $\omega_1$-cocovering holds between $V$ and a Sacks generic extension.

LEMMA 2.10. *Let $\mathbb{P} \in V$ be a partial ordering and suppose that the $V$-regular cardinal $\kappa$ is uniformly least such that forcing with $\mathbb{P}$ adds a new subset to $\kappa$. Let*

*G be $\mathbb{P}$-generic over $V$. Then at least one of the following fails between $V$ and $V[G]$:*

- *$\kappa$-cocovering and*
- *$\kappa$-covering for subsets of $(\kappa^{<\kappa})^V$.*

PROOF. Suppose not. In $V$, let $f : [\kappa]^{<\kappa} \to \lambda$ be a bijection, where $\lambda = \kappa^{<\kappa}$. Suppose $a \in V[G] \setminus V$ and $a \subseteq \kappa$. Then $a$ is unbounded in $\kappa$ and $a \cap \alpha \in V$, for $\alpha < \kappa$ by our choice of $\kappa$. Using $\kappa$-covering, let $b \in V$ be such that $b \supseteq \{f(a \cap \alpha) : \alpha < \kappa\}$ and $|b|^V = \kappa$. Let $g \in V$ be a bijection $g : b \to \kappa$. Then $g``\{f(a \cap \alpha) : \alpha < \kappa\}$ is unbounded in $\kappa$. Using $\kappa$-cocovering, let $c \in V$ be an unbounded subset of this subset of $\kappa$. Then $a = \bigcup_{\gamma \in c} f^{-1}(g^{-1}(\gamma))$, contradicting that $a \notin V$. ⊣

LEMMA 2.11. *In $V$, let $\kappa$ be regular and let $\mathbb{P}$ be a separative partial ordering of uniform density $\kappa$. If $G$ is $\mathbb{P}$-generic over $V$, then $\kappa$-cocovering fails between $V$ and $V[G]$.*

PROOF. Work in $V$. Let $\langle p_i : i < \kappa\rangle$ enumerate a dense subset $D$ of $\mathbb{P}$. For convenience, assume that each element of $D$ appears $\kappa$ times in this enumeration. Define $\langle q_i : i < \kappa\rangle$ by recursion: Choose $q_i \leq p_i$ such that $q_i \in D$ and, for all $j < i$, either $q_i \mid q_j$ or $q_i$ properly extends $q_j$. Such a condition $q_i$ exists because $\{q_j : j < i\}$ is not dense below $p_i$.

Let us make two observations. First, using that each element of $D$ occurs $\kappa$ times in the enumeration $\langle p_i : i < \kappa\rangle$, it is easy to see that

$$\{p \in \mathbb{P} : \exists i \geq \beta\ (p \leq q_i)\}$$

is dense, for each $\beta < \kappa$.

Secondly, if $p \in \mathbb{P}$, we can choose $j$ to be least such that $p_j \leq p$. Then $q_j \leq p$. If $i > j$, then either $q_i \mid q_j$ or $q_i$ is a proper extension of $q_j$. In either case, $p \not\leq q_i$. Using that $\mathbb{P}$ is separative, it follows that $p \not\Vdash \check{q}_i \in \mathring{G}$.

By these two observation, if $G$ is $\mathbb{P}$-generic over $V$, then $a = \{i < \kappa : q_i \in G\}$ witnesses that $\kappa$-cocovering fails between $V$ and $V[G]$. ⊣

The following corollary is derived from the main theorem of this paper in §3. (See the proof just before Lemma 3.6.) That theorem uses a two variable form of cocovering (the principle $\pitchfork_{\lambda,<k}$ stated §1) that allows us to eliminate the hypotheses on cardinal arithmetic in $V$ that appear in the statement of Corollary 2.12.

COROLLARY 2.12. *Let $\kappa$ be regular in $V$ and assume the following:*

- *$\kappa^{<\kappa} = \kappa$ and $2^\kappa = \kappa^+$ in $V$ and*
- *both $\kappa$ and $\kappa^+$-cocovering hold between $V$ and $W$.*

*Then every subset of $\kappa^+$ that lies in $W$ is set generic over $V$ for a partial ordering of $V$-cardinality less than $\kappa$.*

Note the following consequence of Corollary 2.12: If in $V$ there are unboundedly many regular $\kappa$ such that $\kappa^{<\kappa} = \kappa$ and $2^\kappa = \kappa^+$, and if $\lambda$-cocovering holds between $V$ and $W$, for all sufficiently large $\lambda$, then every element of $W$ is set generic over $V$. If also $\lambda$-covering holds, for all sufficiently large $\lambda$, then $W$ itself is a set generic extension of $V$ (see Theorem 4.3). Example 2.7 (the example with Prikry sequences) can be extended to show that the cocovering hypothesis alone is not sufficient to get that $W$ is a set generic extension of $V$.

It is open whether the hypotheses on cardinal arithmetic in $V$ can be eliminated in Corollary 2.12 in favor of some covering hypothesis.

QUESTION 2.13. Let $\kappa$ be regular in $V$ and assume the following hold between $V$ and $W$:

- $\kappa$-covering for subsets of $(2^{\kappa^{<\kappa}})^V$; and
- $\kappa$ and $\kappa^+$-cocovering.

Must every subset of $\kappa^+$ that lies in $W$ be set generic over $V$ for a forcing of $V$-cardinality less than $\kappa$?

Nothing in our work to this point suggests that both $\kappa$ and $\kappa^+$-cocovering should be necessary for Corollary 2.12. In fact, both are necessary even to get the weaker result that every subset of $\kappa$ in $W$ is generic for a partial ordering of cardinality less than $\kappa$ in $V$. That $\kappa$-cocovering is necessary is obvious: Let $V$ be a model of GCH and let $W$ be the result of adding a Cohen subset to $\kappa$ over $V$. But what about $\kappa^+$?

EXAMPLE 2.14. A pair of models in which the GCH holds in $V$;

- $\kappa$-cocovering holds, $\kappa$-covering holds, and $\kappa^+$-cocovering fails; and
- there exists a subset of $\kappa$ in $W$ that is not generic for a forcing of cardinality less than $\kappa$ in $V$.

Let $\kappa$ be ineffable in $V$. Let $\mathbb{P} = \prod_{\alpha<\kappa} \mathbb{Q}(\omega_\alpha^+)$, where $\mathbb{Q}(\nu)$ is the forcing to add a Cohen subset to $\nu$. Then $\mathbb{P}$ is a cardinal and cofinality preserving forcing over $V$ and has uniform density $\kappa^+$. It follows that $\kappa$-covering holds and $\kappa^+$-cocovering fails between $V$ and any $\mathbb{P}$-generic extension. If $G$ is $\mathbb{P}$-generic, then $a = \{\langle\langle\alpha,\delta\rangle\rangle : \exists p \in G\; p_\alpha(\delta) = 1\}$ is a subset of $\kappa$ that is not $\mathbb{Q}$-generic, for any $\mathbb{Q} \in V$ of cardinality less than $\kappa$. All that is left is to see that $\kappa$-cocovering holds between $V$ and $V[G]$. This is where the ineffability of $\kappa$ comes in.

Work in $V$. Let $\mathring{f}$ be a term for a function from $\kappa$ to $\kappa$, and let $p \in \mathbb{P}$. We seek $p' \leq p$ and an unbounded $S \subseteq \kappa$ such that for each $\alpha \in S$, there exists $\beta$ such that $p' \Vdash \mathring{f}(\check{\alpha}) = \check{\beta}$. By extending tails of $p$ in $\kappa$ many steps, if necessary, we may assume that for each $\alpha < \kappa$, then exists $\beta^\alpha$ and $r^\alpha \in \prod_{\delta\leq\alpha} \mathbb{Q}(\omega_\delta^+)$ such that $r^\alpha \leq p \restriction (\alpha+1)$ and $r^\alpha \cup p \restriction [\alpha+1, \kappa) \Vdash \mathring{f}(\check{\alpha}) = \check{\beta}^\alpha$. Code $r^\alpha$ as a subset $A_\alpha$ of $\omega_\alpha$. Because $\kappa$ is ineffable, there exists an unbounded $S \subseteq \kappa$ such that $A_\alpha = A_\beta \cap \omega_\alpha$, for all $\alpha < \beta$ in $S$. Then $r^\beta \restriction (\alpha+1) = r^\alpha$, for $\alpha < \beta$ in $S$. So $p' = \bigcup_{\alpha\in S} r^\alpha$ is as required.

QUESTION 2.15. Assume the GCH in $V$, or better yet, assume $V = L$. Does there exist a countably generated $\mathbb{P}$ such that $\mathbb{P}$ has uniform density $\omega_2$ and $\omega_1$-cocovering holds between $V$ and any $\mathbb{P}$-generic extension?

**§3. $\pitchfork_{\lambda,<\kappa}$ and set genericity without GCH assumptions.** Let us continue to observe the notational convention that $V$ is a standard model of ZFC and that $W$ is an outer model of $V$ such that $(W; V) \models$ ZFC.

It is easy to see (and was implicit in the last example of §2) that $\kappa$-cocovering is equivalent to the following statement: *Given a function $F : \kappa \to \kappa$ that lies in $W$, there exists a function $f \in V$ such that $f \subseteq F$ and $\bigcup \mathrm{dom}(f) = \kappa$.* In view of Example 2.7 (the example using Prikry sequences), one might suspect that a better form of cocovering is *$\kappa$-sequence cocovering*: *Given a function $F : \kappa \to$ OR that lies in $W$, there exists $f \in V$ such that $f \subseteq F$ and $\bigcup \mathrm{dom}(f) = \kappa$.* In addition to $\kappa$-cocovering, this principle implies a certain amount of covering. (In fact, if $\kappa$-sequence cocovering holds for all infinite $V$-regular $\kappa$, then $V = W$. But we do not need this fact in this paper.) In an effort to distinguish the roles of these characters, we chose in §2 not to work with $\kappa$-sequence cocovering. However, as Lemma 3.3 indicates, with $\pitchfork_{\lambda,<\kappa}$ we are implicitly working with some sequence cocovering.

Let $\kappa$ be regular in $V$ and let $\lambda \geq \kappa$. Recall from §1 that $\pitchfork_{\lambda,<\kappa}$ is the following statement:

$$(\pitchfork_{\lambda,<\kappa})\quad \begin{array}{l}\text{For each } F : [\lambda]^{<\kappa} \to \lambda \text{ in } W,\\ \text{there exists } f \in V \text{ such that } f \subseteq F \text{ and } \bigcup \mathrm{dom}(f) = \lambda.\end{array}$$

REMARK 3.1. Restricting to functions $F : [\lambda]^{<\kappa} \cap V \to \lambda$ results in a principle equivalent to $\pitchfork_{\lambda,<\kappa}$ because $\mathrm{dom}(f) \subseteq V$ in its conclusion.

Note that, like $<\kappa$-covering and $\kappa$-cocovering, $\pitchfork_{\lambda,<\kappa}$ holds for generic extensions by forcing of size less than $\kappa$:

LEMMA 3.2. *Suppose that $W = V[G]$, where $G$ is $\mathbb{P}$-generic over $V$ and $|\mathbb{P}|^V < \kappa$. Then $\pitchfork_{\lambda,<\kappa}$ holds, for all $\lambda \geq \kappa$.*

PROOF. Let $\mathring{F}$ be a term for a function from $[\lambda]^{<\kappa}$ to $\lambda$ in $V[G]$. For each $p \in \mathbb{P}$, let

$$f_p = \{(a, \beta) : a \in [\lambda]^{<\kappa} \cap V \text{ and } p \Vdash \mathring{F}(\check{a}) = \check{\beta}, \text{ for some } \beta\}.$$

It suffices to see that $\{p \in \mathbb{P} : \bigcup \mathrm{dom}(f_p) = \lambda\}$ is dense in $\mathbb{P}$. Were this set not dense, we could choose $\bar{p}$ below which it is empty. For $p \leq \bar{p}$, let $\gamma_p < \lambda$ be least such that $\gamma_p \notin \bigcup \mathrm{dom}(f_p)$. Set $a = \{\gamma_p : p \leq \bar{p}\}$. Then no condition below $\bar{p}$ forces a value for $\mathring{F}(\check{a})$, which is absurd. ⊣

Next, let us observe some connections between $\pitchfork_{\lambda,<\kappa}$ and cocovering.

LEMMA 3.3. *Assume that $\kappa$ is regular in $V$ and that $\pitchfork_{\lambda,<\kappa}$ holds.*

(a) *If $\kappa \leq \eta \leq \lambda$ and $\mathrm{cf}^V(\eta) \geq \kappa$, then $\eta$-cocovering holds.*

(b) *$\kappa$-sequence cocovering holds for $H \in W$ such that $H : \kappa \to \lambda$.*

PROOF. For (a), let $a$ be an unbounded subset of $\eta$ in $W$. Define $F : [\lambda]^{<\kappa} \to \lambda$ by $F(s) = \min(a \setminus \sup(s \cap \eta))$, if $\sup(s) < \eta$; let $F(s) = 0$, otherwise. (We are proving $\eta$-cocovering, from which it follows that $\mathrm{cf}^W(\eta) = \mathrm{cf}^V(\eta) \geq \kappa$. So the "otherwise" case never actually occurs.) Let $f \in V$ be such that $f \subseteq F$ and $\bigcup \mathrm{dom}(f) = \lambda$. Then $b = \mathrm{rng}(f) \setminus \{0\}$ is an unbounded subset of $a$.

For (b), let $H : \kappa \to \lambda$ lie in $W$. Set $a = \{\langle\langle \gamma, H(\gamma)\rangle\rangle : \gamma < \kappa\}$. Let $\eta$ be least such that $|a \cap \eta|^W = \kappa$. By (a), there exists $b \in V$ such that $b$ is unbounded in $a \cap \eta$. Set $h = \{(\gamma, \delta) : \langle\langle \gamma, \delta\rangle\rangle \in b\}$. Then $h \in V$ and $h \subseteq H$, and $\bigcup \mathrm{dom}(h) = \kappa$, since $|h| = \kappa$. ⊣

LEMMA 3.4. *Let $\kappa$ be regular in $V$. Then $\pitchfork_{\kappa,<\kappa}$ is equivalent to $\kappa$-cocovering.*

PROOF. We must see that $\kappa$-cocovering implies $\pitchfork_{\kappa,<\kappa}$. (The converse is handled by (a) of Lemma 3.3.) Suppose that $F : [\kappa]^{<\kappa} \to \kappa$ lies in $W$. Set $a = \{\langle\langle \gamma, F(\gamma)\rangle\rangle : \gamma < \kappa\}$. Let $b \in V$ be an unbounded subset of $a$ and set $f = \{(\gamma, \delta) : \langle\langle \gamma, \delta\rangle\rangle \in b\}$. Then $f \subseteq F$ and $\bigcup \mathrm{dom}(f) = \kappa$. ⊣

The principles $\pitchfork_{\lambda,<\kappa}$ get stronger as the gap between $\kappa$ and $\lambda$ gets larger:

LEMMA 3.5. *Assume that $\kappa'$ and $\kappa$ are regular in $V$, and that $\kappa' \leq \kappa \leq \lambda \leq \lambda'$. Then $\pitchfork_{\lambda',<\kappa'}$ implies $\pitchfork_{\lambda,<\kappa}$.*

PROOF. To see $\pitchfork_{\lambda,<\kappa'} \Rightarrow \pitchfork_{\lambda,<\kappa}$: Suppose that $F : [\lambda]^{<\kappa} \to \lambda$ lies in $W$. Let $F' = F \restriction [\lambda]^{<\kappa'}$. Let $f \in V$ be such that $f \subseteq F'$ and $\bigcup \mathrm{dom}(f) = \lambda$. Then $f$ is as required.

To see $\pitchfork_{\lambda',<\kappa} \Rightarrow \pitchfork_{\lambda,<\kappa}$: Let $F : [\lambda]^{<\kappa} \to \lambda$ lie in $W$. Define $F' : [\lambda']^{<\kappa} \to \lambda'$ by setting $F'(a) = F(a \cap \lambda)$. Let $f' \in V$ be such that $f' \subseteq F'$ and $\bigcup \mathrm{dom}(f') = \lambda'$. Note that if $a, b \in \mathrm{dom}(f')$ and $a \cap \lambda = b \cap \lambda$, then $f'(a) = F'(a) = F(a \cap \lambda) = F(b \cap \lambda) = F'(b) = f'(b)$. It follows that $f = \{(a \cap \lambda, f'(a)) : a \in \mathrm{dom}(f)\}$ is as required. ⊣

The main theorem of this paper is proved in §8:

SET FORCING THEOREM. *In $V$, let $\kappa$, $\mu$, and $\lambda$ be cardinals such that*

- $\omega_1 \leq \kappa < \mu \leq \lambda$,
- *$\kappa$ is regular,*
- $\mathrm{cf}(\mu) > \kappa$, *and*
- $\lambda^{<\mu} = \lambda$.

*Assume that*

- *$<\kappa$-covering for subsets of $\lambda$ and*
- $\pitchfork_{\lambda,<\kappa}$

*hold between $V$ and $W$. Then every subset of $\mu$ in $W$ is set generic over $V$ for a forcing of $V$-cardinality less than $\kappa$.*

We can now derive Corollary 2.12 from the Set Forcing Theorem.

PROOF OF COROLLARY 2.12. Let $\kappa$ be regular in $V$ and assume the following:

- $\kappa^{<\kappa} = \kappa$ and $2^\kappa = \kappa^+$ in $V$ and
- both $\kappa$ and $\kappa^+$-cocovering hold between $V$ and $W$.

In $V$, set $\mu = \lambda = \kappa^+$. Then $\mathrm{cf}^V(\mu) > \kappa$ and $(\lambda^{<\mu})^V = \lambda$. By $\kappa$-cocovering, we know that $\kappa$ is regular in $W$. Because $\lambda = \kappa^+$, $\kappa^+$ and $\kappa$-covering for subsets of $\lambda$ is trivial. As observed in Lemma 2.9, $\kappa$-covering for subsets of $\lambda$ implies $<\kappa$-covering for subsets of $\lambda$. By Lemma 3.6 below, we also have $\pitchfork_{\kappa^+,<\kappa}$. By the Set Forcing Theorem it follows that every subset of $\kappa^+$ that lies on $W$ is set generic over $V$ for a forcing of $V$-cardinality less than $\kappa$. ⊣

LEMMA 3.6. *Let $\kappa$ be regular in $V$. Assume*

- $(\kappa^{<\kappa})^V = \kappa$;
- *$\kappa^+$-covering for subsets of $(2^\kappa)^V$; and*
- *both $\kappa$ and $\kappa^+$-cocovering.*

*Then $\pitchfork_{\kappa^+,<\kappa}$ holds.*

PROOF. In $V$, set $\mu = 2^\kappa$ and let $\langle f_\beta : \beta < \mu\rangle$ enumerate all functions $f$ such that $\mathrm{dom}(f) \subseteq [\alpha]^{<\kappa}$, $\mathrm{rng}(f) \subseteq \alpha$, and $\bigcup \mathrm{dom}(f) = \alpha$, for some $\alpha < \kappa^+$. This is possible since $\kappa^{<\kappa} = \kappa$ and $2^\kappa = \mu$.

In $W$, fix some $F : [\kappa^+]^{<\kappa} \to \kappa^+$. Set

$$A = \left\{\alpha < \kappa^+ : \text{ if } a \in \left([\alpha]^{<\kappa}\right)^V, \text{ then } F(a) \in \alpha\right\}.$$

Then $A$ is unbounded in $\kappa^+$ by $(\kappa^{<\kappa})^V = \kappa$ and that $\kappa^+$ is regular (in $W$ by $\kappa^+$-cocovering). If $\alpha \in A$, let $F_\alpha : [\alpha]^{<\kappa} \to \alpha$ by $F_\alpha(a) = F(a)$, if $a \in ([\alpha]^{<\kappa})^V$; otherwise, set $F_\alpha(a) = 0$. From $\kappa$-cocovering, we have $\pitchfork_{\kappa,<\kappa}$. Using this, there exists $\beta(\alpha) < \mu$ such that $f_{\beta(\alpha)} \subseteq F_\alpha$. Set $B = \{\beta(\alpha) : \alpha \in A\}$. Then $|B|^W = \kappa^+$, since $\bigcup \mathrm{dom}(f_{\beta(\alpha)}) = \alpha$, for $\alpha \in A$ (so $\beta(\alpha) \neq \beta(\alpha')$ when $\alpha \neq \alpha'$). By $\kappa^+$-covering for subsets of $\mu$, there exists $C \in V$ such that $C \supseteq B$ and $|C|^V = \kappa^+$. We may assume that $C \subseteq \mu$. Let $g \in V$ be a bijection $g : C \to \kappa^+$. Then $g"B$ is an unbounded subset of $\kappa^+$. Let $D \in V$ be an unbounded subset of $g"B$. Set $f = \bigcup_{g(\beta) \in D} f_\beta$. Then $f \in V$.

Also $f \subseteq F$: If $g(\beta) \in D$, then $\beta \in B$, so $\beta = \beta(\alpha)$ where $f_{\beta(\alpha)} \subseteq F_\alpha$. Because $\mathrm{dom}(f_{\beta(\alpha)}) \subseteq V$, it follows that $f_\beta \subseteq F$.

Finally, note that $\bigcup \mathrm{dom}(f) = \kappa^+$: If $\gamma < \kappa^+$, then there exists $\alpha \in A$ such that $\alpha > \gamma$ and $g(\beta(\alpha)) \in D$, because $|D| = \kappa^+$. Then $f_{\beta(\alpha)} \subseteq f$, so $\gamma \in \bigcup \mathrm{dom}(f)$. ⊣

The two principal hypotheses of the Set Forcing Theorem are $\pitchfork_{\lambda,<\kappa}$ and $<\kappa$-covering for subsets of $\lambda$. Below, Lemma 3.8 shows that if $\lambda$ is not too much larger than $\kappa$ —there are no limit cardinals between $\kappa$ and $\lambda$— then the second of these hypotheses follows from the first. First, though, let us illustrate the utility of Lemma 3.8 by using it to draw the following corollary of the Set Forcing Theorem.

COROLLARY 3.7. *In $V$, let $\kappa$ be a regular uncountable cardinal and set $\lambda = 2^\kappa$. Assume that there are no limit cardinals (of $V$) between $\kappa$ and $\lambda$. If $\pitchfork_{\lambda,<\kappa}$ holds, then every subset of $\kappa^+$ in $W$ is set generic over $V$ for a forcing of $V$-cardinality less than $\kappa$.*

PROOF. In the statement of the Set Forcing Theorem, set $\mu = \kappa^+$ and $\lambda = 2^\kappa$. The hypothesis that $<\kappa$-covering for subsets of $\lambda$ holds follows from Lemma 3.8. ⊣

Let $\kappa^{n+}$ be the $n$th cardinal successor of $\kappa$. Using Lemma 3.3(a), note that $\pitchfork_{\kappa^{n+},<\kappa}$ implies $\eta$-cocovering, for $\kappa \leq \eta \leq \kappa^{n+}$.

LEMMA 3.8. *Let $\kappa$ be regular in $V$ and set $\lambda = (\kappa^{n+})^V$, for some $n \in \omega$. Assume that $\eta$-cocovering holds, for $\kappa \leq \eta \leq \lambda$. Then $<\kappa$-covering for subsets of $\lambda$ holds.*

PROOF. The proof is essentially trivial by induction on $n$. If $n = 0$, then $\lambda = \kappa$. And $<\kappa$-covering for subsets of $\kappa$ is trivial.

Let $n > 0$. By the cocovering hypotheses, every $V$-cardinal from $\kappa$ through $\lambda$ is a $W$-cardinal. If $a \in W$ is a subset of $\lambda$ having $W$-cardinality less than $\kappa$. Then $\alpha = \sup(a) < \lambda$, since $\lambda$ is regular. Project $a$ to a subset of $\kappa^{(n-1)+}$ using a bijection with $\alpha$ that lies in $V$. By induction, cover the projection with a set in $V$ of cardinality less than $\kappa$. Then apply the inverse of the bijection to cover $a$. ⊣

**§4. Applications.** Let us continue to observe the notational convention that $V$ is a standard model of ZFC and that $W$ is an outer model of $V$ such that $(W; V) \models$ ZFC.

**4.1. Set generic outer models.** Our first application is to give a combinatorial characterization of those outer models $W$ such that $W = V[G]$, where $G$ is $\mathbb{P}$-generic over $V$ for some forcing $\mathbb{P} \in V$.

In the early 1970s several such characterizations were proved (see [B]). At the end of this section, we connect our Theorem 4.2 to these.

As remarked before, *$\kappa$-covering* is a strong property: A set of $W$-cardinality $\kappa$ is required to be covered by a set of $V$-cardinality $\kappa$. Let $\kappa$ be an infinite cardinal in $W$. A generally weaker and more familiar principle might be called

**Jensen $\kappa$-covering.** *If $a \in W$ is a set of ordinals, then there exists $b \in V$ such that $b \supseteq a$ and $|b|^W \leq \max(\kappa, |a|^W)$.*

It is clear that $\lambda$-covering, for all $W$-cardinals $\lambda \geq \kappa$, implies Jensen $\kappa$-covering. In contrast, simply collapsing $(\kappa^+)^V$ to cardinality $\kappa$ causes $\kappa$-covering to fail. However, in the presence of $\lambda$-cocovering, for all $V$-regular $\lambda \geq \kappa$, the two are equivalent.

REMARK 4.1. If every $V$-cardinal $\lambda \geq \kappa$ is a cardinal in $W$, then the following are equivalent:

- Jensen $\kappa$-covering
- $\lambda$-covering, for all $\lambda \geq \kappa$

THEOREM 4.2. *The following are equivalent:*

(a) *$W$ is a set generic extension of $V$.*
(b) *There exists a $W$-cardinal $\nu$ such that*
- *Jensen $\nu$-covering and*
- *$\pitchfork_{\lambda,<\nu}$, for $\lambda \geq \nu$,*

*hold between $V$ and $W$.*

PROOF. In light of the previous remark, ((a) $\Rightarrow$ (b)) follows from Lemmas 2.5 and 3.2. We must prove ((b) $\Rightarrow$ (a)).

Note that $W$ and $V$ have the same cardinals and cofinalities greater than or equal to $\nu$ because $\pitchfork_{\lambda,<\nu}$ implies $\eta$-cocovering, for $V$-regular $\eta$ such that $\nu \leq \eta \leq \lambda$. Thus $\lambda$-covering holds, for $\lambda \geq \nu$, by Remark 4.1. Let $\kappa > \nu$ be $V$-regular. Then $<\kappa$-covering follows from $\lambda$-covering, for $\nu \leq \lambda < \kappa$.

In $W$, let $\mu \geq \kappa^{<\kappa}$ be such that $\mathrm{cf}^V(\mu) > \kappa$. Let $\langle x_\delta : \delta < \mu\rangle$ enumerate $[\kappa]^{<\kappa}$, perhaps with repetitions. Let $\lambda$ be a $V$-cardinal such that $\lambda^{<\mu} = \lambda$ in $V$. Set $a = \{\langle\langle\delta, \xi\rangle\rangle : \xi \in x_\delta\}$. Then $a \subseteq \mu$. By the Set Forcing Theorem, there exists $\mathbb{P} \in V$ of $V$-cardinality less than $\kappa$ and $G \in W$ such that $G$ is $\mathbb{P}$-generic over $V$ and $a \in V[G]$.

Note that $\lambda$-cocovering holds between $V[G]$ and $W$, for all $V[G]$-cardinals $\lambda$ : $V[G]$ and $W$ have the same sets of cardinality less than $\kappa$, because $a \in V[G]$, and $\lambda$-cocovering holds between $V$ and $W$, for $\lambda \geq \kappa$. Similarly, $\lambda$-covering holds between $V[G]$ and $W$, for all $\lambda$. Hence $V[G] = W$ by Lemma 2.8. ⊣

With a little assumption on the cardinal arithmetic of $V$, we get a characterization using cocovering in place of $\pitchfork_{\lambda,<\nu}$.

THEOREM 4.3. *Assume that in $V$ there exist unboundedly many regular $\kappa$ such that $\kappa^{<\kappa} = \kappa$ and $2^\kappa = \kappa^+$. The following are equivalent:*

(a) *$W$ is a set generic extension of $V$.*
(b) *There exists a $W$-cardinal $\nu$ such that*
- *Jensen $\nu$-covering and*
- *$\lambda$-cocovering, for $\lambda \geq \nu$,*

*hold between $V$ and $W$.*

PROOF. The proof is the same as for Theorem 4.2, using Corollary 2.12 in place of the Set Forcing Theorem. ⊣

We can use reflection in place of the assumption that covering and cocovering hold on a tail of cardinals. For example, let $S$ be the following

$(W; V)$-definable class:

$$\kappa \in S \quad \text{iff } \kappa : \text{is inaccessible in } W;$$
$$\kappa \text{ and } \kappa^+\text{-cocovering hold; and}$$
$$(2^\kappa)^V = \kappa^+.$$

Say that a class of ordinals is *$\Sigma_n$-stationary* if it meets every $\Sigma_n$-definable club class of ordinals.

THEOREM 4.4. *If $S$ is $\Sigma_3$-stationary in $(W; V)$, then $W$ is a set generic extension of $V$.*

PROOF. If $S$ is $\Sigma_3$-stationary, then there exists $\kappa \in S$ such that $(W_\kappa; V_\kappa) \prec_2 (W; V)$. Let $a \subseteq \kappa$ code $W_\kappa$. By Corollary 2.12, $a$ is $\mathbb{P}$-generic over $V$ for a forcing $\mathbb{P} \in V_\kappa$. Thus $(W_\kappa; V_\kappa) \models \forall x\, \exists \mathring{x} \in V^{\mathbb{P}}\, (x = \mathring{x}^G)$. It follows that $W = V[G]$. ⊣

As mentioned at the beginning of this section, chacterizations of set generic outer models have been known for a long time.

DEFINITION. The following principles are due to Vopěnka and Hájek. Let $\kappa$ be a cardinal of $V$.

$\mathrm{Bd}_{V,W}(\kappa)$ : If $a \in W$ is a subset of $V$,
then there exist $b \in W$ and $c \in V$ such that
$a = \bigcup b$ and $c \supseteq b$ and $|c|^V < \kappa$.

$\mathrm{Apr}_{V,W}(\kappa)$ : If, for some ordinal $\alpha$, $f : \alpha \to \mathrm{OR}$ lies in $W$,
then there exists a function $g \in V$ with domain $\alpha$ such that
$|g(\alpha)|^V < \kappa$ and $f(\delta) \in g(\delta)$, for all $\delta < \alpha$.

It is an easy exercise to verify that if $\kappa$ is regular in $V$, then $\mathrm{Apr}_{V,W}(\kappa)$ is equivalent to the following statement: *If $\lambda \geq \kappa$ and $f : \lambda \to \lambda$ lies in $W$, then in $V$ there exists a club subset $K \subseteq [\lambda]^{<\kappa}$ of $a$ such that $f''a \subseteq a$.*

I believe the first characterization of set generic extensions is the following theorem of Vopěnka and Hájek.

THEOREM 4.5 (Vopěnka and Hájek). *Let $\kappa$ be a cardinal in $V$. The following are equivalent*:

- *$W$ is a set generic extension of $V$ by a forcing of $V$-cardinality less than $\kappa$.*
- $\mathrm{Bd}_{V,W}(\kappa)$

Soon afterwards Bukovský [B] found another, appealing characterization:

THEOREM 4.6 (Bukovský). *Let $\kappa$ be a cardinal in $V$. The following are equivalent*:

- *$W$ is a set generic extension of $V$ via a $\kappa$-c.c. forcing.*
- $\mathrm{Apr}_{V,W}(\kappa)$

The proof of Bukovský's Theorem requires that the Power Set Axiom holds in $W$. For example, if we let $W$ be the result of adding OR-many Cohen reals to $V$ (using a finite support product), then $\mathrm{Apr}_{V,W}(\omega_1)$ holds, even though $W$ is not a set generic extension of $V$. Consequently, we cannot expect to derive the hypotheses of our characterization, Theorem 4.2, directly from $\mathrm{Apr}_{V,W}(\kappa)$.

On the other hand, Lemma 4.7 shows that the two hypotheses of Theorem 4.2 can be derived easily from $\mathrm{Bd}_{V,W}(\kappa)$. Thus Vopěnka's Theorem can be obtained as a corollary of Theorem 4.2.

LEMMA 4.7. *Let $\kappa$ be regular in $V$. Assume $\mathrm{Bd}_{V,W}(\kappa)$. Then*

(a) *$ᗰ_{\lambda,<\kappa}$ and*

(b) *$<\kappa$-covering for subsets of $\lambda$*

*hold for all $\lambda \geq \kappa$.*

PROOF. For (a), let $F : [\lambda]^{<\kappa} \to \lambda$ lie in $W$. Let $b \in W$ and $c \in V$ be such that $F \restriction ([\lambda]^{<\kappa})^V = \bigcup b$ and $c \supseteq b$ and $|c|^V < \kappa$. It suffices to see that there exists $f \in b$ such that $\bigcup \mathrm{dom}(f) = \lambda$. Suppose not. In $V$, let $c' = \{f \in c : \bigcup \mathrm{dom}(f) \neq \lambda\}$. Then $b \subseteq c'$. For $f \in c'$, let $\gamma_f$ be least in $\lambda \setminus \bigcup \mathrm{dom}(f)$. Let $s = \{\gamma_f : f \in c'\}$. Then $s \in [\lambda]^{<\kappa}$, but $s \notin \mathrm{dom}(F)$, which is absurd.

For (b), proceed by induction on $\lambda$. Let $A \in W$ be a subset of $\lambda$ such that $|A|^W < \kappa$. We may assume that $A$ is unbounded in $\lambda$ and that $\lambda$ is a cardinal in $V$. Let $\langle \alpha_\delta : \delta < \eta \rangle \in W$ be increasing and cofinal in $\lambda$. Then $\eta < \kappa$. We may assume that some element of $A$ lies between $\alpha_\delta$ and $\alpha_{\delta+1}$, for each $\delta < \eta$.

By induction, choose $B_\delta \in V$ such that $B_\delta \supseteq A \cap \alpha_\delta$ and $|B_\delta|^V < \kappa$. We may assume that $B_\delta \subseteq \alpha_\delta$. Set $a = \{B_\delta : \delta < \eta\}$. Using $\mathrm{Bd}_{V,W}(\kappa)$, let $b \in W$ and $c \in V$ be such that $a = \bigcup b$ and $c \supseteq b$ and $|c|^V < \kappa$. Note that if $x \in b$, then $x \subseteq a$, so $\mathrm{ot}\{\sup(B) : B \in x\} \leq \eta$. By our choice of $\langle \alpha_\delta : \delta < \eta \rangle$ it follows that $|x|^V \leq \eta < \kappa$.

There are two cases to consider. First, suppose that there exists $x \in b$ such that $|x|^V = \eta$. Then $\bigcup x \supseteq A$. Since $|x|^V < \kappa$ and $|B_\delta|^V < \kappa$, for each $B_\delta \in x$, it follows that $|\bigcup x|^V < \kappa$, as required for $<\kappa$-covering.

In the second case, suppose that every $x \in b$ has $V$-cardinality less than $\kappa$. By thinning $c$, if necessary, we may assume that every $x \in c$ has the same property. Similarly, if $B \in x \in b$, then $|B|^V < \kappa$, so we may assume that this holds for all $B \in x \in c$, as well. Then $A \subseteq \bigcup a \subseteq \bigcup\bigcup b \subseteq \bigcup\bigcup c$. And $\bigcup\bigcup c = \bigcup_{x \in c} \bigcup_{B \in x} B$ is, in $V$, a union of fewer than $\kappa$ sets, each of cardinality less than $\kappa$. ⊣

**4.2. Supercompact cardinals.** In deference to standard notation, let us write "$\mathcal{P}_\kappa\lambda$" for $[\lambda]^{<\kappa}$ in this section.

LEMMA 4.8. *Working in $(W; V)$, suppose that*

- *$D$ is a fine filter on $\mathcal{P}_\kappa\lambda$, that*
- *$(\mathcal{P}_\kappa\lambda)^V \in D$, and that,*

• *for each $f : \mathcal{P}_\kappa\lambda \to \lambda$, there exists $A \in D$ such that $f \restriction A \in V$.*

*Then $(\mathcal{P}_\kappa\lambda)^V = \mathcal{P}_\kappa\lambda \cap V$ and both $\pitchfork_{\lambda,<\kappa}$ and $<\kappa$-covering for subsets of $\lambda$ hold between $V$ and $W$.*

PROOF. Let us begin by checking $<\kappa$-covering for subsets of $\lambda$. Suppose that $a \in \mathcal{P}_\kappa\lambda$. Then $\hat{a} = \{b \in \mathcal{P}_\kappa\lambda : b \supseteq a\} \in D$, since $D$ is fine. Consequently, there exists $b \in \hat{a} \cap (\mathcal{P}_\kappa\lambda)^V$, as required for $<\kappa$-covering.

From $<\kappa$-covering for subsets of $\lambda$, it follows that $(\mathcal{P}_\kappa\lambda)^V = \mathcal{P}_\kappa\lambda \cap V$.

Finally, we must check $\pitchfork_{\lambda,<\kappa}$. Let $F : \mathcal{P}_\kappa\lambda \to \lambda$ lie in $W$. Let $A \in D$ be such that $f = F \restriction A \in V$. If $\gamma < \lambda$, then $\widehat{\{\gamma\}} \in D$, so $A \cap \widehat{\{\gamma\}} \neq \emptyset$. Because $\gamma < \lambda$ was arbitrary, it follows that $\bigcup \mathrm{dom}(f) = \lambda$. ⊣

If $D$ is a $\kappa$-complete ultrafilter on $\mathcal{P}_\kappa\lambda$, let $j_D$ be the associated ultrapower embedding.

DEFINITION. Assume that $D \in V$ is a normal fine $\kappa$-complete ultrafilter on $(\mathcal{P}_\kappa\lambda)^V$. In $(W; V)$ define

$$\widehat{D} = \{A \in \mathcal{P}_\kappa\lambda : \exists B \in D \; B \subseteq A\}.$$

Recall from §1 that $D$ *automatically extends* in $W$ if,

• in $W$, $\widehat{D}$ is a normal fine $\kappa$-complete ultrafilter on $\mathcal{P}_\kappa\lambda$ and

• $j_{\widehat{D}} \restriction V = j_D$.

REMARK 4.9. If $W = V[G]$, where $G$ is $\mathbb{P}$-generic over $V$, for some partial ordering $\mathbb{P} \in V_\kappa$, then any normal fine $\kappa$-complete ultrafilter $D \in V$ on $(\mathcal{P}_\kappa\lambda)^V$ automatically extends. Theorems 4.10 and 4.11 provide a converse to this observation.

THEOREM 4.10. *In $V$, let $\lambda \geq \mu > \kappa$ be uncountable cardinals such that $\kappa$ is regular, $\mathrm{cf}(\mu) > \kappa$, and $\lambda^{<\mu} = \lambda$. If there exists a normal fine $\kappa$-complete ultrafilter $D \in V$ on $(\mathcal{P}_\kappa\lambda)^V$ that automatically extends in $W$, then each subset of $\mu$ that lies in $W$ is $\mathbb{P}$-generic over $V$, for some forcing $\mathbb{P} \in V_\kappa$.*

PROOF. The plan is to verify the hypotheses of Lemma 4.8 in order to get $<\kappa$-covering for subsets of $\lambda$ and $\pitchfork_{\lambda,<\kappa}$. Then Theorem 4.10 follows from the Set Forcing Theorem.

By the definition of $\widehat{D}$, we know that $(\mathcal{P}_\kappa\lambda)^V \in \widehat{D}$.

Factor $j_D$ and $j_{\widehat{D}}$ as $j_D = k \circ e$ and $j_{\widehat{D}} = \hat{k} \circ \hat{e}$, where

$$V \xrightarrow{e} \mathrm{Ult}_D(V) \xrightarrow{k} N \qquad \text{and} \qquad W \xrightarrow{\hat{e}} \mathrm{Ult}_{\widehat{D}}(W) \xrightarrow{\hat{k}} M$$

and $e$ and $\hat{e}$ are canonical embeddings and $k$ and $\hat{k}$ are transitive collapses.

Work in $(W; V)$. Let $f : \mathcal{P}_\kappa\lambda \to \lambda$. Then $\hat{k}([f]_{\widehat{D}}) \in j_{\widehat{D}}(\lambda)$. Because $k$ is onto $\mathrm{OR}^V = \mathrm{OR}^W$, there exists $[g]_D \in \mathrm{Ult}_D(V)$ such that $k([g]_D) = \hat{k}([f]_{\widehat{D}})$. Because $j_D(\lambda) = j_{\widehat{D}}(\lambda)$, we may assume that $g : (\mathcal{P}_\kappa\lambda)^V \to \lambda$. Because $D$ and

$\widehat{D}$ are normal,

$$k([g]_D) = j_D(g)(j_D``\lambda) \quad \text{and} \quad \hat{k}([f]_{\widehat{D}}) = j_{\widehat{D}}(f)(j_{\widehat{D}}``\lambda).$$

Thus $j_{\widehat{D}}``\lambda = j_D``\lambda \in j_{\widehat{D}}(\{a \in (\mathcal{P}_\kappa\lambda)^V : g(a) = f(a)\})$. Consequently, there exists $A \in \widehat{D}$ such that $f \restriction A = g \restriction A$. Let $B \in D$ be such that $B \subseteq A$. Then $f \restriction B = g \restriction B \in V$.

We conclude that $<\kappa$-covering for subsets of $\lambda$ and $\pitchfork_{\lambda,<\kappa}$ hold, using Lemma 4.8. The Set Forcing Theorem then provides the desired conclusion. ⊣

THEOREM 4.11. *Suppose that for every $\lambda \geq \kappa$ there exists a normal fine $\kappa$-complete ultrafilter $D \in V$ on $(\mathcal{P}_\kappa\lambda)^V$ that automatically extends in $W$. Then there exists a forcing $\mathbb{P} \in V_\kappa$ and a filter $G$ such that $G$ is $\mathbb{P}$-generic over $V$ and $W = V[G]$.*

PROOF. Note that $|W_\kappa|^W = \kappa$, since $\kappa$ is inaccessible in $W$. Let $a \in W$ be a subset of $\kappa$ coding $W_\kappa$. By Theorem 4.10, there exists $\mathbb{P} \in V_\kappa$ and a $G \in W$ such that $G$ is $\mathbb{P}$-generic over $V$ and $a \in V[G]$. Thus

$$(W_\kappa; V_\kappa) \models \forall x\, \exists \mathring{x} \in V^{\mathbb{P}}\ (x = \mathring{x}^G).$$

It suffices to see that $(W_\kappa; V_\kappa) \prec_2 (W; V)$. This follows from the observation that, given any $b \in W$, there exists $j_{\widehat{D}} : (W; V) \to (M; N)$ with critical point $\kappa$ and $b \in (W_{j_{\widehat{D}}(\kappa)})^M$ and $N \subseteq V$. (Choose $\lambda$ sufficiently large and let $\widehat{D}$ be the automatic extension of a normal fine $\kappa$-complete ultrafilter $D \in V$ on $(\mathcal{P}_\kappa\lambda)^V$. Then $N \subseteq V$ because $j_{\widehat{D}} \restriction V = j_D$ and $j_D : V \to N \subseteq V$.) ⊣

REMARK 4.12. The analog of Theorem 4.10 for measures on $\kappa$ is false. More precisely, assume that $D \in V$ is a normal $\kappa$-complete ultrafilter on $\kappa$. In $(W; V)$ define

$$\widehat{D} = \{A \subseteq \kappa : \exists B \in D\ B \subseteq A\}.$$

Say that $D$ *automatically extends* in $W$ if,

- in $W$, $\widehat{D}$ is a normal $\kappa$-complete ultrafilter and
- $j_{\widehat{D}} \restriction V = j_D$.

Again, if $W = V[G]$, where $G$ is $\mathbb{P}$-generic over $V$ for some forcing $\mathbb{P} \in V_\kappa$, then any normal $\kappa$-complete ultrafilter on $\kappa$ in $V$ automatically extends in $W$. However, we have this

PROPOSITION 4.13. *In $V$, assume the GCH and let $D$ be a normal $\kappa$-complete ultrafilter on $\kappa$. There exists a set generic extension $V[G]$ of $V$ (via a forcing of $V$-cardinality $\kappa^+$) such that*

- *$D$ automatically extends in $V[G]$, but*
- *there exists a subset of $\kappa$ in $V[G]$ that is not generic for a forcing of cardinality less than $\kappa$ in $V$.*

PROOF. In $V$, let $D$ be a normal $\kappa$-complete ultrafilter on $\kappa$. Let $\mathbb{P}$ be the forcing of Example 2.14. That is, $\mathbb{P} = \prod_{\alpha<\kappa} \mathbb{Q}(\omega_\alpha^+)$, where $\mathbb{Q}(\nu)$ is the forcing to add a Cohen subset to $\nu$.

Let $G$ be $\mathbb{P}$-generic over $V$. By the arguments of Example 2.14, we have

(1) there exists a subset of $\kappa$ in $V[G]$ that is not generic for a forcing of cardinality less than $\kappa$ in $V$, and

(2) if $f : \kappa \to V$ lies in $V[G]$, then there exists $A \in D$ such that $f \restriction A \in V$.

Let us use (2) to show that $D$ automatically extends in $V[G]$.

To see that $\widehat{D}$ is an ultrafilter, let $A \in V[G]$ be a subset of $\kappa$. Let $f : \kappa \to 2$ be its characteristic function. Let $B \in D$ be such that $f \restriction B \in V$. Because $D$ is normal in $V$, there exists $C \in D$ such that $C \subseteq B$ and $f$ is constant on $C$. Then either $C \subseteq A$ or $C \subseteq (\kappa \setminus A)$.

Similarly, it is easy to see that $\widehat{D}$ is normal and $\kappa$-complete.

Finally, we must check that $j_{\widehat{D}} \restriction V = j_D$. For this, it suffices to see that

$$\mathrm{Ult}_D(V)^V \cong \mathrm{Ult}_{\widehat{D}}(V)^{V[G]}.$$

Define $F : \mathrm{Ult}_D(V)^V \to \mathrm{Ult}_{\widehat{D}}(V)^{V[G]}$ by

$$F\big([f]_D\big) = [f]_{\widehat{D}},$$

for $f : \kappa \to V$ that lie in $V$. Note that $F$ is well-defined because $D \subseteq \widehat{D}$. Similarly, note that $F$ is an isomorphic embedding. Finally, observe that $F$ is onto: If $f : \kappa \to V$ lies in $V[G]$, then there exists $A \in D$ such that $f \restriction A \in V$. Let $g \in V$ be such that $g : \kappa \to V$ and $g \supseteq f \restriction A$. Then $F([g]_D) = [g]_{\widehat{D}} = [f]_{\widehat{D}}$. ⊣

**§5. Superdensity reduction.** In this and subsequent sections, we work under the following assumptions:

- $\mathbb{P}$ is either an element of $V$ or a $V$-definable partial ordering. Assume, without loss of generality, that $\mathbb{P}$ is separative and is closed under finite meets. That is, any finite subset of $\mathbb{P}$ has a greatest lower bound,
- $\mathbb{Q} \subseteq \mathbb{P}$ is, in $V$, a set-complete Boolean algebra with $\mathbb{0}$ removed, and $\mathbb{Q}$ is complete in $\mathbb{P}$. That is, if $q \in \mathbb{Q}$, then there exists an element $-q \in \mathbb{Q}$ such that, for all $p \in \mathbb{P}$, $p$ is incompatible with $q$ iff $p \leq -q$. Also, if $d \subseteq \mathbb{Q}$ is a set, then there exists a condition $\bigvee d \in \mathbb{Q}$ such that, for all $p \in \mathbb{P}$, $p \leq \bigvee d$ iff $d$ is predense with respect to $p$. The subordering $\mathbb{Q}$ is also perhaps a proper class.
- $G$ is a filter on $\mathbb{P}$ that is internally generic over $V$. That is,
  - if $p \in G$ and $\bar{p} \geq p$, then $\bar{p} \in G$, and
  - if $p$ and $p'$ lie in $G$, then there exists $p'' \in G$ extending both $p$ and $p'$, and $G$ is internally generic, that is,
  - if $d \in V$ is a set of conditions in $\mathbb{P}$, then either $G \cap d \neq \emptyset$ or there exists $p \in G$ that is incompatible with every element of $d$.

As far as application in this paper go, we might as well assume that $\mathbb{P} = \mathbb{Q}$ and dispense with one or the other name. In [S] it is shown that every "sufficiently non-minimal" outer model of $V$ is an extension that is generic with respect to the language of set theory. The forcing that witnesses this has the structure described in our Assumptions. Another application of the main theorem of this paper is to factor that forcing.

In this section,

- suppose that $\mathbb{R} \subseteq \mathbb{Q}$ is a set and $\mathbb{R}$ is closed under $\bigwedge$. That is, if $x \subseteq \mathbb{R}$ and there exists $p \leq q$, for all $q \in x$, then $\bigwedge x \in \mathbb{R}$.

Define $h_{\mathbb{R}} : \mathbb{P} \to \mathbb{R}$ by

$$h_{\mathbb{R}}(p) = \bigwedge\{r \in \mathbb{R} : p \leq r\}.$$

Say that $e \subseteq \mathbb{R}$ is *superdense* in $\mathbb{R}$ with respect to $\bar{p} \in \mathbb{P}$ if, given any $p \leq \bar{p}$ in $\mathbb{P}$, there exists $q \in e$ such that $q \sim p$ and $q \leq h_{\mathbb{R}}(p)$.

Note that if $e$ is superdense with respect to $\bar{p}$, then $e$ is predense with respect to $\bar{p}$, that is $\bar{p} \leq \bigvee e$.

Let

$$\Pi(\mathbb{R}) = \{\bigwedge x : x \subseteq \mathbb{R} \cup \{-r : r \in \mathbb{R}\}\}.$$

Then $\mathbb{R} \subseteq \Pi(\mathbb{R})$, since $\bigwedge\{r\} = r$, and $\Pi(\mathbb{R})$ is closed under $\bigwedge$. Let $\Pi^2(\mathbb{R}) = \Pi(\Pi(\mathbb{R}))$.

Granted that $\mathbb{R}$ is a set, so are $\Pi(\mathbb{R})$ and $\Pi^2(\mathbb{R})$, and so forth. Iterating the $\Pi$-operation and taking unions at limits generates the smallest set-complete Boolean algebra that is complete in $\mathbb{P}$ and extends $\mathbb{R}$. At successor stages, the ordering is closed under $\bigwedge$; at limits, under complementation. In general, the process iterates through the ordinals without closing off. The plan for our proof is to use $\pitchfork_{\lambda,<\kappa}$ and $<\kappa$-covering for subsets of $\lambda$ to find fixed points in this process.

Recall the principle $\pitchfork_{\lambda,<\kappa}$: If $W$ is an outer model of $V$ and $F : [\lambda]^{<\kappa} \to \lambda$ lies in $W$, then there exists $f \in V$ such that $f \subseteq F$ and $\bigcup \mathrm{dom}(f) = \lambda$.

Superdensity Reduction — Lemma 5.1. *In $V$, let $\kappa \geq \omega_1$ be a regular cardinal and let $\lambda \geq \kappa$ be a cardinal. Assume that $\pitchfork_{\lambda,<\kappa}$ holds between $V$ and $V[G]$. Suppose that $e \in V$ is a subset of $\mathbb{R}$ such that $|e|^V \leq \lambda$ and $e$ is superdense in $\mathbb{R}$ with respect to $\tilde{p} \in G \cap \Pi(\mathbb{R})$. Then there exists $d \in [e]^{<\kappa}$ and $p^* \in G \cap \Pi^2(\mathbb{R})$ such that $d$ is superdense in $\mathbb{R}$ with respect to $p^*$.*

Proof. Let $\kappa$, $\lambda$, $G$, $e$, and $\tilde{p}$ be as in the statement of the lemma. Note first that it suffices to prove the lemma without the requirement that $p^* \in \Pi^2(\mathbb{R})$:

Claim 5.2. It suffices to find $d \in [e]^{<\kappa}$ and $p^* \in G$ such that $d$ is superdense in $\mathbb{R}$ with respect to $p^*$.

Proof of the Claim. Set $p = h_{\Pi^2(\mathbb{R})}(p^*)$. Then $p \in G \cap \Pi^2(\mathbb{R})$. We must see that $d$ is superdense in $\mathbb{R}$ with respect to $p$. Let $p' \leq p$. Then

$p^* \sim h_{\Pi(\mathbb{R})}(p')$, since otherwise $p^* \leq -h_{\Pi(\mathbb{R})}(p') \in \Pi^2(\mathbb{R})$, and then $p \leq -h_{\Pi(\mathbb{R})}(p')$, contradicting that $p' \leq p$. Let $p''$ extend both $p^*$ and $h_{\Pi(\mathbb{R})}(p')$. Because $d$ is superdense in $\mathbb{R}$ with respect to $p^*$, there exists $q \in d$ such that $q \sim p''$ and $q \leq h_{\mathbb{R}}(p'')$. Now $h_{\mathbb{R}}(p') \geq h_{\mathbb{R}}(p'')$ because $h_{\Pi(\mathbb{R})}(p') \geq p''$. Hence $q \leq h_{\mathbb{R}}(p')$. And $q \sim p'$, since otherwise $p' \leq -q$. In this case $h_{\Pi(\mathbb{R})}(p') \leq -q$. But $h_{\Pi(\mathbb{R})}(p') \sim q$ since $p'' \leq h_{\Pi(\mathbb{R})}(p')$ and $p'' \sim q$. ⊣

In $V$, let

$$\langle q_\alpha : \alpha < \lambda \rangle$$

be an enumeration of $e$.

For a contradiction (to $\pitchfork_{\lambda,<\kappa}$), assume that for all $d \in [e]^{<\kappa}$ and all $p^* \in G$, $d$ is not superdense in $\mathbb{R}$ with respect to $p^*$. We have that

$$(\dagger) \qquad \begin{cases} \text{if } p^* \in G \text{ and } d \in [e]^{<\kappa}, \\ \text{then there exists } p \leq p^* \text{ such that,} \\ \text{for all } q \in d, \text{ either } p \mid q \text{ or } q \not\leq h_{\mathbb{R}}(p). \end{cases}$$

The plan is to find a "simple" term $\mathring{F}$ for the function $F : ([\lambda]^{<\kappa})^V \to \lambda$ defined in $V[G]$ by

$$F(s) = \text{the least } \beta \text{ such that } q_\beta \in G \text{ and } \forall \gamma \in s\ (q_\gamma \in G \to q_\gamma \not\leq q_\beta)$$

and argue from (†) that there is no $f \in V$ such that $f \subseteq F$ and $\bigcup \mathrm{dom}(f) = \lambda$. This contradicts $\pitchfork_{\lambda,<\kappa}$. We must be careful carrying out this plan because we are assuming only that $G$ is internally generic.

For $s \in [\lambda]^{<\kappa}$, set

$$B_s = \{q \in \Pi(\mathbb{R}) : \exists \beta\ (q \leq q_\beta \wedge \forall \gamma \in s\ (q \sim q_\gamma \to q_\gamma \not\leq q_\beta))\}.$$

Claim 5.3. $B_s \cap G \neq \emptyset$

Proof. For a contradiction, suppose that $G \cap B_s = \emptyset$. Then $\bigwedge_{q \in B_s} -q \in G$, since otherwise $\bigvee B_s \in G$. In this case, the set of conditions $B_s$ would be predense with respect to this condition in $G$, and we would have $B_s \cap G \neq \emptyset$.

Let $p^* \in G$ extend both $\tilde{p}$ and $\bigwedge_{q \in B_s} - q$. By (†) there exists $p' \leq p^*$ such that

$$(*) \qquad \forall \gamma \in s\ (p' \sim q_\gamma \to q_\gamma \not\leq h_{\mathbb{R}}(p')).$$

Now $p' \leq p^* \leq \tilde{p}$, hence by the superdensity of $e$, there exists $q_\beta$ such that $p' \sim q_\beta$ and $q_\beta \leq h_{\mathbb{R}}(p')$. Let $p$ extend both $p'$ and $q_\beta$. Set $q = h_{\Pi(\mathbb{R})}(p)$. Then $q \leq q_\beta$, since $p \leq q_\beta \in \mathbb{R}$ and $q = h_{\Pi(\mathbb{R})}(p)$.

Fix $\gamma \in s$ and suppose that $q_\gamma \leq q_\beta$. We maintain that $q \mid q_\gamma$. This will show that $q \in B_s$, which is impossible since $p \leq q$ and $p \leq \bigwedge_{q \in B_s} - q$. Now $q_\gamma \leq q_\beta \leq h_{\mathbb{R}}(p')$. Thus $p' \mid q_\gamma$ by (*). So $p \leq -q_\gamma$ since $p \leq p'$. But $-q_\gamma \in \Pi(\mathbb{R})$, so $q \leq -q_\gamma$ since $q = h_{\Pi(\mathbb{R})}(p)$. Thus $q \mid q_\gamma$. ⊣

If $q \in B_s$, let $\beta(q)$ be the least $\beta$ witnessing that $q \in B_s$. Then for $q \in B_s$, we have the following:

(1) $q \leq q_{\beta(q)}$;

(2) $\forall \gamma \in s\ (q \sim q_\gamma \rightarrow q_\gamma \not\leq q_{\beta(q)})$; and

(3) $\forall \beta' < \beta(q)\ (q \leq q_{\beta'} \rightarrow \exists \gamma \in s\ (q \sim q_\gamma \wedge q_\gamma \leq q_{\beta'}))$.

CLAIM 5.4. Suppose that $q \in B_s$ and $q' \leq q$, where $q' \in \Pi(\mathbb{R})$. Then $q' \in B_s$ and $\beta(q') \leq \beta(q)$.

PROOF. Set $\beta = \beta(q)$. Then $q' \leq q \leq q_\beta$ and $\forall \gamma \in s\ (q \sim q_\gamma \rightarrow q_\gamma \not\leq q_\beta)$. It follows that $\beta$ witnesses that $q' \in B_s$. ⊣

In $V$, for $s \in [\lambda]^{<\kappa}$ and $\beta < \lambda$, set

$$p_{s\beta} = \bigvee \{q : q \in B_s \text{ and } \beta(q) = \beta\} \wedge \bigwedge \{-q : q \in B_s \text{ and } \beta(q) < \beta\},$$

if this set Boolean combination is non-$\mathbb{0}$. Otherwise $p_{s\beta}$ is undefined.

Note that if $\beta \neq \beta'$ and both $p_{s\beta}$ and $p_{s\beta'}$ are defined, then $p_{s\beta} \mid p_{s\beta'}$.

Set

$$\mathring{F} = \{((s, \beta)\check{}, p_{s\beta}) : s \in [\lambda]^{<\kappa} \text{ and } \beta < \lambda \text{ and } p_{s\beta} \text{ is defined}\}.$$

Then $\mathring{F}$ is a Shoenfield term in $V^{\mathbb{Q}}$.

CLAIM 5.5. $\mathring{F}^G : ([\lambda]^{<\kappa})^V \rightarrow \lambda$.

PROOF. Fix $s \in ([\lambda]^{<\kappa})^V$. Since $p_{s\beta} \mid p_{s\beta'}$ when $\beta \neq \beta'$ (and both $p_{s\beta}$ and $p_{s\beta'}$ are defined), there exists at most one $\beta$ such that $p_{s\beta} \in G$. It follows that $\mathring{F}^G$ is a function.

Let $\beta$ be least such that $\beta = \beta(q)$, for some $q \in B_s \cap G$. Then $p_{s\beta} \in G$. Hence $(s, \beta) \in \mathring{F}^G$. ⊣

Because we assume only that $G$ is an internally generic filter on $\mathbb{P}$, many statements may hold in $V[G]$ that are not forced by any condition in $G$. On the other hand, $\mathring{F}$ is a term of a special form: Say that a Shoenfield term $\mathring{z}$ of the forcing language for $\mathbb{P}$ is a *simple* $\mathbb{Q}$*-term* for a subset of $a \in V$ if there exist sets of conditions $F_c \subseteq \mathbb{Q}$, for $c \in a$, such that

$$\mathring{z} = \{(\check{c}, q) : c \in a \text{ and } q \in F_c\}.$$

CLAIM 5.6. Assume that $\mathring{z}$ is a simple $\mathbb{Q}$-term for a subset of $a \in V$ and that $b \subseteq \mathring{z}^G$, where $b \in V$. There exists $p \in G$ such that $F_c$ is predense with respect to $p$, for all $c \in b$.

PROOF. Suppose not. Then $\{-\bigvee F_c : c \in b \text{ and } \bigvee F_c \neq \mathbb{1}\}$ is predense with respect to $p$, for all $p \in G$. Because $G$ is internally generic, $-\bigvee F_c \in G$, for some $c \in b$. But then $c \notin \mathring{z}^G$. ⊣

The following claim contradicts that $⋔_{\lambda,<\kappa}$ holds between $V$ and $V[G]$, completing the proof of the Superdensity Reduction Lemma.

CLAIM 5.7. If $f \in V$ and $f \subseteq \mathring{F}^G$, then there exists $\gamma < \lambda$ such that $\gamma \notin \bigcup \mathrm{dom}(f)$.

PROOF. By the previous claim, there exists $p \in G$ such that $F_{(s,f(s))} = \{p_{s\,f(s)}\}$ is predense with respect to $p$, for all $s \in \mathrm{dom}(f)$. Thus $p \leq p_{s\,f(s)}$, for all $s \in \mathrm{dom}(f)$. Now $p_{s\,f(s)} \leq \bigvee\{q \in B_s : \beta(q) = f(s)\}$, and each condition in this set extends $q_{f(s)}$ by the definition of $\beta(q)$. It follows that $p \leq q_{f(s)}$.

We may assume that $p \leq \tilde{p}$, since $G$ is a filter. Using that $e$ is superdense in $\mathbb{R}$ with respect to $\tilde{p}$, let $\gamma$ be such that $q_\gamma \sim p$ and $q_\gamma \leq h_{\mathbb{R}}(p)$. We claim that $\gamma \notin \bigcup \mathrm{dom}(f)$.

Fix $s \in \mathrm{dom}(f)$. Let $p'$ extend both $p$ and $q_\gamma$. Let $q \in B_s$ be such that $\beta(q) = f(s)$ and $q \sim p'$. (Recall that $p \leq \bigvee\{q \in B_s : \beta(q) = f(s)\}$.) Then $q \sim q_\gamma$ because $q \sim p' \leq q_\gamma$. And $q_\gamma \leq q_{f(s)}$ because $q_\gamma \leq h_{\mathbb{R}}(p)$ and $p \leq q_{f(s)}$. By the definition of $\beta(q) = f(s)$, we know that $\forall \gamma \in s\ (q \sim q_\gamma \rightarrow q_\gamma \not\leq q_{f(s)})$. It follows that $\gamma \notin s$. Since $s \in \mathrm{dom}(f)$ was arbitrary, we conclude $\gamma \notin \bigcup \mathrm{dom}(f)$, as required to finish the proof of the Claim and, with it, the proof of the Superdensity Reduction Lemma. ⊣

**§6. Iterated reduction.** In this section, we continue to work under the following assumptions:

- $\mathbb{P}$ is either an element of $V$ or a $V$-definable partial ordering that is separative and is closed under finite meets.
- $\mathbb{Q} \subseteq \mathbb{P}$ is, in $V$, a set-complete Boolean algebra with $\mathbb{0}$ removed, and $\mathbb{Q}$ is complete in $\mathbb{P}$.
- $G$ is a filter on $\mathbb{P}$ that is internally generic over $V$.

Recall that if $W$ is an outer model of $V$ and $\kappa < \lambda$ are cardinals of $V$, then *$<\kappa$-covering for subsets of $\lambda$* holds provided that, given $x \subseteq \lambda$ in $W$ with $|x|^W < \kappa$, there exists $y \in V$ such that $y \supseteq x$ and $|y|^V < \kappa$.

ITERATED SUPERDENSITY REDUCTION — LEMMA 6.1. *In $V$, let $\kappa \geq \omega_1$ be regular, let $\lambda \geq \mu > \kappa$ be cardinals such that $\lambda^{<\mu} = \lambda$. Let $\langle q_\alpha : \alpha < \mu \rangle$ be a sequence of conditions in $\mathbb{Q}$. For $\alpha < \mu$, there exists $E^\alpha \subseteq \mathbb{Q}$ and $e^\alpha_p \subseteq \mathbb{Q}$, for $p \in E^\alpha$, satisfying the following:*

- *$E^\alpha$ is an antichain (perhaps not maximal) and $|E^\alpha| \leq \lambda$.*
- *Each $p \in E^{\alpha+1}$ meets $E^\alpha$. More generally, let $E^\beta \restriction p = \{q \in E^\beta : q \sim p\}$. If $p \in E^\alpha$, then $|E^\beta \restriction p| < \kappa$, for $\beta < \alpha$.*
- *$e^\alpha_p$ is predense in $\mathbb{P}$ with respect to $p$ and $|e^\alpha_p| < \kappa$.*
- *$e^\alpha_p$ is superdense with respect to $p$ in $\Pi(R)$, where*

$$R = \bigcup \left\{ e^\beta_{p'} : \beta < \alpha \text{ and } p' \in E^\beta \restriction p \right\} \cup \{q_\beta : \beta < \alpha\}.$$

- *Let $G$ be an internally $\mathbb{P}$ generic filter over $V$. Assume that $⋔_{\lambda,<\kappa}$ and $<\kappa$-covering for subsets of $\lambda$ holds between $V$ and $V[G]$. Then $G \cap E^\alpha \neq \emptyset$,* *for all $\alpha < \mu$. Furthermore, there exists $p^* \in G$ such that $|E^\alpha \restriction p^*|^V < \kappa$, for all $\alpha < \mu$.*

Before getting started on the construction of the $E^\alpha$'s and $e_p^\alpha$'s, let us prove a lemma that lets us improve predense sets to antichains.

LEMMA 6.2. *Suppose that $P \subseteq \mathbb{Q}$ is a set. There exists an antichain $A \subseteq \mathbb{Q}$ such that*

- $|A| \leq |P|$;
- *every condition in $A$ meets $P$; and*
- *if $p \in P$, then $A$ is predense with respect to $p$.*

PROOF. Let $\langle r_i : i < \theta \rangle$ enumerate $P$. Define $\langle a_i : i < \theta' \rangle$, for some $\theta' < \theta^+$, by recursion. At stage $i$, if there exists $k < \theta$ such that $r_k \sim \bigwedge_{j<i} - a_j$, let $k_i < \theta$ be least with this property and set $a_i = r_{k_i} \wedge \bigwedge_{j<i} - a_j$. Otherwise, set $\theta' = \theta$. Set $A = \{a_i : i < \theta'\}$.

By construction $A$ is an antichain and is predense with respect to any condition in $P$. Any condition meeting $A$ meets $P$ because $a_i \leq r_{k_i} \in P$. Finally, $|\theta'| \leq \theta$, since $i \mapsto k_i$ is a one-to-one function from $\theta'$ into $\theta$. (If $j < i$ and $k_j = k_i$, then $a_i = r_{k_i} \wedge \bigwedge_{\delta<i} - a_\delta \leq r_{k_j} \wedge \bigwedge_{\delta<j} - a_\delta = a_j$, which is absurd.) ⊣

The construction of the $E^\alpha$'s and $e_p^\alpha$'s will be by recursion on $\alpha < \mu$. At limit stages in the construction, the following lemma will be useful. This lemma also establishes that, if the construction can be carried out, then there exists $p^* \in G$ such that $|E^\alpha \restriction p^*|^V < \kappa$, for all $\alpha < \mu$.

LEMMA 6.3. *Let $\alpha \leq \mu$ be a limit ordinal and assume that $E^\beta$ and $e_{p'}^\beta$ are as required, for $\beta < \alpha$ and $p' \in E^\beta$. There exists $p \in G$ such that $|E^\beta \restriction p| < \kappa$, for all $\beta < \alpha$.*

PROOF. Work in $V$. Let $\eta = \mathrm{cf}(\alpha)$ and let $\langle \beta_\delta : \delta < \eta \rangle$ be increasing and cofinal in $\alpha$. Let $\langle q_i^\delta : i < \theta^\delta \rangle$ be a one-to-one enumeration of $E^{\beta_\delta}$, where $\theta^\delta \leq \lambda$.

In $V[G]$, set

$$x = \{\langle\langle \delta, i \rangle\rangle : \delta < \eta \text{ and } i < \theta^\delta \text{ and } q_i^\delta \in G\}.$$

This set lies in $V[G]$ because $G$ is amenable to $V[G]$. Furthermore, for each $\delta < \eta$, there exists a unique $i$ such that $\langle\langle \delta, i \rangle\rangle \in x$.

Recall that $<\kappa$-covering for subsets of $\lambda$ holds by hypothesis, thus $\alpha$ has cofinality less than $\kappa$ in $V[G]$ iff it has cofinality less than $\kappa$ in $V$.

If $\eta < \kappa$, then $|x|^{V[G]} = |\eta|^{V[G]} < \kappa$. Using $<\kappa$-covering, let $y \in V$ be a superset of $x$ such that $|y|^V < \kappa$.

Working in $V$, set

$$p = \bigwedge_{\delta<\eta} \bigvee \{q_i^\delta : \langle\langle \delta, i \rangle\rangle \in y \text{ and } i < \theta^\delta\}.$$

Then $p \in G$. (Otherwise $-p \in G$, contradicting our choice of $x$.) We maintain that $|E^\beta \restriction p| < \kappa$, for all $\beta < \alpha$. Fix $\beta < \alpha$ and let $\delta < \eta$ be least such that $\beta_\delta \geq \beta$. Then

$$E^\beta \restriction p \subseteq \bigcup_{\langle\langle \delta, i\rangle\rangle \in y} E^\beta \restriction q_i^\delta.$$

And $|E^\beta \restriction q_i^\delta| < \kappa$ by hypothesis. Hence $|E^\beta \restriction p| < \kappa$, as required for the lemma.

If $\eta \geq \kappa$, then $|x|^{V[G]} = \eta$. By $\pitchfork_{\lambda,<\kappa}$, we have $\pitchfork_{\lambda,<\eta}$, so, by Lemma 3.3(b), we have $\eta$-sequence cocovering for $H \in V[G]$ such that $H : \eta \to \lambda$. Let $y \in V$ be such that $|y|^V = \eta$ and $y \subseteq x$. Set

$$p = \bigwedge_{\langle\langle \delta, i\rangle\rangle \in y} q_i^\delta.$$

Then $p \in G$, again because $-p \in G$ contradicts our choice of $x$. If $\beta < \alpha$, let $\delta$ be least such that $\beta_\delta \geq \beta$ and $\langle\langle \delta, i\rangle\rangle \in y$, for some $i$. Then $E^\beta \restriction p \subseteq E^\beta \restriction q_i^\delta$. Hence $|E^\beta \restriction p| < \kappa$, as required for the lemma. ⊣

**The construction.** Work in $V$. Begin by setting $E^0 = \{\mathbb{1}\}$ and $e^0_{\mathbb{1}} = \{\mathbb{1}\}$.

If $\alpha > 0$, we first define an antichain $A \subseteq \mathbb{Q}$, then extend conditions in $A$ to get the antichain $E^\alpha$. At the same time, we define $e^\alpha_p$, for $p \in E^\alpha$.

If $\alpha = \beta + 1$, set $A = E^\beta$.

If $\alpha < \mu$ is a limit ordinal, let

$$p \in P \quad \text{iff } p = \bigwedge_{\beta<\alpha} \bigvee x_\beta \neq \mathbb{0}, \text{ for some } \langle x_\beta : \beta < \alpha\rangle$$
$$\text{such that } x_\beta \subseteq E^\beta \text{ and } |x_\beta| < \kappa.$$

Then $|P| \leq (\lambda^{<\kappa})^{<\mu} = \lambda$, since $\lambda^{<\mu} = \lambda$. By the previous lemma, $P \cap G \neq \emptyset$. Using the Lemma 6.2, let $A$ be an antichain in $\mathbb{Q}$ such that every condition in $A$ meets $P$, $A$ is predense with respect to each condition in $P$, and $|A| \leq |P| \leq \lambda$.

Now that we have defined the antichain $A$, we can proceed to define $E^\alpha$ and $e^\alpha_p$, for $p \in E^\alpha$. Fix $\bar{p} \in A$ and set

$$\mathbb{R} = \Pi\left(\bigcup\left\{e^\beta_{p'} : \beta < \alpha \text{ and } p' \in E^\beta \restriction \bar{p}\right\} \cup \{q_\beta : \beta < \alpha\}\right).$$

Then $\mathbb{R} \subseteq \mathbb{Q}$ and $\mathbb{R}$ is closed under $\bigwedge$. Because $\bar{p}$ meets $P$, we know that $|E^\beta \restriction \bar{p}| < \kappa$, for all $\beta < \alpha$. By induction, we know that $|e^\beta_{p'}| < \kappa$, for $p' \in E^\beta$. And $\alpha < \mu$. It follows that $|\mathbb{R}| \leq 2^{<\mu} \leq \lambda^{<\mu} = \lambda$.

Define $D_{\bar{p}} \subseteq [\mathbb{R}]^{<\kappa}$ by

$$D_{\bar{p}} = \left\{d \in [\mathbb{R}]^{<\kappa} : d \text{ is superdense in } \mathbb{R} \text{ with respect to some } p \in \Pi^2(\mathbb{R})\right\}.$$

For $d \in D_{\bar{p}}$, set

$$p(d) = \bigvee \{p \in \Pi^2(\mathbb{R}) : d \text{ is superdense in } \mathbb{R} \text{ with respect to } p\}.$$

Finally, set

$$P_{\bar{p}} = \{p(d) : d \in D_{\bar{p}}\}$$

Note that $|P_{\bar{p}}| \leq \lambda$ because $|[\mathbb{R}]^{<\kappa}| \leq \lambda$.

CLAIM 6.4. *If $\bar{p} \in G \cap A$, then $P_{\bar{p}} \cap G \neq \emptyset$.*

PROOF. Trivially $\mathbb{R}$ is superdense in $\mathbb{R}$ with respect to $\bar{p}$. Set $\tilde{p} = h_{\Pi(\mathbb{R})}(\bar{p})$. Then $\tilde{p} \in G \cap \Pi(\mathbb{R})$. By the Superdensity Reduction Lemma, there exists $p^* \in G \cap \Pi^2(\mathbb{R})$ and $d \in [\mathbb{R}]^{<\kappa}$ such that $d$ is superdense in $\mathbb{R}$ with respect to $p^*$. Then $p^* \leq p(d) \in P_{\bar{p}}$. So $p(d) \in G \cap P_{\bar{p}}$. ⊣

Apply the Lemma 6.2 to get an antichain $A_{\bar{p}} \subseteq \mathbb{Q}$ such that every condition in $A_{\bar{p}}$ meets $P_{\bar{p}}$, and $A_{\bar{p}}$ is predense with respect to any condition in $P_{\bar{p}}$, and $|A_{\bar{p}}| \leq |P_{\bar{p}}| \leq \lambda$.

Note that if $P_{\bar{p}} \cap G \neq \emptyset$, then $A_{\bar{p}} \cap G \neq \emptyset$ because $G$ is internally generic and $A_{\bar{p}}$ is a set of conditions that is predense with respect to any condition in $P_{\bar{p}} \cap G$

For $p \in A_{\bar{p}}$, set $e_p^\alpha = d$, where $d \in D_{\bar{p}}$ is chosen so that $p \leq p(d)$. Then $e_p^\alpha$ is superdense in $\mathbb{R}$ with respect to $p$.

Finally, set $E^\alpha = \bigcup_{\bar{p} \in A} A_{\bar{p}}$. This completes the construction.

**§7. The factor lemma.** Let us continue to work with the following assumptions:

- $\mathbb{P}$ is either an element of $V$ or a $V$-definable partial ordering that is separative and is closed under finite meets.
- $\mathbb{Q} \subseteq \mathbb{P}$ is, in $V$, a set-complete Boolean algebra with $\mathbb{0}$ removed, and $\mathbb{Q}$ is complete in $\mathbb{P}$.
- $G$ is a filter on $\mathbb{P}$ that is internally generic over $V$.

Working in $V$, if $Q \subseteq \mathbb{Q}$, let $\mathbb{B}(Q)$ be the set-complete (in $\mathbb{Q}$) Boolean subalgebra generated by $Q$. In general, no set in $V$ is dense in $\mathbb{B}(Q)$, even if $Q$ is a set in $V$. If $p \in \mathbb{P}$, let

$$\mathbb{B}_p(Q) = \{p \wedge q : q \in \mathbb{B}(Q) \text{ and } p \wedge q \neq \mathbb{0}\}.$$

Then $\mathbb{B}_p(Q)$ is the smallest subclass of $\mathbb{Q}$ that contains $\{p \wedge q : q \in Q\}$ and is closed under $\bigwedge$ and complements relative to $p$. That is, if $q \in \mathbb{B}_p(Q)$ and $p \wedge -q \neq \mathbb{0}$, then $p \wedge -q \in \mathbb{B}_p(Q)$.

FACTOR LEMMA 7.1. *In $V$, let $\kappa$, $\mu$, and $\lambda$ be cardinals such that*

- $\omega_1 \leq \kappa < \mu \leq \lambda$,
- *$\kappa$ is regular,*
- $\mathrm{cf}(\mu) > \kappa$, *and*
- $\lambda^{<\mu} = \lambda$.

*Let $Q \subseteq \mathbb{Q}$ have cardinality at most $\mu$. Assume that*

- *$<\kappa$-covering for subsets of $\lambda$ and*
- $ф_{\lambda,<\kappa}$

*hold between $V$ and $V[G]$. There exists $p^* \in G$ such that $\mathbb{B}_{p^*}(Q)$ has a dense subset of cardinality less than $\kappa$.*

PROOF. Let $\langle q_\alpha : \alpha < \mu \rangle$ be an enumeration of $Q$ in $V$. Let $E^\alpha$, $e^\alpha_p$, and $p^* \in G$ be as provided by the Iterated Superdensity Reduction Lemma.

Note first that there exists $p \in G$ such that $E^\alpha$ is predense with respect to $p$, for all $\alpha < \mu$: Because $G$ is internally generic, it contains a condition $p$ deciding $d = \{-\bigvee E^\alpha : \alpha < \mu\}$. But $G \cap d = \emptyset$, since $G \cap E^\alpha \neq \emptyset$, for all $\alpha$. Let $p \in G$ be incompatible with every element of $d$. Then $p \leq -\bigvee d = \bigwedge_{\alpha<\mu} \bigvee E^\alpha$.

Assume, then, that the condition $p^* \in G$ provided by Iterated Superdensity Reduction has the property that $E^\alpha$ is predense with respect to $p^*$, for all $\alpha \leq \mu$. Set

$$\begin{aligned} d_\alpha &= \bigcup\nolimits_{p \in E^\alpha \restriction p^*} e^\alpha_p; \\ R_\alpha &= \bigcup\nolimits_{\beta<\alpha} d_\beta \cup \{q_\beta : \beta < \alpha\}; \\ \mathbb{R}_\alpha &= \Pi(R_\alpha); \\ \mathbb{R}_{<\alpha} &= \bigcup\nolimits_{\beta<\alpha} \mathbb{R}_\beta; \\ \mathbb{R}^*_\alpha &= \{p^* \wedge r : r \in \mathbb{R}_\alpha\}; \text{ and} \\ \mathbb{R}^*_{<\alpha} &= \bigcup\nolimits_{\beta<\alpha} \mathbb{R}^*_\beta. \end{aligned}$$

Note that

(1) If $\alpha < \mu$, then $d_\alpha$ is superdense in $\mathbb{R}_\alpha$ with respect to $p^*$. This is because $E^\alpha$ is predense with respect to $p^*$ and $e^\alpha_p$ is superdense in $\mathbb{R}_\alpha$ with respect to $p$.

(2) Suppose that $\alpha \leq \alpha' < \mu$. If $q \in d_\alpha$ is compatible with $p^*$, then there exists $q' \in d_{\alpha'}$ such that $q' \leq q$, since $d_{\alpha'}$ is superdense in $\mathbb{R}_{\alpha'}$ with respect to $p^*$.

(3) $|d_\alpha| < \kappa$

(4) If $\alpha \leq \beta$, then $\mathbb{R}_\alpha \subseteq \mathbb{R}_\beta$. Also $\mathbb{R}_{<\alpha} \subseteq \mathbb{R}_\alpha$.

To render our notation less cumbersome, let $h_\alpha = h_{\mathbb{R}_\alpha}$, that is, for $p \in \mathbb{P}$, set

$$h_\alpha(p) = \bigwedge\{r \in \mathbb{R}_\alpha : p \leq r\}.$$

CLAIM 7.2. If $\alpha \leq \mu$ is a limit ordinal, then $\mathbb{R}^*_{<\alpha}$ is closed under complements relative to $p^*$. That is, if $r \in \mathbb{R}_{<\alpha}$ and $p^* \wedge -r \neq \mathbb{0}$, then $p^* \wedge -r \in \mathbb{R}^*_{<\alpha}$.

PROOF. Suppose that $r \in \mathbb{R}_\beta$, where $\beta < \alpha$. It suffices to see that

$$(*) \qquad p^* \wedge -r = p^* \wedge \bigwedge\{-q : q \leq r \text{ and } q \in d_{\beta+1}\}$$

because $\bigwedge\{-q : q \leq r \text{ and } q \in d_{\beta+1}\} \in \Pi(R_{\beta+2}) = \mathbb{R}_{\beta+2} \subseteq \mathbb{R}_{<\alpha}$.

On the one hand, if $q \leq r$, then $-r \leq -q$. Hence

$$-r \leq \bigwedge\{-q : q \leq r \text{ and } q \in d_{\beta+1}\}.$$

So $(*)$ holds with "$\leq$" in place of "$=$".

For the converse relation, suppose for a contradiction that

$$p^* \wedge \bigwedge\{-q : q \leq r \text{ and } q \in d_{\beta+1}\} \not\leq -r.$$

Then there exists $p$ such that

$$(**) \qquad p \leq p^* \wedge r \wedge \bigwedge\{-q : q \leq r \text{ and } q \in d_{\beta+1}\}.$$

Now $d_{\beta+1}$ is superdense in $\mathbb{R}_{\beta+1}$ with respect to $p^*$, so there exists $q \in d_{\beta+1}$ be such that $q \sim p$ and $q \leq h_{\beta+1}(p)$. Then $q \leq r$ because $p \leq r \in \mathbb{R}_\beta$. But then $p \leq -q$ by $(**)$, contradicting that $q \sim p$, and completing the proof of the Claim. ⊣

Claim 7.3. If $\alpha \leq \mu$ is a limit ordinal and $q \in \mathbb{R}_\alpha$, then $q = \bigwedge_{\beta<\alpha} h_\beta(q)$.

Proof. Since $h_\beta(q) \geq q$, it is evident that $q \leq \bigwedge_{\beta<\alpha} h_\beta(q)$. For the converse relation, let $x, y \subseteq R_\alpha$ be such that

$$q = \bigwedge_{r\in x} - r \wedge \bigwedge_{r\in y} r.$$

If $r \in R_\alpha$, then $r \in R_\beta$, for some $\beta < \alpha$. Then $-r \in \Pi(R_\beta) = \mathbb{R}_\beta$, provided $-r \neq \mathbb{0}$. Thus

- if $r \in y$, then $h_\beta(q) \leq r$, for all sufficiently large $\beta < \alpha$, and
- if $r \in x$, then $h_\beta(q) \leq -r$, for all sufficiently large $\beta < \alpha$.

Hence $\bigwedge_{\beta<\alpha} h_\beta(q) \leq \bigwedge_{r\in x} - r \wedge \bigwedge_{r\in y} r = q$. ⊣

Our goal is to see that $\mathbb{R}^*_{<\mu}$ has a dense subset of cardinality less than $\kappa$. On the road to that conclusion, let us prove

Claim 7.4.

(a) $\mathbb{R}^*_{<\mu}$ is $<\kappa$-c.c.
(b) $\mathbb{R}^*_\alpha$ is $<\kappa$-c.c., for all $\alpha < \mu$.

Proof. Note first that (b) follows from (a): If $\alpha < \mu$, then $\mathbb{R}^*_\alpha$ is closed under $\bigwedge$. Thus two conditions in $\mathbb{R}^*_\alpha$ are compatible in $\mathbb{R}^*_\alpha$ iff they are compatible in $\mathbb{R}^*_\beta$, for all $\beta \geq \alpha$. Consequently an antichain of cardinality $\kappa$ in $\mathbb{R}^*_\alpha$ would be an antichain in $\mathbb{R}^*_{<\mu}$, contradicting part (a).

To prove part (a), suppose that $A^* \subseteq \mathbb{R}^*_{<\mu}$ is an antichain (perhaps not maximal) and $|A^*| \leq \kappa$. Our goal is to see that $|A^*| < \kappa$. Let $A \subseteq \mathbb{R}_{<\mu}$ be such that $A^* = \{p^* \wedge r : r \in A\}$ and $|A| \leq \kappa$ (though $A$ may not be an antichain). Choose $\alpha < \mu$ large enough that $A \subseteq \mathbb{R}_\alpha$.

Fix $a \in A^*$. Then $a \leq p^*$. Because $d_\alpha$ is superdense with respect to $p^*$, there exists $q \in d_\alpha$ such that $q \sim a$ and $q \leq h_\alpha(a)$. Then $q \wedge p^* \neq \mathbb{0}$ because $q \sim a \leq p^*$. Let $r \in A$ be such that $a = p^* \wedge r$. Then $q \leq r$, since $q \leq h_\alpha(a)$. So $p^* \wedge q \leq p^* \wedge r$.

Because $a \in A^*$ was arbitrary, this shows that

$$\forall a \in A^* \; \exists q \in d_\alpha \; (p^* \wedge q \neq \mathbb{0} \text{ and } p^* \wedge q \leq a).$$

Because $|d_\alpha| < \kappa$ and $A^*$ is an antichain, this shows that $|A^*| < \kappa$, completing the proof of the Claim. ⊣

Recall that $d_\alpha = \bigcup_{p \in E^\alpha \restriction p^*} e^\alpha_p$. Set

$$d^*_\alpha = \{p^* \wedge r : r \in d_\alpha \text{ and } p^* \sim r\}.$$

CLAIM 7.5.

(a) If $\alpha < \mu$ is a limit ordinal and $\mathrm{cf}(\alpha) \geq \kappa$, then there exists $\beta < \alpha$ such that $d^*_\beta$ is dense in $\mathbb{R}^*_\alpha$.

(b) There exists $\beta < \mu$ such that $d^*_\beta$ is dense in $\mathbb{R}^*_{<\mu}$.

PROOF. First note that part (b) follows from part (a): Assume part (a). Recall that $\mathrm{cf}(\mu) > \kappa$. By Fodor's Lemma, there exists $\beta < \mu$ such that $d^*_\beta$ is dense in $\mathbb{R}^*_\alpha$ for $\alpha$ unbounded in $\mu$. But $\mathbb{R}^*_{<\mu} = \bigcup_{\alpha<\mu} \mathbb{R}^*_\alpha$ and $\mathbb{R}^*_\alpha \subseteq \mathbb{R}^*_\beta$, for $\alpha \leq \beta$.

Turning to the proof of part (a), begin by noting that it suffices to show that

(∗) there exists $\beta < \alpha$ such that each element of $d^*_\alpha$ is extended by some element of $d^*_\beta$.

To see that (∗) suffices, suppose $r \in \mathbb{R}_\alpha$ is compatible with $p^*$. We must find $q' \in d_\beta$ such that $p^* \wedge q' \leq p^* \wedge r$. Let $q \in d_\alpha$ be such that $q \leq h_\alpha(p^* \wedge r)$ and $q \sim p^* \wedge r$. Then $q \leq r$, since $r \in \mathbb{R}_\alpha$. And $p^* \wedge q \in d^*_\alpha$. Let $q' \in d_\beta$ be such that $p^* \wedge q' \leq p^* \wedge q$. Then $p^* \wedge q' \leq p^* \wedge r$.

For a contradiction, suppose that (∗) fails. Then

$$\forall \beta < \alpha\ \exists q \in d_\alpha\ \forall r \in d_\beta\ (p^* \wedge r \not\leq p^* \wedge q).$$

Because $|d_\alpha| < \kappa$ and $\mathrm{cf}(\alpha) \geq \kappa$, it follows that there exists $q \in d_\alpha$ such that for $\beta$ unbounded in $\alpha$, no condition in $d^*_\beta$ extends $q$. Because every condition in $d_\beta$ is extended by some condition in $d_{\beta'}$ when $\beta' \geq \beta$, it follows that

$$\forall \beta < \alpha\ \forall r \in d_\beta\ (p^* \wedge r \not\leq p^* \wedge q).$$

Now $q = \bigwedge_{\gamma<\alpha} h_\gamma(q)$. Thus

$$\forall \beta < \alpha\ \forall r \in d_\beta\ \exists \gamma < \alpha\ (p^* \wedge r \not\leq p^* \wedge h_\gamma(q)).$$

Let $\langle \gamma_i : i < \kappa \rangle$ be an increasing sequence of ordinals less than $\alpha$ such that for $i < j < \kappa$,

$$(\dagger) \qquad \forall r \in d_{\gamma_i}\ (p^* \wedge r \not\leq p^* \wedge h_{\gamma_j}(q)).$$

Using that $d_{\gamma_i}$ is superdense in $\mathbb{R}_{\gamma_i}$ with respect to $p^*$, let $r_i \in d_{\gamma_i}$ be such that $r_i \sim p^* \wedge q$ and $r_i \leq h_{\gamma_i}(p^* \wedge q)$. Then $r_i \wedge -h_{\gamma_{i+1}}(q) \sim p^*$ by (†). Set

$$q_i = p^* \wedge r_i \wedge -h_{\gamma_{i+1}}(q).$$

If $i < j < \kappa$, then $q_j \leq r_j \leq h_{\gamma_j}(q)$, but $q_i \leq -h_{\gamma_{i+1}}(q) \leq -h_{\gamma_j}(q)$. Thus $A^* = \{q_i : i < \kappa\}$ is an antichain in $\mathbb{R}^*_\alpha$, contradicting the previous Claim and completing the proof of the current one. ⊣

Trivially $\mathbb{R}^*_\alpha$ is closed under $\bigwedge$, for all $\alpha$. We have seen that if $\alpha$ is a limit ordinal, then $\mathbb{R}^*_{<\alpha}$ is closed under complements relative to $p^*$. We can now show

CLAIM 7.6. If $\alpha \leq \mu$ and $\mathrm{cf}(\alpha) \geq \kappa$, then $\mathbb{R}^*_{<\alpha}$ is closed under $\bigwedge$.

PROOF. If $x \subseteq \mathbb{R}^*_{<\alpha}$ and $\bigwedge x \neq \mathbb{0}$, then $\bigwedge x \in \mathbb{R}^*_\alpha$. Let $\beta < \alpha$ be such that $d^*_\beta$ is dense in $\mathbb{R}^*_\alpha$. Set

$$(*) \qquad q = \bigvee \{r \in d^*_\beta : r \leq \bigwedge x\}.$$

Then $q \in \mathbb{R}^*_{<\alpha}$. Certainly $q \leq \bigwedge x$. To see $\bigwedge x \leq q$, for a contradiction suppose that $p = -q \wedge \bigwedge x \neq \mathbb{0}$. Now $p \leq p^*$, so $p^* \wedge h_\alpha(p) \in \mathbb{R}^*_\alpha$. Also $h_\alpha(p) \leq -q \wedge \bigwedge x$, since $x \subseteq \mathbb{R}^*_{<\alpha}$ and $q \in \mathbb{R}^*_{<\alpha}$. Let $r \in d^*_\beta$ be such that $r \leq p^* \wedge h_\alpha(p)$. Then $r \leq \bigwedge x$, hence $r \leq q$ by $(*)$. But also $r \leq -q$. This contradiction completes the proof of Claim 7.6. ⊣

We are now ready to finish the proof of the Set Forcing Lemma. By definition, $\mathbb{B}(Q)$ is the smallest subclass of $\mathbb{Q}$ that contains $Q$ and is closed under $\bigwedge$ and non-$\mathbb{0}$ complements. Thus $\mathbb{R}_{<\mu} \subseteq \mathbb{B}(Q)$; hence $\mathbb{R}^*_{<\mu} \subseteq \mathbb{B}_{p^*}(Q)$.

For the converse inclusion, recall that $\mathbb{B}_{p^*}(Q)$ is the smallest subclass of $\mathbb{Q}$ that contains $\{p^* \wedge q : q \in Q\}$ and is closed under $\bigwedge$ and complements relative to $p^*$. By construction $\{p^* \wedge q : q \in Q\} \subseteq \mathbb{R}^*_{<\mu}$ and we have seen $\mathbb{R}^*_{<\mu}$ has these two properties.

Thus $\mathbb{B}_{p^*}(Q) = \mathbb{R}^*_{<\mu}$. There exists $\beta < \mu$ such that $d^*_\beta$ is dense in $\mathbb{R}^*_{<\mu}$. And $|d^*_\beta| < \kappa$. ⊣

## §8. The set forcing theorem.

SET FORCING THEOREM 8.1. *Let $W$ be an outer model of $V$ and assume that $(W; V)$ satisfies ZFC. In $V$, let $\kappa$, $\mu$, and $\lambda$ be cardinals such that*

- $\omega_1 \leq \kappa < \mu \leq \lambda$,
- *$\kappa$ is regular,*
- $\mathrm{cf}(\mu) > \kappa$, *and*
- $\lambda^{<\mu} = \lambda$.

*Assume that*

- *$<\kappa$-covering for subsets of $\lambda$ and*
- $\pitchfork_{\lambda,<\kappa}$

*hold between $V$ and $W$. Then every subset of $\mu$ in $W$ is set generic over $V$ for a forcing of $V$-cardinality less than $\kappa$.*

PROOF. The theorem follows from the Factor Lemma once we have represented $W$ as an internally class generic extension of $V$. This is more or less folklore and has appeared elsewhere in various forms. We include a proof here because we need to take advantage of a detail.

Because it is just as easy, we give a proof that applies even to $W$ and $V$ that do not satisfy Choice. This requires the additional assumption that every set in $W$ is the range of a function in $W$ with domain in $V$. This extra hypothesis is trivial if $W$ satisfies Choice. However, it is necessary: If $W = V[G]$ and $\mathring{a}^G \in W$, then $f : \mathrm{dom}(\mathring{a}) \to \mathring{a}^G$ given by $f(\mathring{b}) = \mathring{b}^G$ is a function onto $\mathring{a}^G$ with domain in $V$.

LEMMA 8.2. *Let $W \supseteq V$ be a countable and standard with $V \cap \mathrm{OR} = W \cap \mathrm{OR}$. Assume that both $V$ and $(W; V)$ satisfy* ZF. *Assume also that every set in $W$ is the range of a function in $W$ with domain in $V$. There exists a $V$-definable partial ordering $\mathbb{Q}$ and a ($W$-amenable) filter $G$ such that $G$ is internally $\mathbb{Q}$-generic over $V$ and $W = V[G]$. Furthermore, $(V[G]; V, \mathbb{P}, G) \models$* ZF *and, in $V$, $\mathbb{Q}$ is a set-complete Boolean algebra with $\mathbb{0}$ removed.*

PROOF. Using a simple class forcing definable over $W$, construct a $W$-amenable class function $F : V \to W$ such that

(1) $\mathrm{rng}(F) = W$;
(2) if $a \in F(b)$, then $a = F(c)$, for some $c \in b$; and
(3) $(W; V, F) \models \mathrm{ZF}$.

Ordered by reverse functional extension, conditions in this forcing are functions $f \in W$ such that

(a) $\mathrm{dom}(f) \in V$,
(b) $\mathrm{rng}(f)$ is transitive, and
(c) $f$ satisfies requirement (2).

These conditions are $<\!\infty$-closed. It is easy to check that, given $a \in W$, conditions $f$ with $a \subseteq \mathrm{rng}(f)$ are dense. Let $F$ be given by a filter sufficiently generic to insure property (c).

For future reference, note that, given any $f$ as in (a)–(c), we may assume that $f \subseteq F$, by choosing the generic $F$ below the condition $f$.

Let $\mathcal{L}$ be the language with the binary relation symbol $\in$ and a constant symbol $\dot{a}$, for each $a \in V$. Work in the smallest infinitary fragment (with equality) including all Boolean combinations of atomic formulas of $\mathcal{L}$ that lie in $V$ and contain only finitely many distinct variables. Let $T$ consist of the axioms

$$\forall x \left( x \in \dot{a} \to \bigvee_{b \in a} x = \dot{b} \right),$$

for $a \in V$. Letting $\widehat{W} = (W; F(a))_{a \in V}$, we have that $\widehat{W} \models T$.

Working in $V$, let $\mathbb{Q}$ consist of all quantifier-free infinitary sentences $q$ of $\mathcal{L}$ such that $T \cup \{q\}$ is consistent. Order $\mathbb{Q}$ by $p \geq q$ iff $T \vdash q \to p$. Modded out by the equivalence relation "$p \leq q$ and $q \leq p$", $\mathbb{Q}$ is a set-complete Boolean algebra with $\mathbb{0}$ removed.

Working in $(W; V, F)$, set $G = \{q \in \mathbb{Q} : \widehat{W} \models q\}$. Then $G$ is a filter on $\mathbb{Q}$ and is internally generic: If $d \in V$ is predense then $T \vdash \bigvee d$, so $\widehat{W} \models \bigvee d$. Hence $G \cap d \neq \emptyset$.

Clearly $V[G] \subseteq W$ because $G$ is definable over $(W; V, F)$. Conversely, in $V$ define a Shoenfield term $\mathring{a}$ for each $a \in V$ by

$$\mathring{a} = \{(\mathring{b}, \{\dot{b} \in \dot{a}\}) : b \in a\}.$$

Then $\mathring{a}^G = F(a)$. Hence $W \subseteq V[G]$. This completes the proof of the lemma. ⊣

Returning to the proof of the theorem, suppose that $a \in W$ is a subset of $\mu$. Let $F$ in the proof of the previous lemma be chosen so that $f \subseteq F$, where $f : \mu + 1 \to W$ by $f \restriction \mu = \mathrm{id} \restriction \mu$ and $f(\mu) = a$. Let $G$ be the internally $\mathbb{Q}$-generic filter provided by $F$. In $V$, set

$$Q = \{\{\dot{\alpha} \in \dot{\mu}\} : \alpha < \mu\}.$$

Then $Q \subseteq \mathbb{Q}$ is a set of $\mu$ conditions. By the Factor Lemma, there exists $p^* \in G$ such that, in $V$, $\mathbb{B}_{p^*}(Q)$ has a dense subset $\mathbb{D}$ of cardinality less than $\kappa$. Then $G \cap \mathbb{D}$ is $\mathbb{D}$-generic over $V$ and $a = F(a) = \mathring{\mu}^G \in V[G \cap \mathbb{D}]$. ⊣

REFERENCES

[B] L. Bukovský, *Characterization of generic extensions of models of set theory*, ***Fundamenta Mathematicae***, vol. 83 (1973), no. 1, pp. 35–46.

[J] T. Jech, ***Set theory, third millennium edition***, Monographs in Mathematics, Springer, 2002.

[S] M. Stanley, *Outer models and genericity*, ***The Journal of Symbolic Logic***, vol. 68 (2003), no. 2, pp. 389–418.

MATH DEPARTMENT
SAN JOSE STATE
SAN JOSE, CA 95192, USA
*E-mail*: stanley@math.sjsu.edu

# ABSTRACT VERSUS CONCRETE COMPUTABILITY: THE CASE OF COUNTABLE ALGEBRAS

J. V. TUCKER AND J. I. ZUCKER[†]

**Abstract.** We consider the distinction between abstract computability, in which computation is independent of data representations, and concrete computability, in which computations are dependent on data representations. The distinction is useful for current research in computability theories for continuous data and uncountable structures, including topological algebras and higher types. The distinction is also interesting in the seemingly simple case of discrete data and countable algebras. We give some theorems about equivalences and inequivalences between abstract models (*e.g.*, computation with 'while' programs) and concrete models (*e.g.*, computation via numberings) in the countable case.

**§1. Introduction.** By a *computability theory* we mean a theory of functions and sets that are definable using a model of computation. By a *model of computation* we mean a model of some general method of calculating the value of a function or of deciding, or enumerating, the elements of a set. We allow the functions and sets to be made from any kind of data.

With this terminology, Classical Computability Theory on the set ℕ of natural numbers is made up of many computability theories. The computable functions and computably enumerable sets are definable by scores of models of computation, based on scores of ideas about machines, programs, algorithms, specifications, rewriting systems, and calculi. It was an important early discovery, in 1936, that different models of computation can be shown to define the same classes of functions and sets. The fact that diverse computability models lead to the same classes of functions and sets on ℕ is the main pillar supporting the Church-Turing Thesis, which gives the classical theory on ℕ its extraordinary unity.

Starting in the 1940s, computability theories have been created for other special sets of data, including:

- higher types over the natural numbers,
- ordinals and set hierarchies,

[†]The research of the second author was supported by a grant from the Natural Sciences and Engineering Research Council of Canada.

**Logic Colloquium '03**
Edited by V. Stoltenberg-Hansen and J. Väänänen
Lecture Notes in Logic, 24

- real numbers and Banach spaces, and
- higher types over real numbers.

More generally, computability theories have been created for axiomatically defined classes of structures, such as

- groups, rings, fields,
- equational and first order theories of universal algebras,
- topological and metric spaces and algebras,
- domains, and
- categories.

Over the years, however, the classification and the proofs of equivalences of models of computation — and, hence, the search for generalised Church-Turing-type Theses and the theoretical unity they represent — have proved much more difficult to achieve for these data types. Cases exist where different models of computation that were of equal conceptual value have been shown not to be equivalent. For example, in higher types, there is Tait's theorem that the fan functional on total functions is recursively continuous but not computable in Kleene's schemes S1–S9 [Tai62, Nor80]. Another more recent example, in computability on real numbers, is the non-equivalence of the models of computability studied in Computable Analysis [PER89, Wei00] and in the theory developed in [BSS89, BCSS98].

There has been much research activity in the above areas in the past decade. For example, Computable Analysis has been greatly extended by mathematicians and computer scientists using *competing* models. Of course, seen from the point of view of Classical Computability Theory, this exciting and rapid growth of Computable Analysis is also rather messy.

In this paper, we seek a general explanation for this phenomenon of non-equivalent computability theories by focussing on the treatment of the data in models of computation. We classify computability theories into two types by introducing the following two concepts.

In an *abstract computability theory* the computations are *independent* of *all* the representations of the data. Computations are *uniform* over all representations and are isomorphism invariants.

In a *concrete computability theory* the computations are *dependent* on *some* representation of the data. Computations are *not* uniform, and different representations can yield different results. Computations are not isomorphism invariants.

Typical of abstract models of computation are models based on abstract ideas of *program*, *equation*, *scheme*, or *logical formula*. Typical of concrete models of computation are those based on concrete ideas of *coding*, *numbering*, or *data representations* using numbers or functions.

Over the years, many people have considered particular problems where abstract and concrete theories can be compared. Whilst the distinction can hardly be new, it is rarely made even in particular cases. One example of the distinction is that of Dag Normann's *internal* ($\approx$ abstract) and *external* ($\approx$ concrete) in higher type computability [Nor82].

For instance, studying examples and analysing algorithms are important in investigating computation on topological algebras. This is usually done using pseudocode, a high-level informal algorithmic language. Typically, the high-level programming and specification languages are designed using abstract models, and low-level implementations are designed using concrete models. It is to be expected there is a need for both abstract and concrete models, and an understanding of their relationship.

Now the distinction of abstract versus concrete is also directly relevant to the seemingly stable and unified classical world of computability on countable algebras. We will consider the distinction in this special case.

First, in Section 2, we describe the basic concepts in more detail and define the ideas of *soundness*, *adequacy* and *completeness* between abstract and concrete models. We point out some examples of the difference between these kinds of computability theory for topological algebras and higher types. Then, in the rest of the paper, we examine the problem of abstract versus concrete for countable algebras. In Section 3 we introduce our concrete models based on codings or numberings of algebras using the natural numbers. In Section 4 we define our abstract models based on 'while' programs equipped with arrays and non-deterministic choice, which may interpreted over any algebra. In Section 5 we discuss the soundness of the abstract for the concrete, under weak assumptions on countable algebras. In Section 6 we discuss adequacy and completeness and give some simple completeness theorems under stronger assumptions. Finally, we note connections with other work in the countable case (*e.g.*, the Ash-Nerode theorem), and pose some problems.

We wish to thank Ulrich Berger, Dag Normann and Viggo Stoltenberg-Hansen for helpful discussions on the subject of this paper, and an anonymous referee for useful comments.

**§2. Abstract and concrete computability theories.** In this section we elaborate the general ideas given in the Introduction. First, let us make some relevant reflections on Computability Theory.

**2.1. Reflections on computability theory and computer science.** Over the past decade or so, a number of surveys and personal reflections on Computability Theory have appeared. There have been the *Handbook of Computability Theory* [Gri99] and the two-volume *Handbook of Recursive Mathematics* [EGNR98], and chapters on the subject in other Handbooks, for example [AGM92]. The reflection [Soa96] proposed a modernisation or, at least, a "make-over" of the subject. The reflection [Fen02] looks back at the distinction between structure and algorithm in generalised computability theory, debated at the Oslo meetings in the 1970s. Some of the technical development of generalised computability theory of that period was influenced by the reflection [Kre71].

Indeed, encouraged by the symbolism of the year 2000, there have been plenty of surveys and reflections on mathematics. Several propose the idea that *algorithms and computations are returning to prominence throughout mathematics*, enriched by the abstract mathematics of the 20th century [Sma98, Gro98] and encouraged by the new ubiquity of mathematics in the contemporary world [Bou01]. The return to prominence of computation in mathematics can be explained by reflecting on the origins of abstraction and structure in 19th century mathematics. But it is, of course, more accurate to say that what helps make the idea sweet enough for many to swallow is the ever growing, all pervasive influence of computing.

The world is a lot more mathematical than it was, thanks to Computer Science.

This is good news for Computability Theory. Computability Theory is a pure mathematical subject that can be said to have *led* the early development of Computer Science. Since its inception, Computability Theory has provided deep conceptual insights, new algorithmic and programming ideas and techniques, and mathematical theories on which to found software technology. For example, in the standard courses on computability one can find the origins of the following *core* ideas of Computer Science: universal computer; recursion; lambda calculus; rewrite systems, formal specification of computations; higher types; decision problems and their classification; data representations; data type implementations; computational complexity; and so on.

From the standpoint of contemporary Computer Science, the mathematical investigations in Computability Theory that led to these fundamental ideas can be seen as brilliant theoretical speculations whose intellectual and practical value is rising daily. Quite simply, Computability Theory is pure mathematics at its best!

Today, computability retains this important speculative role in our quest to understand the big ideas of algorithm, data, specification, program, machine, and, not surprisingly, many computability theorists — including ourselves — are working as computer scientists.

Now, in Computer Science, it is *obvious* that a computation is fundamentally dependent on its data. By a *data type* we mean (i) data, together with (ii) some primitive operations on data. Often we also have in mind the ways these data are (iii) axiomatically specified and (iv) represented or implemented. To compute, we think hard about what forms of data the user may need, how we might model the data in designing a system — where some high level but formal understanding is important — and how we might implement the data in some favoured programming language.

Seen from Computer Science, we propose that

*A computability theory should be focussed equally on the data and the algorithms.*

Now this idea may be difficult to appreciate if one works in one of the classical computability theories of the natural numbers, for data representations rarely seem to be an important topic there. Although the translation between Turing machines and register machines involve data transformations, these can be done on an *ad hoc* basis.

However, representations are *always* important. Indeed, representations are a subject in themselves. This is true even in the simple cases of discrete data forming countable sets, from Gödel numberings of syntax to general numberings of countable sets and structures. From the beginning there has been a great interest in the differences and non-equivalences between numberings, for example in Mal'cev's theory of computability on sets and structures [Mal71b]. The study of different notions of reduction and equivalence of numberings, and the space or spectrum of numberings, has had a profound influence on the theory and has led to quite remarkable results, such as Goncharov's Theorem and its descendants [EGNR98, SHT99a]. These notions also include the idea of invariance under computable isomorphisms, prominent in computable algebra, starting in [FS56]. In the general theory of computing with numbered structures that are countable, there was always the possibility of computing relative to a reasonable numbering that was standard in some sense. For example, earlier work on the word problem for groups, such as [MKS76], and on computable rings and fields, such as [Rab60], did not need to worry about numberings in order to develop fine theorems. Indeed, some authors dispensed with explicit numberings altogether and worked with algebras of natural numbers, thus removing what we consider to be the basic notion of representation.

We see the importance of representations even more clearly when computing with continuous data forming uncountable sets. For example, in computing with real numbers, it has long been known that care must be taken over the representations: if one represents the reals by decimal expansions then one cannot even compute multiplication. But if one chooses the Cauchy sequence representations then a great deal is computable [BH02].

Now there is a chasm between computing with discrete data in countable sets and with continuous data in uncountable sets. Today, it is a prominent problem to understand computability on continuous data because of the possible "new" role of computation in mathematics (*e.g.*, algebra and analysis), and because in computer science there are problems in understanding exact computation with real numbers, and high level programming languages for continuous data. Indeed, in physics there are applications to computing by experiment (*e.g.*, quantum computation) and the debate on the existence of non-computable physical systems; see the survey in [BT04].

We need a unification of the many specialised computability theories that can handle arbitrary data and enable us to integrate discrete and continuous data in specifications, computation and proofs. This could have applications in understanding digital and analogue data in hybrid computing and communication systems; first order and higher order data in specifications, programs and verifications; and discrete and continuous data in simulation, and experiments with physical systems.

**2.2. Abstract and concrete models for computing.** We present a number of *informal* ideas about computing functions and sets on *any* kind of data.

DEFINITION (Model of computation). A *model of computation* is a mathematical model of some general method of calculating the values of functions, or deciding, or enumerating, the elements of sets. A *computability theory* is a theory of functions and sets that are definable using a model of computation.

In an *abstract model of computation* the idea is to compute strictly within a set $D$ of data using programs based on some primitive operations on $D$. The choice of operations on the data creates a structure which defines how data is created and tested in programs. Here is an attempt to define this.

DEFINITION (Abstract model of computation).

(a) An *abstract model of computation* for computing with a set $D$ of data consists of
  (i) a structure $A$ containing the data $D$;
  (ii) a set $\boldsymbol{Prog}(\Sigma)$ of programs based on the signature $\Sigma$ of the structure, naming the primitive functions and relations of $A$; and
  (iii) a semantics

$$[\![\_]\!]^A : \boldsymbol{Prog}(\Sigma) \to \boldsymbol{Func}(D)$$

  that defines partial functions on $D$.

(b) A partial function $f : D \to D$ is *computable* (according to this model) if there exists a program $P \in \boldsymbol{Prog}(\Sigma)$ that defines it, *i.e.*,

$$f = [\![P]\!]^A.$$

(c) A subset $S$ of $D$ is *semicomputable* if it is the domain of a computable partial function.

The intention is that since the programs can use only the functions and predicates of the structure, the programs are at exactly the same level of abstraction as the primitive operations of the structure, as far as the data is concerned. In fact, we expect the programs to be independent of the representations of the data and to generate computations uniformly over all representations: a program $P \in \textbf{\textit{Prog}}(\Sigma)$ can be run on different structures of common signature $\Sigma$. In an abstract model of computation, we may expect that the programs and computations are invariant under isomorphisms. In symbols:

**Invariance Principle.** Let $\phi : A \to B$ be a $\Sigma$-isomorphism of structures. For all $a \in A$,

$$\phi([\![P]\!]^A(a)) = [\![P]\!]^B(\phi(a)).$$

Thus, an abstract model of computation is "abstract" in exactly the same way that an abstract algebra is "abstract": it ignores the "nature" of the elements and concentrates on the properties of operations on the elements. This explains the choice of the name *abstract*.

The structures we have in mind include:

- particular single-sorted structures (*e.g.*, strings, natural numbers, integers, reals);
- classes of single sorted structures (*e.g.*, groups, rings and fields);
- particular, and classes of, finitely sorted structures (*e.g.*, hereditarily finite sets and vector spaces, modules, normed spaces, metric spaces);
- particular, and classes, of infinitely sorted structures (*e.g.*, functions of all finite types).

In a *concrete model of computation*, the idea is to implement computations in a set $D$ of data using computations on another "simpler" set $R$ of data, using a "simpler" computability theory on $R$. Here is an attempt to define this.

Definition (Concrete model of computation). A *concrete model of computation* for computing with a set of data $D$ consists of

(i) a representation of the data in $D$ by data from another set $R$ via a mapping $\alpha : R \to A$,

(ii) a computability theory $\textbf{\textit{Comp}}(R)$ on $R$.

A partial function $f : D^m \rightharpoonup D$ is *computable with respect to* $\alpha$ if there is a computable function $\varphi \in \textbf{\textit{Comp}}(R)$ that tracks $f$ in $R$ in the sense that, for appropriate $r \in R$,

$$f(\alpha^m(r)) = \alpha(\varphi(r));$$

equivalently, if the following diagram commutes:

$$\begin{array}{ccc} D^m & \xrightarrow{\ f\ } & D \\ \uparrow{\scriptstyle \alpha^m} & & \uparrow{\scriptstyle \alpha} \\ R^m & \xrightarrow[\ \varphi\ ]{} & R \end{array}$$

In a concrete model the computations are dependent on some choice of data representation $\alpha$. Computations need not be isomorphism invariant. In fact they need not be uniform, and different representations can yield different results. An important task is to define when one representation reduces, or is equivalent to, another.

A concrete model of computation is "concrete" in exactly the same way as a specific concrete algebra is "concrete". The term "concrete" is chosen as an opposite of "abstract".

In most examples, the representation $R$ is constructed from, or can be reduced to, the natural numbers $\mathbb{N}$ and functions and functionals on $\mathbb{N}$, and the computability theory $\boldsymbol{Comp}(R)$ is some form of classical computability theory on $\mathbb{N}$ or $\mathbb{N} \to \mathbb{N}$.

**2.3. Soundness, adequacy and completeness.** In general, what is the relationship between abstract and concrete computability models with a common set of data $D$?

Let $\boldsymbol{AbstComp}_A(D)$ be the set of functions on the data set $D$ that are computable in an abstract model of computation associated with a structure $A$ containing $D$. Let $\boldsymbol{ConcComp}_\alpha(D)$ be the set of functions on $D$ that are computable in a concrete model of computation with representation $\alpha$.

DEFINITION (Soundness). An abstract model of computation $\boldsymbol{AbstComp}_A(D)$ is *sound* for a concrete model of computation $\boldsymbol{ConcComp}_\alpha(D)$ if

$$\boldsymbol{AbstComp}_A(D) \subseteq \boldsymbol{ConcComp}_\alpha(D).$$

DEFINITION (Adequacy). An abstract model of computation $\boldsymbol{AbstComp}_A(D)$ is *adequate* for a concrete model of computation $\boldsymbol{ConcComp}_\alpha(D)$ if

$$\boldsymbol{ConcComp}_\alpha(D) \subseteq \boldsymbol{AbstComp}_A(D).$$

DEFINITION (Completeness). An abstract model of computation

$$\boldsymbol{AbstComp}_A(D)$$

is *complete* for a concrete model of computation $\boldsymbol{ConcComp}_\alpha(D)$ if it is both sound and adequate, *i.e.*,

$$\boldsymbol{AbstComp}_A(D) = \boldsymbol{ConcComp}_\alpha(D).$$

In general, when comparing abstract and concrete models of computation for a common set $D$ of data, we must compare a structure $A$ and a representation $\alpha$.

In the abstract model, the choice of $\Sigma$-structure $A$ and the set of programs $\boldsymbol{Prog}(\Sigma)$ will determine what is, and is not, computable on $D$. Change the operations or programming constructs, and the model changes. In the concrete model, the choice of representation $\alpha : R \to D$, and the computability model $\boldsymbol{Comp}(R)$ on $R$, will determine what is, and is not, computable on $D$.

In comparing them, first we ask if the primitive operations on the structure $A$ of the abstract model are computable using the concrete model. If not then the abstract model cannot be sound for the concrete model. If the answer is yes, then it is sensible to ask about soundness and adequacy. We then meet the following problem. Let $ConcRep(D)$ be some class of concrete representations of the form $\alpha : R \to D$.

**General Completeness Problem.**

$$\text{Is } AbstComp_A(D) = \bigcap_{\alpha \in ConcRep(D)} ConcComp_\alpha(D)?$$

Now, the abstract model relies *exclusively* on these primitive operations of $A$. However, the representation $\alpha$ allows *all* the properties of the *particular* structure $R$ to be used in defining computations on $D$. Thus, we may expect that in practice,

> *if the primitive operations are concretely computable, then computing with a concrete representation R of the data of D enables more functions on these data to be computed than computing with abstract programs based solely on the operations of a structure A containing these data.*

For example, when $A$ is some finite or countably infinite abstract structure *without* an ordering and $R$ is $\mathbb{N}$, then a concrete model of computation is able to use the total ordering of the data representations in $\mathbb{N}$ as the basis of a global search operator, which is then computable in the concrete model but may not be in the abstract model.

One reason for designing an abstract model is to abstract away from the details of data representations and to define computations that are *uniform* over a class of representations, i.e., to design an abstract model that is sound for the concrete model. What about adequacy and completeness? These properties we might expect to be possible in special cases, where we can limit the class of representations, or allow access to low-level operations in the high-level programs. But high-level languages that capture all the algorithms that may be needed are clearly desirable. Designing high-level languages (= abstract models) that can do this is a familiar problem.

**2.4. Many-sorted signatures and algebras.** We give a short introduction on many-sorted algebras. More details may be found in any of [TZ99, TZ00, TZ04, TZ05].

Given a signature $\Sigma$ with finitely many *sorts* $s, \ldots$ and *function symbols*

(1) $$\mathsf{F} : s_1 \times \cdots \times s_m \to s,$$

a $\Sigma$-algebra $A$ consists of a carrier $A_s$ for each $\Sigma$-sort $s$, and a function

$$\mathsf{F}^A : A_1 \times \cdots \times A_m \to A_s$$

for each $\Sigma$-function symbol as in (1). In general, functions $F^A$ are assumed to be partial. Special further possible properties of $\Sigma$ and the $\Sigma$-algebra $A$ are:

(1) **Standard algebras.** $\Sigma$ and $A$ are *standard* if they contain the sort bool of *booleans* and the corresponding carrier $A_{\mathsf{bool}} = \mathbb{B} = \{\mathtt{t}, \mathtt{f}\}$, together with the standard boolean and boolean-valued operations, including the conditional at all sorts, and equality at certain sorts ("equality sorts");

(2) **N-standard algebras.** $\Sigma$ and $A$ are *N-standard* if they contain the sort nat of *natural numbers* and the corresponding carrier $A_{\mathsf{nat}} = \mathbb{N} = \{0, 1, 2, \dots\}$, together with the standard arithmetical operations of zero, successor, equality and order on $\mathbb{N}$.

In this paper we will always assume (1), *i.e.*,

Assumption 2.1 (Standardness assumption). All signatures $\Sigma$ and $\Sigma$-algebras $A$ are standard.

We will also usually assume N-standardness, but will state that assumption explicitly.

In any case, any standard signature $\Sigma$ can be N-*standardised* to a signature $\Sigma^N$ by adjoining the sort nat and the standard arithmetical operations listed in (2) above. Correspondingly, any standard $\Sigma$-algebra $A$ can be N-*standardised* to an algebra $A^N$ by adjoining the carrier $\mathbb{N}$ together with the corresponding standard functions.

(3) **Array algebras.** We also consider *array signatures* $\Sigma^*$ and *array algebras* $A^*$, which are formed from N-standard signatures $\Sigma$ and algebras $A$ by adding, for each sort $s$, an *array sort* $s^*$, with corresponding carrier $A_s^*$ consisting of all arrays or finite sequences over $A_s$, together with certain standard array operations, as specified in [TZ00, §2.7] or (in an equivalent but simpler version) in [TZ99, §2.4].

**2.5. Computability theories on many-sorted algebras.** There are many abstract computability theories on algebras. Some have their beginnings in

(a) generalisations of computability theory, *e.g.*, prime and search computability of Moschovakis [Mos69a, Mos69b], definability of Montague [Mon68], and the finite algorithmic procedures of Friedman [Fri71] and Shepherdson [She73];

(b) the theory of programs (*e.g.*, flow charts, program schemes, 'while' programs).

A huge number of models of computation and computability theories have been classified, and there many equivalence theorems between disparate models. Most models are based on ideas about programming, but some are based on ideas about definability and specification. The literature has been surveyed in [She85, TZ88, TZ00].

Abstract models of computability theories on algebras have decent mathematical theories, inspired by, but distinct from, the theory of computable functions on the natural numbers. In particular, the abstract case is well understood, and there are stable proposals for generalised Church-Turing Theses [TZ00].

The primary method of modelling concrete representations for computing on algebras is by means of numberings of the form

$$\alpha : \Omega_\alpha \to A \quad (\Omega_\alpha \subseteq \mathbb{N})$$

for all, or part, of an algebra $A$, and in which the operations, or their restrictions, are recursive on codes. If $A$ is countable, $\alpha$ can be taken to be surjective.

The problem of "abstract versus concrete" for countable algebras is examined in the following sections. We will choose the 'while' language and its extensions as our abstract model of computation, and find conditions under which

$$\mathit{WhileCC}^*(A) = \bigcap_{\alpha \in \mathit{ConcRep}(A)} \mathit{Comp}_\alpha(A).$$

**2.6. Example: Computability theories on topological many-sorted algebras.** Consider many-sorted algebras equipped with a topology and continuous operations. Analysis makes heavy use of such algebraic structures, through topological groups and vector spaces, Banach spaces, Hilbert spaces, $C^*$ algebras, and many more. These many-sorted topological algebras specify

(i) some basic operations;
(ii) normal forms for the algebraic representation of elements (*e.g.*, using bases);
(iii) structure-preserving operators (*i.e.*, homomorphisms such as linear operators);
(iv) approximations, through inner products, norms, metrics and topologies.

We can create abstract computability theories by simply applying the abstract models of §2.4 to these algebras. However, thanks to the method of approximation (iv), we obtain two classes of functions: the *computable* functions and the *computably approximable* functions, defined as follows:

DEFINITION. Let $A$ be an algebra with metric $\mathrm{d} : A^2 \to \mathbb{R}$. A function $f : A \to A$ is *computably approximable* in an abstract model if there is an abstract program $P \in \mathbf{Prog}(\Sigma)$ such that

$$\mathrm{d}(f(a), [\![P]\!]^A(n, a)) < 2^{-n}$$

for all $a \in A$.

Despite the simplicity of this abstract approach, most computable analysis is done using *concrete models* of computation. In the case of concrete computability, there have been a number of general approaches to the analysis and classification of metric and topological structures:

- *Effective metric spaces* [Cei59, Mos64];
- *Computable sequence structures for Banach spaces* [PER89];

- *Type 2 enumerations* [Wei00];
- *Algebraic domain representations* [SHT88, SHT95];
- *Continuous domain representations* [Eda97];
- *Numbered spaces* [Spr98].

In fact, for certain basic topological algebras, most of these concrete computability theories have been shown to be essentially equivalent [SHT99b].

The comparison of abstract and concrete models of computation for topological algebras is a promising research area.

First, we must look at the primitive operations of the algebras. In the concrete models computable functions are continuous. Thus tests, like equality and order, which are total functions with discontinuities, cannot be primitive operations. If the discontinuities are removed and the functions are made partial then the functions become continuous and may be allowed as operations. For example, the continuous partial form of equality $=_p \,:\, A^2 \to \mathbb{B}$ is defined by

$$(x =_p y) \simeq \begin{cases} \uparrow & \text{if } x = y \\ \text{ff} & \text{otherwise.} \end{cases}$$

Thus, partial algebras play an essential role.

Next, the *computably approximable functions* have been found to be necessary to bridge the gap between abstract and concrete in the case of total functions on the real numbers [TZ99]. Furthermore, it has become clear that multivaluedness is necessary to bridge the abstract and concrete for general classes of topological algebras. In [TZ04] there is a detailed comparison using our 'while'-array language with nondeterministic countable choice construct

$$\mathsf{x} := \mathsf{choose}\ \mathsf{z} : b(\mathsf{z}, \mathsf{x}, \mathsf{y})$$

(see Section 4). Soundness is proved in considerable generality, but completeness demands a number of special hypotheses, though there is no shortage of examples. These equivalence results have been extended to include local coverings of metric algebras in [TZ05]. We show that for effectively locally uniformly continuous functions, and a wide class of metric algebras, approximation by 'while'-array programs with countable choice is equivalent to a simple general concrete computational model based on effective metric spaces.

Briefly, let us consider the case of the real numbers. To compute on the set $\mathbb{R}$ of real numbers with an abstract model of computation, we have to choose a structure $A$ in which $\mathbb{R}$ is a carrier set. There are infinitely many choices of operations with which to make an algebra, and so there are infinitely many choices of classes of computable functions. Thanks to the theory of computable functions on many-sorted algebras (§2.4), all these classes of abstractly computable functions on $\mathbb{R}$ will have decent mathematical theories. For example, computability on rings and fields of reals, with and without orderings, forms the basis of abstract computability theories in [BCSS98], which have been extended to complexity theory. A problem with these particular abstract theories is that, when interpreted on $\mathbb{R}$, equality and order are assumed total, violating continuity (see remarks above). Earlier such theories have been given by Herman and Isard [HI70] and Engeler [Eng68].

In contrast, to compute on the set $\mathbb{R}$ of real numbers with a concrete model of computation, we choose an appropriate concrete representation of the set $\mathbb{R}$, such as computable Cauchy sequences. The study of the computability of the reals began with [Tur36], but was only later taken up in a systematic way, in [Ric54, Lac55a, Lac55b, Lac55c, Grz55, Grz57] for example.

For a special partial metric algebra $\mathcal{R}_p^N$ over $\mathbb{R}$, with continuous partial operations of equality and order, and a special total metric algebra $\mathcal{R}_t^N$ over $\mathbb{R}$, with a continuous total division operation of reals by naturals, (both algebras containing the sort of naturals as well as reals), the following can be shown [TZ05, §4.4]:

THEOREM. *Let* $f : \mathbb{R}^n \to \mathbb{R}^m$ *be a total function that is effectively locally uniformly continuous on a standard open exhaustion of* $\mathbb{R}^n$. *Then the following are equivalent*:

(i) $f$ *is Grzegorczyk-Lacombe computable on* $\mathbb{R}$;
(ii) $f$ *is effectively trackable on the computable reals* $\mathbb{R}_c$;
(iii) $f$ *is effectively locally* $\mathbb{Q}$*-polynomial approximable on* $\mathbb{R}$;
(iv) $f$ *is locally* 'while' *approximable on* $\mathcal{R}_t^N$;
(v) $f$ *is locally* 'while'*-array approximable on* $\mathcal{R}_t^N$;
(vi) $f$ *is locally* 'while'*-array with countable choice approximable on* $\mathcal{R}_p^N$.

There is much to explore on the borderline between abstract and concrete computability: notions of approximation, limit processes, nondeterminism and multivaluedness. There are the important results of Brattka [Bra96, Bra97] that show that by strengthening a fundamental abstract computability model (relations defined by primitive recursion, minimalisation and a limit operation) it is possible to characterise a fundamental concrete computability model (relations defined by type 2 enumerability). The implications of Brattka's results for other abstract models need further investigation.

**2.7. Computability theories on higher types.** The computation of functionals on higher types over the natural numbers $\mathbb{N}$ has been analysed in many different abstract and concrete computability theories the classification of which has been, and continues to be, a hugely important undertaking. Higher type computations are technically complicated. One must consider total and partial functionals on total and partial functions as arguments, which may or may not be continuous. See the concise survey [Nor99] for an introduction and the invaluable historical survey [Lon04] for a full description.

Abstract models for computable partial higher type functionals begin to be studied in earnest with the Kleene schemes S1–S9, which were first proposed in [Kle59] for computing on all the total functionals, and later reinterpreted on the total continuous functionals. These schemes have an enumeration scheme (S9) involving algorithms — but not data — and were refined by other equivalent abstract models based on the $\lambda$-calculus, in [Pla66] and [Mol77], for the hereditarily monotone functionals. A basic abstract model of computation on higher types is a simply typed $\lambda$-calculus with fixed points. An example in computer science is the language PCF and its extensions, which are abstract models of functional programming.

Different concrete models were also introduced by Kleene and Kreisel [Kle59, Kre59] for total functionals with total arguments. Kleene's *recursively continuous functionals* are the total functionals on total functions that are continuous and have recursive associates. Other concrete models include the *hereditarily recursive continuous functions* and the *hereditarily effective operators*. Ershov proved the latter are equivalent, thus generalising the Kreisel-Lacombe-Schoenfield Theorem to all finite types. However, they are incomparable with the recursively continuous functionals.

The comparison of the early models of higher type computation encountered problems of comparing abstract and concrete models. Up to type 2, Kleene's S1–S9 computability is complete for his notion of recursively continuous. Indeed, at all finite types, Kleene's S1–S9 computability is sound for the recursively continuous functionals. However at type 3, we have Tait's theorem that the fan functional at type 3 is recursively continuous but not computable by Kleene's S1–S9. Thus, completeness fails at type 3 [Tai62, Nor80].

However the first bridging theorems between abstract and concrete were proved for partial (rather than total) continuous functionals defined on domains. Computable functions on domains are defined to have recursively enumerable sets of approximations [SHLG94]; they generalise the recursively continuous functionals.

In the remarkable study [Plo77] of the language PCF, it was shown that whilst PCF was sound for all the computable partial functionals over the domain of naturals, it was not complete. However, PCF augmented with a parallel conditional and existential operator was complete for the computable partial functionals.

Berger conjectured that if one restricts to total functions on total arguments then the parallel constructs are not needed and PCF is complete w.r.t. the computable total continuous functionals [Ber93]. This striking fact was proved by Normann [Nor00].

Although PCF and S1–S9 are equivalent over partial continuous functionals, this result does not contradict Tait's Theorem, because in that case S1–S9 are interpreted over arguments that are total, or total continuous, functionals (*cf.* scheme S8), whereas (as stated above) the domain semantics of PCF is defined over partial arguments.

Recently, computability in higher types has also been studied over the real numbers. Escardo [Esc96] shows that a language Real PCF, containing parallel extensions, is complete with the computable partial functionals over the interval domain representation of the reals. However, the corresponding version of Berger's Conjecture is open.

Other examples of abstract versus concrete models at higher type can be extracted from [Lon04]. An overview on totality, continuity, computability can be found in [Ber02].

**§3. Concrete computation: Numbering of algebras.** In this section we assume $A$ is a countable N-standard partial $\Sigma$-algebra.

**3.1. Numberings.**

DEFINITION 3.1 (Numbering). A *numbering* of $A$ is a family

$$\alpha = \langle \alpha_s : \Omega_{\alpha,s} \twoheadrightarrow A_s \mid s \in \mathit{Sort}(\Sigma)\rangle$$

of surjective maps $\alpha_s : \Omega_{\alpha,s} \twoheadrightarrow A_s$, for some family

$$\Omega_\alpha = \langle \Omega_{\alpha,s} \mid s \in \mathit{Sort}(\Sigma)\rangle$$

of sets $\Omega_{\alpha,s} \subseteq \mathrm{N}$, the *code sets* of $\alpha$; we say that $A$ is *numbered* by $\alpha$. If $u = s_1 \times \cdots \times s_m$, we write $\Omega^u_\alpha =_{df} \Omega_{s_1} \times \cdots \times \Omega_{s_m}$. We assume that $\alpha_{\mathsf{nat}} = id_{\mathrm{N}}$ and that $\alpha_{\mathsf{bool}} : 0 \mapsto \mathfrak{f},\ n(\neq 0) \mapsto \mathfrak{t}$.

Let $\alpha : \Omega \twoheadrightarrow A$ be a fixed numbering of $A$.

DEFINITION 3.2 (Kernel). The *kernel* of $\alpha$ is the family of relations

$$\equiv_\alpha = \{\equiv_{\alpha,s} \subseteq \mathrm{N}^2 \mid s \in \mathit{Sort}(\Sigma)\}$$

where for all $\Sigma$-sorts $s$

$$\equiv_{\alpha,s} = \{(k,l) \in \Omega^2_{\alpha,s} \mid \alpha_s(k) = \alpha_s(l)\}.$$

DEFINITION 3.3 (Tracking function). Let $f : A^u \rightharpoonup A_s$ and $\varphi : \Omega^u_\alpha \rightharpoonup \Omega_{\alpha,s}$. Then $\varphi$ is called an *$\alpha$-tracking function for $f$* if the following diagram commutes:

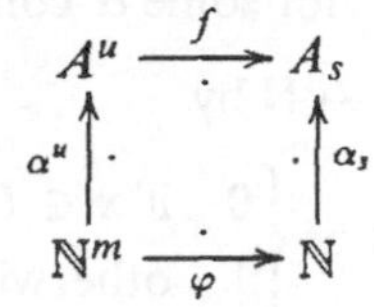

in the sense that for all $k \in \Omega^u_\alpha$

$$f(\alpha^u(k))\downarrow \Longrightarrow \varphi(k)\downarrow \in \Omega_{\alpha,s} \wedge f(\alpha^u(k)) = \alpha_s(\varphi(k))$$

and

$$f(\alpha^u(k))\uparrow \Longrightarrow \varphi(k)\uparrow.$$

Note that a possible alternative definition of "$\alpha$-tracking function" would replace the second clause above by

$$f(\alpha^u(k))\uparrow \Longrightarrow \varphi(k)\uparrow \vee \varphi(k)\downarrow\notin \Omega_{\alpha,s}.$$

However the completeness proof for ***WhileCC****** computability given in [TZ04, Thm B] requires this (apparently) stronger definition of $\alpha$-tracking function.

Note also that nothing is said about the behaviour of $\varphi$ off $\Omega^u_\alpha$.

**3.2. Computable functions and semicomputable sets.**

DEFINITION 3.4 ($\alpha$-computability and $\alpha$-semicomputability).

(a) A function on $A$ is *$\alpha$-computable* if it has a computable (*i.e.*, partial recursive) $\alpha$-tracking function.
(b) A relation on $A$ is *$\alpha$-computable* if its characteristic function is $\alpha$-computable.
(c) A relation on $A$ is *$\alpha$-semicomputable* if it is the domain of an $\alpha$-computable function.

We write $\boldsymbol{Comp}_\alpha(A)$ for the set of $\alpha$-computable functions on $A$, and $\boldsymbol{SComp}_\alpha(A)$ for the set of $\alpha$-semicomputable relations on $A$.

We say that $\Omega_\alpha$ is *recursive* (or *r.e.*, or *co-r.e.*) if $\Omega_{\alpha,s}$ is recursive (etc.) for each $\Sigma$-sort $s$. There is a similar meaning for $\equiv_\alpha$ being recursive (etc.).

LEMMA 3.5. *Let $U \subseteq A^u$. Then $U$ is $\alpha$-semicomputable* $\Longleftrightarrow$

(2) $\qquad (\alpha^u)^{-1}[U] = S \cap \Omega^u_\alpha \quad$ *for some r.e.* $S \subseteq \mathbb{N}$.

PROOF. ($\Rightarrow$) Suppose $U$ is $\alpha$-semicomputable. Then $U = \boldsymbol{dom}(f)$, where $f : A^u \rightharpoonup A_s$ is $\alpha$-computable. Let $\varphi : \Omega^u_\alpha \rightharpoonup \Omega_{\alpha,s}$ be a computable tracking function for $f$. Let $S = \boldsymbol{dom}(\varphi)$. Then $S$ is r.e., and it is easy to check that

$$(\alpha^u)^{-1}[U] = S \cap \Omega^u_\alpha.$$

($\Leftarrow$) Suppose (2). We must show:

(3) $\qquad U = \boldsymbol{dom}(f) \quad$ for some $\alpha$-computable $f$.

Define $f : A^u \rightharpoonup \mathbb{N}$ and $\varphi : \mathbb{N}^m \rightharpoonup \mathbb{N}$ by

$$f(x) \simeq \begin{cases} 0 & \text{if } x \in U \\ \uparrow & \text{otherwise} \end{cases}$$

and

$$\varphi(k) \simeq \begin{cases} 0 & \text{if } k \in S \\ \uparrow & \text{otherwise.} \end{cases}$$

Then, since $S$ is r.e., $\varphi$ is computable. Also it is easy to check that $\varphi$ is an $\alpha$-tracking function for $f$, thus proving (3). ⊣

COROLLARY 3.6. *Suppose $\Omega_\alpha$ is r.e. Then for $U \subseteq A^u$,*

$$U \text{ is } \alpha\text{-semicomputable} \iff (\alpha^u)^{-1}[U] \text{ is r.e.}$$

**3.3. Computability of numberings, code sets and kernels.** We are interested in various computability aspects of $\alpha$, namely $\alpha$-computability of the $\Sigma$-operations, and computability or semicomputability of the *code set* and *kernel*.

DEFINITION 3.7 ($\Sigma$-effective algebra under $\alpha$). The partial algebra $A$ is $\Sigma$-*effective* under $\alpha$, or *$\alpha$-$\Sigma$-effective*, if for every $F \in$ ***Func***$(\Sigma)$, $F^A$ is $\alpha$-computable.

We also describe this by saying that $\alpha$ *has computable $\Sigma$-operations*.

Note that in [TZ04, TZ05] the concepts defined in Definitions 3.3, 3.4 and 3.7 were called *strict $\alpha$-tracking*, *strict $\alpha$-computability* and *strict $\Sigma$-effectivity* of $\alpha$ respectively.

DEFINITION 3.8 (Effective algebra under $\alpha$). The partial algebra $A$ is *effective* under $\alpha$, or *$\alpha$-effective*, if $A$ is $\Sigma$-effective under $\alpha$, and also $\Omega_\alpha$ is r.e.

DEFINITION 3.9 (Computable algebra under $\alpha$). (a) The partial algebra $A$ is *computable* (or *semicomputable*, or *cosemicomputable*) under $\alpha$, or *$\alpha$-computable* (or *$\alpha$-semicomputable*, or *$\alpha$-cosemicomputable*), if $A$ is $\alpha$-effective, and also, for all $\Sigma$-sorts $s$ there is a recursive (or r.e., or co-r.e.) relation $R \subseteq \mathbb{N}^2$ such that

$$R \cap \Omega^2_{\alpha,s} = \equiv_{\alpha,s}.$$

(b) $A$ is said to be *computable* (etc.) if it is computable (etc.) under some numbering.

We also say that $\alpha$ is a *$\Sigma$-effective* (or *effective*, or *computable*, etc.) numbering of $A$ if $A$ is $\Sigma$-effective (etc.) under $\alpha$.

Let $\Sigma$-***EffNum***$(A)$ and ***CompNum***$(A)$ be the set of all $\Sigma$-effective and computable (respectively) numberings of $A$.

Note that the following are equivalent:

(i) $\alpha$ is $\alpha$-computable by $id_{\Omega_\alpha}$, which is computable on $\mathbb{N}$,

(ii) $\Omega_\alpha$ is r.e.

This is clear from the diagram:

$$\begin{array}{ccc} \mathbb{N} & \xrightarrow{\ \alpha_s\ } & A_s \\ {\scriptstyle (\alpha_{nat}=)id_{\mathbb{N}}}\uparrow & & \uparrow{\scriptstyle \alpha_s} \\ \mathbb{N} & \xrightarrow[\ id_{\Omega_{\alpha,s}}\ ]{} & \Omega_{\alpha,s} \end{array}$$

LEMMA 3.10. *Suppose $\Omega_\alpha$ is r.e., i.e., $\Omega_{\alpha,s}$ is r.e. for all $s$. For all $s$, let $\nu_s$ be a total recursive function with range $\Omega_{\alpha,s}$. Let $\beta : \mathbb{N} \twoheadrightarrow A$ be the numbering of $A$ defined by*

$$\beta =_{df} \alpha \circ \nu =_{df} \{\alpha_s \circ \nu_s \mid s \in \textbf{\textit{Sort}}(\Sigma)\}.$$

*Then*

(a) *If a function* $f : A^u \to A_s$ *is* $\alpha$*-tracked by* $\varphi$*, then it is* $\beta$*-tracked by the function* $\psi$ *defined as follows:*

$$\begin{array}{ccc} A^u & \xrightarrow{\;f\;} & A_s \\ \uparrow{\scriptstyle \alpha^u} & & \uparrow{\scriptstyle \alpha_s} \\ \Omega^u_\alpha & \xrightarrow[\varphi]{\cdot} & \Omega_{\alpha,s} \\ \uparrow{\scriptstyle \nu_u} & & \uparrow{\scriptstyle \nu_s} \\ \mathbb{N}^m & \xrightarrow[\psi]{\cdot} & \mathbb{N} \end{array}$$

(b) *For any* $\Sigma$*-sort* $s$*, if*

$$\equiv_{\alpha,s} = R \cap \Omega_{\alpha,s}$$

*for some computable (or semicomputable, or cosemicomputable) relation* $R$ *on* $\mathbb{N}$*, then* $\equiv_{\beta,s}$ *is computable (or semicomputable, or cosemicomputable, respectively).*

PROOF. For (a), the algorithm for $\psi$ is: with *input* $k \in \mathbb{N}^m$, compute $\varphi(\nu_u(k))$. (If and) when this converges, find an $l$ such that $\nu_s(l) = \varphi(\nu_u(k))$ (which exists, by surjectivity of $\nu$ on $\Omega$). *Output* such an $l$.

Part (b) is proved by noting that $\Omega_{\beta,s} = \mathbb{N}$. ⊣

COROLLARY 3.11. *Suppose* $\Omega_\alpha$ *is r.e. Then, with* $\beta$ *defined as in Lemma* 3.10:

(a) *If* $A$ *is* $\alpha$*-effective, then* $A$ *is* $\beta$*-effective.*

(b) *If* $A$ *is computable (or semicomputable, or cosemicomputable) under* $\alpha$*, then* $A$ *is also computable (or semicomputable, or cosemicomputable) under* $\beta$*.*

REMARK 3.12. Hence if $A$ is $\alpha$-effective, or $\alpha$-computable, or $\alpha$-(co)semicomputable, then we can assume without loss of generality that $\Omega_{\alpha,s} = \mathbb{N}$ for all $s$.

Further, under this assumption on $\Omega_\alpha$, the definition (3.9) of computability (etc.) of an algebra can be simplified to:

*A is* $\alpha$*-computable (or* $\alpha$*-semicomputable, or* $\alpha$*-cosemicomputable) iff A is* $\alpha$*-effective, and also* $\equiv_\alpha$ *is recursive (or r.e., or co-r.e.).*

LEMMA 3.13. (Canonical extension of numbering to N-standard and array algebra). *A numbering* $\alpha$ *of* $A$ *can be canonically extended to numberings* $\alpha^N$ *of* $A^N$*, and* $\alpha^*$ *of* $A^*$*, such that if* $A$ *is* $\Sigma$*-effective, effective, computable, or (co-)semicomputable under* $\alpha$*, then so are* $A^N$ *under* $\alpha^N$*, and* $\alpha^*$ *under* $A^*$*.*

**§4. Abstract computation: the *While* language and its extensions.** Again, we assume that $A$ is a countable N-standard $\Sigma$-algebra.

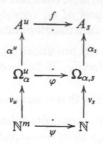

**4.1. The *WhileCC** programming language.** The programming language *WhileCC*($\Sigma$) ('CC' for "countable choice") is an extension of *While*($\Sigma$) [TZ00, §3] with an extra 'choose' rule of term formation. The idea of the semantics for 'choose' is to select *all* possible implementations satisfying a given property. The complete description of its syntax and semantics, as well as motivation for it, are given in [TZ04, TZ05]. Here we give a brief review.

The language *WhileCC* has a 'choose' construct in the context of an assignment statement, which has one of three forms:

(i) $\mathsf{x} := t$ (simultaneous assignment),
(ii) $\mathsf{x} := \mathsf{choose}\ \mathsf{z} : b(\mathsf{z}, \dots)$,
(iii) $\mathsf{x} := \mathsf{choose}\ \mathsf{z} : P(\mathsf{z}, \dots)$,

where z is a variable of sort nat, and in (ii) $b(\mathsf{z}, \dots)$ is a *boolean term*, and in (iii) $P(\mathsf{z}, \dots)$ is a *semicomputable predicate* of z (and other variables), *i.e.*, the halting set of a boolean-valued *WhileCC* procedure $P$ with z among its input variables.

Thus the semantics of *WhileCC* is given by a many-valued function.

In [TZ04] an algebraic operational semantics is given for *WhileCC*, whereby a *WhileCC* procedure $P : u \to v$ has a meaning in an N-standard $\Sigma$-algebra $A$:

$$P^A : A^u \to \mathcal{P}_\omega^+(A^v \cup \{\uparrow\})$$

where $\mathcal{P}_\omega^+(X)$ is the set of all countable *non-empty* subsets of $X$, and '$\uparrow$' represents a divergent computation. This is also written in "multivalued function" notation:

$$P^A : A^u \rightrightarrows^+ A^{v^\uparrow}$$

where $X^\uparrow$ denotes $X \cup \{\uparrow\}$.

**4.2. *WhileCC** computable functions and semicomputable sets.**

DEFINITION 4.1. (a) A partial function on $A$ is *WhileCC**($\Sigma$) computable on $A$ if it is *WhileCC*($\Sigma$) computable on $A^*$.

(b) *WhileCC*($A$) is the class of all *WhileCC*($\Sigma$) computable partial functions on $A$.

(c) *WhileCC**($A$) is the class of all *WhileCC**($\Sigma$) computable partial functions on $A$.

LEMMA 4.2. *If $A$ is total then*
(a) *WhileCC*($A$) = *While*($A$).
(b) *WhileCC**($A$) = *While**($A$).

PROOF. We can implement $\mathsf{choose}\ \mathsf{z} : b(\mathsf{z})$ as $\mu\mathsf{z} : b(\mathsf{z})$, since by totality, for each value $\mathsf{n} = 0, 1, 2, \dots$ of z, computation of $b(\mathsf{n})$ converges to a value $\mathbb{t}$ or $\mathbb{f}$. ⊣

DEFINITION 4.3 (*WhileCC*-semicomputability). (a) The *halting set* of a *WhileCC* procedure $P : u \to v$ on $A$ is the set

$$\{x \in A^u \mid P^A(x) \backslash \{\uparrow\} \neq \emptyset\}.$$

(b) A set is ***WhileCC-semicomputable*** on $A$ if it is the halting set on $A$ of a ***WhileCC*** procedure.

(c) ***WhileCC*** *semicomputability* is defined similarly.

(d) ***SWhileCC***$(A)$ is the class of ***WhileCC***-semicomputable relations on $A$.

(e) ***SWhileCC****$(A)$ is defined similarly.

If $R$ is the halting set of the procedure $P$, then a *code* or *Gödel number* of $R$ is given by a code of $P$.

**4.3. Closure under projections and countable unions.**

DEFINITION 4.4 (Minimal carrier). A carrier $A_s$ of an algebra $A$ is said to be *$\Sigma$-minimal* if every element of it is the value of a closed $\Sigma$-term of sort $s$.

DEFINITION 4.5 (TEP). (a) The *term evaluation representing function* on $A$ relative to a tuple of variables $\mathsf{x} : u$ and $\Sigma$-sort $s$ is the function

$$\boldsymbol{te}^A_{\mathsf{x},s} : \ulcorner \boldsymbol{Term}_{\mathsf{x},s} \urcorner \times A^u \to A_s$$

where $\boldsymbol{Term}_{\mathsf{x},s}$ is the class of $\Sigma$-terms of sort $s$ with variables among $\mathsf{x}$ only, and for any term $t \in \boldsymbol{Term}_{\mathsf{x},s}$ and $a \in A^u$, $\boldsymbol{te}^A_{\mathsf{x},s}(\ulcorner t \urcorner, a)$ is the value of $t$ when $\mathsf{x}$ is assigned the value $a$.

(b) The partial algebra $A$ has the *term evaluation property* (*TEP*) if for all $\mathsf{x}$ and $s$, the function $\boldsymbol{te}^A_{\mathsf{x},s}$ is ***While***$(A)$ computable.

LEMMA 4.6. *The function* $\boldsymbol{te}^A_{\mathsf{x},s}$ *is always* ***While****$(A)$ *computable.*

LEMMA 4.7 (Closure under projection).

(a) ***SWhileCC***$(A)$ *and* ***SWhileCC****$(A)$ *are closed under projection off* nat, *i.e., existential quantification over* $\mathbb{N}$.

(b) *If* $A_s$ *is minimal, then* ***SWhileCC****$(A)$ *is closed under projection off sort* $s$, *i.e., existential quantification over* $A_s$.

(c) *If* $A_s$ *is minimal and* $A$ *has TEP, then* ***SWhileCC***$(A)$ *is closed under projection off sort* $s$, *i.e., existential quantification over* $A_s$.

PROOF. (a) Note that $\exists n R(x,n) \Longleftrightarrow R(x, \mathsf{choose}\ n : R(x,n))$.

(b) Note that $\exists y R(x,y) \Longleftrightarrow \exists n R(x, \boldsymbol{te}^A_s(n, \langle\rangle))$.

(c) Like (b). Use Lemma 4.6. ⊣

LEMMA 4.8 (Closure under effective countable unions). *If* $A$ *has TEP, then* ***SWhileCC****$(A)$ *is closed under effective countable unions.*

PROOF. From TEP there follows a Universal Function Theorem for ***WhileCC****: namely there is a ***WhileCC**** function

$$\mathsf{Univ}^A_{u\to\mathsf{bool}} : \mathbb{N} \times A^u \to \mathbb{B}$$

which is universal for $\boldsymbol{Proc}_{u\to\mathsf{bool}}$ on $A$, in the sense that for all $P \in \boldsymbol{Proc}_{u\to\mathsf{bool}}$ and $x \in A^u$,

$$\mathsf{Univ}^A_{u\to\mathsf{bool}}(\ulcorner P \urcorner, x) \simeq P^A(x).$$

This can be proved by an adaptation of the methods of [TZ00] (*cf.* [Jia03]). Now let $f : \mathbb{N} \to \mathbb{N}$ be a total computable numbering of a sequence of codes of ***WhileCC**** semicomputable relations $R_n \subseteq A^u$ ($n = 0, 1, 2, \ldots$). We may assume w.l.o.g. that the procedures enumerated by $f$ all have range type bool. Then

$$\begin{aligned}\bigcup_n R_n &= \{x \in A^u \mid \exists n\ P_{f(n)}(x)\downarrow\}\\ &= \{x \in A^u \mid \exists n\ \mathsf{Univ}^A_{u\to\mathsf{bool}}(f(n), x)\downarrow\}\end{aligned}$$

which is in ***SWhileCC**** by Lemma 4.7(a). ⊣

**4.4. Locality of computation.** The locality of computation theorem, proved for the deterministic ***While*** language in [TZ00, §3.8], also applies to ***WhileCC****. The proof, in broad lines, follows the proof in [TZ00]. (For $X \subseteq A$, we write $\langle X\rangle^A_v$ for the retract to $A^v$ of the $\Sigma$-subalgebra of $A$ generated by $X$.)

LEMMA 4.9 (Locality for terms). *If* $t : s$ *and* $var(t) \subseteq \mathsf{x}$ *then*

$$[\![t]\!]^A\sigma \subseteq \langle\sigma[\mathsf{x}]\rangle^A_v.$$

PROOF. By structural induction on $t$, as in [TZ00, §3.8]. The interesting new case is $t \equiv \mathsf{choose}\ \mathsf{z} : b$. ⊣

LEMMA 4.10 (Locality for computation trees). *If* ***var***$(S) \subseteq \mathsf{x} : u$, *then for all* $n$

$$\textit{\textbf{CompTreeStage}}^A(S, \sigma, n)[\mathsf{x}] \subseteq \langle\sigma[\mathsf{x}]\rangle^A_u.$$

Here the l.h.s. means the set of $u$-tuples $\sigma'[\mathsf{x}]$ for all states $\sigma'$ in

$$\textit{\textbf{CompTreeStage}}^A(S, \sigma, n),$$

*i.e.*, the first $n$ stages of the computation tree of statement $S$ at state $\sigma$ [TZ04, §4.2(*e*)].

PROOF. By induction on $n$. ⊣

From this follows:

LEMMA 4.11 (Locality for procedures). *If* $P^A : u \to v$ *is a* ***WhileCC****$(\Sigma)$ *procedure, then for all* $a \in A^u$,

$$P^A(a)\backslash\{\uparrow\} \subseteq \langle a\rangle^A_v.$$

THEOREM 4.12 (Locality for functions). *If* $f : A^u \rightharpoonup A_s$ *is* ***WhileCC**** *computable on* $A$, *then for any* $a \in A^u$, *if* $f(a)\downarrow$ *then*

$$f(a) \in \langle a\rangle^A_s.$$

**§5. A general soundness theorem.** Again, $A$ is a countable partial $\Sigma$-algebra. However we need not assume N-standardness of $A$ in this section.

**5.1. Soundness theorem.**

THEOREM 5.1 (Soundness of *WhileCC** computation). *Suppose $A$ is $\Sigma$-effective under $\alpha$. Then*

$$\boldsymbol{WhileCC}^*(A) \subseteq \boldsymbol{Comp}_\alpha(A).$$

This is Theorem $A_0$ in [TZ04, §7]. A complete proof is given there. Here we present, as an indication of the proof, the last part of the main lemma (Lemma Scheme 7.3.1($g$)) from which the theorem easily follows.

LEMMA 5.2 (Tracking of procedure evaluation). *For a specific triple of lists of variables* a : $u$, b : $v$, c : $w$, *let* $\boldsymbol{Proc}_{\mathsf{a,b,c}}$ *be the class of all* ***WhileCC*** *procedures of type* $u \to v$, *with declaration* 'in a out b aux c'. *The (many-valued) procedure evaluation function localised to this declaration:*

$$\boldsymbol{PE}^A_{\mathsf{a,b,c}} : \boldsymbol{Proc}_{\mathsf{a,b,c}} \times A^u \rightrightarrows^+ A^{v^\dagger}$$

*defined by*

$$\boldsymbol{PE}^A_{\mathsf{a,b,c}}(P, a) = P^A(a),$$

*is $\alpha$-tracked by a computable function*

$$\boldsymbol{pe}^A_{\mathsf{a,b,c}} : \ulcorner \boldsymbol{Proc}_{\mathsf{a,b,c}} \urcorner \times \Omega^u_\alpha \to \Omega^v_\alpha,$$

*i.e., the following diagram commutes:*

$$\begin{array}{ccc}
\boldsymbol{Proc}_{\mathsf{a,b,c}} \times A^u & \xrightarrow{\ \boldsymbol{PE}^A_{\mathsf{a,b,c}}\ } & +A^{v^\dagger} \\
\Big\uparrow {\scriptstyle (\boldsymbol{enum}, \alpha^u)} & & \Big\uparrow {\scriptstyle \alpha^v} \\
\ulcorner \boldsymbol{Proc}_{\mathsf{a,b,c}} \urcorner \times \Omega^u_\alpha & \xrightarrow[\ \boldsymbol{pe}^A_{\mathsf{a,b,c}}\ ]{} & \Omega^v_\alpha
\end{array}$$

*in the sense that*

$$\boldsymbol{pe}^A_{\mathsf{a,b,c}}(\ulcorner P \urcorner, k) \downarrow l \Longrightarrow \alpha(l) \in \boldsymbol{PE}^A_{\mathsf{a,b,c}}(P, \alpha(k)),$$
$$\boldsymbol{pe}^A_{\mathsf{a,b,c}}(\ulcorner P \urcorner, k) \uparrow \Longrightarrow \ \uparrow \in \boldsymbol{PE}^A_{\mathsf{a,b,c}}(P, \alpha(k)).$$

Here $\boldsymbol{pe}^A_{\mathsf{a,b,c}}$ is a combination "tracking function" and "selection function". We can think of $\boldsymbol{pe}^A_{\mathsf{a,b,c}}$ as giving one possible implementation of $\boldsymbol{PE}^A_{\mathsf{a,b,c}}$.

PROOF OF THEOREM 5.1. Suppose $f : A^u \rightharpoonup A_s$ is ***WhileCC**** computable on $A$. Then there is a deterministic ***WhileCC**** procedure

$$P : u \to s$$

such that for all $x \in A^u$,

$$f(x) \downarrow y \Longrightarrow P^A(x) = \{y\},$$
$$f(x) \uparrow \Longrightarrow P^A(x) = \{\uparrow\}.$$

Hence by Lemma 5.2 (substituting a suitable constant for $\ulcorner P \urcorner$) there is a computable (partial) function

$$\varphi : \Omega_\alpha^u \overset{\cdot}{\rightarrow} \Omega_{\alpha,s}$$

which $\alpha$-tracks $f$, as required.

Note that in applying Lemma 5.2 to prove Theorem 5.1, we are implicitly using the canonical extension $\alpha^*$ of $\alpha$ given by Lemma 3.13. ⊣

**5.2. Applications.**

COROLLARY 5.3. *Suppose $A$ is $\Sigma$-effective under $\alpha$. Then for any relation on $A$,*

$$\textit{WhileCC}^*\text{-semicomputability} \Longrightarrow \alpha\text{-semicomputability}.$$

PROOF. By the Soundness Theorem 5.1. ⊣

Let '$=^A$' be the family of equality relations $\langle =_s^A \mid s \in \mathbf{Sort}(\Sigma)\rangle$.

COROLLARY 5.4. (a) *If $=^A$ is **WhileCC**$^*$-computable, then $\alpha$ has a computable kernel.*

(b) *If $=^A$ is **WhileCC**$^*$-semicomputable, then $\alpha$ has a semicomputable kernel.*

DEFINITION 5.5. (a) A ***DE*** *(disjunctive-existential) $\Sigma$-formula* is one of the form

$$\bigvee_{i=0}^{\infty} \exists \mathsf{y}_i \, b_i(\mathsf{x}, \mathsf{y}_i)$$

*i.e.*, an infinite disjunction of a *computable sequence* of existentially quantified $\Sigma$-booleans, $\mathsf{x} \equiv (\mathsf{x}_1, \ldots, \mathsf{x}_k)$ and $\mathsf{y}_i \equiv (\mathsf{y}_{i1}, \ldots, \mathsf{y}_{il_i})$.

(b) $\boldsymbol{DE}(A)$ is the class of $\boldsymbol{DE}(\Sigma)$ definable relations on $A$.

THEOREM 5.6 (Invariance of $\alpha$-semicomputability). *Let $A$ be a countable $\Sigma$-algebra. Then*

$$\boldsymbol{DE}(A) \subseteq \bigcap_{\alpha \in \Sigma\text{-}\boldsymbol{EffNum}(A)} \boldsymbol{SComp}_\alpha(A).$$

.

PROOF. Straightforward. To deal with evaluation in $A$ of the sequences $\mathsf{y}_i$ of bound variables (of finite but unbounded length), we need term evaluation on $A^*$, which is ***While****-computable on $A^N$ [TZ00, §4.7], and hence (by Lemma 3.13 and the Soundness Theorem 5.1) $\alpha^*$-computable. ⊣

It follows from Engeler's Lemma [TZ00, §§5.11/12] and the above theorem that

*projective **While** semicomputability*

$\Longrightarrow$ ***DE**-definability*

$\Longrightarrow$ *$\alpha$-semicomputability for all $\alpha \in \Sigma$-**EffNum**$(A)$.*

The converse directions, *i.e.*, finding "reasonable" side conditions under which the above arrows can be reversed, remain to be investigated.

**§6. Examples of completeness theorems.** We begin with an example of incompleteness of abstract w.r.t. concrete computation, and then find conditions on the algebra which guarantee completeness.

**6.1. An example of incompleteness.** Consider the single-sorted algebra

$$A = (\mathbb{N}, 0, \mathsf{pd})$$

where pd is the predecessor function on $\mathbb{N}$: $\mathsf{pd}(0) = 0$, $\mathsf{pd}(n+1) = n$.

This algebra is clearly computable (in the sense of Definition 3.9), since it is computable under *e.g.* the identity numbering $\boldsymbol{id} : \mathbb{N} \to \mathbb{N}$. For a more general kind of concrete representation of $A$ we have the following incompleteness result. Let $\mathsf{suc} : \mathbb{N} \to \mathbb{N}$ be the successor function, $\mathsf{suc}(n) = n + 1$.

THEOREM 6.1. *For any computable numbering $\alpha$ of $A$, we have*

$$\mathsf{suc} \in \boldsymbol{Comp}_\alpha(A) \qquad \textit{but} \qquad \mathsf{suc} \notin \boldsymbol{WhileCC}^*(A)$$

*and hence*

$$\boldsymbol{WhileCC}^*(A) \subsetneq \bigcap_{\alpha \in \boldsymbol{CompNum}(A)} \boldsymbol{Comp}_\alpha(A).$$

PROOF. First, suppose $\mathsf{suc} \in \boldsymbol{WhileCC}^*(A)$. Then by the locality of computation theorem (4.12), we would have $n + 1 \in \langle n \rangle^A$. But

$$\langle n \rangle^A = \{0, 1, \dots, n\},$$

and so $\mathsf{suc} \notin \boldsymbol{WhileCC}^*(A)$.

Next, let $\alpha : \Omega_\alpha \to \mathbb{N}$ be a computable numbering of $A$ with 0 $\alpha$-coded by $\mathsf{c} \in \Omega_\alpha$ and pd recursively $\alpha$-tracked by $\varphi : \Omega_\alpha \to \Omega_\alpha$. Define $\psi : \mathbb{N} \rightharpoonup \mathbb{N}$ by

$$\psi(k) \simeq \text{some } l\ [l \in \Omega_\alpha \wedge l \not\equiv_\alpha \mathsf{c} \wedge \varphi(l) \equiv_\alpha k].$$

Since $\Omega_\alpha$ is r.e. and $\equiv_\alpha$ is recursive, $\psi$ is recursive. Since $\varphi : \Omega_\alpha \to \Omega_\alpha$, it follows that

$$\psi : \Omega_\alpha \to \Omega_\alpha$$

and for all $k, l \in \Omega_\alpha$:

$$\begin{aligned} \psi(k) \downarrow l &\Rightarrow l \not\equiv_\alpha \mathsf{c} \wedge \varphi(l) \equiv_\alpha k \\ &\Leftrightarrow \alpha(l) \neq 0 \wedge \mathsf{pd}(\alpha(l)) = \alpha(k) \\ &\Leftrightarrow \alpha(k) + 1 = \alpha(l). \end{aligned}$$

So $\varphi$ recursively $\alpha$-tracks suc. ⊣

Call the above sort of naturals (with 0 and predecessor) $\mathsf{nat}^-$. Note that essentially the same counterexample to completeness can be constructed on $\mathsf{nat}^-$ even if we $\mathbb{N}$-standardise $A$, *i.e.*, adjoin the standard sort nat of naturals with 0 and successor, in addition to $\mathsf{nat}^-$.

**6.2. Sections, equality and completeness.** We turn to conditions which guarantee completeness, *i.e.*, a sort of converse of the Soundness Theorem 5.1.

We assume again that $A$ is an N-standard partial $\Sigma$-algebra.

DEFINITION 6.2 (Sections). A *section* of $\alpha$ is a right inverse of $\alpha$, *i.e.*, a family

$$\hat{\alpha} = \langle \hat{\alpha}_s \mid s \in \boldsymbol{Sort}(\Sigma) \rangle$$

of mappings

$$\hat{\alpha}_s : A_s \longrightarrow \mathbb{N}$$

such that

$$\alpha \circ \hat{\alpha} = \boldsymbol{id}_A, \tag{4}$$

*i.e.*, for all $s \in \boldsymbol{Sort}(\Sigma)$,

$$\alpha_s \circ \hat{\alpha}_s = \boldsymbol{id}_{A_s}.$$

Note from (4) that (by the property of left and right inverses) $\alpha$ is onto (which we already knew) and $\hat{\alpha}$ is 1-1. So for all $a \in A$, $\hat{\alpha}(a)$ selects an element of $\alpha^{-1}(\{a\})$ (which is not empty, since $\alpha$ is onto).

Thus a section $\hat{\alpha}$ of $\alpha$ always exists, by the Axiom of Choice. The interesting question is: when does $\alpha$ have a *computable* section?

LEMMA 6.3. *Suppose $\alpha$ is **While*** computable. Then the following are equivalent*:

(1) *$\alpha$ has a **WhileCC*** computable section $\hat{\alpha}$,*
(2) *$=^A$ is **WhileCC*** computable,*
(3) *$=^A$ is **WhileCC*** semicomputable.*

PROOF. (1)⇒(2): Assume (1). Then for all $a, b \in A_s$:

$$a = b \Longleftrightarrow \hat{\alpha}_s(a) = \hat{\alpha}_s(b) \quad \text{in } \mathbb{N},$$

which gives a ***WhileCC**** decision procedure for $=^A$.

(2)⇒(3): trivial.

(3)⇒(1): Suppose $=^A_s$ is ***WhileCC**** semicomputable. Define $\hat{\alpha}_s : A_s \to \mathbb{N}$ as follows. With *input* $a \in A_s$, the *output* is given by

$$\mathsf{choose\ k} : \alpha_s(\mathsf{k}) = a,$$

where we are choosing an item satisfying a ***WhileCC****-semicomputable predicate. ⊣

Note that in the proof of (1)⇒(2), ***WhileCC**** computability of $\alpha$ was not used.

THEOREM 6.4 (Adequacy of ***WhileCC**** computation). *Suppose $\alpha$ is **While*** computable. Then any of the conditions* (1), (2) *or* (3) *of Lemma* 6.3 *implies*:

$$\boldsymbol{Comp}_\alpha(A) \subseteq \boldsymbol{WhileCC}^*(A).$$

PROOF. Assume any of (1), (2) or (3) of Lemma 6.3. (We will actually use (1).) Let $f$ be $\alpha$-computable on $A$ by a recursive $\varphi$:

$$\begin{array}{ccc} A^u & \xrightarrow{\ f\ } & A_s \\ {\scriptstyle \alpha^u}\uparrow\downarrow{\scriptstyle \hat{\alpha}_u} & & \uparrow{\scriptstyle \alpha_s} \\ \Omega^u_\alpha & \xrightarrow[\varphi]{} & \Omega_{\alpha,s} \end{array}$$

Then

$$f = \alpha_s \circ \varphi \circ \hat{\alpha}_u$$

which is ***WhileCC****** computable. ⊣

Note that $\Sigma$-effectivity of $A$ under $\alpha$ is not assumed in the proof of adequacy (Theorem 6.4) or Lemma 6.3. It is, however, assumed for the reverse inclusion, soundness (Theorem 5.1).

Combining Theorems 5.1 (soundness) and 6.4 (adequacy) we have

THEOREM 6.5 (Completeness of ***WhileCC****** computation, Version 1). *Suppose $A$ is $\Sigma$-effective under $\alpha$, and $\alpha$ is **While*** computable. Then any of the conditions* (1), (2) *or* (3) *of Lemma* 6.3 *implies:*

$$\boldsymbol{Comp}_\alpha(A) = \boldsymbol{WhileCC}^*(A).$$

Combining this with Lemma 6.3 provides another formulation of completeness:

THEOREM 6.6 (Completeness of ***WhileCC****** computation, Version 2). *Suppose $A$ is $\Sigma$-effective under $\alpha$, and $\alpha$ is **WhileCC*** computable. Then the following are equivalent*:

(1) *$\alpha$ has a **WhileCC*** computable section,*
(2) *$=^A$ is **WhileCC*** computable,*
(3) *$=^A$ is **WhileCC*** semicomputable,*
(4) ***Comp***$_\alpha(A) =$ ***WhileCC****$(A)$ *and $\alpha$ has a semicomputable kernel,*
(5) ***Comp***$_\alpha(A) =$ ***WhileCC****$(A)$ *and $\alpha$ has a computable kernel.*

Clearly, if $A$ satisfies the assumptions of this theorem, and also has equality at all sorts in its signature, then

$$\boldsymbol{Comp}_\alpha(A) = \boldsymbol{WhileCC}^*(A).$$

For our next corollary, we suppose that $A$ is a *finitely generated $\Sigma$-algebra*, say

$$A = \langle \mathrm{c}_1^A, \dots, \mathrm{c}_p^A \rangle_A.$$

Then $A$ has a *canonoical numbering* $\alpha_{\mathrm{c}}$, defined as the composition

$$\mathrm{N} \xrightarrow{\ \gamma\ } T(\Sigma, \mathrm{x}) \xrightarrow{\ te_c^A\ } A$$

where $\mathsf{x} \equiv (\mathsf{x}_1,\ldots,\mathsf{x}_p)$ is a tuple of variables of the same type as $\mathsf{c} \equiv (\mathsf{c}_1,\ldots,\mathsf{c}_p)$, $T(\Sigma,\mathsf{x})$ is the set of $\Sigma$-terms generated from $\mathsf{x}$, $\gamma$ is a *standard effective numbering* of $T(\Sigma,\mathsf{x})$, and $te^A_{\mathsf{c}}$ is the evaluation of terms in $T(\Sigma,\mathsf{x})$ determined by the assignment $\langle \mathsf{x} \mapsto \mathsf{c}^A \rangle$. By [TZ00, Corollary 4.7], $\alpha_{\mathsf{c}}$ is ***For***$^*$ (and hence ***While***$^*$) computable. This yields the following Corollary of Theorem 6.6.

COROLLARY 6.7 (Completeness for algebras with canonical numberings). *Suppose $A$ is a finitely generated $\Sigma$-algebra, and is $\Sigma$-effective under the canonical numbering $\alpha_{\mathsf{c}}$. Then the following are equivalent:*

(1) *$\alpha_{\mathsf{c}}$ has a **WhileCC**$^*$ computable section,*
(2) *$=^A$ is **WhileCC**$^*$ computable,*
(3) *$=^A$ is **WhileCC**$^*$ semicomputable,*
(4) *$\mathbf{Comp}_{\alpha_{\mathsf{c}}}(A) = \mathbf{WhileCC}^*(A)$ and $\alpha_{\mathsf{c}}$ has a semicomputable kernel,*
(5) *$\mathbf{Comp}_{\alpha_{\mathsf{c}}}(A) = \mathbf{WhileCC}^*(A)$ and $\alpha_{\mathsf{c}}$ has a computable kernel.*

A similar result holds when $A$ is generated not by a finite set of $\Sigma$-constants, as above, but by an infinite set $\mathsf{c}_0, \mathsf{c}_1, \mathsf{c}_2, \ldots$, where the function $n \mapsto \mathsf{c}_n$ is in $\Sigma$.

**6.3. Invariance, definability and the Ash-Nerode theorem.** In concrete computability theories based on numbered structures, invariance questions of the following kind arise:

**Invariance Problem.** Let $A$ be a computable $\Sigma$-algebra. What are

$$\bigcap_{\alpha \in CompNum(A)} \mathbf{Comp}_\alpha(A) \quad \text{and} \quad \bigcap_{\alpha \in CompNum(A)} \mathbf{SComp}_\alpha(A)?$$

Viewed from the theory of numbered algebras, our work on soundness and completeness can be seen as trying to answer these sorts of questions using abstract computability theories.

However, recalling the role of the arithmetic hierarchy for computability theories on $\mathbb{N}$, it is also natural to ask if definability in logical languages based on $\Sigma$ can characterise the computability of functions and sets on a numbered algebra $A$. Definability problems of this kind, first considered in [Mal71a], can provide *some* answers to the Invariance Problem. Not surprisingly, very special properties of $A$ are needed for completeness, and it seems that to progress we must consider either

(i) particular structures, such as matrix groups, or
(ii) structures satisfying stringent conditions that are hard to satisfy.

We will discuss one source of illumination of the latter kind.

Now, thanks to soundness, we already have the general observation (Theorem 5.6) that, for any countable $A$, effective infinite disjunctions of existentially quantified $\Sigma$-booleans are $\alpha$-semicomputable under every $\Sigma$-effective numbering $\alpha$ of $A$.

We will look at the Ash-Nerode Theorem which contains strong sufficient conditions on $A$ that imply the converse and, hence, a completeness result. The Ash-Nerode Theorem was first proved in [AN81]; a new account can be found in Chapter 11 of [AK00]. We modify the original definitions, in keeping with the theory of numberings.

First, the kind of invariance Ash-Nerode studied is captured by the following definition.

DEFINITION 6.8. Let $A$ be a $\Sigma$-algebra computable under a numbering $\alpha : \Omega_\alpha \to A$. Let $R \subseteq A^k$. Then $R$ is *intrinsically semicomputable* if for any $\Sigma$-algebra $B$ computable under $\beta : \Omega_\beta \to B$, and any $\Sigma$-isomorphism $\phi : A \to B$, $\phi(R)$ is $\beta$-semicomputable.

Note that, if $R$ is intrinsically semicomputable, then (taking $B = A$, $\beta = \alpha$, and $\phi = I_A$), it follows that $R$ is $\alpha$-semicomputable.

It is easy to connect this notion with the Invariance Problem:

LEMMA 6.9. *Let $A$ be an $\alpha$-computable $\Sigma$-algebra, and $R \subseteq A^k$. Then the following are equivalent*:

(1) *$R$ is intrinsically semicomputable,*

(2) *$R$ is semicomputable in every computable numbering of $A$, i.e.,*

$$R \in \bigcap_{\alpha \in CompNum(A)} SComp_\alpha(A).$$

The special condition of the Ash-Nerode Theorem is based on the decidability of this property:

DEFINITION 6.10. Let $A$ be a $\Sigma$-algebra computable under $\alpha$, and $R \subseteq A^k$. The *satisfiability problem outside R* is the following decision problem: For any finitary existentially quantified $\Sigma$-boolean

$$\exists \mathsf{y}\, b(\mathsf{x}, \mathsf{y}) \quad (\mathsf{x} \equiv (\mathsf{x}_1, \ldots, \mathsf{x}_k),\ \mathsf{y} \equiv (\mathsf{y}_1, \ldots, \mathsf{y}_l),\ l > 0)$$

and any $a \in A^k$, is there $b \in A^l$ such that

$$b \notin R \text{ and } A \models b[a, b]?$$

Note that if the existential diagram of $(A, R)$ is $\alpha$-decidable, then the satisfiability problem outside $R$ is $\alpha$-decidable.

The *Ash-Nerode Theorem* says that if the satisfiability problem outside $R$ is $\alpha$-decidable, then $R$ is intrinsically semicomputable if, and only if, $R$ can be expressed as a $DE(\Sigma)$-formula (see Definition 5.5). Combining this with Lemma 6.9, we have

THEOREM 6.11. *Let $A$ be a $\Sigma$-algebra computable under $\alpha$ and let $R \subseteq A^k$. Suppose the satisfiability problem outside $R$ is $\alpha$-decidable. Then the following are equivalent*:

(1) *R is intrinsically semicomputable;*
(2) *R is semicomputable in every computable numbering of A, i.e.,*

$$R \in \bigcap_{\alpha \in CompNum(A)} SComp_{\alpha}(A);$$

(3) *R is expressible as* $\boldsymbol{DE}(\Sigma)$*-formula:*

$$x \in R \iff \bigvee\!\!\!\!\bigvee_{i=0}^{\infty} \exists y_i \boldsymbol{b}_i(x, y_i).$$

The equivalence of (2) and (3) is a form of completeness theorem.

## REFERENCES

[AGM92] S. Abramsky, D. Gabbay, and T. Maibaum (editors), ***Handbook of logic in computer science***, Oxford University Press, 1992, In 5 volumes, 1992–2000.

[AK00] C.J. Ash and Julia Knight, ***Computable structures and the hyperarithmetical hierarchy***, Elsevier, 2000.

[AN81] C.J. Ash and A. Nerode, ***Intrinsically recursive relations***, ***Proceedings of the conference on aspects of effective algebra, Monash university*** (J.N. Crossley, editor), U.D.A. Book Co., Steel's Creek, Victoria, Australia, 1981, pp. 26–41.

[BT04] E.J. Beggs and J.V. Tucker, *Computations via experiments with kinematic systems*, Technical Report 5-2004, Department of Computer Science, University of Wales, Swansea, March 2004.

[Ber93] U. Berger, *Total sets and objects in domain theory*, ***Annals of Pure and Applied Logic***, vol. 60 (1993), pp. 91–117.

[Ber02] ———, *Computability and totality in domains*, ***Mathematical Structures in Computer Science***, vol. 12 (2002), pp. 455–467.

[BCSS98] L. Blum, F. Cucker, M. Shub, and S. Smale, ***Complexity and real computation***, Springer-Verlag, 1998.

[BSS89] L. Blum, M. Shub, and S. Smale, *On a theory of computation and complexity over the real numbers: NP-completeness, recursive functions and universal machines*, ***Bulletin of the American Mathematical Society***, vol. 21 (1989), pp. 1–46.

[Bou01] J.P. Bourguignon, *A basis for a new relationship between mathematics and society*, ***Mathematics unlimited — 2001 and beyond*** (B. Engquist and W. Schmid, editors), Springer-Verlag, 2001, pp. 171–188.

[Bra96] V. Brattka, *Recursive characterisation of computable real-valued functions and relations*, ***Theoretical Computer Science***, vol. 162 (1996), pp. 45–77.

[Bra97] ———, *Order-free recursion on the real numbers*, ***Mathematical Logic Quarterly***, vol. 43 (1997), pp. 216–234.

[BH02] V. Brattka and P. Hertling, *Topological properties of real number representations*, ***Theoretical Computer Science***, vol. 284 (2002), no. 2, pp. 241–257.

[Cei59] G.S. Ceitin, *Algebraic operators in constructive complete separable metric spaces*, ***Doklady Akademii Nauk SSSR***, vol. 128 (1959), pp. 49–52.

[Eda97] A. Edalat, *Domains for computation in mathematics, physics and exact real arithmetic*, ***The Bulletin of Symbolic Logic***, vol. 3 (1997), pp. 401–452.

[Eng68] E. Engeler, ***Formal languages: Automata and structures***, Markham Publishing Co, 1968.

[EGNR98] Y.L. Ershov, S.S. Goncharov, A.S. Nerode, and J.B. Remmel (editors), ***Handbook of recursive mathematics***, Elsevier, 1998, In 2 volumes.

[Esc96] M. Escardó, *PCF extended with real numbers*, ***Theoretical Computer Science***, vol. 162 (1996), pp. 79–115.

[Fen02] J.E. Fenstad, *Computability theory: Structure or algorithms*, ***Reflections on the foundations of mathematics: Essays in honor of Solomon Feferman*** (W. Sieg, R. Sommer, and C. Talcott, editors), Lecture Notes in Logic, vol. 15, Association for Symbolic Logic, 2002, pp. 188–213.

[Fri71] H. Friedman, *Algebraic procedures, generalized Turing algorithms, and elementary recursion theory*, ***Logic colloquium '69*** (R.O. Gandy and C.M.E. Yates, editors), North Holland, 1971, pp. 361–389.

[FS56] A. Fröhlich and J. Shepherdson, *Effective procedures in field theory*, ***Philosophical Transactions of the Royal Society London***, vol. (A) 248 (1956), pp. 407–432.

[Gri99] E. Griffor (editor), ***Handbook of computability theory***, North Holland, 1999.

[Gro98] M. Gromov, *Possible trends in mathematics in the coming decades*, ***Notices of the American Mathematical Society***, vol. 45 (1998), pp. 846–847.

[Grz55] A. Grzegorczyk, *Computable functions*, ***Fundamenta Mathematicae***, vol. 42 (1955), pp. 168–202.

[Grz57] ———, *On the defintions of computable real continuous functions*, ***Fundamenta Mathematicae***, vol. 44 (1957), pp. 61–71.

[HI70] G.T. Herman and S.D. Isard, *Computability over arbitrary fields*, ***Journal of the London Mathematical Society (2)***, vol. 2 (1970), pp. 73–79.

[Jia03] W. Jiang, ***Universality and semicomputability for non-deterministic programming languages over abstract algebras***, M.Sc. Thesis, Department of Computing & Software, McMaster University, 2003, Technical Report CAS-03-03-JZ, Department of Computing & Software, McMaster University.

[Kle59] S.C. Kleene, *Countable functionals*, ***Constructivity in mathematics*** (A. Heyting, editor), North Holland, 1959, pp. 81–100.

[Kre59] G. Kreisel, *Interpretation of analysis by means of constructive functionals of finite type*, ***Constructivity in mathematics*** (A. Heyting, editor), North Holland, 1959, pp. 101–128.

[Kre71] ———, *Some reasons for generalizing recursion theory*, ***Logic colloquium '69*** (R.O. Gandy and C.M.E. Yates, editors), North Holland, 1971, pp. 139–198.

[Lac55a] D. Lacombe, *Extension de la notion de fonction récursive aux fonctions d'une ou plusieurs variables réelles, I*, ***Comptes Rendus Mathématique. Académie des Sciences. Paris***, vol. 240 (1955), pp. 2470–2480.

[Lac55b] ———, *Extension de la notion de fonction récursive aux fonctions d'une ou plusieurs variables réelles, II*, ***Comptes Rendus Mathématique. Académie des Sciences. Paris***, vol. 241 (1955), pp. 13–14.

[Lac55c] ———, *Extension de la notion de fonction récursive aux fonctions d'une ou plusieurs variables réelles, III*, ***Comptes Rendus Mathématique. Académie des Sciences. Paris***, vol. 241 (1955), pp. 151–153.

[Lon04] J. Longley, *Notions of computability at higher types I*, ***Proceedings of the ASL logic colloquium, Paris, 2000***, AK Peters, 2004, To appear.

[MKS76] W. Magnus, A. Karass, and D. Solitar, ***Combinatorial group theory***, Dover, 1976.

[Mal71a] A.I. Mal'cev, *Closely related models and recursively perfect algebras*, ***The metamathematics of algebraic systems. A. I. Malcev, collected papers: 1936–1967***, North Holland, 1971, pp. 255–261.

[Mal71b] ———, *Constructive algebras I*, ***The metamathematics of algebraic systems. A. I. Malcev, collected papers: 1936–1967***, North Holland, 1971, pp. 148–212.

[Mol77] J. Moldestad, ***Computations in higher types***, Lecture Notes in Mathematics, vol. 574, Springer-Verlag, 1977.

[Mon68] R. Montague, *Recursion theory as a branch of model theory*, ***Logic, methodology & philosophy of science III*** (B. van Rootselaar and J.F. Staal, editors), North Holland, 1968, pp. 63–86.

[Mos64] Y.N. Moschovakis, *Recursive metric spaces*, ***Fundamenta Mathematicae***, vol. 55 (1964), pp. 215–238.

[Mos69a] ———, *Abstract first order computability I*, ***Transactions of the American Mathematical Society***, vol. 138 (1969), pp. 427–464.

[Mos69b] ———, *Abstract first order computability II*, ***Transactions of the American Mathematical Society***, vol. 138 (1969), pp. 465–504.

[Nor80] D. Normann, ***Recursion on the countable functionals***, Lecture Notes in Mathematics, vol. 811, Springer-Verlag, 1980.

[Nor82] ———, *External and internal algorithms on the continuous functionals*, ***Patras logic symposium*** (G. Metakides, editor), Studies in Logic, vol. 109, North Holland, 1982, pp. 137–144.

[Nor99] ———, *The continuous functionals*, ***Handbook of computability theory*** (E. Griffor, editor), North Holland, 1999.

[Nor00] ———, *Computability over the partial continuous functionals*, ***The Journal of Symbolic Logic***, (2000), pp. 1133–1142.

[Pla66] R.A. Platek, ***Foundations of recursion theory***, Ph.D. thesis, Department of Mathematics, Stanford University, 1966.

[Plo77] G.D. Plotkin, *LCF considered as a programming language*, ***Theoretical Computer Science***, vol. 5 (1977), pp. 223–255.

[PER89] M.B. Pour-El and J.I. Richards, ***Computability in analysis and physics***, Springer-Verlag, 1989.

[Rab60] M. Rabin, *Computable algebra, general theory and the theory of computable fields*, ***Transactions of the American Mathematical Society***, vol. 95 (1960), pp. 341–360.

[Ric54] H.G. Rice, *Recursive real numbers*, ***Proceedings of the American Mathematical Society***, vol. 5 (1954), pp. 784–791.

[She73] J.C. Shepherdson, *Computations over abstract structures: serial and parallel procedures and Friedman's effective definitional schemes*, ***Logic colloquium '73*** (H.E. Rose and J.C. Shepherdson, editors), North Holland, 1973, pp. 445–513.

[She85] ———, *Algebraic procedures, generalized turing algorithms, and elementary recursion theory*, ***Harvey Friedman's research on the foundations of mathematics*** (L.A. Harrington, M.D. Morley, A. Ščedrov, and S.G. Simpson, editors), North Holland, 1985, pp. 285–308.

[Sma98] S. Smale, *Mathematical problems for the next century*, ***The Mathematical Intelligencer***, vol. 20 (1998), pp. 7–15.

[Soa96] R.L. Soare, *Computability and recursion*, ***The Bulletin of Symbolic Logic***, vol. 2 (1996), pp. 284–321.

[Spr98] D. Spreen, *On effective topological spaces*, ***The Journal of Symbolic Logic***, vol. 63 (1998), pp. 185–221.

[SHLG94] Stoltenberg-Hansen, I. Lindström, and E. Griffor, ***Mathematical theory of domains***, Cambridge University Press, 1994.

[SHT88] V. Stoltenberg-Hansen and J.V. Tucker, *Complete local rings as domains*, ***The Journal of Symbolic Logic***, vol. 53 (1988), pp. 603–624.

[SHT95] ———, *Effective algebras*, ***Handbook of logic in computer science*** (S. Abramsky, D. Gabbay, and T. Maibaum, editors), vol. 4, Oxford University Press, 1995, pp. 357–526.

[SHT99a] ———, *Computable rings and fields*, ***Handbook of computability theory*** (E. Griffor, editor), North Holland, 1999.

[SHT99b] ———, *Concrete models of computation for topological algebras*, ***Theoretical Computer Science***, vol. 219 (1999), pp. 347–378.

[Tai62] W.W. Tait, *Continuity properties of partial recursive functionals of finite type*, Unpublished notes, 1962.

[TZ88] J.V. Tucker and J.I. Zucker, ***Program correctness over abstract data types, with error-state semantics***, CWI Monographs, vol. 6, North Holland, 1988.

[TZ99] ———, *Computation by 'while' programs on topological partial algebras*, ***Theoretical Computer Science***, vol. 219 (1999), pp. 379–420.

[TZ00] ———, *Computable functions and semicomputable sets on many-sorted algebras*, ***Handbook of logic in computer science*** (S. Abramsky, D. Gabbay, and T. Maibaum, editors), vol. 5, Oxford University Press, 2000, pp. 317–523.

[TZ04] ———, *Abstract versus concrete models of computation on partial metric algebras*, ***ACM Transactions on Computational Logic***, vol. 5 (2004), no. 4, pp. 611–668.

[TZ05] ———, *Computable total functions on metric algebras, universal algebraic specifications and dynamical systems*, ***Journal of Algebraic and Logic Programming***, vol. 62 (2005), pp. 71–108.

[Tur36] A.M. Turing, *On computable numbers, with an application to the Entscheidungsproblem*, ***Proceedings of the London Mathematical Society***, vol. 42 (1936), pp. 230–265, With correction, *ibid.*, 43, 544–546, 1937. Reprinted in *The Undecidable*, M. Davis, ed., Raven Press, 1965.

[Wei00] K. Weihrauch, ***Computable analysis: An introduction***, Springer-Verlag, 2000.

DEPARTMENT OF COMPUTER SCIENCE
UNIVERSITY OF WALES SWANSEA
SINGLETON PARK, SWANSEA SA2 8PP, WALES
*E-mail*: J.V.Tucker@swansea.ac.uk

DEPARTMENT OF COMPUTING AND SOFTWARE
MCMASTER UNIVERSITY
HAMILTON, ONTARIO L8S 4K1, CANADA
*E-mail*: zucker@mcmaster.ca

LECTURE NOTES IN LOGIC

## General Remarks

This series is intended to serve researchers, teachers, and students in the field of symbolic logic, broadly interpreted. The aim of the series is to bring publications to the logic community with the least possible delay and to provide rapid dissemination of the latest research. Scientific quality is the overriding criterion by which submissions are evaluated.

Books in the Lecture Notes in Logic series are printed by photo-offset from master copy prepared using LaTeX and the ASL style files. For this purpose the Association for Symbolic Logic provides technical instructions to authors. Careful preparation of manuscripts will help keep production time short, reduce costs, and ensure quality of appearance of the finished book. Authors receive 50 free copies of their book. No royalty is paid on LNL volumes.

Commitment to publish may be made by letter of intent rather than by signing a formal contract, at the discretion of the ASL Publisher. The Association for Symbolic Logic secures the copyright for each volume.

The editors prefer email contact and encourage electronic submissions.

## Editorial Board

## Editorial Policy

1. Submissions are invited in the following categories:
i) Research monographs  iii) Reports of meetings
ii) Lecture and seminar notes  iv) Texts which are out of print
Those considering a project which might be suitable for the series are strongly advised to contact the publisher or the series editors at an early stage.

2. Categories i) and ii). These categories will be emphasized by Lecture Notes in Logic and are normally reserved for works written by one or two authors. The goal is to report new developments quickly, informally, and in a way that will make them accessible to non-specialists. Books in these categories should include

– at least 100 pages of text;
– a table of contents and a subject index;
– an informative introduction, perhaps with some historical remarks, which should be accessible to readers unfamiliar with the topic treated;

In the evaluation of submissions, timeliness of the work is an important criterion. Texts should be well-rounded and reasonably self-contained. In most cases the work will contain results of others as well as those of the authors. In each case, the author(s) should provide sufficient motivation, examples, and applications. Ph.D. theses will be suitable for this series only when they are of exceptional interest and of high expository quality.

Proposals in these categories should be submitted (preferably in duplicate) to one of the series editors, and will be refereed. A provisional judgment on the acceptability of a project can be based on partial information about the work: a first draft, or a detailed outline describing the contents of each chapter, the estimated length, a bibliography, and one or two sample chapters. A final decision whether to accept will rest on an evaluation of the completed work.

3. Category iii). Reports of meetings will be considered for publication provided that they are of lasting interest. In exceptional cases, other multi-authored volumes may be considered in this category. One or more expert participant(s) will act as the scientific editor(s) of the volume. They select the papers which are suitable for inclusion and have them individually refereed as for a journal. Organizers should contact the Managing Editor of Lecture Notes in Logic in the early planning stages.

4. Category iv). This category provides an avenue to provide out-of-print books that are still in demand to a new generation of logicians.

5. Format. Works in English are preferred. After the manuscript is accepted in its final form, an electronic copy in LaTeX format will be appreciated and will advance considerably the publication date of the book. Authors are strongly urged to seek typesetting instructions from the Association for Symbolic Logic at an early stage of manuscript preparation.

# Lecture Notes in Logic

From 1993 to 1999 this series was published under an agreement between the Association for Symbolic Logic and Springer-Verlag. Since 1999 the ASL is Publisher and A K Peters, Ltd. is Co-publisher. The ASL is committed to keeping all books in the series in print.

Current information may be found at `http://www.aslonline.org`, the ASL Web site. Editorial and submission policies and the list of Editors may also be found above.

Previously published books in the *Lecture Notes in Logic* are:

1. *Recursion Theory.* J. R. Shoenfield. (1993, reprinted 2001; 84 pp.)

2. *Logic Colloquium '90; Proceedings of the Annual European Summer Meeting of the Association for Symbolic Logic, held in Helsinki, Finland, July 15–22, 1990.* Eds. J. Oikkonen and J. Väänänen. (1993, reprinted 2001; 305 pp.)

3. *Fine Structure and Iteration Trees.* W. Mitchell and J. Steel. (1994; 130 pp.)

4. *Descriptive Set Theory and Forcing: How to Prove Theorems about Borel Sets the Hard Way.* A. W. Miller. (1995; 130 pp.)

5. *Model Theory of Fields.* D. Marker, M. Messmer, and A. Pillay. (First edition, 1996, 154 pp. Second edition, 2006, 155 pp.)

6. *Gödel '96; Logical Foundations of Mathematics, Computer Science and Physics; Kurt Gödel's Legacy. Brno, Czech Republic, August 1996, Proceedings.* Ed. P. Hajek. (1996, reprinted 2001; 322 pp.)

7. *A General Algebraic Semantics for Sentential Objects.* J. M. Font and R. Jansana. (1996; 135 pp.)

8. *The Core Model Iterability Problem.* J. Steel. (1997; 112 pp.)

9. *Bounded Variable Logics and Counting.* M. Otto. (1997; 183 pp.)

10. *Aspects of Incompleteness.* P. Lindstrom. (First edition, 1997. Second edition, 2003, 163 pp.)

11. *Logic Colloquium '95; Proceedings of the Annual European Summer Meeting of the Association for Symbolic Logic, held in Haifa, Israel, August 9–18, 1995.* Eds. J. A. Makowsky and E. V. Ravve. (1998; 364 pp.)

12. *Logic Colloquium '96; Proceedings of the Colloquium held in San Sebastian, Spain, July 9–15, 1996.* Eds. J. M. Larrazabal, D. Lascar, and G. Mints. (1998; 268 pp.)

13. *Logic Colloquium '98; Proceedings of the Annual European Summer Meeting of the Association for Symbolic Logic, held in Prague, Czech Republic, August 9–15, 1998.* Eds. S. R. Buss, P. Hájek, and P. Pudlák. (2000; 541 pp.)

14. *Model Theory of Stochastic Processes.* S. Fajardo and H. J. Keisler. (2002; 136 pp.)

15. *Reflections on the Foundations of Mathematics; Essays in Honor of Solomon Feferman.* Eds. W. Seig, R. Sommer, and C. Talcott. (2002; 444 pp.)

16. *Inexhaustibility; A Non-exhaustive Treatment.* T. Franzén. (2004; 255 pp.)

17. *Logic Colloquium '99; Proceedings of the Annual European Summer Meeting of the Association for Symbolic Logic, held in Utrecht, Netherlands, August 1–6, 1999.* Eds. J. van Eijck, V. van Oostrom, and A. Visser. (2004; 208 pp.)

18. *The Notre Dame Lectures.* Ed. P. Cholak. (2005, 185 pp.)

19. *Logic Colloquium 2000; Proceedings of the Annual European Summer Meeting of the Association for Symbolic Logic, held in Paris, France, July 23–31, 2000.* Eds. R. Cori, A. Razborov, S. Todorčević, and C. Wood. (2005; 408 pp.)

20. *Logic Colloquium '01; Proceedings of the Annual European Summer Meeting of the Association for Symbolic Logic, held in Vienna, Austria, August 1–6, 2001.* Eds. M. Baaz, S. Friedman, and J. Krajíček. (2005, 486 pp.)

21. *Reverse Mathematics 2001.* Ed. S. Simpson. (2005, 401 pp.)

22. *Intensionality.* Ed. R. Kahle. (2005, 265 pp.)

23. *Logicism Renewed: Logical Foundations for Mathematics and Computer Science.* P. Gilmore. (2005, 230 pp.)

24. *Logic Colloquium '03; Proceedings of the Annual European Summer Meeting of the Association for Symbolic Logic, held in Helsinki, Finland, August 14–20, 2003.* Eds. V. Stoltenberg-Hansen and J. Väänänen. (2006; 407 pp.)

T - #0461 - 101024 - C0 - 229/152/23 - PB - 9781568812946 - Gloss Lamination